D1447884

TEXT–BOOK OF ADVANCED

MACHINE WORK

PREPARED FOR

STUDENTS IN TECHNICAL, MANUAL TRAINING,
AND TRADE SCHOOLS, AND FOR THE
APPRENTICE AND THE MACHIN-
IST IN THE SHOP

BY

ROBERT H. SMITH

MASSACHUSETTS INSTITUTE OF TECHNOLOG

797 pages, 886 Illustrations

EIGHT EDITION,
REVISED AND ENLARGED

Advanced Machine Work

Advanced Machine Work
reprinted from an edition published in 1925 by
Industrial Education Book Co.

Copyright © 1984 by Lindsay Publications, Bradley IL

ISBN 0-917914-23-6

5 6 7 8 9 0

WARNING

Remember that the materials and methods described here are from another era. Workers were less safety conscious then, and some methods may be downright dangerous. Be careful! Use good solid judgement in your work. Lindsay Publications has not tested these methods and materials and does not endorse them. Our job is merely to pass along to you information from another era. Safety is your responsibility.

Write for a catalog of other unusual books available from:

Lindsay Publications
PO Box 12
Bradley IL 60915-0012

A SECTIONAL BOOK

This book is built in Sections and paged as follows:

Sections	Pages
1	101 to 167
2	201 to 242
3	301 to 356
4	401 to 440
5	501 to 548
6	601 to 636
7	701 to 732
8	801 to 832
9	901 to 938
10	1001 to 1049
11	1101 to 1190
12	1201 to 1287
13	1301 to 1345
14	1401 to 1446
15	1501 to 1524

Page Numbering of This Book
Patented, July 24, 1917

FOREWORD

TO

PRINCIPLES OF MACHINE WORK
AND
ADVANCED MACHINE WORK

Text-books. — To teach any subject rapidly, good text-books are a necessity. For the study of languages, mathematics, physics, chemistry, etc., in the class-room and the laboratory, excellent text-books are obtainable scientifically arranged to lead the student progressively and rapidly through elementary and advanced principles.

Lack of Text-Books on Machine Work. — In this Age of Machinery, teachers, students, apprentices, machine operators and all those who are interested in the art and science of machine construction have been handicapped by the lack of text-books comparable with those that aid the student and teacher in other subjects.

Need of Text-books on Machine Building. — To meet the urgent demand for such text-books, the author has prepared and published two books — the text-book of *The Principles of Machine Work* and the text-book of *Advanced Machine Work* — that beginners might have the advantages of text-books as in the older subjects and be able to acquire in a short time, the fundamental and the advanced principles of machine building, logically, systematically and progressively. These books supersede all the other editions of his books on machine work.

" **Principles of Machine Work** " describes the metals and materials used in machine construction, and names the tools and instruments and schedules of operations and hand processes. It treats of Heat Treatment of Steel; Hardening and Tempering Carbon Tool Steels and High-speed Steels; Oxy-Acetylene Welding and Cutting, Lead Burning; Thermit Welding, Electric Welding, Brazing and Soldering; Pipe Fitting; Laying Out Work; Hand and Pneumatic Chipping, Riveting

7

and Drilling, Filing and Scraping; Lacing Belts, Alining Shafting and Installing Machines; Drills and Drilling; Taps and Tapping; Speed Lathes, Hand Tools and Hand Turning.

"**Advanced Machine Work**" treats of Engine Lathe Work; Cutting Tools; Measuring; Turning; Fitting; Threading; Chucking; Reaming; Mandrels or Arbors; Curve Turning and Forming; Inside Calipers and Inside Micrometers; Boring and Inside Threading; Brass Finishing; Broaching; Drilling Jigs; Boring, Boring Bars and Boring Machines; Eccentric Turning; Nurling; Cylindrical, Internal, Surface and Cutter Grinding; Planing; Milling; Spur, Spiral, Worm and Bevel Gear Cutting; Generating Gears; Toolmaking; Spiral Milling; The Plug and Button Methods of Locating Holes of Precision in Jigs and Fixtures; Sine Bar; Relieving or Backing Off Cutters, Taps, Counterbores and Hobs; Inspection and Limit System.

Books in Sectional Form. — These text-books are divided into Sections. Each Section is complete in itself.

Illustrations. — Machines, mechanisms and tools, problems and processes, methods and operations are graphically illustrated by perspective and mechanical drawings.

The drawings have been made especially for these books and are so clearly marked with letters, words, and figures, that many of them are self-explanatory of the operations and processes which they represent and tell things far better than could be told by words.

Schedules of Operations. — To secure efficiency in teaching or manufacturing, it is necessary to be equipped with a well-defined plan of attack for the problem in hand. The schedules of operations in these text-books provide the student and teacher with a complete plan in table form, for the rapid production of standard and typical· problems in machine construction. They name the time necessary to complete the work, the materials, machines, speeds, feeds, tools, jigs, and fixtures for each consecutively numbered or lettered operation, and the accompanying illustration is numbered and lettered to correspond.

These schedules of operations have been used by the **author** for many years in his classes and are the methods now used in all modern manufacturing efficiency systems.

To the Student. — These text-books constitute a complete treatise on the indispensable principles and processes of modern machine-shop practice for the production of machine parts, machines and tools. They tell how to do things with that theory which connects principles and practice and no person can build machines or superintend the construction of machinery without consciously or unconsciously understanding and applying these principles. The basic operations and processes have been reduced to problem form. The study and practice of these problems and methods supplemented by lectures, demonstrations, and instruction, will give the student not only an excellent training in machine work, but also a broader training by teaching him to plan methods of doing things, to study his movements and avoid wasting steps and motions, and thus to conserve his time and his energy, thereby obtaining mental and physical precision and scientific efficiency.

To the Instructor these text-books will be of great assistance by furnishing an organized course. They will supplement the lectures, enrich the individual instruction, and supply information and answers to the innumerable questions of the students, thereby conserving the instructor's time and enabling the students to work with greater efficiency and dispatch.

To obtain the greatest benefit, the use is urged of as many of the problems as time and conditions will permit, and in the order given in the books, so far as the equipment of special machines will allow. These books will be found valuable in experimental work and in building apparatus and machines, as the schedules of operations and the processes give the correct methods for making machine parts, tools, and for building machines.

In fact, the schedules of problems and processes are complete and condensed lessons in scientific efficiency, and teach the shortest, quickest and easiest way to obtain results.

These books do not teach casual ideas but *scientific principles* developed by thirty years' study of the subject. These *powerful* lessons enable the student to make the most of his efforts and time without waste of energy.

To the Machine Operator. — To the man who has been trained on one machine, or to perform but few operations, these text-books open the door of opportunity to a broader training, increased efficiency, and *greater earning power.*

To the Apprentice. — These books will be of great value to apprentices and to young machinists for any problem which may arise in the machine shop, as neither the superintendent, foremen, toolmakers nor machinists have the time to instruct the apprentices in the principles and processes presented.

An apprentice or young machinist may have worked a long time in the shop without seeing or doing a particular kind of work, for shop work depends on commercial requirements. When new work comes to such a man he will find the method of procedure so clearly set forth in these books that he will have but little difficulty in following directions.

As these principles and processes can be applied to all machine work, their faithful study by the apprentice or young machinist will not only increase his knowledge, develop high efficiency and rapid and accurate workmanship, but will also make him more valuable to his employer during his period of learning by largely increasing his power of production.

Furthermore, these text-books teach the apprentice to read and understand technical literature. They train him to translate printed matter into intelligent action. This is a great acquisition, for in no other way can an apprentice keep abreast of everything technical and scientific. It is self-evident that the training we have so briefly outlined will tend to increase his earning power and to fit him for a better position at the close of his apprenticeship.

To the Machinist. — To machinists who have served their apprenticeships in the small machine shops, with limited equipment and range of work; to those who were trained in the large machine shops well equipped with improved machinery, but where intensive methods of manufacturing and repetitional production may have limited their opportunities to acquire that broad fundamental training so necessary to future success; and to *all* machinists who wish to fit themselves for better positions by further study, these books will be an aid and an inspiration. Even to those who have had a superior training, such as managers, superintendents and foremen, they will be valuable as works of reference.

The Increased Efficiency obtained at the Massachusetts Institute of Technology by use of these text-books, and the kind reception given them by the technical press, schools and shops, teachers and students, apprentices and machinists, have shown the need of text-books on machine work and justified their production.

The increasing number of state universities, technical, trade, and manual training schools that are adopting these books as a required text is evidence that the want they meet is widespread.

Grateful Acknowledgment. — To the teachers and educators, manufacturers and engineers, foremen and mechanics, associates, and other friends in all parts of the country, who have kindly assisted with information, help and encouragement, I take this opportunity to express my indebtedness and appreciation.

R. H. S.

Boston, U. S. A., *March* 1, 1919.

CONTENTS

Section 1

LATHE WORK

Section 3

LATHE WORK

Section 4

LATHE WORK

Section 5

LATHE WORK

Section 6

DRILLING JIGS, BORING BARS, ECCENTRIC TURNING

18 CONTENTS

Section 7

CYLINDRICAL GRINDING (EXTERNAL) INTERNAL GRINDING

Section 8

SURFACE GRINDING   CUTTER GRINDING

Section 9

PLANING

Section 10

MILLING

Section 11

GEAR CUTTING

Section 12

TOOL MAKING

Section 13

INSPECTION, LIMIT SYSTEM OF MANUFACTURING, INSPECTION OF MACHINE PARTS

Section 14

INSPECTION OF MACHINE TOOLS, PRECISION MEASURE-
MENT OF THREADS, HEAVY DUTY TURNING AND
BORING

SAFETY WARNINGS

TO

AVOID ACCIDENTS WHEN OPERATING MACHINES, TOOLS, OR APPLIANCES

Guards and safety appliances. — Guard dangerous parts of power-driven machines properly, according to modern safety standards, and to comply with the laws of the State or Municipality. Gears, pulleys, belts, couplings, ends of shafts having keyways and other revolving or reciprocating dangerous parts, are some of the mechanisms that are generally required to be guarded to a height of six feet above the floor. Safety set screws and collars are required. Guards should be removed only for repairing or adjusting the machine, then replaced before operating the machine.

Danger Notice. — When engine or motor is stopped do not begin work on the line shaft, jackshafts, countershafts, overhead pulleys or belts without first notifying the engineer or the man in charge of the motor.

Light and sanitation. — As factors of safety, efficiency, and health, rooms should be kept clean, well-lighted, and have a large volume of circulating air.

Operating machines. — Do not start or attempt to operate a machine until its mechanism and its attending dangers are fully explained by a qualified person.

Lack of knowledge is the cause of more accidents by machines than carelessness. Most accidents are avoidable.

Oiling overhead bearings. — Never expose yourself to the dangers of revolving shafting, pulleys, and gears, by oiling overhead shaft bearings, countershafts, loose pulleys or clutches while in motion.

Clothing. — Wear a suitable working uniform, preferably overalls and jumper, with jumper inside overalls. The working uniform should be washed at regular intervals. Badly soiled clothing is unsanitary, unhygienic and inefficient and may cause infection in scratches, cuts or wounds. Loose or torn clothing, particularly loose or torn sleeves, or a flowing necktie, is dangerous. Never wear gloves when operating a machine except when absolutely necessary.

Cleaning, adjusting, repairing, and oiling machines. — *Stop the machine* before cleaning, adjusting, or repairing. If a dangerous part of a machine must be oiled while running, use an oil can with a long spout. The belt or clutch shipper (shifter) should be *locked*, or otherwise fastened, when adjusting or repairing a dangerous part of a machine. Serious accidents have occurred by some person starting a machine while another person was adjusting or repairing it.

Do not lean against a machine.

Belting. — Do not pinch or try to hold a belt when pushing it on or off a pulley. Keep the fingers straight, stiff and together, and act quickly.

To put belts on or off overhead pulleys, use a pole with a suitable shifting device or hook at the end. It is dangerous to climb a ladder to put on an overhead belt in the midst of other running belts. Ladders should have safety feet.

Gearing. — Do not clean running gears. Cotton waste, loose clothing, or the end of the belt you may be wearing, is quickly caught in running gears and thereby may draw in and crush the fingers, the hand or the arm, or cause injury to the body. Running gears are among the most dangerous of mechanisms. All gearing should be well guarded.

Engine and speed lathes. — Revolve a lathe one complete revolution before starting it by power to see that there is ample clearance between work, dog, chuck, and carriage. When filing or polishing work in an engine or speed lathe, see that the sleeves are tight or rolled up. Keep hand and arm above the rotating dog and work.

Do not put your hand or fingers on revolving work, spindles, pulleys, gears, chucks, or shafts.

Chucks. — Hold a chuck firmly when screwing it on or off the nose of the lathe spindle, otherwise it is liable to drop on the ways of the lathe and injure the hand. Hold small chuck with the hand. Medium and larger chucks are held by resting the chuck on the right forearm while rotating the spindle, by means of the belt, with the left hand. For very large chucks, have an assistant, or use a crane or tackle. Before starting the lathe, be sure to fasten the work into the chuck firmly and *remove chuck wrench.*

Drilling with drill press or speed lathe. — Never try to hold a short piece of work with the hand while drilling as it is liable to be jerked away suddenly and cause a painful and often a serious accident by cutting or bruising the hand.

Work that is to be drilled with a drill press should be fastened firmly in a vise, clamped to the table or screwed to a suitable fixture. Work that is to be drilled in a speed lathe should be held with some suitable device such as a hand vise or a lathe dog supported by the Tee rest.

A torn sleeve or a flowing necktie may be quickly caught by a high-speed spindle of a sensitive or multiple-spindle drilling machine and cause a serious accident.

Milling cutters. — Do not brush the chips from a revolving milling cutter with the fingers, or waste. If need be, use a suitable brush.

A loose or torn sleeve or cotton waste, may quickly catch in a revolving milling cutter and draw in and crush the fingers or the hand.

Grinding wheels. — Rotate a grinding wheel by pulling belt by hand, and examine the wheel carefully to be sure that it is not broken before starting by power. To avoid vibration and bursting, keep the wheel true with a dresser. Do not run the wheel faster than the speed recommended. Grinding wheels must be well guarded.

To grind tools or work held on a rest, clamp the rest firmly and closely to the wheel. When grinding with a dry wheel, protect the eyes by wearing safety goggles. When grinding with an unprotected grinding wheel, stand a little to one side to avoid the danger of flying emery and breaking wheel.

Planers and shapers. — Note the position of jigs or fixtures and work on table, in relation to the cutting tool, before starting a planer. Note also the position of the table dogs, to avoid the danger of the tool cutting into the work or the fixture that may be on the table.

Do not leave loose pieces, such as bolts, nuts, tools, or fixtures, on the table when it is running as they may shake off and drop into and break the gearing inside the bed.

Do not use the space inside the bed, as a receptacle for tools and reach for these tools while the table is running; this has caused many serious accidents. To prevent such accidents, the space inside of bed should be covered with a suitable guard.

Metal chips and shavings. — Do not remove chips or shavings with the hand. Remove cast-iron chips which break into small fragments with a brush. Steel turnings, steel planer chips, and steel drilling chips, which are long shavings, may cause a bad cut if an attempt is made to remove them with the hand. They may be broken with the crooked end of a scratch awl or bent rod, and then brushed off the machine.

Hand tools. — Never use a file without a handle. Handles of files, hammers, and other tools, should be tight and free from splints.

Chipping. — In hand or pneumatic chipping, always chip *toward* the wall, screen, or shield, and *away* from other persons.

Repeated blows on the head of a chisel, or similar tool will develop burrs. They should be ground off or the chisel head dressed and annealed to avoid flying chips due to the brittle burrs breaking off. When others are chipping in the same room, it is best to protect the eyes with safety goggles.

Babbitt, lead, solder. — When handling or pouring molten babbitt, lead, solder, or other metal, protect the eyes with safety goggles or a helmet.

Avoid water, moisture, or even a slight dampness on bench, in ladle, or on other tools, or on floor or the ground; for if molten metal is dropped or spattered on wet objects, it may cause a dangerous explosion and serious injury.

As a necessary measure of prevention, dry tongs, pokers, ladles, or other tools by preheating them in the forge, furnace, or by torch, before putting them into molten metal.

To pour babbitt bearings, or to pour lead or other soft metal into molds, cracks, crevices, or foundations, first carefully dry the parts by preheating with a torch.

Bedding on foundations. — Type metal is often used for bedding on foundations, for engines, pumps, power hammers, and other classes of heavy machinery; also for bedding castings, eye beams, etc., on concrete or granite foundations.

To remove dampness and to avoid an explosion, powdered rosin may be sprinkled over the surfaces where it is not convenient to preheat the work with a torch. The molten metal is poured from the side preferably through a spout. The powdered rosin will burn out ahead of the molten metal and remove the moisture or dampness.

The spaces into which molten metal is to be poured should have large vents to allow the hot gases to escape.

Various cold materials are used for bedding structural work, base plates, or castings on foundations *without danger of explosion:*

Grout which may consist of cement alone or cement and fine sand of equal parts and sufficient water to make the mixture pour; also a wet mixture of sal ammoniac and iron filings; and sheet lead.

Grout or molten sulphur is used to anchor bolt heads in holes in foundations.

Autogenous, thermit and electric Welding. — Protect the eyes with suitable goggles (scientifically colored) when welding by the autogenous, thermit, or electric processes. For electric arc welding, protect the face as well as the eyes by a helmet which is made for the purpose.

Welding by any process that requires an extremely high temperature will cause serious and often permanent injury to the eyes and in some cases, to the skin, if they are not protected.

FIRST AID SUGGESTIONS

First Aid Equipment. — Keep a standard medical kit,* splints, stretcher, disinfectants, sterilized gauze and bandages in some accessible place, and where possible have hot water available. Designate and instruct some person who has natural qualifications to render first aid, and order all others when injured to report at once to him. If the injury appears serious or the patient seems to be profoundly affected by the accident, seek medical advice immediately. Make a thorough inspection of the injury. Think carefully but act promptly. Keep cool as self-control is one of the first essentials in the person rendering aid.

Blood poisoning. — Treat and dress immediately **every** scratch, cut, wound, bruise or burn to avoid infection or blood poisoning which might incapacitate a person from work for weeks or months. The slightest scratch is worthy of attention. Remember that soap and water cleansing of the parts injured is a good preliminary measure, but should be followed by disinfection.

Hands that are to treat the wound should be washed — surgically clean. A neglect of cleanliness may cost the patient's life.

To disinfect and relieve pain wash with a weak solution of carbolic acid and water (1 part to 40) and rinse with warm water. Iodine (tincture) with alcohol, half and half, is another disinfectant highly recommended as a local antiseptic and should be in every kit. Paint the wound thoroughly with the tincture of iodine. Cover wound with sterilized gauze then bind with a clean bandage. An open wound should not be sealed by tight bandaging or by collodion. The inspection of such a wound should be made at least every 24 hours and a physician consulted at the first signs of inflammation.

Attention. — Other emergency disinfectants are alcohol, weak sulphur-naphthol, gasoline, shellac or soap and water.

* A standard "First Aid Kit" such as recommended by the National Safety Council will be found very satisfactory.

For burns put a teaspoonful of bicarbonate of soda into a pint of hot water and bathe or use carron oil (linseed oil and lime water). For muscle sprains and injuries to joints apply hot or cold water, keeping the injured parts at rest. In the event of fracture place the limb where it cannot be jarred and support it on a splint made soft by the use of cotton wadding, a pillow or any pliable material available and send for a physician.

Medical and dental departments in shops and factories. — No person can work efficiently with steel or emery in his eye, with the toothache or while suffering pain in any form. If the size of a plant justifies, medical and dental departments should be provided, with a physician, dentist and nurse, on duty, so that attention may be given injuries at once in order that men can return promptly to work.

ADVANCED MACHINE WORK

SECTION 1

LATHE WORK

Engine Lathes. Electrically-driven Machine Tools. Truing and Alining Centers. Center Holes. Cutting Tools. Lathe Tools for Cast Iron. Grinding Lathe Tools. Setting and Using Outside Calipers. Cutting Speeds, Cut-Meter, and Feeds. Lubricants for Cutting Tools. Inspecting and Measuring Material (Stock).

ENGINE LATHES

1. Evolution of the lathe. — The lathe is the most general and useful of all machine tools and is used to produce cylindrical surfaces.

The date of its origin is lost in antiquity. The first lathes consisted of two short posts driven into the ground, and a nail driven into each formed the centers on which the work revolved, operated by a rope, treadle and sapling, or lath, and from the latter name the term *lathe* is derived.

To Henry Maudslay of England, belongs the credit of inventing the slide rest and applying it to the lathe about 1794; and later, to other machines. Planing machines came next, and did for plane surfaces what the lathe had done for cylindrical surfaces. Then followed milling machines, grinding machines, screw machines, gear-cutting machines, etc. The improvements in machine tools during the past fifty years have been greater than in all the preceding years.

2. Swing of lathes.—A lathe is designated by its swing and total length of bed. A 14" × 6' engine lathe will swing fourteen inches in diameter over ways, but will only swing about 8" over a rise and fall rest, and about 10" over a plain or compound rest. In length it will turn six feet less the combined length of head and footstock. A 6' bed will turn about 35" between centers.

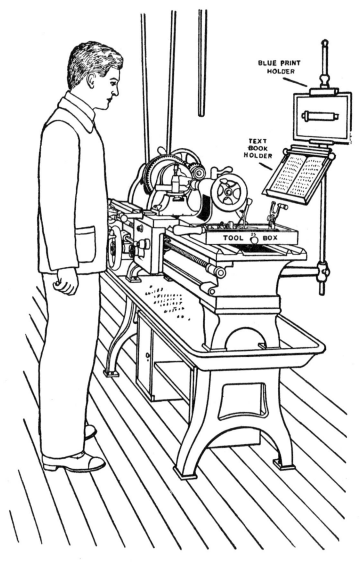

BLUE PRINT
HOLDER

TEXT
BOOK
HOLDER

TOOL ⊙ BOX

FIG. 1. — CORRECT POSITION AT ENGINE LATHE FOR TURNING.

3. Classes of lathes. — Lathes are divided into many classes, some of which are designed especially for the work performed upon them, as the wheel lathe, axle lathe, pulley lathe, turret lathe, bench lathe, jeweler's lathe, etc., and for general work the engine lathe, which when supplied with special attachments is called a tool-maker's lathe.

4. The engine lathe, see Figs. 1, 2, is supplied with hand and power long. (*longitudinal*) and cross feeds, is arranged for screw cutting for which a lēad screw is provided, and is usually constructed with back gears. The cutting tool is held in a tool post which is clamped to the tool block and the whole mounted on a carriage.

CONE HEADSTOCK, COUNTERSHAFT DRIVE, BELT OR
GEAR FEED

ESSENTIAL PARTS, FIG. 2.

5. *A — Bed.*

B and *B'* — Legs fastened to bed by cap screws, to floor by lag screws.

C and *C'* — Front ways of two pairs of *V* ways, planed and scraped.

D — Headstock bolted to ways.

E — Footstock or tailstock; position adjustable.

F and *F'* — Bolts for clamping footstock to ways *C*.

G — Carriage, two parts, movable on ways.

H — Saddle; carries tool mechanism.

H' — Apron; carries feed mechanism.

6. *Footstock.*

May be set over for taper turning.

I — Front screw of a pair for adjustment of upper part of footstock; back screw not visible.

J — Spindle.

K — Handle operating footstock spindle.

L — Binder for clamping *J*.

M — Dead center.

N — Oil well and oiler for dead center.

7. *Carriage.*

O — Tool post.

O' — Screw for fastening cutting tool.

P — Slide rest, rise and fall type (or elevating rest).

Q — Handle for adjusting height of tool.

R — Thread stop, used when cutting screw threads.

S — Handle for operating long. feed by hand.

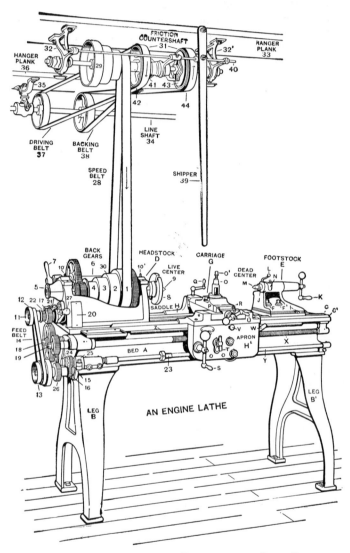

FIG. 2.— 14-Inch Engine Lathe, Countershaft, Line Shaft, and Belt Connections,

T — Knob for operating long. feed by power.

U — Handle for operating cross feed by hand.

V — Knob for operating cross feed by power.

W — Lever for operating split nut (half-nuts) inside apron *H'*.

X — Lead screw engaged by split nut when cutting threads.

Y — Feed shaft.

Z — Feed rack.

8. *Headstock.*

1, 2, 3, 4 — Steps on headstock cone. Belt on 1, slowest speed; on 4, fastest speed.

5 — Thrust bearing and end adjustment.

6 — Back gears.

7 — Lever for throwing 6 " in " or " out."

8 — Face plate, slotted to receive dog.

9 — Live center.

10 and 10' — Oil holes for live spindle.

9. *Feed.*

11 — Stud on feed spindle; transmits motion from lathe spindle to carriage for turning by 12, 13, 14, 15, 16 to feed shaft *Y*; for screw cutting by 17, 18, 19 to lead screw *X*.

A set of change gears is supplied for screw cutting and gear feed.

12, 13 — Feed cones.

14 — Feed belt.

20 — Index plate of gears, for screw cutting.

21 — Supplementary radial arm to carry two gears fixed on sleeve, for compounding change gears for

fine or coarse thread; serves to connect 22 on 11 to 18, and thence to *X*.

22 — Gear.

10. *Automatic Feed Stop.*

23 — Automatic stop sleeve.

24 — Clutch (23 and 24 used to stop carriage automatically at desired point).

25 — Clamping bolt (swinging 13 outward tightens feed belt).

26 — Gear feed. Remove belt 14, swing 13 until 26 meshes with 18. By different combinations of gears a large variety of feeds is obtainable. Six cone belt feeds are provided by interchanging 16 and 26.

27 — Reversing lever; reverses feed mechanism in headstock.

11. *Countershaft (friction type) and Line Shaft.*

28 — Speed belt.

29 — Countershaft cone pulley.

30 — Headstock cone pulley.

31 — Countershaft mechanism (consists of shaft, cone pulley, pulley for driving forward belt, pulley for driving backward belt, and clutch mechanism).

32 and 32' — Hangers bolted to hanger plank.

33 — Hanger plank.

34 — Line shaft; drives 31.

35 — Line shaft hanger.

36 — Hanger plank.

37 — Driving belt, 37 to 42; drives lathe forward.

38 — Backing belt, 38 to 44; drives lathe backward.

39 — Shipper pole, pivoted to 33.

40 — Shipper rod; controls friction-clutch mechanism.

41 — Expanding clutch; engages pulley, driving lathe forward.

42 — Driving pulley.

43 — Expanding clutch; engages pulley, driving lathe backward.

44 — "Backing" pulley.

(To run lathe "forward," push shipper to left; clutch 41 engages 42; "backward," push shipper to right; clutch 43 engages 44.)

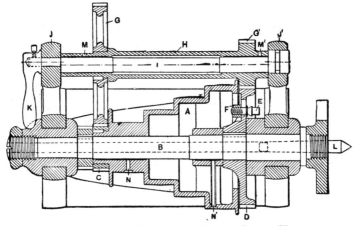

Fig. 3. — Horizontal Section of Engine-Lathe Headstock.

12. Back Gears and Headstock.
(Fig. 3.)

Back gears are used to reduce speed and increase power of machine. Ratio is about 10 to 1.

A — Cone pulley; running fit on spindle.

B — Spindle.

C — Gear fast to cone pulley.

D — Gear keyed to spindle B.

E — Slide bolt to fasten A to D.

F — Slide nut.

G and G' — Back gears fast on sleeve.

H — Sleeve; running fit on shaft.

I — Eccentric shaft.

J and J' — Brackets, part of headstock casting.

K — Lever to rotate shaft I, throwing back gear "in" or "out."

L — Live center; taper fit in B.

M and M' — Oil holes (oiled before using back gears).

N and N' — Oil holes (oiled before using back gears).

To Operate Back Gears.

For direct cone drive, slide nut F is in slot in cone A, back gears "out." To use back gears, drop bolt E and secure, throw lever

K forward. To obtain direct cone speed again, throw lever back, loosen *E* and revolve cone until *F* engages slot in *A*. Tighten *E*.

Double and Triple Back Gears.
To obtain a greater reduction of speed, lathes are built double or triple back geared.

Information. — Modern engine lathes have a micrometer dial upon the cross-feed screw which is convenient for fine adjustments and to use in conjunction with micrometer calipers. Each graduation gives a cut one-thousandth in depth which reduces the work two-thousandths in diameter. See Micrometer Calipers, p. 207.

Attention. — Gear guards are often provided to prevent accident and to keep dirt and chips from gear teeth.

13. A typical lathe apron. Fig. 4. — The apron of a lathe carries the greater part of the feed mechanism. Fig. 4 shows a lathe apron and the three distinct mechanisms, the long. feed, cross feed, and screw-cutting feed. The first is used for moving carriage back and forth along bed for turning; the second for moving cross slide in and out for squaring; the third, the lead screw and split nut, for moving carriage along bed for cutting screw threads.

14. To change direction of feeds. — The reversing mechanism to change direction of rotation of feed shaft may be in the headstock and operated by lever 27, Fig. 2, or in the lathe apron. See Inspection of an Engine Lathe, p. 1402.

LONG. FEED. CROSS FEED. LEAD SCREW

15. *Long. Feed.*

A — Long. feed handle.
B — Pinion.
C — Spur gear.
D — Sliding stud.
E — Sliding pinion, on inner end of stud *D*.
F — Feed rack, fastened to under side of bed.
G — Splined feed shaft.
H — Feather-keyed worm held in bracket *H'*.

J — Worm gear.
K — Friction clutch to connect *J* and *L*.
L — Pinion fast to *K*.
M — Knob controlling clutch *K*. Hand long. feed is obtained by rotating *A* which drives through *B*, *C*, *E* to *F*. Power long. feed is obtained from *G*. Clutch *K*, shown out of action, is thrown in, which causes *G* to drive through *H*, *J*, *L*, *C*, *E* to *F*.

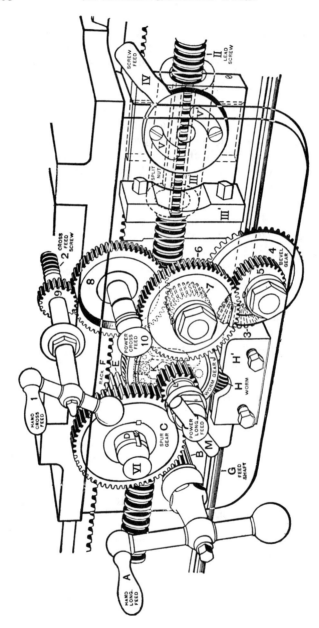

FIG. 4. — LATHE APRON SHOWN TRANSPARENTLY.

16. *Cross Feed.*

1 — Cross-feed handle.

2 — Operating screw, in nut under cross slide.

3 — Bevel pinion feather-keyed to *G.*

4 — Bevel gear.

5 — Pinion fast to 4.

6 — Driving gear.

7 — Pinion fast to 6.

8 — Gear always in mesh with 7.

9 — Cross-feed pinion.

10 — Knob controlling position of 8. Hand cross feed is obtained by rotating handle 1.

Power cross feed is obtained by meshing 8 with 9 by means of knob 10.

17. *Lead Screw and Split Nut.*

II — Lead screw.

III — Split nut.

III' — Split nut bracket.

IV — Lever to operate split nut III.

V and V' — Cams closing split nut III.

VI — Knob to disengage *E* and *F* when screw cutting. To operate for screw cutting, pull out knob VI, throw IV downward, closing split nut III.

There are types of lathes, where a splined lead screw performs the combined duty of feed shaft and lead screw, the worm and bevel pinion being driven directly by it.

Attention. — Care should be taken not to have both long. feed and lead screw thrown in at the same time. Some lathes are fitted with devices which will prevent this.

18. Care of machines and small tools. — When through using a machine, clean it, first with a brush, then wipe with cotton waste.

Wrenches, handles, bolts, straps, and fixtures should be put away in their proper places so that they may be found when wanted. All tools out on checks should be returned to the tool room as soon as possible.

19. Lathe box or tray for tools and work. — Do not place tools, work, or other metallic objects on the ways of a machine, as they would scar and affect their truth. A wooden box or tray should be supplied for the tools or small work. It may be placed on the ways near the end of the lathe. See Fig. 1. Large work should be placed on a bench, truck, or the floor.

20. Attachments for lathes. — Among the attachments for lathes are the taper attachment, compound rest, steady rest, follower rest, and attachments for milling and grinding.

Fig. 5.—Engine Lathe with a Rapid Change-Gear Mechanism.

21. Stops for duplicating sizes. — In addition to the thread stop some lathes are equipped with long. and cross-feed stops. After the first piece is turned or threaded to size, the back stop is set to check movement of cross slide. The carriage or long. stop is used for shoulders and lengths. By aid of these stops a number of pieces can be duplicated.

22. Lathe with a rapid change-gear mechanism for threads and feeds. — Some lathes, as in Fig. 5, are equipped with a rapid system of change gears by which different threads or feeds are obtained quickly.

ESSENTIAL PARTS.

A — Change gears in cone form, on end of lead screw.

B — Handle operating change gears.

C — Compound gears.

D — Handle operating compound gears.

E — Index plate, giving positions for handles B and D to cut a desired thread.

F — Sector to carry gear for cutting special thread.

G — Lever to reverse carriage.

H and K — Automatic carriage stops. Carriage stop (invisible) is located on back ways under letter L.

23. Cutting a screw using a rapid system of change gears. — To cut a screw of five threads per inch, find 5 on index E and place handle B in notch and hole under it; place handle D under hole 3 on compound gear box, which is indicated in the third column on the same line as 5.

For threads given on the index, no change of gears is necessary. Gears may be calculated for other threads and applied as on the ordinary lathe. The feed for turning is four times threads per inch expressed in turns per inch of tool travel.

24. For threading short screws the carriage may be reversed by moving reversing lever G up or down, depending on whether a right or left thread is being cut; or the automatic stops H and K may be used.

ELECTRICALLY-DRIVEN MACHINE TOOLS

25. Arrangement of machine tools for electric drive. — They may be group-driven by electricity by using a constant-speed motor to drive group line shaft, or individually driven by attaching a constant or variable-speed motor to each machine. Constant-speed motors are used on machines that need but little speed variation, or on machines that have a large variety of mechanical speed changes.

26. Motor-driven engine lathe, Fig. 6. **Variable speed motor** *A* gives wide speed variation. To start lathe, close

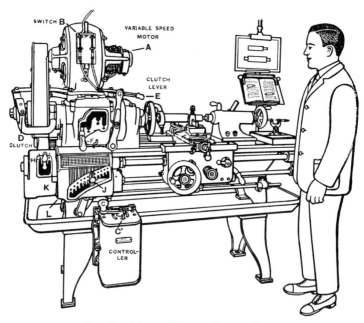

FIG. 6.— MOTOR-DRIVEN ENGINE LATHE.

switch *B* with controller handle *C* on the "off" position as shown at *C'*. Move controller handle *C* clockwise to obtain desired speed. Throw in clutch *D* by means of lever *E*.

To reverse motor or to run the spindle backward, remove bolt on controller and turn handle in opposite direction.

The gear headstock consists of a cone of gears on spindle controlled by lever *F* and back gears, giving eight spindle speeds with power shaft running at constant speed. Back gear drive is controlled by positive clutch lever *G*.

Thirty-six different threads and feeds are obtainable by means of levers *H* and *J* in gear boxes *K* and *L*, respectively.

TRUING AND ALINING CENTERS

27. Center gage *A*, Fig. 7, is used for defining angles of 60°. The large notch is used for testing lathe centers, as *B*; notches *C* and *D* for testing and setting outside threading

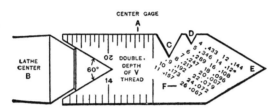

FIG. 7. — TESTING ANGLE OF LATHE CENTER

tools, and *E* for inside threading tools. At *F* is a table of double depths of Sharp *V* threads for determining diameter of tap drills by subtracting number in thousandths opposite pitch from diameter of tap.

28. Requirements for successful use of engine lathes. — Engine lathes should be accurate enough to turn, bore and face straight and true, and be equipped with an accurate lead screw for screw cutting.

To produce accurate work requires that both live and dead centers should be true and in accurate alinement. Furthermore, lathes wear, causing loss of alinement and looseness of working parts that must be detected and corrected.

29. Lathe centers. — The dead center is hardened and tempered. The live center may or may not be hardened.

30. To test truth of live center of any lathe. — Move footstock until dead center is close to live center, run lathe at highest speed, look for error. Move lathe tool close to revolving live center or use test indicator.

**31.   To true engine-lathe cen-
ters.   Figs. 8 and 9.**

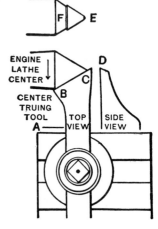

FIG. 8. — LOCATING CENTER IN   FIG. 9. — TRUING ENGINE-LATHE
        HEADSTOCK SPINDLE.   CENTER.

SCHEDULE OF OPERATIONS
Live Center.

Speed: — *For 12" to 16" engine lathe, 3d speed, back gears out.*

1. Remove center
2. Clean hole and center with waste.
3. Insert center with lines coincident *A, B*, Fig. 8.
4. Drive center lightly with lead hammer.
5. Start lathe; if still out of true, use center-truing tool *A*, Fig. 9. *BC* shows cutting edge, *D* the clearance.
6. Fasten tool lightly at height of center.

7. Adjust edge to fit center.
8. Clamp tool tightly.
9. Run lathe at moderate speed.
10. Operate both long. and cross feeds slowly by hand.
11. Test with center gage. Readjust tool, if necessary, until center fits gage.
12. Turn point *E* at a more · obtuse angle.   Portion *F* may be cut away in advance with cutting-off tool, to facilitate truing center.

Attention. — Centers may be trued by means of a compound rest. In an emergency a right side tool may be used back of the center and the lathe run backward.

SCHEDULE OF OPERATIONS *Concluded*
Dead Center.

1. Remove center.	6. File center with 8″ mill file.
2. Anneal.	7. Polish slightly with fine
3. Insert in live spindle.	emery cloth held under file.
4. True similarly to live center.	8. Reharden.
5. Run lathe at high speed.	9. Temper to light straw color.

32. To grind hardened and soft lathe centers, Fig. 10.

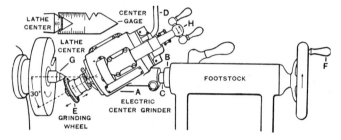

FIG. 10. — TRUING CENTER WITH ELECTRIC CENTER GRINDER.

SCHEDULE OF OPERATIONS

1. Mount center grinder *A*, Fig. 10, in footstock spindle *B*.

One end of shank or arbor *C*, fits taper hole in spindle *B*, and the other end fits hole in grinder *A*. Clean center and spindle before inserting.

2. Start lathe, 3d or 4th speed.

3. Connect electric cable *D* to a socket, and start grinder.

4. Feed revolving grinding wheel *E* with footstock handle *F* to lightly touch revolving lathe center *G*. Then grind by taking light cuts, moving grinding wheel slowly back and forth along center by rotating handle *H*.

Attention. — When necessary to grind both centers, grind dead center first, this leaves live center true to axis of rotation.

Warning. — Both wheel and center must revolve in same direction.

Information. — Electric center grinders are obtainable for direct (D. C.), alternating (A. C.), or for either (universal) currents.

33. To set dead center in alinement to turn straight. Figs. 11 and 12.

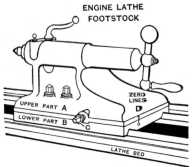

ENGINE LATHE
FOOTSTOCK

ZERO LINES

UPPER PART A

LOWER PART B

C

D

LATHE BED

FIG. 11. — SETTING DEAD CENTER IN ALINEMENT. APPROXIMATE METHODS.

SCHEDULE OF OPERATIONS

34.   *Two Approximate Methods.*

I. Unclamp footstock, Fig. 11, and move upper part *A* upon *B* by screws *C* and *C'* (*C'* not shown) until zero lines at *D* coincide.

II. Move footstock until dead center is close to live center, adjust screws *C* and *C'*, and aline centers by sight.

35.   *Accurate Method.*

Material, use piece that you are working on after it is rough-turned $\frac{1}{64}''$ to $\frac{1}{32}''$ large, or trial piece the same length

1. Set dead center by Approximate Method, No. I or No. II.		
2. Mount work on centers, Fig. 12.	Engine lathe, 12'' to 16''.	Dog, copper under set screw.
3. Take short light cut as shown at *A*, Fig. 12, .002'' to .003'' in diameter, estimated, or use dial on cross-feed handle.	3d speed, or 50 F.P.M. Fine power feed—140 to 1''.	For cast iron or brass, round-nose tool. For steel or wrought iron, diamond-point tool or holder and cutter 35° rake.
4. Take work off center.		
5. Run carriage back near dead center without moving cross-feed handle.		
6. Remount work.		
7. Take short light cut as shown dotted at *B*.		
8. Measure *A* and *B* with micrometer.		Micrometer.
9. If not alike, adjust foot-stock and repeat operations.		

Attention. — In setting the lathe for straight turning or fitting, the diameter at *A*, Fig. 12, must be either equal to *B* or a fraction of a thousandth of an inch larger than *B*. If *A* is larger than *B*, the error can be corrected by filing; if smaller, the error *can not* be corrected and the work is spoiled.

While it is best to have a lathe set straight, an error of a thousandth of an inch either way is permissible on such work as turning pulleys, flanges, and gear blanks.

Note. — A test indicator and parallel mandrel, preferably of the length of the work to be turned, may also be used to set a lathe to turn straight. See Test Indicators, pp. **12**10–**12**13.

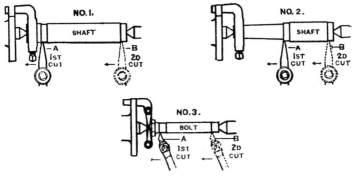

FIG. 12. — ACCURATE METHOD OF SETTING DEAD CENTER IN ALINEMENT FOR DIFFERENT KINDS OF WORK.

CENTER HOLES

36. Center holes are made in the ends of material (stock) to fit lathe centers by locating and drilling small holes, then counter-sinking with a 60° countersink, *A* and *B*, Fig. 13.

The countersink should be large enough to provide ample bearing to prevent excessive wear, and should be in proportion to the diameter of work.

The drilled hole must be deeper than countersink to provide a reservoir for oil and to prevent lathe centers from bottoming, as at *C*, Fig. 14, as this would injure centers and

cause work to run out of true. Center holes must be wiped
clean before mounting on centers. Chips in center holes *D*,
Fig. 15, will spoil both work and center.

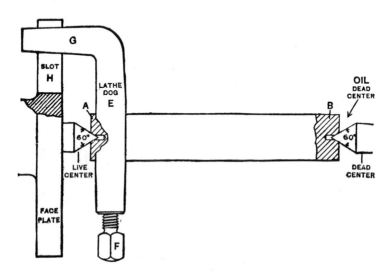

FIG. 13. — STOCK CORRECTLY CENTERED AND PROPERLY MOUNTED ON
LATHE CENTERS.

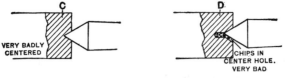

FIG. 14. — INCORRECT
CENTERING.

FIG. 15. — STOCK CARELESSLY
MOUNTED ON CENTERS.

37. To drive the work, dog *E*, Fig. 13, is fastened to work
by screw *F*, and tail *G* must be loose in face-plate slot *H*.

38. Table of center-hole dimensions. Fig. 16. — These center-hole dimensions provide good bearings for ordinary lathe work to resist tool pressure.

PLAN TO HAVE THE CENTER HOLES
THE GIVEN DIMENSIONS
WHEN WORK IS FINISHED TO EXACT LENGTH

Fig. 16.

Diameter of shaft.	Diameter of counter- sinks.	Drill size.		Depth of hole = $B + \frac{1}{16}''$.
		Nearest 64th.	Gage No.*	
A.	B.	C.		D.
$\frac{1}{8}-\frac{3}{16}$	$\frac{5}{64}$	$\frac{3}{64}$	56	$\frac{9}{64}$
$\frac{1}{4}-\frac{7}{16}$	$\frac{3}{32}$	$\frac{1}{16}$	52	$\frac{5}{32}$
$\frac{1}{2}-\frac{11}{16}$	$\frac{1}{8}$	$\frac{1}{16}$	52	$\frac{3}{16}$
$\frac{3}{4}-\frac{15}{16}$	$\frac{5}{32}$	$\frac{3}{32}$	42	$\frac{7}{32}$
$1-1\frac{3}{16}$	$\frac{3}{16}$	$\frac{3}{32}$	42	$\frac{1}{4}$
$1\frac{1}{4}-1\frac{7}{16}$	$\frac{7}{32}$	$\frac{3}{32}$	42	$\frac{9}{32}$
$1\frac{1}{2}-1\frac{11}{16}$	$\frac{1}{4}$	$\frac{1}{8}$	31	$\frac{5}{16}$
$1\frac{3}{4}-1\frac{15}{16}$	$\frac{9}{32}$	$\frac{1}{8}$	31	$\frac{11}{32}$
$2-2\frac{7}{16}$	$\frac{5}{16}$	$\frac{1}{8}$	31	$\frac{3}{8}$
$2\frac{1}{2}-2\frac{15}{16}$	$\frac{11}{32}$	$\frac{1}{8}$	31	$\frac{13}{32}$
$3-3\frac{7}{16}$	$\frac{3}{8}$	$\frac{5}{32}$	22	$\frac{7}{16}$
$3\frac{1}{2}-3\frac{15}{16}$	$\frac{7}{16}$	$\frac{5}{32}$	22	$\frac{1}{2}$
$4-4\frac{7}{16}$	$\frac{1}{2}$	$\frac{5}{32}$	22	$\frac{9}{16}$
$4\frac{1}{2}-4\frac{15}{16}$	$\frac{9}{16}$	$\frac{5}{32}$	22	$\frac{5}{8}$
$5-5\frac{7}{16}$	$\frac{5}{8}$	$\frac{5}{32}$	22	$\frac{11}{16}$
$5\frac{1}{2}-6$	$\frac{11}{16}$	$\frac{3}{16}$	13	$\frac{3}{4}$

Attention. — If drill and countersink are within one size they will answer.

* Twist drill and steel wire gage. See *Principles of Machine Work*

39. Countersinks or **center reamers,** Figs. 17, 18, are made to an angle of 60°. They may be made any desired diameter and with straight shanks, as shown, to be held in a chuck, or with taper shanks to be inserted in taper collet or machine spindle. They may have several cutting lips as in Fig. 17, or but a single cutting lip as in Fig. 18.

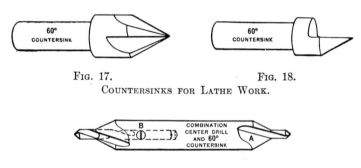

FIG. 17.   FIG. 18.
COUNTERSINKS FOR LATHE WORK.

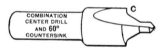

FIG. 19. — COMBINED DRILL AND COUNTERSINK.

Combination center drill and countersink, Fig. 19, produces a countersink central with drilled hole. At end *A* drill and countersink are one piece, while at end *B* a hole is made in countersink and a center drill inserted and held by a set screw.

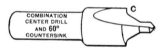

FIG. 20. — COMBINED DRILL, COUNTERSINK, AND COUNTERBORE.

Fig. 20 shows a combination center drill and countersink of the solid type as at end *A*, Fig. 19, and in addition a counterboring lip as at *C* for rounding the corners of the countersink. Combination countersinks of this type are used for centering mandrels, milling machine arbors and any other work that is mounted and remounted on centers frequently. Their use also facilitates the squaring of work.

40. **Hand method of finding the center,** drilling and countersinking. Figs. 21, 22, 23, 24, 25, 26.

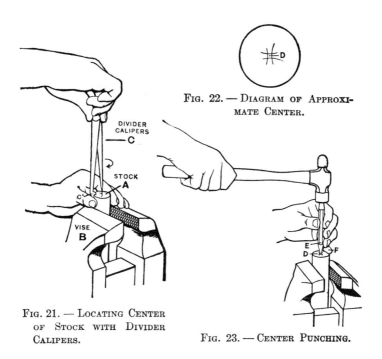

FIG. 22. — DIAGRAM OF APPROXIMATE CENTER.

FIG. 21. — LOCATING CENTER OF STOCK WITH DIVIDER CALIPERS.

FIG. 23. — CENTER PUNCHING.

SCHEDULE OF OPERATIONS

I. Grip stock A, Fig. 21, in vise B. Smooth ends with file and rub chalk on ends.

II. Describe arcs with divider calipers C (shoulder C' placed on edge of stock) from four points with radius equal to about one-half diameter, as at Fig. 22.

III. Locate center D, Figs. 22, 23, by eye with center punch E, Fig. 23. Steady with finger F and strike with hammer as shown.

IV. Mount on bench or lathe centers, revolve and test its truth with chalk near each end. If too much out of true, set over center punch mark as at G, Fig. 24. Repeat if necessary.

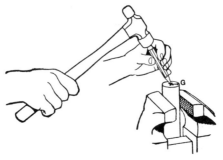

Fig. 24. — Setting Over Punch Mark.

SCHEDULE OF OPERATIONS *Concluded*

V. Enlarge center punch marks with heavier blows before drilling.

VI. Use speed lathe, 3d or 4th speed. Place drill *A* in drill chuck *B*, Fig. 25, and one end of stock *C* on dead center. Support with hand; start lathe, and drill one end, reverse work and drill the other end.

Attention.—Withdraw work occasionally to let out chips, and oil into hole. Drill cast iron or brass dry; on steel or wrought iron, use oil.

VII. 2d or 3d speed. Place countersink *D* in drill chuck *E*, Fig. 26, hold stock *F* as before, start lathe and countersink one end to desired size, reverse work and countersink the other end.

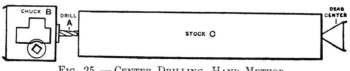

Fig. 25. — Center Drilling, Hand Method.

Fig. 26. Countersinking, Hand Method.

41. To remove or anneal broken center drill. — If the broken piece of drill does not drop out of hole with the aid of a scratch awl, strike stock with hammer; if it still remains, the work must be annealed, and the hole redrilled.

42. Machine method of finding the center, drilling and countersinking. — In Fig. 27 is a centering machine used to drill and countersink center holes accurately in shafts of any length.

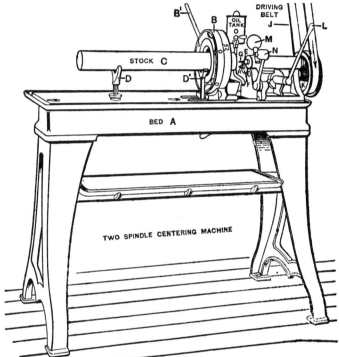

FIG. 27. — CENTER DRILLING AND COUNTERSINKING, MACHINE METHOD.

CENTERING MACHINE SCHEDULE OF PARTS AND OPERATIONS

A — Bed bolted to floor.

B and B′ — Universal chuck and operating lever.

C — Shows manner of holding stock to be centered.

D and D′ — Supports.

E — Drill spindle.

F — Countersink spindle.

G — Center drill.

H — Countersink.

J — Driving belt.

K — Gear mechanism which drives E and F at different speeds.

L — Handle operating both spindles.

M—Ball lever for tipping head to bring either spindle into aline-ment with stock.

N — Stock or work stop for uniform drilling.

O — Oil tank.

Warning. — The end of stock must be filed approximately flat and the drill started slowly or the drill will be liable to break.

43. Straightening shafting, rods, and bolts. — As soon as stock is centered, it is mounted on centers and revolved by hand and its eccentricity tested with chalk. If it is short or rigid, it may have to be straightened with a hammer on the anvil; but if slender, it may be straightened on lathe centers or with straightening press.

44. To test and straighten centered shafts in a lathe. — Unturned work that is centered is tested by rotating it in lathe and marking with chalk. For finished work, use copper tool held in tool post or a test indicator shown at *A*, Fig. 28.

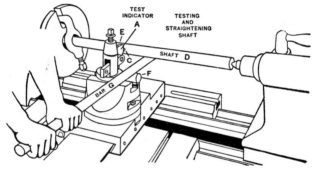

Fig. 28. — Testing and Straightening Shaft in Lathe.

Shank *B* is held in tool post; the cross slide is fed inward until feeler *C* touches revolving shaft *D*, when pointer *E* will indicate error in thousandths of an inch. With piece *F* for fulcrum and bar *G*, the shaft is straightened. Sometimes it

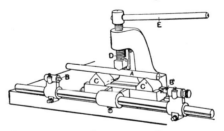

Fig. 29. — Straightening Shaft with Straightening Press.

is necessary to peen shaft by a few light blows of hammer on upper side, struck while shaft is pressed upward.

45. Straightening press, Fig. 29.—If shaft A is centered, it is tested by mounting on centers B and B'; if not centered, it is tested by sighting along its length and marking with chalk or metal workers' crayon (soapstone). It is placed on supports C and C' with its high side under screw D and pressure is applied with screw D and lever E.

CUTTING TOOLS

46. All cutting tools may be considered primarily as wedges driven into the material to separate it. A thin-edge tool cuts more easily, because it generates less friction, distorts the chips less and gives a greater freedom to their removal. The edge must be thick enough to carry a heavy cut at a suitable speed, and have a point of sufficient width to stand the heat generated by friction. Excessive heat will soon destroy the point of the tool. This limits the cutting speeds. A portion of the heat is conducted away through the work to the air and from point of tool to body and to air by direct radiation, and on tenacious metal by the lubricant. See Lubricants for Cutting Tools, pp. 148–150.

47. Rake, clearance, and cutting angle defined. — *Rake* is applied to angle of upper surface and *clearance* to angle of lower surface. The angle included between these surfaces is the *cutting* angle, or angle of keenness.

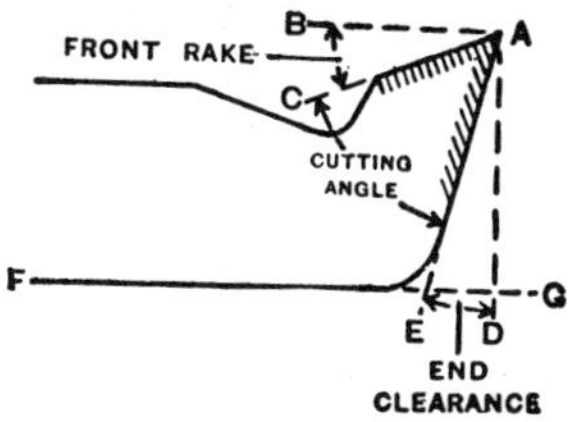

48. Front rake, end clearance, and cutting angle. — Fig. 30 is a side view of a square-nose tool partially cut by a plane giving a section CAE. Through point A lines AB and AD are drawn parallel and perpendicular to the base line FG. The front-rake angle

FIG. 30. — DIAGRAM OF FRONT RAKE, CLEARANCE, AND CUTTING ANGLE.

is BAC, positive when below and negative when above AB. The end-clearance angle is EAD. The cutting angle is CAE.

49. Side rake, side clearance, and cutting angle. — Fig. 31 is an end view of a right-side tool partially cut by a plane,

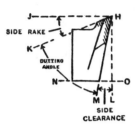

giving section KHM. Through point H lines HJ and HL are drawn parallel and perpendicular to the base of tool NO. The side-rake angle is JHK. The side-clearance angle MHL. The cutting angle is KHM.

50. The keenness of a cutting edge is increased or decreased by increasing or decreasing the rake angle.

FIG. 31.—DIAGRAM OF SIDE RAKE, SIDE CLEARANCE, AND CUTTING ANGLE.

Cast iron may be cut successfully with a cutting angle of from 60° to 75° ; steel and wrought iron, with a cutting angle of from 40° to 50°. See Brass Finishing, pp. 533–537.

51. Front and side rake combined. — For clearness the front rake and side rake are shown on separate tools, but some turning tools will cut more effectively if the top face is given a combination front and side rake in varying degrees to suit the nature of the work, as the diamond-point tool, Fig. 44.

52. "Rough square" and "rough turn." — Terms used to name the operations of removing the surplus material from a piece of metal by one or more roughing cuts preparatory to finishing.

53. "Finish square" and "finish turn." — Terms used to name the final finishing cuts which reduce any piece of metal to required size.

LATHE TOOLS FOR CAST IRON

54. Lathe tools are made of carbon steel and high-speed steel. See Holders and Cutters, pp. 306–309, 312, 314 and *Principles of Machine Work*.

55. A chart of forged lathe tools is shown in Fig. 32. For 10″ to 12″ lathes, they are made $\frac{3}{8}″ \times \frac{3}{4}″$ in section and 7″ in length; for 14″ to 16″ lathes, $\frac{1}{2}″ \times 1″$ section, 9″ in length. Other sizes in proportion. After being forged it is best to file them to proper shape before they are hardened.

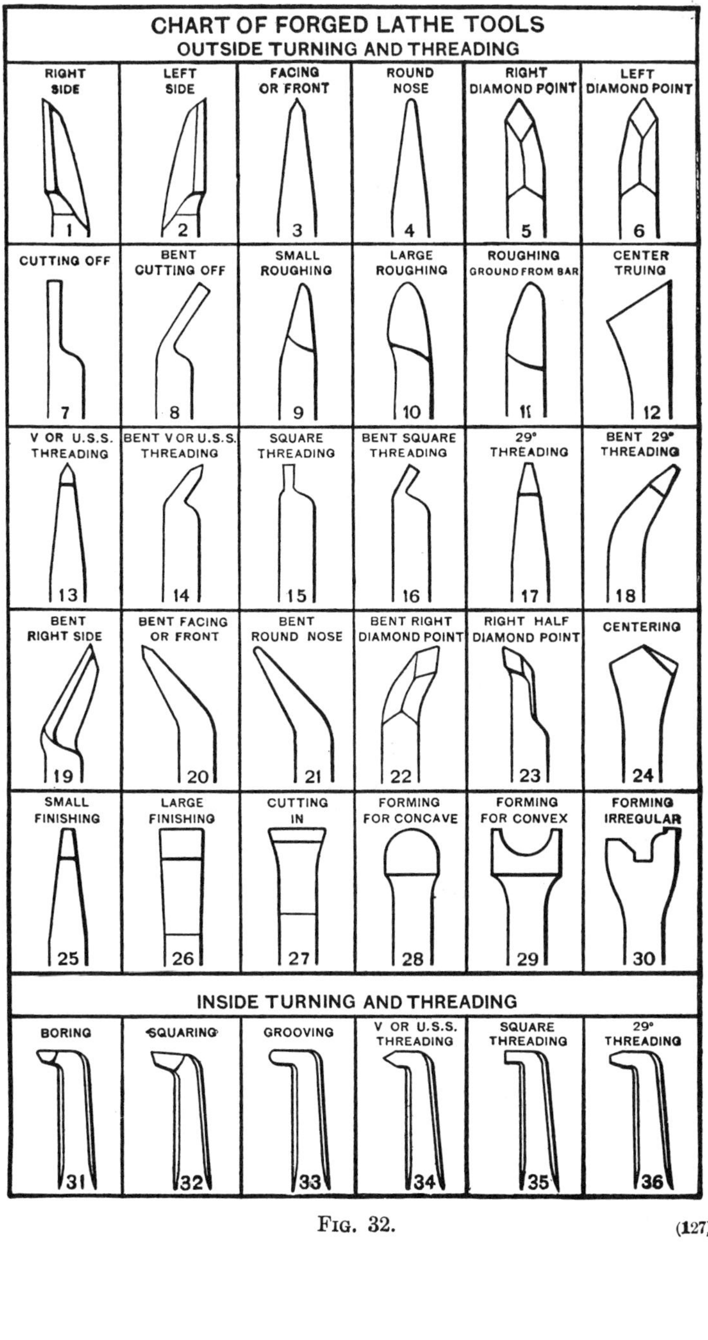

CHART OF FORGED LATHE TOOLS
OUTSIDE TURNING AND THREADING

RIGHT SIDE	LEFT SIDE	FACING OR FRONT	ROUND NOSE	RIGHT DIAMOND POINT	LEFT DIAMOND POINT
1	2	3	4	5	6
CUTTING OFF	BENT CUTTING OFF	SMALL ROUGHING	LARGE ROUGHING	ROUGHING GROUND FROM BAR	CENTER TRUING
7	8	9	10	11	12
V OR U.S.S. THREADING	BENT V OR U.S.S. THREADING	SQUARE THREADING	BENT SQUARE THREADING	29° THREADING	BENT 29° THREADING
13	14	15	16	17	18
BENT RIGHT SIDE	BENT FACING OR FRONT	BENT ROUND NOSE	BENT RIGHT DIAMOND POINT	RIGHT HALF DIAMOND POINT	CENTERING
19	20	21	22	23	24
SMALL FINISHING	LARGE FINISHING	CUTTING IN	FORMING FOR CONCAVE	FORMING FOR CONVEX	FORMING IRREGULAR
25	26	27	28	29	30

INSIDE TURNING AND THREADING

BORING	SQUARING	GROOVING	V OR U.S.S. THREADING	SQUARE THREADING	29° THREADING
31	32	33	34	35	36

FIG. 32.

(127)

56. Right and left tools. — A *right* tool cuts from right to left and a *left* tool from left to right. Tools are understood to be *right* unless otherwise designated.

57. The angles of lathe cutting tools. — In order to select the proper tool and prepare the correct cutting angle, the student should consider the kind and, if possible, the hardness of the metal and whether for taking a roughing or finishing cut. If the metal is very hard, the tool must be ground to a less acute cutting angle, the cutting speed reduced, or both.

58. Height of tool and tool block. — Various devices are used to regulate the height of the point of the tool. On

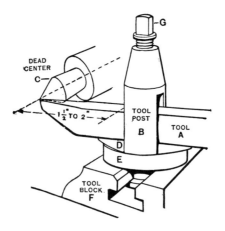

Fig. 33. — Setting Tool Height of Centers — Plain Rest.

small lathes the rise and fall rest operated by an elevating screw is perhaps the most common. Fig. 33 shows a plain rest. The point of tool *A* is adjusted in tool post *B* to height of dead center *C* by a tilting action of circular wedge *D* in concave washer *E* and the shank clamped by screw *G*.

59. Round-nose tool. — Fig. 34 shows a small round-nose tool used for roughing and finishing cast iron or brass. Face
A has no rake, but the sides have 10° clearance. The point is about
$\frac{1}{8}''$ thick. B is the cutting edge and CBD the clearance angle. Too little clearance will cause the tool to ride on the work and too much will weaken the cutting edge. When dull, grind end B and a little on top A.

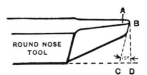

FIG. 34.— ROUND-NOSE TOOL FOR CAST IRON.

60. To square with round-nose tool. — The scale or skin on cast iron is very hard, and the round-nose tool is used to rough square ends and remove surplus stock, as in Fig. 35. Cast iron is machined dry. See Lubricants for Cutting Tools, pp. 148–150.

The work is mounted on centers and the lathe run at proper speed; arrow 1 shows direction of rotation and arrow 2 direction of cut. The long. feed handle is held firmly with one hand, while the tool is fed with the other operating the cross-feed handle.

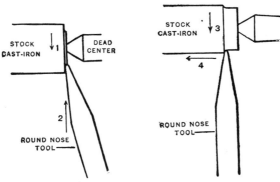

FIG. 35. — ROUGH SQUARING CAST IRON.

FIG. 36. — ROUGH TURNING CAST IRON.

61. To turn with round-nose tool. — For light rough turning on small diameters and for finish turning with fine feed, a round-nose tool may be used to advantage on cast iron, as in

Fig. 36. Arrow 3 shows direction of rotation of work and arrow 4 direction of cut. It is sometimes necessary to slant the tool to the left to turn close to a shoulder or dog, but the tool must be clamped extra firm or it may draw into the work and turn the diameter too small.

In Fig. 37 is shown a large round-nose tool for turning or facing large work. It is ground to shape from the bar and given side rake as at A, to give freedom to removal of chips.

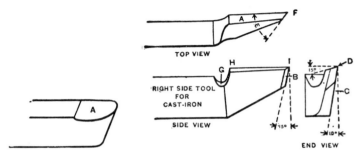

FIG. 37. — TOOL FOR HEAVY   FIG. 38. — SIDE TOOL FOR SQUARING
CUTS, ROUGH SQUARING   CAST IRON.
OR TURNING CAST IRON.

62. Side tool. — For squaring or facing the ends of shafts, shoulders, etc., a right-side tool, Fig. 38, is used. The tool has end clearance B, 15°; side clearance C, 10°; and side rake D, 15°. The angle E for point F is 60°. It is forged hollow at G, to facilitate grinding. Grinding is done on top A and end B with a little on side C. Cutting edge HI should be kept horizontal.

Fig. 39 shows a right-side tool suitable for heavy work.

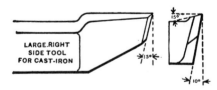

FIG. 39. — SIDE TOOL FOR HEAVY CUTS, SQUARING CAST IRON.

63. To square an end with side tool. — A side tool is set at the height of the center, as in Fig. 40. On diameters not larger than ⅜″, edge AB is set at right angles to the axis of work, so as to square the whole end at one cut. For large diameters the point should " drag " a little, as at A, Fig. 41,

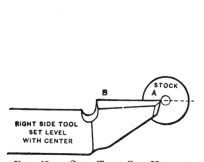
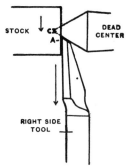

Fig. 40. — Side Tool Set Height of Centers.

Fig. 41. — Finish squaring Cast Iron.

for both roughing and finishing cuts. The point A being slightly the deepest, the tool when carried from center to circumference will produce a smooth surface, provided the tool is properly hardened, tempered, and ground, and the speed and feed are correct. For some purposes, especially in squaring compositions of brass, a side tool is fed inward.

64. To remove burr around countersink. — To remove the burr that remains around the countersink after taking the finishing cut, feed the point of tool up to surface of work and close to dead center; then unclamp binder and relieve dead center slightly with right hand and at same moment slightly feed tool inward with left hand, which will remove burr; then simultaneously feed tool outward and dead center back in place.

65. Grooved dead center for squaring. — The extra operation of removing burr around countersink when squaring may be avoided and time saved by using a grooved dead center A, Fig. 42. As point of tool B may be started or terminated in groove, no burr remains.

66. To square a shoulder with side tool. — To turn a portion
of a piece of stock and square the shoulder, as in Fig. 43, it is
marked as at *A*, then the cut taken to *B*, and the shoulder
squared to mark *A*.   The side tool is fed inward to touch the

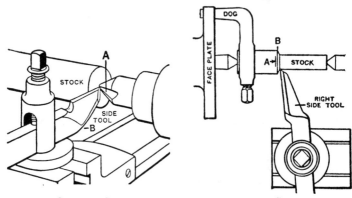

Fig. 42. — Grooved Dead Center   Fig. 43. — Squaring
         for Squaring.   Shoulder.

stock.   A moderately fast speed is used and the long. feed is
fed slowly with one hand, while the cross feed is held firmly
with the other; when the cut is carried far enough, the long.
feed is held firmly and the cross feed fed outward.

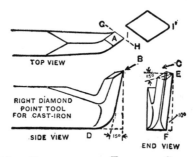

Fig. 44. — Diamond-point Tool for Cast Iron.

67. Right diamond-point tool. — Fig. 44 shows a right
diamond-point tool.   *A* is the top face, which is given a com-

bination front and side rake, as indicated by arrows *B* and *C*. Side clearance *EF* is 10°, but for a very coarse feed should be more.

Cast iron of small diameter may be turned by tools without rake, but for large diameters and heavy cuts a combination side and front rake of about 15° is effective. The cutting is done by edge *GH* and point *I*, which should be rounded, as shown enlarged at *I'*, to strengthen it and produce a smoother cut. The tool is ground on the top face *A*, and if necessary, a little on the side faces.

68. Height of lathe turning tools in relation to axis of work.— The point of taper turning and threading tools must be set at

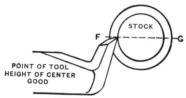

FIG. 45. — DIAMOND-POINT TOOL SET HEIGHT OF CENTERS. GOOD.

height of center, as at *FG*, Fig. 45. The student may apply this rule at all times and obtain good results.

69. Evil effects of setting a tool too low or too high. — A tool point set below the center *FG*, as exaggerated in Fig. 46, increases the clearance and decreases front rake, will not cut

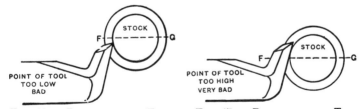

FIG. 46. — DIAMOND-POINT TOOL SET BELOW CENTERS. BAD.

FIG. 47. — DIAMOND-POINT TOOL SET ABOVE CENTERS. BAD.

properly, and will dull quickly. A tool point too high above center *FG*, exaggerated in Fig. 47, reduces the clearance, will ride on the work and soon destroy itself by friction.

70. Theoretical height of turning tools for straight work. —
In Fig. 48 is shown the theoretically correct height to set
the point of a tool, which increases its keenness and gives the
greatest support to its cutting edge. This height is at the
tangent point A of line BC, and is located by drawing line DE
through center of work at 90° to line BC. As this gives no
clearance, the tool in practice is set slightly below this point.

After a little training, one is able to set the tool point at
the most suitable height in relation to center FG for any
diameter of work.

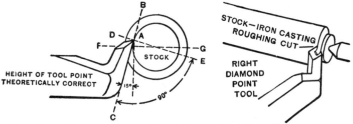

FIG. 48. — DIAMOND-POINT TOOL SET   FIG. 49. — ROUGH TURNING
THEORETICAL HEIGHT. GOOD.   CAST IRON, COARSE FEED.

71. Rough turning cast iron. — Fig. 49 shows a diamond-
point tool taking a roughing cut by power long. feed on a
cast-iron piece mounted on centers in an engine lathe. The
chips from cast iron break off in small fragments.

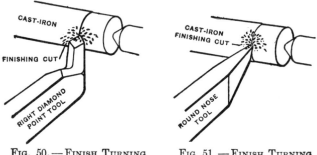

FIG. 50. — FINISH TURNING   FIG. 51. — FINISH TURNING
CAST IRON, FINE FEED.   CAST IRON, FINE FEED.

72. To finish turn cast iron. — Fig. 50 shows diamond-
point and Fig. 51 round-nose tool taking finishing cuts ($\frac{1}{64}''$)

with fine feed. Fig. 52 shows small square-nose tool and Fig. 53, large square-nose tool taking finishing cuts (.010") with coarse feeds.

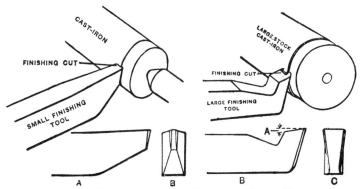

FIG. 52.— FINISH TURNING CAST IRON, MEDIUM FEED.

FIG. 53. — FINISH TURNING CAST IRON, COARSE FEED.

GRINDING OR SHARPENING LATHE TOOLS

73. Grinding wheels for sharpening lathe tools are generally made by the silicate process from natural or manufactured abrasives. The numbered size of the grain is usually about 30 and the grade, $3\frac{3}{4}$. When vitrified wheels are used, the grain size is usually 24 to 36 and the grade, M. Wheels may be from 20" to 24" in diameter, and width of face from $1\frac{1}{2}$" to 2". The surface speed is 4000 to 5000 feet per minute and a good supply of water should be provided. Cup wheels that give a straight line clearance are also used for tool grinding.

Danger. — To avoid accidents, grinding wheels must be kept true with a dresser or diamond. The tool rest must be securely clamped close to the wheel. Do not grind on the side of disk wheels. Never grind short pieces upon these wheels.

Warning. — Always move a tool back and forth slightly, when grinding. This prevents grinding grooves in the surface of the tool, and also avoids the generation of excessive heat and consequent drawing the temper and softening the cutting edge. This movement allows the water to circulate between wheel and tool.

Grinding with too heavy pressure on a dry wheel, or when the water supply is insufficient, will draw the temper and soften the point of the tool.

GRINDING FORGED TOOLS

74. Forged lathe tools are made from bar stock to the approximate shapes, sizes and angles, and then ground to rake and clearance angles. See pp. 125, 126.

To grind forged round-nose tools. — Forged round-nose tools are ground with front rake and end and side-clearance angles. Sharpening requires only regrinding of end clearance angle.

Important. — If round-nose tools are properly forged and ground with side-clearance angles, sharpening will require no regrinding of side clearance.

To grind end clearance on forged round-nose tools. — Fig. 54 shows the general method of grinding end clearances on lathe tools. To obtain this clearance, hold tool *A* as dotted in Fig. 55, with shank *B* upon rest *C*, and with heel *D* of tool

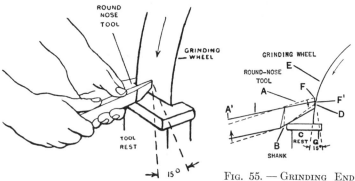

FIG. 54. — GRINDING A ROUND-NOSE TOOL.

FIG. 55. — GRINDING END CLEARANCE ON A ROUND-NOSE TOOL.

just touching grinding wheel *E*. Then raise shank of tool to full line position *A'* or until point *F* of tool touches the wheel, as at *F'*. This preserves the original clearance angle *G*.

To round end of forged round-nose tools. — Place tool *A*, Fig. 56, on rest *B*, and hold end of tool against grinding wheel *C*, as shown in detail, to preserve clearance angle *D*.

With *E* as a center, swing shank of tool back and forth several times, to positions *F* and *F'*, to round end of tool. Angle of swing should be about 100°.

To grind top face or front rake of forged round-nose tools,
Fig. 57. — Hold tool *A* on rest *B* with top face *C* against grind-

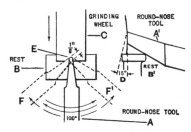

FIG. 56. — GRINDING THE ROUNDED
END OF A ROUND-NOSE TOOL.

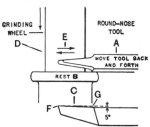

FIG. 57. — GRINDING THE TOP
OF A LATHE TOOL.

ing wheel *D*. Move tool back and forth slightly, as indicated by arrows *E*. Point *F* of tool should be slightly higher than point *G* to form front rake of 5°.

To grind forged side tools. — Forged side tools are ground with front and side rake angles and with end and side clearance angles. Sharpening requires only regrinding of end clearance and side rake.

To grind end clearance on forged side tools. — Place tool *A*, Fig. 58, on rest *B* with end of tool against grinding wheel

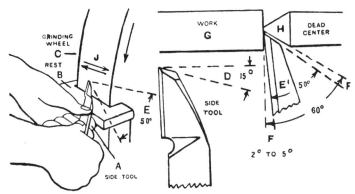

FIG. 58. — GRINDING END CLEARANCE OF FORGED SIDE TOOL.

C. Hold tool in position to give clearance of 15°, as at *D*. Grind point of tool to an angle of 50°, as at *E*, *E'*, to give clearance F, F' between end of work G and dead center H. Move tool back and forth in direction of arrows at J.

To grind front and side rake on forged side tools. — Hold tool A, Fig. 59, on rest B against grinding wheel C with tool tilted to give front rake of 5°, as at D, and side rake of 15° for

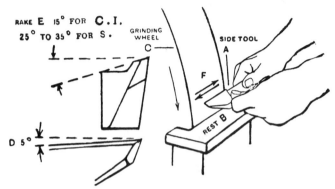

FIG. 59. — GRINDING THE TOP OR RAKE ANGLE OF A SIDE TOOL.

cast iron or from 25° to 35° for steel, as at E. Move tool back and forth in direction of arrows F.

Note. — The front rake of 5° is often called the *back* slope; and the side rake of 15° to 35°, the *side* slope.

To grind side clearance on forged side tools. — Place tool A, Fig. 60, on rest B against grinding wheel C. Hold tool at

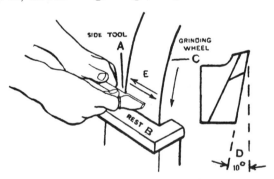

FIG. 60. — GRINDING THE SIDE CLEARANCE OF A SIDE TOOL.

an angle to give side clearance of 10°, as at *D*. Use light
pressure. Move tool back and forth in direction of arrows *E*.

To grind forged diamond-point tools. — Forged diamond-
point tools are ground with front and side-rake angles, and end
and side clearance angles. Sharpening requires only regrinding
of front and side rake.

To grind front and side rake on forged diamond-point tools,
Fig. 61. — Place tool *A* on rest *B* with end of tool against
grinding wheel *C*. Hold tool in position to give front rake

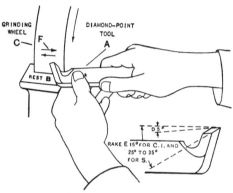

Fig. 61. — Grinding the Top Face or Rake Angle of a Diamond-
Point Tool.

of 5°, as at *D*, and side rake of 15° for cast iron or from 25°
to 35° for steel, as at *E*. Move tool back and forth in direction of arrows *F*.

**To grind side and end clearance on forged diamond-point
tools,** Fig. 62. — Place tool *A* on rest *B* with face *C*, *C'* against
grinding wheel *D*. Hold tool to give side clearance of 10°
to 15° as at *E*. Swing tool from position *F* through 100° to
F', to round end of tool slightly, as at *G*, and to produce end
clearance of 15°, as at *H*. End of tool should appear as in
detail, at *J*. When grinding flat faces, move tool back and
forth slightly, in direction of arrows *K*.

Note.— Round end of tool at *G* $\frac{1}{16}''$ for work $\frac{3}{4}''$ or less in diameter.
Use larger curvature for larger work.

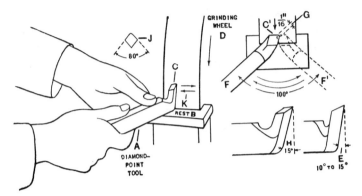

FIG. 62. — GRINDING THE CLEARANCE OF A DIAMOND-POINT TOOL.

GRINDING HIGH–SPEED STEEL INSERTED CUTTERS OR BITS

75. Forged lathe tools, and holders and cutters compared. — Fig. 63 shows a forged side tool at A, and a side tool cutter or

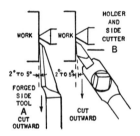

FIG. 63. — SIDE TOOLS.

bit in holder, at B. Both kinds of tools are used for facing or squaring. They are clamped in the positions shown in Fig. 63, at an angle of 2° to 5° from the surface to be finished.

At A, Fig. 64, a forged diamond-point tool is shown, and a diamond-point cutter or bit in holder, at B. They are clamped in the same relative positions for rough or finish turning.

Fig. 65 shows a round-nose cutter or bit and holder, used in place of a forged tool.

The set-up for turning is shown at *A*, and for rough squaring or cutting-in on large work, at *B*.

When forged tools have been ground excessively they require reforging to reform the cutting end. Cutters or bits

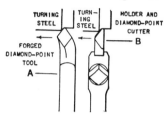

FIG. 64. — TURNING TOOLS.

may be extended in the holders and reground without forging.

Inserted cutters or bits are obtainable in square sections $\frac{3}{16}''$ to $1\frac{1}{8}''$, lengths $1\frac{3}{4}''$ to $8''$; and in round sections $\frac{1}{4}''$ to $\frac{5}{8}''$, lengths $2\frac{3}{8}''$ to $5\frac{1}{2}''$, heat treated or hardened ready for grinding.

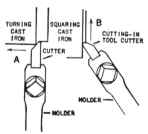

FIG. 65. — CUTTERS GROUND FOR CAST IRON.

A cutter is generally ground to the same rake angles on each end, although the type of point may be different. This permits each end of a cutter to be used on the same kind of material. See pp. 307, 308.

To grind side tool cutter in tool holder. — Side tool cutters are ground with front and side rake angles, and end and side clearance angles.

To sharpen, grind end and side clearance, and rake angle slightly.

Extend cutter in holder. — To avoid grinding holder, it is best to extend cutter or bit a further distance out of holder.

To grind side clearance on side tool cutter, Fig. 66. — Place holder *B* on rest *C* with end of cutter against grinding wheel *D*. Hold tool in position to grind face *E* and produce angle of 35°

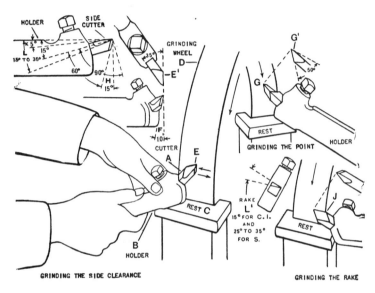

Fɪɢ. 66. — Gʀɪɴᴅɪɴɢ ᴀ Sɪᴅᴇ Tᴏᴏʟ Cᴜᴛᴛᴇʀ ɪɴ Hᴏʟᴅᴇʀ.

with holder, as at *E'*. Then tilt holder slightly and grind to produce side clearance angle of 10°, as at *F*. Grind point of tool to an angle of 50°, as at *G* and *G'*, to give clearance between work and center, and make end clearance angle of 15°, as at *H*.

To grind front and side rake on side tool cutter. — Place holder on rest with cutter against grinding wheel and holder inclined, as at *J*, Fig. 66, to give front rake of 5°, as at *K*, and side rake of 15° for cast iron, or from 25° to 35° for steel, as at *L* and *L'*.

Important. — When sharpening tool, avoid excessive grinding of rake angle *L*, *L'*, which will injure shape of tool.

Information. — The square hole of holder is usually 15°. There-fore, this angle must be taken into consideration if the cutter is ground out of holder, or in holder with parallel hole.

To grind diamond-point cutter in holder. — Diamond-point cutters are ground with front and side rake angles, and end and side clearance angles. Sharpening requires regrinding end and side clearance.

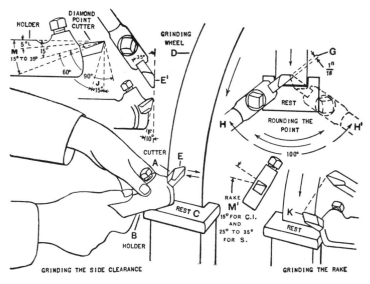

FIG. 67. — GRINDING A DIAMOND-POINT CUTTER IN HOLDER.

To grind side and end clearance on diamond-point cutter. — Place holder B on rest C, Fig. 67, with end of cutter against grinding wheel D. Hold tool in position to grind face E and produce angle of 35° with holder as shown at E'. Then tilt holder slightly and grind to produce side clearance of 10°, as at F. Round point G of tool by swinging holder through angle of 100° from H to H', and at same time grind end clearance angle of 15°, as at J.

To grind front and side rake on diamond-point cutter. — Place holder on rest with cutter against grinding wheel, as at *K*, Fig. 67, to give front rake of 5°, as at *L*, and side rake of 15° for cast iron, or from 25° to 35° for steel, as at *M* and *M'*.

Important. — When sharpening tools, avoid excessive grinding of rake angle *M*, *M'* which will injure shape of tool.

Information. — The hole in this holder is usually 15°. Therefore, this angle must be taken into consideration if the cutter is ground out of holder or in holder, with parallel hole.

To grind high-speed steel or stellite cutters for large work or heavy cuts. — See Stellite, p. 310. The clearance and rake angles are much smaller for stellite cutters than for carbon or high-speed steel tools.

Fasten cutter *A*, Fig. 67a, in special holder *B* provided with a pressure clamp *C* to prevent set screw *D* breaking the cutter.

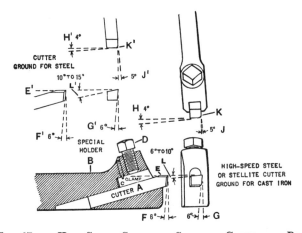

Fig. 67a. — High-Speed Steel or Stellite Cutter or Bit.

Grind cutter either on large grinding wheels or on the face of cupped wheels. Grind cutting edges horizontal to give no rake, as at *E,E'*. End clearance at *F,F'* and side clearance at *G,G'* should not be over 6°. Grind front of tool at angle of 4° with work to give drag clearance, as at *H, H'*. The cutting edge is ground back to an angle of 5°, as at J,J' to give a slight diamond-point tool action. Round corner of tool, as at K, K', to a radius of $\frac{2}{3}$ depth of cut. Grind rake angle of 6° to 10°, as at L for cast iron and 10° to 15° for steel, as at L'.

Tc grind cutters separate from tool holders. — To avoid

FIG. 67b. — SHARPENING CUTTERS IN SPECIAL GRINDING HOLDERS.

cutting into tool holder when grinding, special holders are obtainable, as at A, Fig. 67b. The hole in this type of holder is parallel to the holder and not at an angle of 15° as in regular holders.

Cutters or bits may also be held with the fingers and ground, but this method is not recommended.

76. To grind cutting-off tools. — Extend cutter A, Fig. 67c, if necessary, a short distance out of holder B. Hold

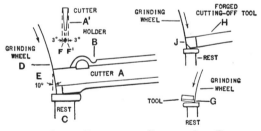

FIG. 67c. — GRINDING A CUTTING-OFF TOOL.

cutter and holder on rest C with end of cutter against grinding wheel D. Grind end clearance of 10°, as at E. The cutters are provided with a side clearance of about 3°, as at FF', and require no grinding. The top face may be ground, as at G. For forged cutting-off tool H, grind end clearance, as at J. The side clearance should be forged correctly but sides may be smoothed by grinding. Grind top face if required, as at G.

77. To grind forged United States Standard or Sharp V threading tool. — Place tool A, Fig. 67d, on rest B with top face of tool against grinding wheel C. Move tool back and forth in direction of arrows D. Grind top of tool flat and parallel with shank.

To bevel end, hold tool E against grinding wheel and grind a bevel F approximately 30° with the shank of the tool. Move tool back and forth in direction of arrows G. Reverse position of tool and grind a second bevel to give a 60°-angle at end of tool. Test angle with gage, as at H. Grinding operations should also give end clearance of 15°, as at J.

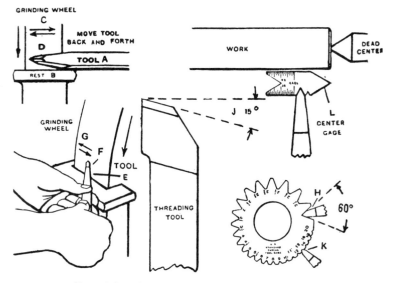

FIG. 67d. — GRINDING A FORGED THREADING TOOL.

Grind point of tool to fit gage for required number of United States Standard threads, as at K. For Sharp V threads this operation is omitted.

To set tool perpendicular to work, use center gage L.

Important. — For sharpening, grind tool on top surface. Do not grind end angle after once correctly obtained.

To grind threading tool cutter in holder. — Threading tool cutters for use with special holders, are obtainable ground to shape, and require sharpening on top face only, see p. 309, No. 13; also see p. 1216.

To grind ordinary cutters, as at A, Fig. 67e, first extend cutter if necessary, in holder B.

Place holder B on rest C with end of tool against wheel D. Grind top of tool flat, as at E, and parallel with top of holder. Move tool back and forth in direction of arrows F.

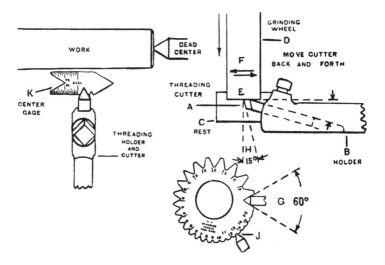

Fig. 67e. — Grinding a Threading Cutter in Holder.

Bevel end of tool to angle of 60° by method shown in **Fig.** 67d. Test angle of tool with gage, as at G. Grinding operations should also give clearance angle of 15°, as at H.

Grind point of tool to fit gage for required number of United States Standard threads, as at J. For Sharp V threads this is not required.

To set tool perpendicular to work, use center gage K.

Important. — For sharpening, grind tool on top surface. Do not disturb end angle after once correctly obtained.

Information. — Square, 29°, and other forms of threading tools, may be made by grinding a cutter or bit to the desired shape.

78. To grind boring or inside turning tools. — Boring tools, either forged or holder and cutter, are ground with rake and clearance similar to those on tools for external work. The end clearance angle is 15° for large holes, as at *A*, Fig. 67f, but should be increased to 20° for small holes to prevent contact at *B*. Side clearance is generally 10°, as at *C*. The front

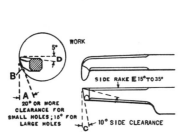

Fig. 67f. — A Forged Boring Tool.

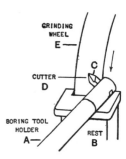

Fig. 67g. — Grinding the Side Clearance on a Boring Tool (Holder and Cutter).

rake is 5°, as at *D*, and the side rake varies from 15° for cast iron to 25° to 35° for steel, as at *E*.

To grind boring or inside turning tools. — To grind side clearance, place forged tool or holder and cutter *A*, Fig. 67g, on rest *B* with face *C* of tool or cutter *D* against grinding wheel *E*. Hold in position to grind side clearance of 10°, as at *C*, Fig. 67f.

To grind end clearance, place forged tool or holder *A*, Fig. 67h, on rest *B* with face *C* of tool or cutter *D* against grinding wheel *E*. Swing tool back and forth from position *A* in direction of arrow *F* to round end of tool, as at *G*, and to produce end clearance angle of 15° or 20°, as at *H*. See also *A*, Fig. 67f.

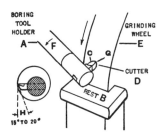

FIG. 67h. — GRINDING END
CLEARANCE ON BORING TOOL.

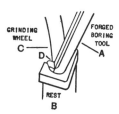

FIG. 67j. — GRINDING A FORGED
BORING TOOL.

Round end of tool G, Fig. 67h, $\frac{1}{16}''$ for work $\frac{3}{4}''$ or less in diameter. Use larger curvature for larger work.

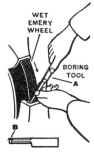

FIG. 67k. — GRINDING A
FORGED BORING TOOL.

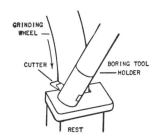

FIG. 67 l. — GRINDING THE
TOP OR RAKE OF A BORING
TOOL (HOLDER AND CUTTER).

To grind front and side rake, place end of shank of boring tool A, Fig. 67j, on rest B. Hold tool inclined as at A, Fig. 67k. Press tool against grinding wheel C, Fig. 67j, as at D, and grind to give front and side rake, as at D and E, Fig. 67f.

Fig. 67 l shows the method of grinding rake angle on boring cutter.

Important. — When sharpening boring tools, grind sparingly the side and end clearance.

79. To grind United States Standard or Sharp V inside or internal threading tools. — Inside or internal threading tools may be forged, as at A, Fig. 67m, and are also obtainable in the form of high-speed steel cutters to fit holders, as at B, and as formed cutters to fit holders at C.

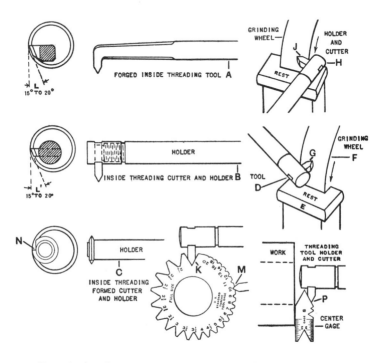

FIG. 67m. — GRINDING AND SETTING INSIDE THREADING TOOLS.

To grind forged tool or cutter and holder, place tool D on rest E and against grinding wheel F.

Hold tool in position and grind bevel at G approximately 30° with cutter. Reverse position of tool, as at H, and grind a second bevel J to give a 60°-angle at end of tool. Test angle with gage, as at K. Grinding operations should also give end clearance of 15° to 20°, as at L and L'. Grind point of tool to fit gage for required number of United States Standard threads, as at M. For Sharp V threads this operation is omitted. Grind top of tool slightly to remove burrs. Allow no rake. In grinding formed cutter, as at C, do not grind sides since cutter is made with end angle accurate. This cutter may be ground repeatedly on top face N without changing shape of cutter.

To set tool perpendicular to work, use center gage, as at P.

Important. — For sharpening, grind tool on top surface. Do not d sturb end angle after once correctly obtained.

80. Universal tool grinder. — Machine method of grinding lathe and planer tools, and duplicating angles.

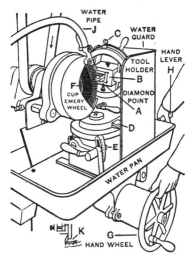

FIG. 68. — DUPLICATING ANGLES OF LATHE AND PLANER TOOLS.

Fig. 68 shows how to grind rake of a right diamond-point tool. Clamp diamond-point A in holder B. Set three graduated circles, C, D, and E, to readings obtained from chart (chart furnished with machine), or obtain setting by trial.

Move tool to radial face of cup-shaped emery wheel F with hand wheel G. Carry tool back and forth across wheel with lever H. Water is supplied from pipe J.

A fourth graduated circle on rear of tool holder as shown in detail at K, is used in grinding bent tools.

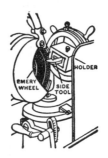

FIG. 69. — GRINDING SIDE TOOL.

To grind side faces, revolve holder B and set dials to give proper angles. To reduce area to be ground, tools may be forged in former blocks, or by hand to more clearance than desired. The method of grinding rake of a right side tool is

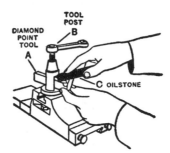

FIG. 70. — OILSTONING LATHE TOOLS.

shown in Fig. 69. When correct angles and settings of tools are obtained, they can be accurately duplicated.

Round-nose or circular forming tools are located centrally with a gage and ground by swinging on a vertical axis. The top is ground in the usual way.

81. To oilstone tools. — Use fine manufactured or hard Arkansas stone about $4'' \times \frac{3}{4}'' \times \frac{3}{4}''$. Clamp diamond-point tool A reversed in tool-post B and apply oilstone C as in Fig. 70, with long strokes. Also oilstone side faces. Use kerosene oil for India, lard or sperm oil for Arkansas, and sperm or lard oil for carborundum stones.

SETTING AND USING OUTSIDE CALIPERS

82. To set outside calipers. — If the light comes from the side, it is best to hold the calipers and rule while adjusting, as in Fig. 71-I.

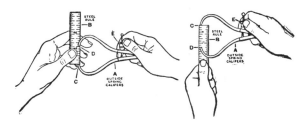

I   II
FROM SIDE–LIGHT   FROM OVER–HEAD LIGHT
FIG. 71 — I, II. — SETTING OUTSIDE CALIPERS.

Place one point of calipers A, Fig. 71-I, against end of rule B and steady with fingers as at C. Adjust other point to desired distance D, by turning thumb nut E.

If light comes from overhead, it is best to hold the calipers and rule in position shown in Fig. 71-II while adjusting. The width of lines on steel rules is from $.002''$ to $.004''$.

Note. — To set calipers accurately to line on rule, avoid a shadow at D by holding rule toward source of light.

Information. — Calipers carefully set to a steel rule, and used with a delicate touch, will measure within $.001''$ (one-thousandth of an inch).

Warning. — A student almost invariably sets outside calipers large and inside calipers small.

When extreme accuracy is required, such as turning work to be fitted, the calipers should be set by a standard plug gage or mandrel, or work of the desired diameter.

83. To measure diameter of lathe work with outside calipers. Fig. 72. — Set calipers F to size. Hold work G stationary and apply calipers at right angles to axis of work as HI, not as JK or LM. Turn work with tool N until calipers will pass over it with a light yet distinct touch, but

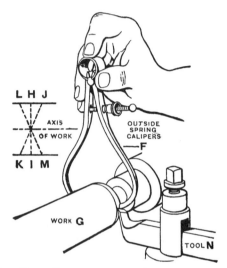

FIG. 72. — MEASURING WITH OUTSIDE CALIPERS.

not hard enough to spring calipers or to sustain their weight. Usually the calipers have to be passed over work a number of times to determine this touch.

84. To adjust the tool to turn work to a desired diameter. Fig. 72.— Move tool N inward at end of work to cut under scale, start lathe and feed tool to cut by hand, then throw in power long. feed and allow a travel of $\frac{1}{8}''$ to $\frac{1}{4}''$. Stop lathe and test diameter with calipers. If correct, continue turning; but if too large, start lathe, release power long. feed and run tool back to end and again slightly advance tool, throw in long. feed and so on until correct diameter is obtained.

Warning. — Hold long. feed firmly with one hand while releasing power feed with the other.

85. To learn to measure accurately with outside calipers. — Set calipers to rule, turn work until calipers will pass over it with a delicate touch, then test with micrometer calipers.

86. To transfer a setting from one pair of calipers to another. Fig. 73. — Set inside calipers A to size of hole, then set outside calipers B by them. Bring lower points of both calipers in contact and steady as at C, then adjust point D

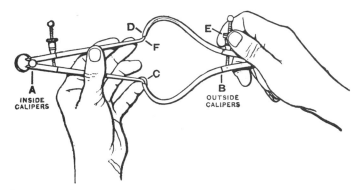

FIG. 73. — TRANSFERRING MEASUREMENT FROM INSIDE TO OUTSIDE CALIPERS.

by nut E until it touches point F. To transfer setting of outside to inside calipers, reverse calipers in hands and adjust.

CUTTING SPEEDS, CUT-METER, AND FEEDS

87. In turning, three things must be considered:

First, the cutting speed in feet per minute, which is controlled by the diameter of work and speed obtained from the table; it is calculated or may be directly measured by a cut-meter.

Second, the depth of cut, one-half the amount that the diameter is reduced.

Third, the feed or amount the tool advances per revolution of work.

88. The cut-meter, *A*, Fig. 74, may be used to measure cutting speed automatically, also speed of drills, milling cutters, etc. It consists of a case *B*, which contains the magnetic mechanism for registration. Scale *C* is calibrated to read the

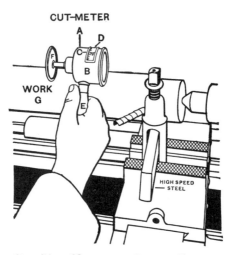

FIG. 74. — MEASURING CUTTING SPEED.

cutting speed in feet per minute; the 0 line is on glass *D*. To use the cut-meter, it is held by handle *E* and wheel *F* pressed against the revolving work *G*.

89. Surface speed attachment for speed indicator. — A rubber-tired wheel *A*, Fig. 75, 6″ in circumference, is slipped over

FIG. 75. — SURFACE SPEED ATTACHMENT FOR SPEED INDICATOR.

point of a speed indicator. See *Principles of Machine Work.* The wheel is pressed against work as at *B*, Fig. 75, and number of revolutions noted in a given time, as one minute. To get surface speed in feet, divide number of revolutions per minute by 2.

90. To find the lathe revolutions, given the cutting speed and diameter of the work: Multiply the cutting speed by 12 to reduce it to inches, then divide by the diameter of the work multiplied by 3.1416.

Example. — The work is 2″ in diameter and it is desired to turn it at 35 feet per minute. How many lathe revolutions are necessary?

Solution. $\dfrac{35 \times 12}{2 \times 3.1416} = 67$ R.P.M.

91. To find the cutting speed, given diameter of the work and lathe revolutions per minute: Multiply diameter of work by 3.1416 and by number of lathe revolutions, then divide by 12.

Example. — The work is 2″ in diameter and makes 67 revolutions per minute in the lathe. What is the cutting speed?

Solution. $\dfrac{2 \times 3.1416 \times 67}{12} = 35$ feet.

92. Eight or ten speeds are possible in an engine lathe. — Place the belt on step of cone that will give the speed nearest to that required.

TABLE OF LATHE CUTTING SPEEDS

MATERIALS.	SPEEDS FOR ROUGHING.	
	CARBON STEEL TOOLS.	HIGH-SPEED STEEL TOOLS.
Cast iron......................	30	60
Steel or wrought iron............	25	50
Carbon steel, annealed...........	20	40
Brass composition...............	95	190

Speed for finishing is 50% to 100% higher than roughing speed.

Speed for filing (brass excepted) equals four times roughing speed.

Speed for filing brass equals three times roughing speed.

Attention. — Roughing cuts are not to reduce diameters more than $\frac{1}{8}$″; and finishing cuts, not more than $\frac{1}{64}$″. Roughing feeds are not to be less than 17 revolutions per 1″ tool travel, and finishing feeds are not to be less than 90 revolutions per 1″ tool travel.

If the above is exceeded, a less speed than that given in table must be used.

For large diameters, use reduced speed, as the strain on the tool is greater.

93. Cutting feeds. — Lathes with belt feed are limited to three changes; lathes with gear feed are not limited.

For heavy work, coarse feed is about 17 lathe revolutions to 1″ of tool travel.

For average work, medium feed is about 38 lathe revolutions to 1″ of tool travel.

For average work, fine feed is about 90 lathe revolutions to 1″ of tool travel.

For tool making, a feed of 200 revolutions is often used.

Attention. — A student may use finer feeds until he acquires some experience; about 80 to 1″ for roughing and 140 to 1″ for finishing.

LUBRICANTS FOR CUTTING TOOLS

94. Some metals are machined dry, others require a lubricant. — Cast iron, with about two exceptions, as tapping and polishing, must be machined dry. Steel and wrought iron, with two exceptions, as machining with side or diamond-point tool (and these are optional), must be machined with a lubricant. A neglect of this may cause destruction of both tool and work. See Table of Lubricants, p. 149.

A lubricant on tenacious metal prevents excessive friction and conducts away heat, thus preserving the point of the tool and producing a smooth finish on the work. It also helps to carry away the chips.

Oil. — " Oil " in tables means lard oil or some oil mixture. Mixtures and compounds are often used as substitutes because of cheapness.

Note. — Never use mineral oils (lubricating oils) for cutting tools. To machine bronze, see p. 149.

95.   TABLE OF LUBRICANTS FOR CUTTING TOOLS.

OPERA-TIONS.	CAST IRON	MACHINE STEEL OR WROUGHT IRON.	CARBON OR HIGH-SPEED STEEL.	COP-PER.	BRASS OR BRONZE.*	ALUMINUM.	LEAD. BAB-BITT.
			METALS.				
Turning. Boring.	Dry.	Dry, oil or soda water.	Dry or oil.	Milk.	Dry.	Kerosene or turpentine.	Dry.
Cutting off. Grooving.	Dry.	Oil or soda water.	Oil or soda water.	Milk.	Dry.	Kerosene or turpentine.	Dry.
Screw cutting.	Dry.	Oil.	Oil.	Milk.	Dry.	Kerosene or turpentine.	Dry.
Threading with dies.	Dry.	Oil.	Oil.	Milk.	Dry.	Kerosene or turpentine.	Oil.
Tapping.	Oil.	Oil.	Oil.				
Drilling. Counter-sinking.	Dry. Dry.	Oil or soda water.	Oil or soda water.	Milk.	Dry.	Kerosene or turpentine.	Oil.
Counter-boring.	Dry.	Oil or soda water.	Oil or soda water.	Milk.	Dry.	Kerosene or turpentine.	Oil.
Chucking.	Dry.	Oil or soda water.	Oil or soda water.	Milk.	Dry.	Kerosene or turpentine.	Dry.
Reaming.	Dry.	Oil.	Oil.	Milk.	Dry.	Kerosene or turpentine.	Dry.
Milling.	Dry.	Oil, soap mixture, or soda water.	Oil, soap mixture, or soda water.	Milk.	Dry.	Kerosene or turpentine.	Dry.
Planing.	Dry.	Dry, oil or soda water.	Dry or oil.	Milk.	Dry.	Kerosene or turpentine.	Dry.
Nurling.	Oil.	Oil.	Oil.	Milk.	Oil.	Kerosene or turpentine.	Dry.
Filing.	Dry.	Dry or oil.	Dry or oil.	Dry or milk.	Dry.	Kerosene or turpentine.	Oil.
Polishing with emery cloth.	Oil.	Oil.	Oil.	Oil.	Oil.	Oil.	Oil.

* *Warning.* — When tapping or using a die on rolled brass (not cast brass), it is often necessary to use lard oil to prevent clogging and ruining **tap or die.**

Soda water (sal soda dissolved in water) is useful in machining steel or wrought iron, and produces what is known as a water finish.

Soap mixture. — ½ lb. sal soda, ½ pt. lard oil, ½ pt. soft soap, 10 qts. water; boil one-half hour. This is good for milling and drilling steel and wrought iron.

96. Drilling extra hard steel such as unannealed carbon steel. — Use a flat drill, with turpentine or kerosene oil as a lubricant.

97. Drilling glass. — Use a very hard flat drill, scratch glass with an old file to remove polish and use turpentine as a lubricant. Drill part way through, then turn over and finish from the other side to avoid chipping surface.

INSPECTING AND MEASURING MATERIAL (STOCK)

98. Inspecting and measuring material (stock). — On receiving a piece of stock, inspect it for imperfections, as blowholes in castings, flaws (cold short) in forging or bars. Stock for work to be hardened must be carbon or high-speed steel. To distinguish high-grade machine steel from tool steel, a trial piece may be tested by the hardening process.

Measure piece to see if it is large enough to finish to dimensions given on the drawing. If the piece be a rough casting or forging, there should be at least ⅛″ surplus stock; if smooth, $\frac{1}{16}$″.

When requested to finish work begun by another, inspect and measure it, and make a note of its condition, that you may not be held responsible for errors not your own.

99. Rough turn all over before finishing. — As metals alter in form when the skin or outside is removed, when possible remove the skin from all surfaces before finishing any part. An exception is made in some classes of lathe work, as a shaft which is squared to exact length before the diameter is roughed out.

100. When and how to oil the bearings of machine tools. —
All bearings must be regularly oiled. Use good lubricating
oil, — machine oil, not lard oil. If the oil does not sink into
the oil holes, they should be cleared out. A few drops of oil
in each bearing is enough, Fig. 76. Flat bearings, as the

FIG. 76. — OILING SPINDLE
OF A MACHINE.

FIG. 77. — OILING WAYS OF
A MACHINE.

ways of a lathe, should be wiped with waste before oiling
and the oil distributed with the fingers, Fig. 77. Oil twice
a day. All automatic oilers should be filled periodically.

**101. Treatment when bearings rough up and machine stops
from lack of oil or too close adjustment.** — First force in a
liberal quantity of oil; if this does not release the bearing,
force in naphtha or benzine and then more oil; if the latter
is not effective, take the bearing apart and smooth the rough
places on the journal by filing and those in the box by scraping or filing. Then wipe clean, oil freely, and put bearing
together and adjust to run loosely for a while.

**102. To prevent rusting or corroding of machine or finished
work,** coat with vaseline or thin oil. When tools or machines
are not in use, coat with a thick oil. The rust should be
removed from a surface before it is oiled or rusting will
continue.

103. To remove rust, first scrape then apply special rust remover,
kerosene,or alcohol; let set a short time; clean with gasolene, wipe dry.

To remove rusted screws or bolts apply rust remover, kerosene or
denatured alcohol to ends of screws or bolts.

Information. — When assembling machine parts apply oil, grease or
graphite to threads, bolts and shafts.

Rust prevented by dipping. — To prevent small tools and instruments rusting, dip in a thin solution of gasolene and vaseline.

SHIFTING, REMOVING AND REPLACING BELTS

104. Importance of a knowledge of belt manipulation. — Machine tools are generally driven by belts running on cone or step pulleys, or single or constant speed pulleys. The belts are usually made from leather but canvas, rubber and steel are used, particularly for non-shifting belts such as over-head belts and belts on motor-driven machine tools.

Belts are fastened in various ways. The common methods are to use rawhide lacing, wire lacing or wire hooks; other metallic fastenings however, are generally utilized where belts have frequently to be shortened. Endless leather belts made by cementing or gluing and endless steel belts are applied to motors, generators, and where belt tighteners are employed, also for heavy over-head belts. See *Principles of Machine Work*.

To operate machines efficiently, it is necessary to be able to shift or change speed, feed, and driving belts quickly, skillfully and safely. Overhead belts sometimes come off when started suddenly or overloaded. Quick and skillful replacement saves a large amount of time and annoyance and often prevents damage or accident.

Important. — Belts and pulleys should be clean and free from grease and dirt; leather belts should not be run where there is moisture without using a preservative. They should not be too tight. For best operation the pulley should be a little wider than the belt.

An idler pulley should not be placed on the tight side of a belt. Avoid the use of vertical belts when possible. Laps that are inclined to open should be cemented at once.

Danger. — Do not wear loose clothing near running belts.

Do not shift the belts of high speed machines by hand when in motion, but use a pole, see pp. 164, 165.

Do not shift a belt by hand, if laced with wire or metal fasteners.

105. Cone or step pulley drives. — On engine lathes, speed or hand lathes, drilling machines, and milling machines, cone or step pulleys are generally used for obtaining the various spindle speeds. The usual position of the cones on the machine is such that the slowest speed is obtained when the belt is on the step nearest the front end of the spindle.

Fig. 78 shows the common arrangement of the lower or headstock step or cone pulley A and the upper or countershaft step or cone pulley B of an engine lathe.   The different speeds are determined and obtained as follows: 4th, 3d, 2d, 1st.

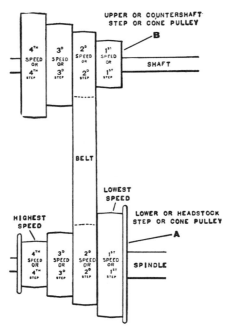

Fig. 78. — Arrangement of Step or Cone Pulleys for Obtaining Different Speeds on Machines.

Thus, the slowest speed is obtained when the belt is on step one (1) or nearest the spindle head; and the highest speed, when belt is on step four (4).

Important. — The various diameters of speed cones are called either steps or speeds, or both.

Information. — The tendency of belts running on cone pulleys to jump to the next step may be prevented, to some extent, by using belts with rounded edges.

106. To shift a lathe speed belt from third speed or step to second speed or step, Figs. 79, 80, and 81.

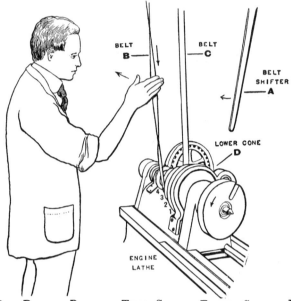

FIG. 79. — PUSHING BELT OFF THIRD STEP TO FOURTH STEP OF LOWER CONE.

I. Push belt off third step of lower cone to fourth step.

Start lathe forward with belt shifter A, Fig. 79.

With fingers of right hand extended and joined, push belt B off third step of lower cone D to fourth step.

Keep hand at nearly height of shoulder and allow belt to slip by the hand.

II. Pull belt from third step of upper cone to second step. Grasp carriage F, Fig. 80, of engine lathe with the right hand for a support.

Then, with the left hand, pull belt C off third step of upper cone E to second step. Keep hand as high as possible and allow belt to slip through hand.

III. Push belt on to second step of lower cone, Fig. 81. With fingers of left hand extended and joined, and starting high on edge of belt B, Fig. 81, push belt on to second step of lower cone D. With left hand push hard sideways and follow edge of belt downward, as shown at G.

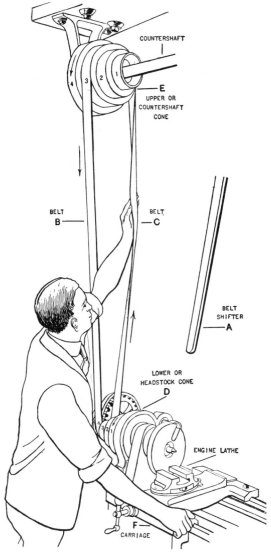

FIG. 80. — PULLING BELT OFF THIRD STEP TO SECOND STEP OF UPPER
CONE.

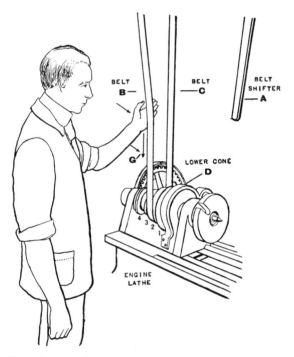

FIG. 81. — PUSHING BELT ON TO SECOND STEP OF LOWER CONE.

Warning. — Do not carry hand down low enough to pinch fingers between cone and belt. Swing fingers outward on nearing end of stroke.

Important. — This method of shifting belts from third to second step also applies to shifting from the second step to the first, or any other consecutive pair to decrease speed, except from the highest speed.

Information. — Ordinarily a belt can be pushed with the left hand on to the third or any other low-speed step on lower cone to correspond to the step on upper cone; but, if the belt is heavy, or tight, it may be necessary to use a wrench handle in the same manner that the hand is used.

107. To shift a lathe speed belt from second speed or step to third speed or step, Figs. 82, 83.

I. Push belt off second step of lower cone to fourth step. Start lathe forward with belt shifter and push belt off second step of lower cone to fourth step by method shown in Fig. 79.

II. Throw or snap belt on to third step of upper cone. Place right hand inside of belt *B*, Fig. 82, and pull belt outward until lower cone *D* is running at a fast speed.

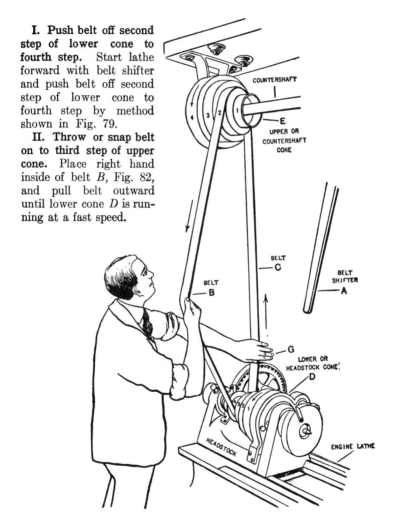

Fig. 82. — First Position in Shifting Belt to Higher Step of Upper Cone.

With fingers of the left hand extended and joined, as at G, start low on edge of belt C and with a very quick movement, a long stroke, and considerable pressure, move hand upward at same speed belt is running; see G', Fig. 83. At end of stroke throw or snap the belt C upward, and at nearly same time, or the fraction of a second later, release pull on belt B with the right hand, as at H, and belt will mount third step of upper cone E.

III. Putting belt on to third step of lower cone. With fingers of the left hand extended and joined start high on edge of belt and push belt on to third step of lower cone by method shown in Fig. 81.

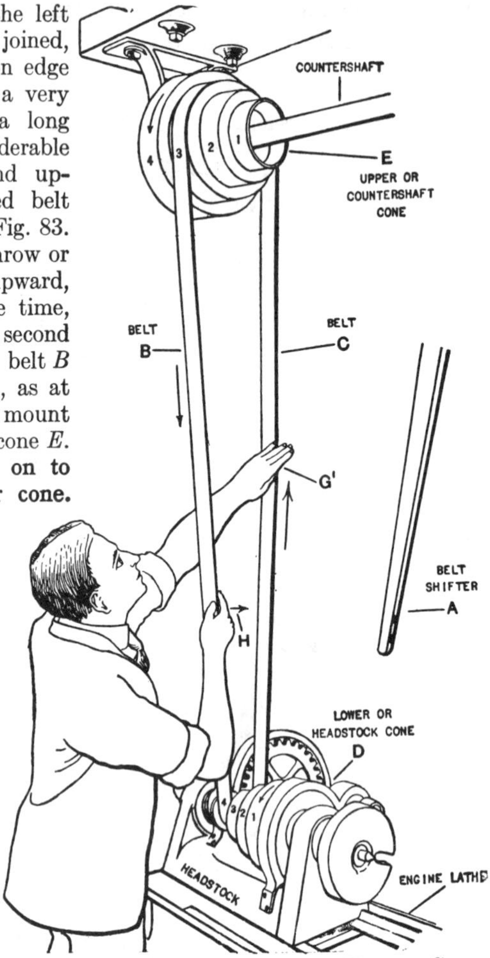

FIG. 83. — SECOND POSITION IN SHIFTING BELT TO HIGHER STEP.

Important. — This method of shifting belts from second to third step also applies to shifting from the first step to the second, or any other consecutive pair, to increase speed, except from the next highest to highest.

Information. — Ordinarily a speed belt can be thrown or snapped on to the second or third step of an upper cone pulley without difficulty, but if unable to do so, use a belt pole, as in Fig. 87.

108. To shift lathe speed belt to fourth or highest speed or step from any other speed or step, Figs. 84, 85. (Most cone-driven machines have four-step cones, but some have three and others five steps.)

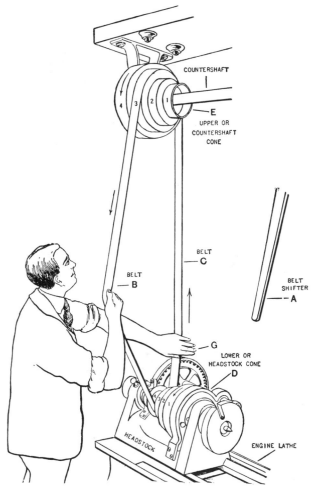

FIG. 84. — FIRST POSITION IN THROWING BELT TO FOURTH STEP OR HIGHEST SPEED OF UPPER CONE.

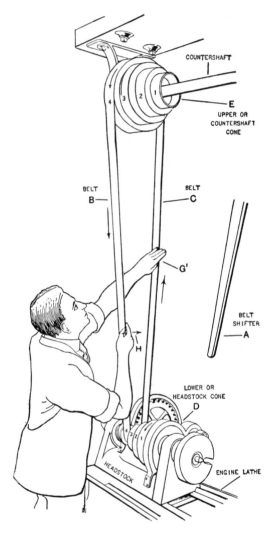

FIG. 85. — SECOND POSITION IN THROWING OR SNAPPING BELT TO FOURTH
STEP OR HIGHEST SPEED OF UPPER CONE.

SCHEDULE OF OPERATIONS

I. Push belt off third or any other step of lower cone to fourth step. Start lathe forward with belt shifter, and push belt off first, second, or third step of lower cone to fourth step by method shown in Fig. 79.

II. Throw or snap belt on to fourth step of upper cone. Place right hand inside of belt *B*, Fig. 84, and pull belt outward until lower cone *D* is running at a fast speed. Then, with fingers of the left hand extended and joined, as at *G*, start low on edge of belt *C* and follow belt upward, as at *G'*, Fig. 85; with a long stroke, considerable pressure, and a very quick movement, move hand upward at same speed belt is running; at end of stroke throw or snap belt *C* upward, and at nearly the same time, or the fraction of a second later, release pull on belt *B* with the right hand, as at *H*, and belt will mount fourth step of upper cone *E*.

If the belt is new, or very old, or large, or very tight so that it will not mount the upper cone readily by hand, use a belt pole as in Fig. 87.

Attention. — If belt comes off the upper cone at any time, keep countershaft running, and push belt on to fourth step of lower cone and then place belt upon first step of upper cone with a belt pole.

109. To pull belt from the fourth speed or step of upper cone, Fig. 86.

SCHEDULE OF OPERATIONS

I. Pull belt off fourth step of upper cone. Start lathe forward with belt shifter *A*, Fig. 86.

Grasp carriage *F* of engine lathe with the right hand for a support, as shown. Then, with the left hand, pull belt *C* off fourth step of upper cone *E* to third or any other desired step. Keep left hand as high as possible and allow belt to slip through hand.

Attention. — A speed belt may be pulled off the third or second step in the same manner as off the fourth step. See Fig. 80.

Information. — If the belt is so tight and stiff that it cannot be pulled from the highest step by hand, use a belt pole in the same manner that the hand is used.

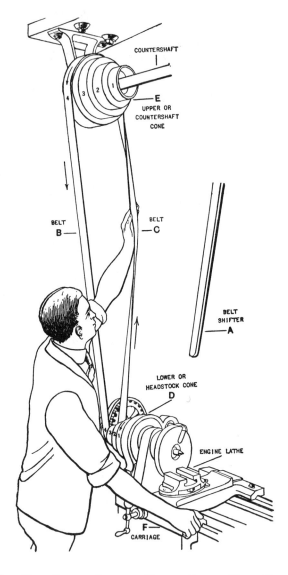

FIG. 86. — PULLING BELT OFF FOURTH STEP OF UPPER CONE.

110. To shift a lathe speed belt on to highest speed or step of upper cone, using a belt pole, Fig. 87.

Start lathe forward with belt shifter *A*, Fig. 87. Place hook *B* of belt pole *C* against up-running edge of belt *D* a few inches below upper cone *E*; press hard and with a quick upward side movement, push or force belt *D* on to fourth step of upper cone *E*.

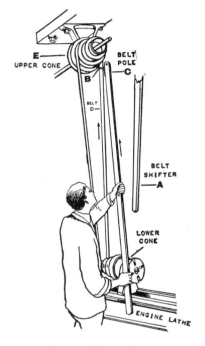

111. Cone belt shifters. — To eliminate the need and danger of shifting large cone belts by hand, cone-belt shifters are obtainable. This attachment is fastened to the headstock and overhead planking, and by movement of the handle, the belt may be readily shifted to any desired step.

FIG. 87. — SHIFTING BELT TO HIGHEST STEP OF UPPER CONE WITH A POLE.

112. To put on and push off overhead belts. — Small and medium overhead belts may be readily put on or pushed off with a belt pole. After some practice a student may become skillful in these movements.

Tight or heavy belts cannot always be put on with a pole. Light belts may be put on by standing on a ladder and pulling the belt on to the pulleys with the hand. Heavy belts are put on by using a cord; but it is dangerous to do this with the pulleys and belts running, and it is advisable to either stop the power or slow down the speed.

113. To put on an overhead belt with a belt pole, driving pulley running, Fig. 88. —

Place belt A upon driven pulley B with belt pole C. Place hook of belt pole inside of belt, as at D, then lift and place belt on bottom of receiving face of the driving pulley E. Then with a quick movement, snap, force or press the belt upon the revolving pulley.

Attention. — If the driving pulley is running in the direction opposite to that shown in Fig. 88, the belt must be put on at the top of the pulley, since the top would be the receiving side.

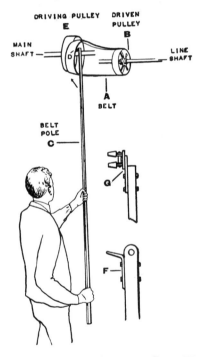

FIG. 88. — PUTTING ON AN OVERHEAD BELT WITH A POLE.

Note. — If the belt slips off the countershaft or driven pulley, have some one hold it on with second pole, or if countershaft is provided with clutch, engage clutch lightly and have lathe cone disconnected from feed or gear shaft.

114. Belt poles *C*, Fig. 88, consist of a large hardwood stick or pole at the upper end of which a flat metal hook *F* is fastened or, preferably, a double-roller hook *G*. Hook *F* may be home-made.

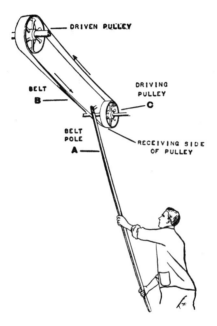

FIG. 89. — PUSHING OFF AN OVERHEAD BELT.

115. To push off an overhead belt with a belt pole, while running, Fig. 89. —

Place belt pole *A* against belt *B* near receiving side of driving pulley *C*. Push hard against edge of belt until the belt drops from the driving pulley. Push belt off driven pulley if required.

Warning. — Never push belt off driven pulley before removing it from driving pulley. If belt accidentally slips off driven pulley and remains on driving pulley, *push it off driving pulley at once.*

116. To put on an overhead belt by hand, driving pulley revolving, Fig. 90. — Place step-ladder *A*, in front of pulley. With the right-hand fingers extended and joined, place belt *B* against the receiving face of the revolving driving pulley *C*. Push belt on pulley as far as possible. With hard hand sidepressure, follow up edge of belt and with a quick movement or snap at end of stroke, force or press the belt onto the revolving driving pulley and instantly remove hand from belt.

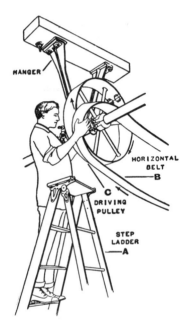

FIG. 90. — PUTTING ON OVERHEAD BELT BY HAND.

Attention. — If the belt is very heavy, it is safer and easier to shut off the power first, then have an assistant start power very slowly while belt is put on, as in Fig. 88.

Warning. — When using a step-ladder be sure to place it so that the pulling strain will not tip it over and cause an accident.

Danger. — Have coat buttoned tightly, sleeves rolled up, and wear no loose clothing near revolving shafting.

117. To put on an overhead belt with a cord, Fig. 91.

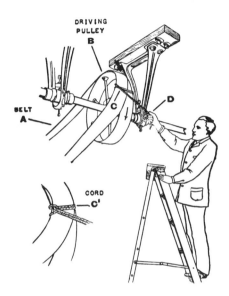

FIG. 91. — PUTTING ON AN OVERHEAD BELT WITH A CORD.

Stop power. Put belt A on driven pulley. Pull belt as far as possible onto receiving face of driving pulley B; loop a light cord around both belt and pulley, as at C,C', and hold belt in place by pulling ends of cord, as at D. Have assistant start power slowly, then pull hard on cord and the belt will go on pulley. When belt is in place, quickly let go of cord and *stop* power. Remove cord; *start* power.

Attention. — If cord should be accidentally drawn from hand, it will separate itself from pulley and drop to the floor.

ADVANCED MACHINE WORK

SECTION 2

LATHE WORK

Time Element and Schedule of Operations. Centering, Squaring, and Straight Turning. Filing Lathe Work. Micrometer and Vernier Calipers. Dimension-Limit System. Fits in Machine Construction with Tables of Allowances. Standard and Limit Gages. Taper Turning and Fitting. Straight Turning and Fitting.

TIME ELEMENT AND SCHEDULE OF OPERATIONS

1. Schedules of operations. — The problems in this book are presented in schedule form. Each operation is given in its logical order together with machines, tools, speeds, feeds, and time. Such schedules are the fundamental and necessary bases of all efficiency systems. They make teaching systematic; learning rapid; and promote industrial efficiency.

A multiple schedule for two or more duplicate pieces is the same as for a single piece, except that each operation is performed on all pieces before beginning the next.

2. The time element in the schedules includes grinding of tools, and is the average time required by an experienced workman for completing the given problem. Students and other beginners will take from 50% to 100% more time on the introductory problems, but as they become familiar with the tools and machines this excess time is reduced to 25% or less, depending on the ability of the student and the efficiency of the equipment and instruction.

3. The student may begin lathe work by turning soft cast iron, as it machines more easily than steel or wrought iron; the cutting angles of the tools are easy to shape, and the material is not expensive. See pp. 102, 203–205.

4. To mount work on lathe centers. — Clean center holes and centers with waste. Fasten dog on work (see p. 118), mount on live center. With left hand under end of work, hold work in line with dead center; move and clamp foot-stock width of tool block from end of work. Oil dead center. With the little finger of left hand to guide the dead center into center hole, screw out spindle with the right hand until there is no end movement. To test adjustment, move tail of dog back and forth; when right a slight resistance is felt by the hand; clamp binder.

Warning. — When taking a heavy cut or rotating work at a high speed, the work will heat and expand and thus bind on the centers. The student should relieve and oil the dead center occasionally. If this is neglected, a *" hot center " will result, which usually destroys both work and center.*

5. To turn work to one diameter from end to end. — First, set tool to take roughing cut and turn approximately one-half the length (Fig. 1). Stop feed, then stop lathe.

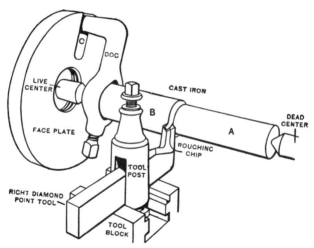

Fig. 1. — Rough Turning Shaft from End to End, First Half.

Take work out of lathe and run carriage back by hand to dead center. Do not disturb cross feed. Fasten dog on turned end A', Fig. 2, with a piece of copper D under set screw to avoid scarring work. Again mount work on centers and turn off half marked *B'*. Regrind tool and follow same method with finishing cut.

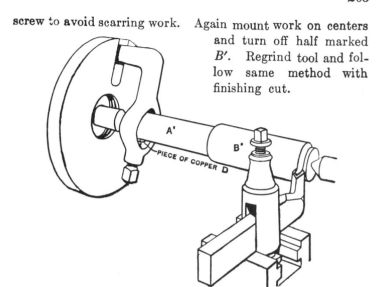

FIG. 2. — ROUGH TURNING SHAFT FROM END TO END, SECOND HALF.

CENTERING, SQUARING AND STRAIGHT TURNING

6. To center, square and turn straight. See Figs. 3 and 4.

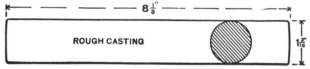

FIG. 3.

Specifications: Material, iron casting $\frac{1}{8}''$ large; weight, 2 lb. 8 oz., Fig. 3, free from visible defects.

Hardness, 28 to 32 (scleroscope).

Oil bearings of lathe with machine oil.

True live center. Set dead center in approximate aline-ment, see p. 116. Carbon-steel cutting tools. See *Exception*, p. 205.

Time: Study drawing and schedule in advance, 10 min. — Oil lathe, 4 min. — Hand center and square, 19 min. — Rough and finish turn, 23 min. — Clean lathe, 3 min. — Total, 59 min. (Machine center and square, 17 min.)

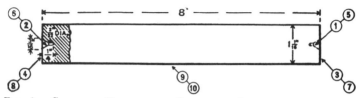

FIG. 4.—SCHEDULE DRAWING OF CENTERING, SQUARING AND STRAIGHT TURNING.

SCHEDULE OF OPERATIONS, MACHINES AND TOOLS

OPERATIONS.	MACHINES, SPEEDS, FEEDS.	TOOLS.
Snag. Center by hand method, pp. **121, 122.** Center holes $\frac{3}{16}''$ diameter, (1), (2).	Speed lathe, 8'' to 12''. 3d or 4th speed, or 1400 R.P.M.*; countersink 3d speed or 700 R.P.M.	Vise, hammer, chisel, file, chalk, dividers or divider calipers, center punch, $\frac{3}{32}''$ drill, 60° countersink, rule.
Mount and adjust on centers. Rough square (3), round nose inward, heavy cut, and side tool outward, light cuts. Reverse work on centers and repeat on (4). Test length with calipers. Take light cuts outward on (4) until work measures $8'' + \frac{1}{64}''$. Remove dog.	Engine lathe, 12'' to 16''. 1st or 2d speed, or 40 F.P.M.† Hand feed.	Dog, round-nose tool, side tool 15° rake, calipers, 12'' steel rule.
Recenter to $\frac{3}{16}''$, (5), (6)	Speed lathe.	Drill, countersink.
Finish square (7) one cut. Reverse work on centers and repeat on (8). Test length with calipers. Take light cuts on (8) until work measures 8''. To remove burr around countersinks, see p. **131.** Test flatness of end with steel rule.	Engine lathe. 2d or 3d speed, or 60 F.P.M. Hand feed.	Side tool 15° rake, calipers, rule.
Rough turn to $1\frac{1}{16}'' + \frac{1}{64}''$, (9) one cut. See Figs. 1, 2. Test with calipers at ends and middle.	1st or 2d speed, or 25 to 40 F.P.M. Medium power feed—80 to 1''‡	Copper under set screw of dog, diamond-point tool 15° rake, small calipers, 3'' steel rule.
Regrind and oilstone tool. Finish turn to $1\frac{1}{16}''$, (10), one cut. See p. **134.**	2d or 3d speed, or 55 F.P.M. Fine power feed — 140 to 1''.	Diamond-point tool 15° rake, or round-nose tool, calipers, rule.
Test with micrometer calipers, at both ends and middle. Limit allowed .001''. Stamp name on end.		Micrometer, steel name stamp, machinists' hammer, vise, copper jaws.
Clean lathe.		Brush and waste.

* R.P.M. = revolutions per minute.   † F.P.M. = feet per minute.

‡ 80 to 1'' = 80 revolutions of work to 1'' of tool travel.

Warning. — Do not recenter work after the diameter has been turned because the center reamer cuts unevenly and throws the work out of true.

Attention. — If necessary, when roughing the tool may be removed and reground, but in finishing it should be prepared to carry its cut without regrinding,

Note. — When squaring, hold long. feed handle firmly with one hand while operating cross feed handle with the other.

Caution. — As a heavy cut may draw the tool inward, caliper occasionally and readjust tool, if need be, to avoid roughing diameter too small.

Exception. — If a schedule says " carbon-steel cutting tools " and it is desired to change to "high-speed steel or stellite," the cutting speed may be increased from 50% to 100%.

If a schedule says " high-speed steel or stellite cutting tools " and it is desired to change to " carbon-steel " the cutting speed *must* be reduced 25% to 50%.

FILING LATHE WORK

7. **Lathe work** is filed to remove tool marks, to make a fit, to produce an exact diameter, and also to prepare the surface for polishing. A small amount of filing improves the condition of the work; excessive or careless filing will destroy its truth.

8. **Mill files** are best for lathe work, files single-cut (sometimes called floats), as at *C*, Fig. 5. A mill bastard is useful for a large variety of lathe filing, but a mill 2d cut is only used for finer classes of work. See *Principles of Machine Work.*

9. **Speed for filing.** — If work is revolved at too high a speed, the file will not bite, but will simply glaze the work and rapidly destroy itself. As a rule, in filing steel and cast iron the lathe may be run at a speed between four and five times faster than that used in taking the roughing cut. Cast iron must be filed dry. See Table of Cutting Speeds, p. 1**4**7. On brass, the rule is from two to three times the roughing speed.

On work $1\frac{1}{2}''$ and less in diameter, run engine lathe at its fastest speed. Small work $\frac{1}{2}''$ and less may be filed in the speed lathe.

Warning. — To avoid a hot center when filing, loosen and oil the dead center before and occasionally after increasing the speed.

10. To hold and to use a file on lathe work. — The work A, Fig. 5, is revolved at a moderately fast speed in the direction of arrow B. Good results are obtained by moving mill file C at right angles to the work as at D. If the work to be filed (A') is a tenacious metal such as machine steel or wrought iron, hold file at an angle of about 10° as at C' but move it at right angles as at D' which increases the cutting angle from 23° to 33°. The stroke should be long but slower

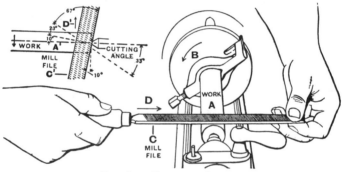

Fig. 5. — Filing in Lathe.

than in vise work in order that the work may make a number of revolutions during each stroke; the pressure should be lighter, as the number of teeth in contact are fewer.

Note. — A file may be moved slightly from left to right; that is, against its tendency to glide on its forward stroke, but not from right to left as it may chatter.

Attention. — Clean the file frequently with a file card to remove chips, or they will scratch the work.

11. To finish radial or side surfaces. — A hand file is sometimes used, but preferably use a scraper. See p. 425.

12. Round and half-round files for lathe work. — Fillets and concave surfaces are generally made with forming tools and scraped with hand tools. To polish such surfaces, they may be prepared by filing with round or half-round files, giving the file on its forward stroke a slight sweep following the curve. See Lubricants for Cutting Tools, pp. 148–150.

MICROMETER CALIPERS

13. The micrometer principle consists of a combination of an accurate screw and graduated head or nut by which fine measurements and adjustments may be obtained. It is applied to feed screws of lathes, planers, milling and grinding machines, etc., and to instruments of precision, as the micrometer caliper, Fig. 6, which consists of frame A and barrel B;

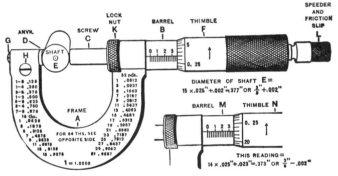

FIG. 6. — MEASURING WITH MICROMETER CALIPER.

in the end of the barrel is a fixed nut through which passes screw C. The caliper is graduated to read in thousandths of an inch, but half, quarter, and tenths of thousandths may be readily estimated.

14. To measure work with micrometer. — Place shaft E between the faces. Rotate thimble F with thumb and finger until a light but distinct contact is obtained. Prove results by moving micrometer up and down on shaft E, or by passing work between micrometer anvil D and screw C with a slight rotative motion.

15. To read micrometer. — Screw C has 40 threads to the inch; the graduations on barrel B are 40 to the inch. One revolution of the screw opens the caliper one-fortieth of an inch, or .025″. Thimble F is graduated into twenty-five parts. Each division when passing the axial line on barrel indicates one-twenty-fifth of one-fortieth of an inch ($\frac{1}{25} \times \frac{1}{40} = \frac{1}{1000}$″).

Reading. — Every fourth division on the barrel is figured 1, 2, 3, 4, etc., and may be read 0.100″, 0.200″, etc.; the figure 3 may be read 0.300″ and the three additional divisions as 0.075″, making 0.375″ on barrel. Then add the two divisions or 0.002″ on the thimble, which makes the complete measurement 0.377″.

Attention. — A student occasionally reads the example at MN backward from zero and obtains .377″ instead of .373″.

16. Adjusting the anvil to correct error. — If zero lines do not coincide when anvil and screw are clean and in contact, adjust anvil D by screw G, first loosening screw H.

17. Lock nut K may be used to clamp the screw and preserve any setting.

18. Speeder and friction slip consists of a ratchet and pawl used as a speeder for the rapid movement of screw, and as a friction slip so that the same pressure of contact may be obtained at every reading.

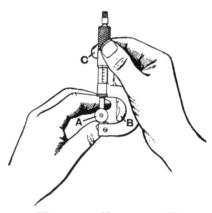

Fig. 7. — Measuring Work with Micrometer Held with One Hand

19. A one-hand method of measuring with a micrometer.
Fig. 7. — Hold work A with left hand and insert third finger
of right hand into frame to steady caliper B. Adjust screw C
to work by rotating the thimble with the thumb and first
finger.

20. To measure work in a lathe with micrometer. Fig. 8.
— Hold frame A over work B and operate thimble C.

Fig. 8. — Measuring Work in the Lathe with Micrometer.

21. Large micrometer and stand. — Micrometer stand A,
in Fig. 9, on bench B, not only protects micrometer C from

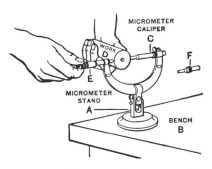

Fig. 9. — Measuring Work on Bench with Micrometer.

injury, but avoids expansion due to the heat of the hand.
To measure, hold work D as shown. This micrometer is of

the interchangeable anvil type. One of the shorter anvils is shown at F.

Standard end-measuring rods and disks are supplied with micrometers measuring more than one inch, to test their accuracy.

22. Decimal equivalents of common fractions. — As micrometers are graduated to read decimally, the common fractions and their decimal equivalents are stamped on the frames. A student should memorize the decimal equivalents of such fractions as $\frac{1}{2}''$, $\frac{1}{4}''$, $\frac{1}{8}''$, $\frac{1}{16}''$, $\frac{1}{32}''$, and $\frac{1}{64}''$.

TABLE OF COMMON FRACTIONS AND DECIMAL EQUIVALENTS

$\frac{1}{64}$						.015625	$\frac{33}{64}$						.515625
	$\frac{1}{32}$					.03125		$\frac{17}{32}$					.53125
$\frac{3}{64}$						.046875	$\frac{35}{64}$						.546875
		$\frac{1}{16}$				.0625			$\frac{9}{16}$				.5625
$\frac{5}{64}$						.078125	$\frac{37}{64}$						.578125
	$\frac{3}{32}$					.09375		$\frac{19}{32}$					.59375
$\frac{7}{64}$						.109375	$\frac{39}{64}$						.609375
			$\frac{1}{8}$			.125				$\frac{5}{8}$			.625
$\frac{9}{64}$						.140625	$\frac{41}{64}$						.640625
	$\frac{5}{32}$					.15625		$\frac{21}{32}$					.65625
$\frac{11}{64}$						.171875	$\frac{43}{64}$						.671875
		$\frac{3}{16}$				.1875			$\frac{11}{16}$				.6875
$\frac{13}{64}$						.203125	$\frac{45}{64}$						.703125
	$\frac{7}{32}$					.21875		$\frac{23}{32}$					.71875
$\frac{15}{64}$						.234375	$\frac{47}{64}$						.734375
				$\frac{1}{4}$		.250					$\frac{3}{4}$		.750
$\frac{17}{64}$						.265625	$\frac{49}{64}$						.765625
	$\frac{9}{32}$					.28125		$\frac{25}{32}$					.78125
$\frac{19}{64}$						.296875	$\frac{51}{64}$						.796875
		$\frac{5}{16}$				.3125			$\frac{13}{16}$				.8125
$\frac{21}{64}$						.328125	$\frac{53}{64}$						.828125
	$\frac{11}{32}$					.34375		$\frac{27}{32}$					.84375
$\frac{23}{64}$						.359375	$\frac{55}{64}$						.859375
			$\frac{3}{8}$			.375				$\frac{7}{8}$			.875
$\frac{25}{64}$						.390625	$\frac{57}{64}$						.890625
	$\frac{13}{32}$					.40625		$\frac{29}{32}$					.90625
$\frac{27}{64}$						.421875	$\frac{59}{64}$						.921875
		$\frac{7}{16}$				.4375			$\frac{15}{16}$				.9375
$\frac{29}{64}$						.453125	$\frac{61}{64}$						.953125
	$\frac{15}{32}$					.46875		$\frac{31}{32}$					.96875
$\frac{31}{64}$						.484375	$\frac{63}{64}$						.984375
					$\frac{1}{2}$	.500						1	1.000000

See Metric Micrometer, p. **133**3.

VERNIER CALIPERS

23. Vernier principle. — Some instruments of precision have in conjunction with the main rule (or scale) a short movable scale, called a vernier, to read fractional parts of the smallest divisions on the main rule.

The vernier is divided into one more or one less divisions than a given number of divisions on the rule. A division on the rule thus differs from a division on the vernier by the fraction shown by 1 divided by the number of divisions on vernier.

Rule A, Fig. 10, is divided into inches, tenths, and fortieths ($\frac{1}{40}'' = .025''$). Vernier B is divided into twenty-five

FIG. 10. — VERNIER PRINCIPLE.

divisions which equal twenty-four divisions on rule A. As each division on rule A is .025" ($\frac{1}{40}''$), each division on vernier B is .024" or .001" less ($\frac{1}{40}'' \times \frac{1}{25}'' = \frac{1}{1000}'' = .001''$).

If the zero line on vernier is set to coincide with the zero line or any line on rule A, the next two lines to the right will differ from each other by .001", and the difference will increase .001" for each division.

24. Example in reading. — *First, read the rule.* Each tenth is read .100" and each fortieth .025". The first line on rule to left of zero line on vernier is distant from zero line on rule .300" and three fortieths, .075", which gives 1.375".

Second, read vernier. Counting to the right, we find that the second line on the vernier B coincides with a line on the rule as indicated by arrows. The distance from the zero line on the vernier to the coinciding lines is .002".

Third, to reading of rule add reading of vernier to obtain complete measurement. Thus: 1.375" + .002" = 1.377".

25. The application of vernier principle to the vernier caliper. — In Fig. 11 each inch of beam C is divided into 40 parts, and vernier D, attached to sliding head E, into 25 parts. To measure work, as shaft F, the caliper is placed over the shaft and the sliding head is moved to bring sliding jaw G and solid jaw H in contact with shaft, clamp K is fastened to beam by thumb screw L. The jaws are made to touch shaft delicately yet distinctly by adjusting nut M, after which head E is fastened to the beam

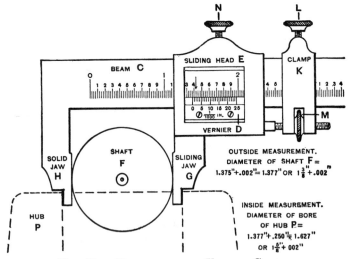

FIG. 11. — MEASURING WITH VERNIER CALIPER.

by thumb screw N. The caliper reads the same as the former example, Fig. 10, that is: 1.377″ or $1\frac{3}{8}″$ + .002″. It is best to use a magnifying glass to read or set a vernier caliper.

26. To obtain inside measurements with a vernier caliper. — To take inside measurements, as the bore of hub P shown dotted, the caliper is read exactly as above, then the width of the points of the jaws, which differ with the size of the caliper is added. For the caliper in Fig. 11 two hundred and fifty thousandths of an inch (0.250″) must be thus added to the reading on the vernier side.

27. Vernier caliper as a caliper square. — The vernier caliper may be used as a caliper square for ordinary outside and inside measurements by reading the back of beam *C*, which is graduated to read to 64ths.

28. A ten-thousandth micrometer. — Micrometer calipers are obtainable to read to the tenth part of a thousandth of an inch. A vernier of ten divisions is marked on the barrel *A*, Fig. 12, and in the space occupied by nine divisions on the thimble *B*. The microm- eter reads .250″+. To

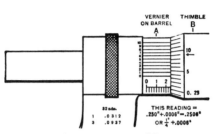

FIG. 12.—A TEN-THOUSANDTH MICROMETER.

read the vernier to obtain the fourth decimal place, locate the line on the vernier, as line 6, which coincides with a line on the thimble, and add .0006″ to reading, as .250″ + .0006″ = .2506″ or ¼ + .0006″.

DIMENSION–LIMIT SYSTEM

29. Drawings giving dimension limits and indicating the measuring tools are given for accurate work, the systems varying. See Limit System, p. 1303.

Certain dimensions may be $\frac{1}{1000}$″ or $\frac{2}{1000}$″ under or over nominal size, and if indicated on the drawing it will save the time used in finishing work with undue accuracy. The extra time taken to make the drawings is saved many times in machining the work.

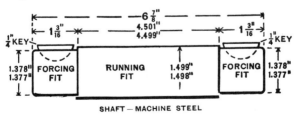

SHAFT — MACHINE STEEL

FIG. 13. — DRAWING SHOWING DIMENSION-LIMIT SYSTEM.

30. A single dimension indicated by a whole number, fraction, or mixed number as 6⅞″, Fig. 13, means that rule and caliper are sufficiently accurate to measure the parts.

31. Double dimensions in decimal form, placed one above the other as $\frac{1.378''}{1.377''}$, Fig. 13, indicate the limit allowed for a required size, and that micrometer, vernier, or limit gage should be used.

32. A dimension allowing no limit is indicated by a single decimal as 8.000″ and means that the greatest possible accuracy must be obtained with the measuring instruments at hand.

Instead of double dimensions, limits are sometimes indicated by plus and minus signs after the nominal dimensions, as 4.125″ + or − .001″.

FITS IN MACHINE CONSTRUCTION WITH TABLES
OF ALLOWANCES

33. Fits in machine construction are most important, and unless made according to requirements, impair the usefulness of the machine. See Dimension-Limit System, pp. 213, 214.

Examples.—In an engine lathe the live spindle of steel is a running fit in bronze or babbitt boxes in the headstock. The footstock spindle of steel is a sliding fit in the cast-iron footstock. The headstock cone is a running fit and the headstock gear a drive fit on spindle. The headstock, footstock, and carriage of cast iron are sliding fits on the ways of the cast-iron bed. See Scraping, *Principles of Machine Work.*

Like metals are generally used for sliding surface fits, *unlike* metals for running fits. Cast iron to cast iron wears well for any fit, but steel to steel will quickly abrade unless hardened and ground.

34. The classes of fits used in machine construction are: running, sliding, driving, forcing and shrinking fits; and taper fits, running and forcing. The hole or bore for duplicate work should be standard. See pp. 221, 222.

35. Running fits vary with the class of work. The amount of looseness varies from .0002″ on fine watch work to $\frac{1}{64}''$ on some classes of cotton and woolen machinery. See Tables, Classes I, II, III. Running fits are often made taper. The adjustment is obtained by moving the spindle along the box, or the box along the spindle.

Running Fits. — Class I

TABLE OF ALLOWANCES AND LIMITS UNDER STANDARD FOR FINE WORK

Hole Diameter, Inches.	Shaft Diameter = Hole Diameter −	Limit Shaft Diameter + or −
.00 to .49	.000625	.000125
.50 to 1.99	.0012	.000375
2.00 to 3.99	.0016	.000625
4.00 to 5.99	.002	.00075
6.00 to 8.00	.002375	.000875

Running Fits. — Class II

TABLE OF ALLOWANCES AND LIMITS UNDER STANDARD FOR AVERAGE WORK WHERE HIGH SPEEDS ARE REQUIRED

Hole Diameter, Inches.	Shaft Diameter = Hole Diameter −	Limit Shaft Diameter + or −
.00 to .49	.001	.00025
.50 to .99	.0015	.0005
1.00 to 1.99	.0019	.000625
2.00 to 2.99	.0023	.00075
3.00 to 3.99	.0027	.00075
4.00 to 5.99	.0035	.001
6.00 to 8.00	.0038	.00125

Running Fits. — Class III

TABLE OF ALLOWANCES AND LIMITS UNDER STANDARD FOR ENGINE WORK WHERE EASY FITS ARE REQUIRED

Hole Diameter, Inches.	Shaft Diameter = Hole Diameter −	Limit Shaft Diameter + or −
.00 to .49	.0015	.0005
.50 to .99	.002	.00075
1.00 to 1.99	.0026	.000875
2.00 to 3.99	.0034	.001125
4.00 to 5.99	.00425	.00125
6.00 to 8.00	.005	.0015

36. Sliding fits such as the footstock spindle of lathes are made similar to running fits except that the final fitting is done by draw-filing. See *Principles of Machine Work.* Many lathe manufacturers grind the spindle slightly large, then force it back and forth in the footstock with a power press, which smooths, straightens, and stretches the hole, eliminates wear to some extent and produces a fine sliding fit

TABLE OF ALLOWANCES AND LIMITS UNDER STANDARD
FOR SLIDING FITS

Hole Diameter, Inches.	Shaft Diameter = Hole Diameter −	Limit Shaft Diameter + or −
.00 to .49	.0005	.00025
.50 to .99	.00075	.00025
1.00 to 1.99	.00125	.00025
2.00 to 3.99	.00175	.00025
4.00 to 5.99	.00225	.00025
6.00 to 8.00	.0035	.0005

37. Driving or drive fits, easy and hard are made by turning the shaft to size with allowance for filing. For easy fits, file or grind until shaft will enter hole about two-thirds length of fit with hand pressure; for hard fits, one-third. The former are used for light-keyed fits and small work; the latter for ordinary work which may be removed for repair.

TABLE OF ALLOWANCES AND LIMITS OVER STANDARD
FOR DRIVING FITS

Hole Diameter, Inches.	Shaft Diameter = Hole Diameter +	Limit Shaft Diameter + or −
.00 to .49	.000375	.000125
.50 to 1.24	.00075	.00025
1.25 to 2.49	.00125	.00025
2.50 to 8.00	.002	.0005

38. Forcing or force fits. — In assembling and erecting machinery, nothing is more important than the proper fitting of the parts that have to be driven or forced together, or driven or forced apart, when making repairs. Examples of forcing fits are gears, couplings, locomotive driving wheels, crank pins, car axles, rod bushing, crank disks, various kinds of bushings, linings into cylinders, various parts of engines, generators and motors, iron bands on wagon-wheel hubs, crank shafts into automobile fly wheels, various parts of built-up cranks, or any two machine parts that have to be joined by forcing one into the other with sufficient power to prevent them ever becoming loose.

TABLE OF ALLOWANCES AND LIMITS OVER STANDARD
FOR FORCING FITS

Hole Diameter, Inches.	Shaft Diameter = Hole Diameter +	Limit Shaft Diameter + or −
.00 to .49	.00075	.00025
.50 to .99	.0015	.0005
1.00 to 1.49	.0025	.0005
1.50 to 1.99	.0035	.0005
2.00 to 2.49	.0045	.0005
2.50 to 3.24	.0055	.0005
3.25 to 3.99	.0065	.0005
4.00 to 4.99	.0075	.0005
5.00 to 5.99	.0085	.0005
6.00 to 8.00	.0095	.0005

Attention. — To avoid abrasion and destruction of work, lubricate both surfaces with machine oil. For heavy work use cylinder oil, white lead, or grease.

39. Taper forcing fits. — For some classes of machinery, such as marine and engine work, taper forcing fits are used. The hole is bored to a taper of .060″ to 1′, and the shaft ground or turned the same but large enough when put together by hand to stop about one-fourth to one-eighth of its length from the desired location, then forced to place. Large taper holes are often scraped true to gages before forcing the shaft into place.

On some classes of work, the hole is made straight and the shaft ground or turned to a taper of about .001″ to 1″, which makes an effective fit.

40. Pressures and allowances for forcing fits. — Formulas have been deduced and tables made giving pressures required to force two parts together; these possess considerable value, but serve only as a guide, for the pressure required depends not only on the difference in diameters but on the diameter and length of fit, smoothness of both parts, material of both shaft and hub, and diameter of hub. The material will often make a difference of from twenty-five per cent to fifty per cent in the pressure for the same allowance. For this reason the method that is generally followed is to make tables of allowances and pressures for each class of fits and for different materials.

41. Forcing presses for forcing fits. — The old method of making forcing fits is with sledge hammers, rams, etc., but such methods are now nearly obsolete except for occasional fits, as the process is difficult, dangerous, and lacks uniformity. The economical, easy, safe, and scientific method is by hydraulic-power or belt-power presses of which there are a great variety both vertical and horizontal, small and large, stationary and portable, to suit all classes of work. Small forcing fits may be made with a mandrel or arbor press.

42. To force screw into bevel gear. Fig. 14. Belt-power forcing press. — This type of press may be used for forcing fits between mandrel press work and hydraulic press work and not requiring a pressure exceeding fifty tons.

Screw A is keyed, lubricated, preferably with linseed oil, and forced into bevel gear B by ram C, to which motion is transmitted by gearing actuated by the driving belt D and controlled by hand wheel E which operates a friction device. The wheel E is used to raise and lower the ram and to control the pressure through its operation of a friction device; the harder the wheel is turned the greater the friction and the greater the pressure produced. The friction only acts on the downward motion.

The plunger F which compresses glycerine in a chamber at the end of ram C, records the pressure in tons per square inch by means of gage G.

Specifications. — The gear in Fig. 14 is steel case-hardened; hole straight, $1\frac{1}{4}''$ diameter; screw, crucible steel; length of fit, $3''$; allowance, and taper, $1.250''$ in diameter at end and $1.254''$ at shoulder (ground); pressure, 8 tons.

43. To force shaft into flange, forcing fit. Fig. 15. Belt-power forcing press. — Shaft A is keyed, oiled, and forced into flange B.

Specifications. — Flange, cast iron; hole $1''$ diameter; shaft, machine steel; length of fit, $1\frac{3}{4}''$, straight; allowance, $.002''$ to $.0025''$, large (ground); pressure, about 3 tons.

Attention. — Condition of surfaces (smooth or rough) will vary the pressure.

44. To force shaft into malleable-iron gear. —
Specifications. — Malleable-iron gear; hole $1\frac{3}{16}''$ diameter; length of fit, $2''$; shaft, crucible steel; allowance and taper, $1.1875''$ at end and $1.1915''$ at shoulder; pressure, 5 tons.

45. To force shaft into machine-steel gear. —
Specifications. — Machine-steel gear; hole $2''$ diameter; length of fit, $5''$; shaft-machine steel; allowance and taper, $2.000''$ at end, $2.004''$ at shoulder; pressure, 13 to 15 tons.

46. Shrinking fits are used to fasten a collar, sleeve, crank pin, crank, or other piece permanently in place. They differ from forcing fits in manner of fitting only. The fit is made by heating the hollow piece until it expands sufficiently to go on the cold shaft easily. The parts must be put together quickly or the heat will expand the shaft and the parts stick hard before they are in place, in which case they should be driven or pressed apart as quickly as possible. When in place, cool slowly with water.

If the proper shrinkage is allowed, it will not be necessary to heat piece above a dull red, 800° F.

For small work the hole may be standard and the amount for shrinkage allowed on shaft.

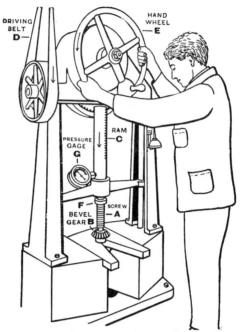

FIG. 14. — FORCING SCREW INTO BEVEL GEAR. FORCING FIT. BELT-POWER FORCING PRESS.

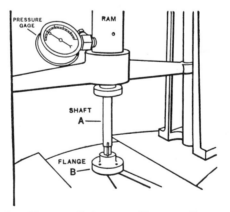

FIG. 15. — FORCING SHAFT INTO FLANGE. FORCING FIT. BELT-POWER FORCING PRESS.

TABLE OF ALLOWANCES AND LIMITS OVER STANDARD FOR
SHRINKING FITS

Hole Diameter, Inches.	Shaft Diameter = Hole Diameter +	Limit Shaft Diameter + or −
.00 to .49	.00075	.00025
.50 to .99	.0015	.0005
1.00 to 1.49	.002	.0005
1.50 to 1.99	.0025	.0005
2.00 to 2.49	.003	.0005
2.50 to 3.24	.0035	.0005
3.25 to 3.99	.004	.0005
4.00 to 4.99	.0045	.0005
5.00 to 5.99	.0055	.0005
6.00 to 8.00	.0075	.0005

47. To shrink tires on wheel centers. — Wheel centers
are turned standard and the allowance is made in boring
the tire.

The allowance for shrinkage in rings or jackets for guns in
the United States naval gun factories varies slightly for different classes of guns, but it is generally about .001″ to the inch.
That is, if the diameter is 12″ the shrinkage will be .012″.

TABLE OF ALLOWANCES FOR LOCOMOTIVE DRIVING WHEEL
TIRES FOR SHRINKING FITS

Wheel Center Diameter, Inches.	Bore of Tire = Wheel Center Diameter −	Wheel Center Diameter, Inches.	Bore of Tire = Wheel Center Diameter −
38.00	.040	56.00	.060
44.00	.047	62.00	.066
50.00	.053	66.00	.070

48. Standard holes. — The holes referred to in tables of
allowances for the various kinds of fits are either bored or bored
and reamed, and are within limits given in table of .00025″.
They are tested by plug and limit gages.

TABLE OF LIMITS OF HOLE DIAMETERS ALLOWABLE UNDER
AND OVER STANDARD

HOLE DIAMETER, INCHES.	LIMIT STANDARD DIAMETER	
	+	−
.00 to 1.24	.00025	.00025
1.25 to 2.49	.00075	.00025
2.50 to 5.99	.001	.0005
6.00 to 8.00	.001	.0075

49. Tables of allowances and limits for standard fits, represent common practice, but are not intended to conform to every case that may arise in fitting. In making allowances for any fit, certain conditions must always be considered. For example, the allowances for forcing fits are for cast-iron hubs twice the diameter of machine-steel shafts, and subject to modifications for different conditions, as the amount of metal surrounding the hole, the length of hole, and the elasticity of the metal.

50. To fit by trial and correction, and by allowance.—When only a few pieces have to be fitted, fit by trial. But for many pieces, fit according to allowance given in tables.

51. To turn and file fits, and to grind fits.—Work that was formerly turned and filed in the lathe is now roughed out with high-speed steel tools and finished in the grinding machine. This process produces work quicker, and truer, cylindrically. See p. 701.

52. To produce standard fits with spring calipers. — Fit ordinary calipers to a plug gage, reference rod or disk which has the proper allowance. Turn and file or grind shaft to fit caliper. Another method is to use limit caliper gages directly on work. See Limit Gages, pp. 223-226.

53. To produce standard fits with micrometers. — Turn work direct to diameter plus double depth of tool marks; the allowance for filing with a fine feed and sharp tools is from .003″ to .004″.

STANDARD AND LIMIT GAGES

54. Gages are instruments of reference for standardizing measurements and for determining dimensions exactly or within limits. See Limit System, p. 1303.

When a piece is machined nearly to size, a gage is invaluable for determining the exact dimension, the tightness or looseness of the fit giving an idea of its size.

55. Standard cylindrical gages, ring and plug. — Fig. 16 represents an accurate subdivision of the Imperial yard.

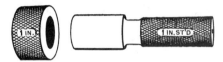

FIG. 16. — STANDARD RING AND PLUG GAGES.

These gages are made within various limits of accuracy, such as .0002″, .0001″, .00005″, .00002″.

56. Caliper gages, Fig. 17. — The measuring faces of the

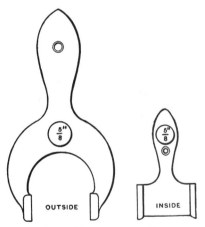

FIG. 17. — CALIPER GAGES.

outside gage are flat, the inside cylindrical. For general use, this gage is preferred to plug and ring gages.

57. Reference disk, Fig. 18, is used for testing and setting calipers.

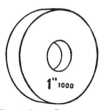

Attention. — As the heat from the hand will enlarge a gage perceptibly, the body of gages for very accurate testing is often covered with a jacket of rubber or wood, or provided with a hole to receive a wooden handle, as rubber and wood are poor conductors of heat.

FIG. 18. — STANDARD REFERENCE DISK.

58. End-measuring rod, Fig. 19, for gaging rings, cylinders, setting ordinary calipers, etc. The ends are sections of true spheres having diameters equal to the length of the rod.

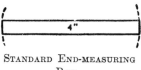

STANDARD END-MEASURING ROD.

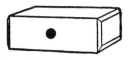

STANDARD END-MEASURE TEST PIECE.

FIG. 19.

59. End-measure test piece, Fig. 19, is used for testing micrometers, caliper gages and setting tools in planer and shaper.

60. Limit gage system of measurement. — In the manufacture of parts for various classes of machines, there are different degrees of accuracy called *limit or tolerance permissible,* depending on the requirements.

To make the parts rapidly and interchangeably, a schedule of permissible variations from nominal sizes should be planned to save time and confusion. See Fits In Machine Construction With Tables Of Allowances, p. 214. To produce work within these limits, different kinds of working or inspection gages are obtainable.

Solid outside and inside limit gages, as at *A* and *B*, Fig. 20, are stamped with dimensions and directions for using.

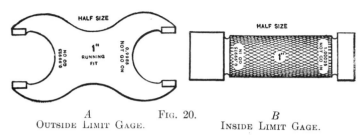

A   Fig. 20.   *B*
OUTSIDE LIMIT GAGE.   INSIDE LIMIT GAGE.

Adjustable limit snap gages are shown at *A*, *B* and *C*, Fig. 21. Each gage is provided with two pairs of contacts and with a

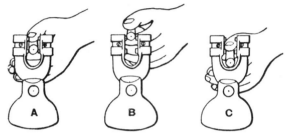

Fig. 21. — INSPECTING WORK WITH ADJUSTABLE LIMIT SNAP GAGES.

setting plug gage covering both limits. The upper contacts are adjusted by set and locking screws to the maximum size, and

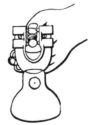

Fig. 22. — INSPECTING WORK WITH ADJUSTABLE LIMIT SNAP-THREAD GAGE.

the lower, to the minimum size. To test the work, place it on the upper contacts and if by its own weight or gravity it passes the upper contacts and rests on the lower, as at *A*, Fig. 21, it is within the limit permissible; if it does not pass the upper, as at *B*, Fig. 21, it is too large. If the work passes both sets of contacts, as at *C*, Fig. 21, it is too small.

61. Adjustable limit snap-thread gage, Fig. 22, with two sets of conical contacts for determining the limits on threaded work, is set and used in the same manner as the plain gages.

62. Special gages. — Various other standard and special gages are obtainable.

Attention. — Care must be exercised to clean and oil gages before using and also not to force them into or over work with undue pressure or they will wear excessively and soon become unreliable.

Information. — To measure the pitch diameter, the conical points are offset one-half the pitch of the thread.

Warning. — Limit snap gages are adjusted to size by set and locking screws, and to detect tampering, recesses in back of the screws may be filled with wax and sealed.

TAPER TURNING AND FITTING

63. Tapers are expressed as so much per unit of length, as 1″ per foot; that is, a piece 1′ in length would be 1″ larger at one end than at the other, as at *A* and *B*, Fig. 23. They

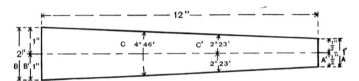

FIG. 23. — DIAGRAM FOR READING TAPERS.

may be expressed as so much per foot from the center line; as, ½″ per foot from center line, ½″ at *A′* and 1″ at *B′*, the same taper as 1″ per foot. They are also expressed in angular measurement as at *C*, angle of 4° 46′, or as at *C′*, 2° 23′, see Table of Tapers and Angles, p. 1188.

64. Tapers, standard and special. — Taper parts are used on nearly all machines. The Morse taper system, approximately ⅝″ per foot (see pp. 233, 234), is used on drills and in drilling-machine and lathe spindles; the Brown & Sharpe taper system, ½″ per foot (see pp. 233, 235), in milling-machine spindles. Both are designated by name and numbers.

The Jarno taper system, .600″ per foot, is used to some extent in lathe and grinding-machine spindles. See pp. 236, 237.

Special tapers are used for spindle boxes, pins, and similar work.

65. Methods of turning taper. — Set over footstock, Fig. 24, or use a taper attachment, Figs. 28, 29. The first method can only be used for outside tapers; the others can be used for either outside or inside tapers.

66. To calculate distance to set over footstock. — Multiply one-half taper per foot (in inches) by whole length of work or mandrel (in feet).

Example. — The Morse Taper No. 3, Fig. 24, is .602″ per foot, length of work 8″. Find amount to set over footstock.

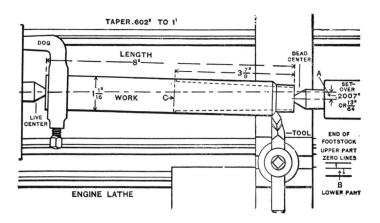

FIG. 24. — TURNING TAPER WITH FOOTSTOCK SET OVER.

Solution. — $\dfrac{.602}{2} \times \dfrac{8}{12} = .2007″$ or $\dfrac{13}{64}″$.

The distance the centers enter the work affects this rule so slightly that it is ignored.

67. To set over footstock calculated distance, unclamp footstock and rotate adjusting screws to move dead center forward .2007″ or $\frac{13}{64}″$, as at *A*, Fig. 24, measuring lines *B* with dividers or rule set to .20″. See Table of Footstock Set-overs for Morse, and Brown & Sharpe Tapers, pp. **2**38, **2**39.

68. To calculate set-over when only length and diameters are given. Fig. 25.

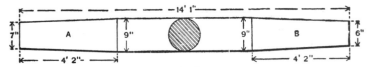

Fig. 25. — Example in Long Taper Turning.

Formula. — $\dfrac{\text{Total length}}{\text{Length taper portion}} \times \dfrac{\text{Difference in diameters}}{2}$
= Set-over.

Example. — To find set-over; total length 14′ 1″; tapered portion 4′ 2″; difference in diameter 9″ − 6″ = 3″.

Solution. —   14′ 1″ = 169″; 4′ 2″ = 50″.
$\frac{169}{50} \times \frac{3}{2}$ = 5.07″ (set-over).

69. To use a pattern to obtain set-over. — If a piece of taper work has to be duplicated, use it as a pattern to obtain approximate set-over by mounting it on centers and using a *test indicator.* See Tapering Pulleys, pp. **4**28, **4**29.

70. To turn, file and fit Morse taper No. 3 (.602″ to 1′), Figs. 26 and 27.

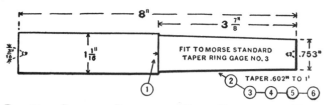

Fig. 26. — Schedule Drawing of Taper Turning and Fitting.

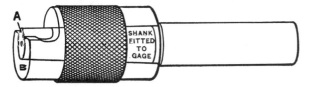

Fig. 27. — Morse Standard Taper Ring Gage, No. 3.

Specifications: Material, iron casting $\frac{1}{8}$″ large; weight, 2 lb. 8 oz. Hardness, 28 to 32 (scleroscope.)
True live center. Set dead center in approximate alinement, see p. **116.**
Carbon-steel cutting tools. See *Exception,* p. **205.**

Time: Study drawing and schedule in advance, 5 min. — Oil machine, 4 min. — Set taper, turn, file, and fit, 45. min. (With blank prepared.) — Clean lathe, 3 min. Total, 57 min. — (Preparation of blank, 40 min. extra.)

SCHEDULE OF OPERATIONS, MACHINES AND TOOLS

OPERATIONS.	MACHINES, SPEEDS, FEEDS.	TOOLS.
Snag, center, rough-square, recenter, finish square. See pp. 2o3, 2o4.	Centering machine, Engine lathe, 12″ to 16″.	Vise, chisel, file, dog, round-nose tool, side tool, 15° rake.
Straight turn. Rough, one cut, finish, one cut. (Or use blank machined to 8″ × 1$\frac{1}{16}$″, Fig. 4.)	....	Diamond-point tool, 15° rake, rule, calipers. Copper under set screw of dog.
Lay off length of taper, (1).	....	Vise, chalk or copper sulphate, rule, scriber.
Set over footstock (forward) to .2007″ or $\frac{13}{64}$″ as in Fig. 24 or use taper attachment. Set tool height of centers.		Rule, dividers. diamond-point tool, 15° rake.
Rough turn $\frac{13}{16}$″ at small end, one cut, (2). When within $\frac{1}{16}$″ of C, Fig. 24, release power feed and use hand feed to within $\frac{1}{64}$″ of C.	1st or 2d speed, or 35 F. P. M. Medium power feed — 80 to 1″.	Calipers, rule.
Grind and oilstone tool. Take a trial cut about .004″ to .005″ (3). Piece will enter gage about 2$\frac{1}{2}$″, enough to determine fit.	3d speed, or 50 F. P. M. Fine power feed — 140 to 1″.	Round-nose tool, Morse taper ring gage, No. 3, Fig. 27.
Clean hole and work. Chalk line along work and test in gage with rotating motion. If chalk line shows contact throughout by even rubbing, taper is correct. If only large end bears, set footstock backward slightly or vice versa, and take another trial cut. Repeat above until taper is correct.	....	Round-nose tool, Morse taper ring gage, No. 3, chalk or Prussian blue.
Rough turn to $\frac{49}{64}$″ one cut, (4), or until it will reach within $\frac{7}{32}$″ of end of gage. Regrind and oilstone tool.	2d or 3d speeds, or 35 to 50 F.P.M. Medium power feed.	Diamond-point tool, 15° rake, calipers, rule.
Finish turn to .757″ (.753″ + .004″) at end, one cut (the .004″ is for filing) or until end of work comes within $\frac{1}{16}$″ of end of gage (5). *Important. — Set footstock back in approximate alinement, see p. 1.16.*	3d speed, or 50 F.P.M. Fine power feed — 140 to 1″.	Round-nose tool, 1″ micrometer.
File lightly all over to remove tool marks, (6). Chalk line on work. Test work in gage, and file bright spots. Continue until fit is uniform and end A is even with end of gage, B, Fig. 27, or .753″.	Engine lathe, 4th speed, or speed lathe, 3d or 4th speed, or 105 F.P.M.	8″ or 10″ mill bastard file, file card, Morse taper ring gage, No. 3, 1″ micrometer, chalk or Prussian blue.

Attention. — Set tool to turn taper at exact height of centers. Take light finishing cuts. Avoid excessive filing. For very accurate work, coat walls of gage with Prussian blue to test taper.

71. To turn taper with taper attachment. Figs. 28, 29.— These attachments are applied to engine lathes to turn out-

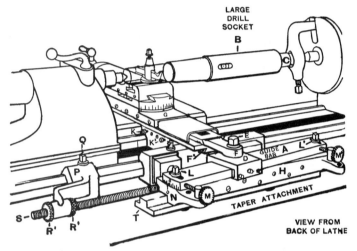

Fig. 28. — Turning a Taper with Taper Attachment.

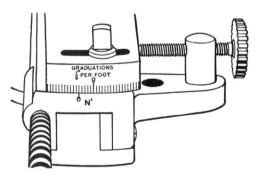

Fig. 29. — Taper Attachment Set to Turn a Taper of $\frac{5}{8}''$ per Foot.

side and inside tapers. The footstock does not have to be set over, and the length of the work does not have to be considered.

SCHEDULE OF OPERATIONS
Figs. 28, 29.

Guide bar *A* is set to turn taper of drill socket *B*. Stub mandrel *C* provides a center for hollow end of socket. Guide bar *A* carries gibbed sliding block *D* which is connected by bolt *E* to extension *F* of supplementary slide *F'* which forms the base of cross slide *G*. Guide bar is swiveled upon its base *H* by unclamping screws *L*, *L'*. Drill socket *B* is to have a Morse taper, No. 4, .623″ or $\frac{5}{8}$″ to 1′.

To Set. — Rotate adjusting screws *M*, *M'* until the graduations on scale at *N* indicate the taper. See *N'*, Fig. 29.

Graduations on one end of bar give a taper of $\frac{1}{8}$″ to 1′, and on the other, $\frac{1}{16}$″ to 1′ or in degrees.

Fasten clamp *P* by bolt *Q* and clamp nuts *R*, *R'* on screw *S*. Sliding block *D* and slide *T* move with carriage. Guide bar *A* and base *H* remain stationary.

Attention. — Take trial cuts and test in taper ring gage or in spindle, and make corrections by adjusting guide bar. See Taper Fitting, pp. 226–230.

Caution. — Take up back lash before each cut by running carriage back at least $\frac{1}{2}$″ beyond end of work to avoid turning a portion at end straight.

Note. — To use lathe for straight work without disturbing taper setting, remove nut *R* and move clamp *P* away.

STRAIGHT TURNING AND FITTING

72. To turn and file a 1″ straight running fit.

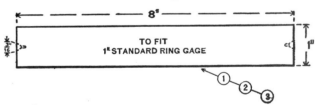

FIG. 30. — SCHEDULE DRAWING OF STRAIGHT TURNING AND FITTING.

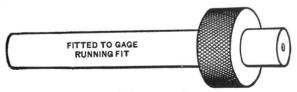

FIG. 31. — 1″ STANDARD RING GAGE.

Specifications: Material, iron casting $\frac{1}{8}''$ large; weight, 2 lb. 8 oz. Hardness, 28 to 32 (scleroscope). True live center. Set dead center in approximate alinement, see p. **116**. Carbon-steel cutting tools. See *Exception*, p. **205**. **Time:** Study drawing and schedule in advance, 5 min. — Oil lathe, 4 min. — Set lathe straight, rough and finish turn, file and fit, 45 min. (With blank prepared.) — Clean lathe, 3 min. Total, 57 min. (Preparation of blank, 40 min. extra.)

SCHEDULE OF OPERATIONS, MACHINES, AND TOOLS

Operations.	Machines, Speeds, Feeds.	Tools.
To Make Fit by Trial and Correction.		
Snag, center, rough square, recenter and finish square. See p. **2**03.	Engine lathe, 12″ to 16″.	Dog, calipers, rule, round-nose tool, side tool, 15° rake.
Rough turn $1'' + \frac{1}{32}''$, one cut, **(1)**, (or use plain part of blank used for taper turning, $4\frac{1}{8}'' \times 1\frac{1}{16}''$, Fig 27).	1st or 2d speed, or 25 F. P. M. Medium power feed — 80 to 1″.	Copper under set screw of dog, d i amond-p o i n t tool, 15° rake, calipers, rule.
Set dead center in accurate alinement to turn straight using this piece or trial piece of equal length. See pp. **1**16, **1**17.	3d speed, or 50 F. P. M. Fine p o w e r f e e d— 140 to 1″.	Round-nose tool, 2″ micrometer.
Rough turn $1'' + \frac{1}{64}''$, one cut **(1)**.	2d or 3d speed, or 50 F. P. M. Medium power feed — 80 to 1″.	Round-nose or diamond-point tool, 15° rake, calipers, rule.
Grind and oilstone tool. Oil mandrel and push in gage. Set calipers to large end close to gage. Take light cuts at end about $\frac{1}{8}''$ in length until calipers fit work slightly harder than on mandrel. Oil and try in gage, Fig. 31. When it enters about $\frac{1}{64}''$ with hand pressure, wipe off oil and take cut from end to end by reversing work, **(2)**, one cut.	2d or 3d speed, or 50 F. P. M. Fine power feed — 140 to 1″.	Round-nose or diamond-point tool, 15° rake, oilstone, calipers, 1″ mandrel, 1″ ring gage.
File a small portion at end. Clean and oil work and gage. Try in gage. Wipe off oil after each trial. Continue fitting over whole length in this manner, **(3)**.	Engine lathe, 4th speed, or speed lathe, 3d or 4th speed, or 110 F. P. M.	8″ or 10″ mill bastard file, file card, 1″ ring gage, machine, or lard oil.

Attention. — Test with calipers frequently. Avoid excessive filing. Test work in both ends of gage. The lack of oil when testing in gage will spoil both work and gage. Reamed holes are usually slightly larger at one end.

73. To make a 1″ running fit by allowance, Figs. 30 and 31.

Turn to $\left\{ \begin{array}{l} 1.002'' \dots\dots\dots\dots\dots\dots\dots\dots\dots\dots\dots\dots\dots\dots\dots \\ 1.003'' \dots\dots\dots\dots\dots\dots\dots\dots\dots\dots\dots\dots\dots\dots\dots \\ \text{and} \end{array} \right\}$ Micrometer calipers

File to $\left\{ \begin{array}{l} .9995''\dots\dots\dots\dots\dots\dots\dots\dots \\ .9990''\dots\dots\dots\dots\dots\dots\dots\dots \end{array} \right\}$ Micrometer calipers

74. Diagram of Morse tapers.

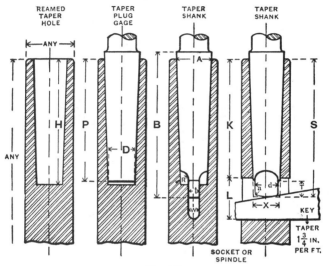

Fig. 32.

75. Diagram of Brown & Sharpe tapers.

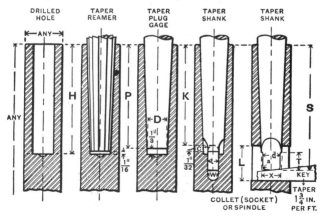

TAPER $\frac{1}{2}$ IN. TO 1 FOOT EXCEPT NO. 10 WHICH IS .5161 PER FOOT

Fig. 33.

76. Table of Morse tapers. For holes in drilling machine spindles, lathe spindles, shanks of drills, sockets, end mills, etc.

Number of Taper	Diam. of Plug at Small End. D	Diam. at End of Socket. A	Shank Whole Length. B	Shank Depth. S	Depth of Hole. H	Standard Plug Depth. P	Tongue Thickness. t	Tongue Length. T	Tongue Rad. of Mill. R	Tongue Diam. d	Tongue Radius. A	Keyway Width. W	Keyway Length. L	End of Socket to Keyway. K	Taper per Foot.	Taper per Inch.	Diam. of Point of Shank. X	Limit for Tongue to Project through Gage.	No. of Key.	Rea. Projects through Gage.
0	.252	.3561	2 11/32	2 7/32	2 1/32	2	5/32	1/4	5/32	.235	.04	.160	9/16	1 15/16	.625	.05208	.2406	.003	0	1/32
1	.369	.475	2 9/16	2 7/16	2 3/16	2 7/16	13/64	3/8	3/16	.343	.05	.213	3/4	2 1/16	.600	.05	.353	.003	1	1/32
2	.572	.700	3 1/8	2 15/16	2 11/16	2 9/16	1/4	7/16	1/4	7/16	.06	.260	7/8	2 1/2	.602	.05016	.5527	.004	2	3/64
3	.778	.938	3 7/8	3 11/16	3 1/4	3 3/16	5/16	9/16	9/32	9/16	.08	.322	1 3/16	3 1/16	.602	.05016	.753	.004	3	1/16
4	1.020	1.231	4 7/8	4 5/8	4 1/8	4 1/16	15/32	5/8	5/16	5/8	.10	.478	1 1/4	3 5/8	.623	.05191	.991	.004	4	1/16
5	1.475	1.748	5 7/8	5 5/8	5 1/8	5 3/16	5/8	3/4	3/8	3/4	.12	.635	1 1/2	4 15/16	.630	.0525	1.4395	.005	5	5/64
6	2.116	2.494	8 1/4	8 1/4	7 3/8	7 1/4	3/4	1 1/8	1/2	1	.15	.760	1 3/4	7	.626	.05216	2.0636	.005	6	5/64
7	2.750	3.270	11 5/8	11 1/4	10 5/8	10	1 1/8	1 3/8	3/4	2 5/8	.18	1.135	2 5/8	9 1/2	.625	.05208	2.685	.005	7	5/64

77. Table of Brown & Sharpe tapers. For holes in milling machine spindles, collets, and shanks of arbors, end mills, etc.

No. of Taper	Diam. of Plug at Small End. (D)	Plug Depth. (P)	Depth of Hole. (H)	Keyway from End of Spindle. (K)	Shank Depth. (S)	Length of Keyway. (L)	Width of Keyway. (W)	Length of Arbor Tongue. (T)	Diameter of Arbor Tongue. (d)	Thickness of Arbor Tongue. (t)	Radius of Tongue Circle. (C)	Radius of Tongue at "A." (A)	Diam. of Arbor at Small End Before it is Tongued. (X)	Limit for Tongue to Project through Test Tool Gage.
1	.20	15/16	1 1/16	15/16	1 3/16	3/8	.135	3/16	.170	1/8	3/16	.030	.189	.003
2	.25	1 3/16	1 5/16	1 1/64	1 1/2	1/2	.166	1/4	.220	5/32	3/16	.030	.237	.003
3	.312	1 1/2	1 5/8	1 15/32	1 7/8	5/8	.197	5/16	.282	3/16	3/16	.040	.296	.003
3	.312	1 3/4	1 7/8	1 23/32	2 1/8	5/8	.197	5/16	.282	3/16	3/16	.040	.296	.003
3	.312	2	2 1/8	1 31/32	2 3/8	5/8	.197	5/16	.282	3/16	3/16	.040	.296	.003
4	.35	1 1/4	1 3/8	1 13/64	1 21/32	11/16	.228	11/32	.320	3/32	5/16	.050	.333	.003
4	.35	1 11/16	1 13/16	1 41/64	2 3/32	11/16	.228	11/32	.320	3/32	5/16	.050	.333	.003
5	.45	1 3/4	1 7/8	1 11/16	2 3/16	3/4	.260	3/8	.420	1/4	5/16	.060	.432	.003
5	.45	2	2 1/8	1 15/16	2 7/16	3/4	.260	3/8	.420	1/4	5/16	.060	.432	.003
5	.45	2 1/4	2 1/4	2 1/16	2 9/16	3/4	.260	3/8	.420	1/4	5/16	.060	.432	.003
6	.50	2 3/8	2 1/2	2 19/64	2 7/8	7/8	.291	7/16	.460	9/32	5/16	.060	.479	.005
6	.50	3 1/4	3 3/8	3 11/32	3 3/4	7/8	.291	7/16	.460	9/32	5/16	.060	.479	.005
7	.60	2 1/2	2 5/8	2 13/32	3 1/16	15/16	.322	15/32	.560	5/16	3/8	.070	.578	.005
7	.60	2 7/8	3	2 25/32	3 13/32	15/16	.322	15/32	.560	5/16	3/8	.070	.578	.005
7	.60	3	3 1/8	2 29/32	3 17/32	15/16	.322	15/32	.560	5/16	3/8	.070	.578	.005
7	.60	4	4 1/8	3 29/32	4 17/32	15/16	.322	15/32	.560	5/16	3/8	.070	.578	.005
8	.75	3 9/16	3 11/16	3 9/16	4 3/8	1	.353	1/2	.710	11/32	3/8	.080	.726	.005
8	.75	4	4 1/8	3 57/64	4 9/16	1	.353	1/2	.710	11/32	3/8	.080	.726	.005
9	.90	4	4 1/8	3 3/4	4 5/8	1 1/8	.385	9/16	.860	3/8	7/16	.100	.874	.005
9	.90	4 1/4	4 3/8	4 1/8	4 7/8	1 1/8	.385	9/16	.860	3/8	7/16	.100	.874	.005
10	1.0446	5	5 1/8	4 27/32	5 23/32	1 1/16	.447	21/32	1.010	7/16	7/16	.110	1.0137	.005
10	1.0446	5 11/16	5 13/16	5 17/32	6 13/32	1 5/16	.447	21/32	1.010	7/16	7/16	.110	1.0137	.005
10	1.0446	5 7/32	6 1/2	6 1/8	6 1/16	1 5/16	.447	21/32	1.010	7/16	7/16	.110	1.0137	.005
11	1.25	5 15/16	6 1/16	5 23/32	6 21/32	1 1/16	.447	21/32	1.210	7/16	1/2	.130	1.247	.005
11	1.25	6 1/4	6 7/8	6 1/2	7 15/32	1 5/16	.447	21/32	1.210	7/16	1/2	.130	1.247	.005
12	1.50	7 1/8	7 1/4	6 15/16	7 15/16	1 1/2	.510	3/4	1.460	1/2	5/8	.150	1.466	.005
13	1.75	7 1/4	7 1/2	7 1/16	8 1/16	1 1/2	.510	3/4	1.710	1/2	5/8	.170	1.716	.010
14	2.00	8 1/4	8 3/8	8 3/32	9 5/32	1 11/16	.572	27/32	1.960	9/16	3/4	.190	1.962	.010
15	2.25	8 3/4	8 7/8	8 17/32	9 21/32	1 11/16	.572	27/32	2.210	9/16	3/4	.210	2.212	.010
16	2.50	9 1/4	9 3/8	9	10 1/4	1 7/8	.635	15/16	2.450	5/8	1	.230	3.458	.010
17	2.75	9 3/4	9 7/8	....	....	....	....	....	....	....	....	....	2.708	....
18	3.00	10 1/4	10 3/8	....	....	....	....	....	....	....	....	....	2.957	....

JARNO TAPERS

78. The Jarno taper system is 0.600″ per foot (1 in 20) for all numbers, and the number of the taper determines all the other dimensions, as all parts are functions of the number. For example, No. 4 taper is $\frac{4}{10}$″ at small end, $\frac{4}{8}$″ at large end, and $\frac{4}{2}$″ (or 2″) in length.

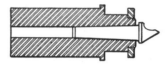

FIG. 34. — JARNO TAPER IN GRINDING-MACHINE SPINDLE.

As there is $\frac{1}{10}$″ in diameter between consecutive numbers at small end and $\frac{1}{8}$″ at large end, the number of a taper can be readily found by rough measurement with rule, or rule and calipers. Some lathe and grinding machine manufacturers have adopted the Jarno taper for holes in spindles, and others retain the Jarno taper per foot but modify the other specifications by changing the diameters or lengths to suit special conditions.

79. Table of Jarno tapers.

Taper per foot = .6 inch. Taper per inch = .050 inch.

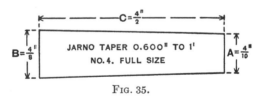

FIG. 35.

Diameter of small end $= \dfrac{\text{No. of taper}}{10}$.

Diameter of large end $= \dfrac{\text{No. of taper}}{8}$.

Length of taper $= \dfrac{\text{No. of taper}}{2}$.

NUMBER.	"A."	"B."	"C."
2	.20	.250	1
3	.30	.375	1½
4	.40	.500	2
5	.50	.625	2½
6	.60	.750	3
7	.70	.875	3½
8	.80	1.000	4
9	.90	1.125	4½
10	1.00	1.250	5
11	1.10	1.375	5½
12	1.20	1.500	6
13	1.30	1.625	6½
14	1.40	1.750	7
15	1.50	1.875	7½
16	1.60	2.000	8
17	1.70	2.125	8½
18	1.80	2.250	9
19	1.90	2.375	9½
20	2.00	2.500	10

80. Diagram of footstock setovers for Morse, and Brown & Sharpe tapers.

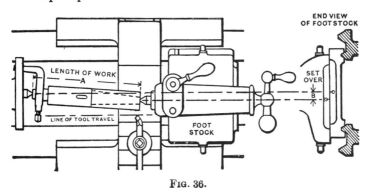

FIG. 36.

81. Table of footstock setovers for Morse, and Brown & Sharpe tapers.

THE AMOUNT IN INCHES OF SETOVER OF FOOTSTOCK GIVEN IN DECIMALS AND IN APPROXIMATE FRACTIONS

LENGTH OF STOCK IN INCHES.	SETOVER FOR MORSE TAPERS.							SETOVER FOR BROWN & SHARPE TAPERS.	
	No. 0.	No. 1.	Nos. 2 and 3.	No. 4.	No. 5.	No. 6.	No. 7.	ALL Nos. EXCEPT No. 10.	No. 10.
2	.0521	.0500	.0502	.0519				.0417	
3	.0781	.0750	.0752	.0779				.0625	
4	.1042	.1000	.1003	.1038	.1050			.0833	
5	.1302	.1250	.1254	.1298	.1313			.1042	.1075
6	.1563	.1500	.1505	.1558	.1575	.1565		.1250	.1290
7	.1823	.1750	.1756	.1817	.1838	.1826		.1458	.1505
8	.2083	.2000	.2007	.2077	.2100	.2087		.1667	.1720
9	.2344	.2250	.2258	.2336	.2363	.2348	.2344	.1875	.1935
10	.2604	.2500	.2508	.2596	.2625	.2608	.2604	.2083	.2150
11	.2865	.2750	.2759	.2855	.2888	.2869	.2865	.2292	.2365
12	.3125	.3000	.3010	.3115	.3150	.3130	.3125	.2500	.2581
13	.3385	.3250	.3261	.3375	.3413	.3391	.3385	.2708	.2796
14	.3646	.3500	.3512	.3634	.3675	.3652	.3646	.2917	.3011
15	.3906	.3750	.3763	.3894	.3938	.3913	.3906	.3125	.3226
16	.4167	.4000	.4013	.4153	.4200	.4173	.4167	.3333	.3441
17	.4427	.4250	.4264	.4413	.4463	.4434	.4427	.3542	.3655
18	.4688	.4500	.4515	.4673	.4725	.4695	.4688	.3750	.3871

Table of footstock setovers for Morse, and Brown & Sharpe tapers. *Concluded.*

THE AMOUNT IN INCHES OF SETOVER OF FOOTSTOCK GIVEN IN DECIMALS AND IN APPROXIMATE FRACTIONS

Length of Stock in Inches	No. 0	No. 1	Nos. 2 and 3	No. 4	No. 5	No. 6	No. 7	All Nos. Except No. 10	No. 10
				Setover for Morse Tapers				Setover for Brown & Sharpe Tapers	
19	.4948	.4750	.4766	.4932	.4988	.4956	.4948	.3958	.4086
20	.5208	.5000	.5017	.5192	.5250	.5217	.5208	.4167	.4301
21	.5469	.5250	.5268	.5451	.5513	.5478	.5469	.4375	.4516
22	.5729	.5500	.5518	.5711	.5775	.5738	.5729	.4583	.4731
23	.5990	.5750	.5769	.5970	.6038	.5999	.5990	.4792	.4946
24	.6250	.6000	.6020	.6230	.6300	.6260	.6250	.5000	.5161
25	.6510	.6250	.6271	.6490	.6563	.6521	.6510	.5208	.5376
26	.6771	.6500	.6522	.6749	.6625	.6782	.6771	.5417	.5591
27	.7031	.6750	.6773	.7009	.7088	.7043	.7031	.5625	.5806
28	.7292	.7000	.7023	.7268	.7350	.7303	.7292	.5833	.6021
29	.7552	.7250	.7274	.7528	.7613	.7564	.7552	.6042	.6236
30	.7813	.7500	.7525	.7788	.7875	.7825	.7813	.6250	.6451
31	.8073	.7750	.7776	.8047	.8138	.8086	.8073	.6458	.6666
32	.8333	.8000	.8027	.8307	.8400	.8347	.8333	.6667	.6881
33	.8594	.8250	.8278	.8566	.8663	.8608	.8594	.6875	.7096
34	.8854	.8500	.8528	.8826	.8925	.8868	.8854	.7083	.7311
35	.9115	.8750	.8779	.9085	.9188	.9129	.9115	.7292	.7526
36	.9375	.9000	.9030	.9345	.9450	.9390	.9375	.7500	.7742

ATTENTION. — Preferably use the decimals if in black face type; if not, use the common fractions as these are nearer correct.

82. Number of revolutions required to obtain surface speeds of from 20 feet to 70 feet per minute.

DIAMETERS FROM $\frac{1}{2}$ INCH TO 24 INCHES

FEET PER MIN.	20	25	30	35	40	45	50	55	60	65	70
DIAM., INCHES.	REVOLUTIONS PER MINUTE										
$\frac{1}{2}$	152	191	229	267	305	344	382	420	458	496	535
$\frac{3}{4}$	101	127	152	178	203	228	254	279	305	330	356
1	76	95	114	133	152	172	191	210	229	248	267
$1\frac{1}{8}$	68	85	102	119	136	153	170	187	204	221	238
$1\frac{1}{4}$	61	76	91	106	122	137	153	168	183	199	213
$1\frac{1}{2}$	50	63	76	89	101	114	127	139	152	165	178
$1\frac{3}{4}$	43	54	65	76	87	98	109	120	131	141	152
2	38	47	57	66	76	86	95	105	114	124	133
3	25	31	38	44	50	57	63	70	76	82	89
4	19	23	28	33	38	43	47	52	57	62	66
5	15	19	22	26	30	34	38	42	45	49	53
6	12	15	19	22	25	28	31	34	38	41	44
7	10	13	16	19	21	24	27	30	32	35	38
8	9	11	14	16	19	21	23	26	28	31	33
9	8	10	12	14	17	19	21	23	25	27	29
10	7	9	11	13	15	17	19	21	22	24	26
11	6	8	10	12	13	15	17	19	20	22	24
12	6	8	9	11	12	14	15	17	19	20	22
13	5	7	8	10	11	13	14	16	17	19	20
14	5	6	8	9	10	12	13	15	16	17	19
15	5	6	7	8	10	11	12	14	15	16	17
16	4	6	7	8	9	10	11	13	14	15	16
17	4	5	6	7	8	10	11	12	13	14	15
18	4	5	6	7	8	9	10	11	12	13	14
19	4	5	6	7	8	9	10	11	12	13	14
20	3	4	5	6	7	8	9	10	11	12	13
21	3	4	5	6	7	8	9	10	10	11	12
22	3	4	5	6	6	7	8	9	10	11	12
23	3	4	4	5	6	7	8	9	10	10	11
24	3	4	4	5	6	7	7	8	9	10	11

83. Number of revolutions required to obtain surface speeds from 75 feet to 125 feet per minute.

DIAMETERS FROM ½ INCH TO 24 INCHES

FEET PER MIN.	75	80	85	90	95	100	105	110	115	120	125
DIAM., INCHES.	REVOLUTIONS PER MINUTE										
½	573	611	649	687	726	764	802	840	878	916	955
¾	381	406	432	457	482	508	535	560	586	612	637
1	286	305	324	344	363	382	401	420	439	458	477
1⅛	255	272	289	306	323	340	357	374	391	408	425
1¼	228	245	260	274	290	306	321	336	351	367	382
1½	191	203	216	229	241	254	266	280	292	306	318
1¾	163	174	185	196	207	218	238	240	250	262	273
2	143	152	162	172	181	191	200	210	219	229	238
3	95	101	108	114	121	127	133	140	146	153	159
4	71	76	81	86	90	95	100	105	109	114	119
5	57	61	64	68	72	76	80	84	87	91	95
6	47	51	54	57	60	63	67	70	73	76	79
7	40	43	46	49	51	54	57	60	62	65	68
8	35	38	40	42	45	47	50	52	55	57	59
9	31	34	36	38	40	42	44	46	48	51	53
10	28	30	32	34	36	38	40	42	44	46	47
11	26	27	29	31	33	34	36	38	40	41	43
12	23	25	27	28	30	31	33	35	36	38	39
13	22	23	25	26	27	29	30	32	33	35	36
14	20	21	23	24	25	27	28	30	31	32	34
15	19	20	21	23	24	25	26	28	29	30	31
16	17	19	20	21	22	23	25	26	27	28	29
17	16	17	19	20	21	22	23	24	25	26	28
18	15	16	18	19	20	21	22	23	24	25	26
19	15	16	17	18	19	20	21	22	23	24	25
20	14	15	16	17	18	19	20	21	22	22	23
21	13	14	15	16	17	18	19	20	20	21	22
22	13	13	14	15	16	17	18	19	19	20	21
23	12	13	14	15	15	16	17	18	19	19	20
24	11	12	13	14	15	16	16	17	18	19	19

Attention. — To calculate any cutting speed, see pp. 145–148.

84. Time required for tool to travel 1″ when feed is $\frac{1}{100}''$ per revolution.

Diam., Inches	20	25	30	35	40	45	50	60	Diam., Inches
	Min Sec	Min Sec	Min Sec	Min Sec	Min Sec	Min Sec	Min Sec	Min Sec	
1/4	0 20	0 16	0 13	0 11	0 10	0 9	0 8	0 7	1/4
5/16	0 25	0 20	0 16	0 14	0 12	0 11	0 10	0 8	5/16
3/8	0 29	0 23	0 20	0 17	0 15	0 13	0 12	0 10	3/8
7/16	0 34	0 27	0 23	0 20	0 17	0 15	0 14	0 11	7/16
1/2	0 39	0 31	0 26	0 22	0 20	0 17	0 16	0 13	1/2
9/16	0 44	0 35	0 29	0 25	0 22	0 20	0 18	0 15	9/16
5/8	0 49	0 39	0 33	0 28	0 25	0 22	0 20	0 16	5/8
11/16	0 54	0 43	0 36	0 31	0 27	0 24	0 22	0 18	11/16
3/4	0 59	0 47	0 39	0 34	0 29	0 26	0 24	0 20	3/4
13/16	1 4	0 51	0 43	0 36	0 32	0 28	0 26	0 21	13/16
7/8	1 9	0 55	0 46	0 39	0 34	0 30	0 28	0 23	7/8
15/16	1 14	0 59	0 49	0 42	0 37	0 33	0 29	0 25	15/16
1	1 19	1 3	0 52	0 45	0 39	0 35	0 31	0 26	1
1 1/8	1 28	1 11	0 59	0 50	0 44	0 39	0 35	0 29	1 1/8
1 1/4	1 38	1 18	1 5	0 56	0 49	0 43	0 39	0 33	1 1/4
1 3/8	1 48	1 26	1 12	1 2	0 54	0 48	0 43	0 36	1 3/8
1 1/2	1 58	1 34	1 18	1 7	0 59	0 52	0 47	0 39	1 1/2
1 5/8	2 8	1 42	1 25	1 13	1 4	0 57	0 51	0 42	1 5/8
1 3/4	2 18	1 50	1 31	1 19	1 9	1 1	0 55	0 46	1 3/4
1 7/8	2 27	1 58	1 38	1 24	1 14	1 5	0 59	0 49	1 7/8
2	2 37	2 6	1 45	1 30	1 19	1 9	1 3	0 52	2
2 1/8	2 47	2 13	1 51	1 35	1 23	1 14	1 7	0 56	2 1/8
2 1/4	2 56	2 21	1 58	1 41	1 28	1 19	1 11	0 59	2 1/4
2 3/8	3 6	2 29	2 4	1 46	1 33	1 23	1 15	1 2	2 3/8
2 1/2	3 16	2 37	2 10	1 52	1 38	1 27	1 19	1 5	2 1/2
2 3/4	3 36	2 53	2 24	2 3	1 48	1 36	1 27	1 12	2 3/4
3	3 55	3 9	2 37	2 15	1 58	1 44	1 34	1 18	3
3 1/4	4 15	3 24	2 50	2 26	2 8	1 53	1 42	1 25	3 1/4
3 1/2	4 35	3 40	3 3	2 37	2 18	2 2	1 50	1 33	3 1/2
3 3/4	4 54	3 56	3 16	2 48	2 27	2 11	1 58	1 38	3 3/4
4	5 14	4 11	3 29	2 59	2 37	2 19	2 6	1 44	4
4 1/4	5 33	4 26	3 42	3 10	2 47	2 27	2 13	1 51	4 1/4
4 1/2	5 52	4 42	3 56	3 21	2 56	2 37	2 21	1 57	4 1/2
4 3/4	6 12	4 58	4 9	3 32	3 6	2 45	2 29	2 3	4 3/4
5	6 32	5 14	4 21	3 44	3 16	2 54	2 37	2 11	5
5 1/2	7 12	5 46	4 47	4 8	3 37	3 11	2 54	2 24	5 1/2
6	7 52	6 18	5 12	4 29	3 56	3 28	3 9	2 37	6
6 1/2	8 33	6 42	5 39	4 52	4 17	3 46	3 25	2 50	6 1/2
7	9 10	7 20	6 9	5 14	4 35	4 4	3 40	3 3	7
7 1/2	9 49	7 51	6 32	5 39	4 54	4 21	3 58	3 16	7 1/2
8	10 28	8 22	6 59	5 58	5 14	4 39	4 11	3 29	8
8 1/2	11 7	8 54	7 25	6 20	5 34	4 56	4 27	3 43	8 1/2
9	11 46	9 25	7 51	6 42	5 53	5 14	4 42	3 55	9
9 1/2	12 25	9 56	8 17	7 5	6 12	5 32	4 58	4 9	9 1/2
10	13 5	10 28	8 43	7 30	6 32	5 49	5 4	4 22	10
10 1/2	13 44	10 59	9 2	7 51	6 52	6 9	5 28	4 35	10 1/2
11	14 29	11 31	9 34	8 13	7 12	6 23	5 45	4 47	11
11 1/2	15 2	12 3	10 2	8 36	7 31	6 41	6 1	5 1	11 1/2
12	15 41	12 35	10 28	8 58	7 50	6 58	6 18	5 14	12
12 1/2	16 21	13 6	10 55	9 21	8 10	7 16	6 34	5 27	12 1/2
13	17 6	13 36	11 26	9 44	8 30	7 34	6 50	5 40	13
13 1/2	17 40	14 7	11 41	10 0	8 49	7 51	7 3	5 53	13 1/2
14	18 20	14 40	12 12	10 27	9 7	8 8	7 20	6 6	14
14 1/2	18 50	15 11	12 40	10 50	9 29	8 26	7 36	6 19	14 1/2
15	19 38	15 41	13 5	11 10	9 49	8 43	7 50	6 32	15
15 1/2	20 10	16 18	13 31	11 35	10 9	9 1	8 9	6 44	15 1/2
16	20 56	16 46	13 58	11 57	10 29	9 18	8 23	6 58	16

Surface Speeds in Feet per Minute.

ADVANCED MACHINE WORK

SECTION 3

LATHE WORK

Lathe Tools for Steel or Wrought Iron. Holders and Cutters. **Turning Steel.** Cutting-off Tools. Threading or Screw Cutting. **Bolt and Nut Making.** Making a Tensile Test Specimen. **Making a Stud.** Table U. S. S. Bolt Heads, etc. International and French Standard Threads. Indexing in Engine Lathe. Making an Engine Lathe Live Center. Automobile Screws and Nuts.

LATHE TOOLS FOR STEEL OR WROUGHT IRON

1. More rake is used on these tools than on those used for cast iron.

2. Right-side tool. — Top face A, Fig. 1, is given a side rake of about 35° and side clearance of 10°. A lubricant may be used, but good results are obtained in squaring and turning small work dry. See Lubricants for Cutting Tools, pp. 148-150.

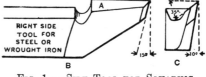

Fig. 1. — Side Tool for Squaring Steel or Wrought Iron.

3. Step method of squaring. — To square a rough end or remove extra length, the step method is used: Set side tool about 95° from axis of work as shown in Fig. 2, which gives a clearance of about 5° for rough squaring. The side tool, Fig. 2, is fed inward about $\frac{1}{16}''$, then fed by hand long. feed. This cuts the first step, as at 1. The process is repeated at 2, and so on to the countersink. Then a continuous roughing cut is taken outward. When a forging or stock for a shaft, or spindle, is extra long, the first end is rough squared

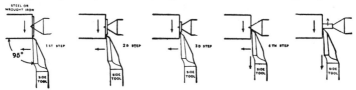

Fig. 2. — Step Method of Squaring.

Fig. 3. — Method of Squaring Large Amount of Stock.

301

in the regular way, and the second end by the step method, as at *A*, Fig. 3. The stem left is chipped and filed off. The ends are recentered and finish squared. If extra length exceeds ⅜″ a cutting-off tool may be used.

For finish squaring set side tool about 93° from axis of work, which gives a clearance of about 3°, and feed outward.

Information. — On some classes of work where length is not important, the ends are chamfered and all squaring omitted or the work is only rough squared and finish squaring omitted.

4. Right diamond-point tool. — For machine steel or wrought iron the top face *A*, Fig. 4, is given a combination of front and side rake of about 35° for roughing or finishing. For carbon steel (annealed) it is usually given less rake, 25° or 30°, and the cutting speed reduced.

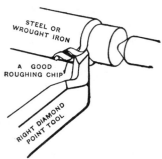

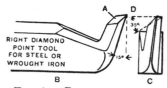

FIG. 4. — DIAMOND-POINT TOOL FOR TURNING STEEL OR WROUGHT IRON.

FIG. 5. — ROUGH TURNING STEEL OR WROUGHT IRON. GOOD.

5. To turn with a diamond-point tool. — Fig. 5 shows a diamond-point tool taking a roughing cut on machinery steel or wrought iron.

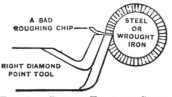

FIG. 6. — ROUGH TURNING STEEL OR WROUGHT IRON. BAD.

6. Steel or wrought-iron chips. — In Fig. 5 a chip cut by a tool with 35° rake, which produces a good surface, is shown, while in Fig. 6 the straight broken chip is cut by a tool without rake, which requires more power and leaves a ragged surface. Only the inexperienced would use a tool without rake for squaring and turning steel or wrought iron.

7. A small roughing tool, substitute for diamond-point, as in Fig. 7, is often used in roughing and finishing steel or

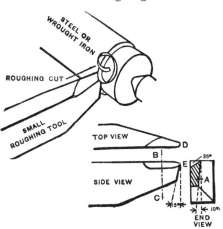

FIG. 7. — ROUGH TURNING STEEL OR WROUGHT IRON.

wrought iron. It may be used for heavy cuts on cast iron. It is given considerable side clearance, see end view, which shows at A a section taken at BC. The point is shaped as shown at D and E.

8. A large roughing tool for steel or wrought iron is shown in Fig. 8.

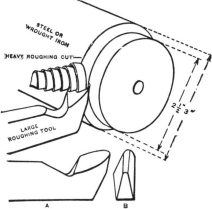

FIG. 8. — ROUGH TURNING STEEL OR WROUGHT IRON.

9. Large roughing tool ground from bar. — Fig. 9 shows a
heavy roughing cut on a piece of tough nickel steel taken

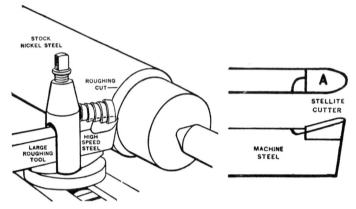

FIG. 9. — ROUGH TURNING STEEL OR WROUGHT IRON.

with a large roughing tool made of high-speed steel and ground
from the bar. See Stellite, p. 310.

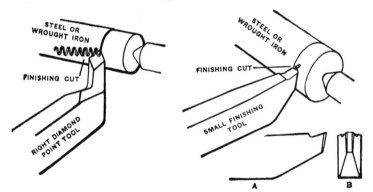

FIG. 10. — FINISHING STEEL
OR WROUGHT IRON.

FIG. 11. — FINISHING STEEL
OR WROUGHT IRON.

10. To finish turn steel or wrought iron. — The diamond-
point tool in Fig. 10, with a fine feed, is best adapted to use
for finishing.   To save time, small and large square-nose finishing tools, Figs. 11, 12, are often used on large work with a lubri-

cant, light cut and coarse feed. The tools drag a little; that is, the back corners are set to cut deeper than the front corners to avoid chattering. See *A*, Fig. 12.

11. Spring finishing tool. — Fig. 13 shows a high-speed steel spring tool taking a light finishing cut with a lubricant on a shaft of nickel steel.

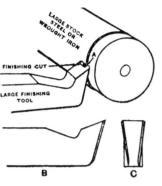

Attention. — Broad-nose tools have a great tendency to chatter and produce a rough corrugated surface caused by the long cutting edge, and made worse by frail, slender work, loose spindle bearings and loose cross slide.

12. Action of spring tools. — Broad-nose finishing tools have

FIG. 12. — FINISHING STEEL OR WROUGHT IRON.

a tendency to dig into the work. The curved portion of tool in Fig. 13 serves as a spring, and when the cutting edge is set at the height of center, it will spring away from rather

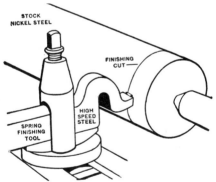

FIG. 13. — FINISHING STEEL OR WROUGHT IRON.

than into the work when a hard spot is encountered as the pivoting point is above the center. The success of spring tools depends upon having the proper amount of spring in proportion to length of cutting edge and diameter of work.

13. A left side tool is shown in Fig. 14. It is ground with the same angles as the right side tool and is used for squaring left shoulders.

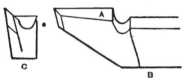

FIG. 14. — LEFT SIDE TOOL FOR STEEL OR WROUGHT IRON.

14. A left diamond-point tool is shown in Fig. 15. It is used for turning from left to right.

15. A right half-diamond point tool is shown in Fig. 16. *A* is its top face. It can be used to turn up to and to square a shoulder.

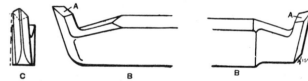

FIG. 15. — LEFT DIAMOND-POINT TOOL FOR STEEL OR WROUGHT IRON.

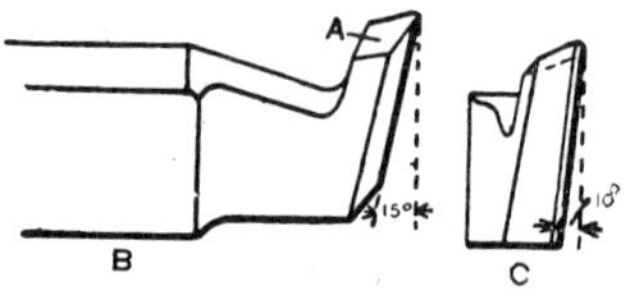

FIG. 16. — HALF DIAMOND-POINT TOOL FOR STEEL OR WROUGHT IRON.

HOLDERS AND CUTTERS

16. Holders with inserted cutters or bits, in Chart, Fig. 22, are displacing forged tools to a large extent. The cutters are high-speed steel or stellite, see p. 310, and obtainable square, round, etc., 2″ to 3″ in length, hardened and ready to grind or in unhardened bars, which are cut by nicking all four sides with a grinding wheel and breaking off in a vise. See *Principles of Machine Work*. The cutting angles are the same as on forged tools, and a student should become familiar with forged tools before using holders and cutters.

17. Holders and cutters for cast iron, Fig. 17. — Cutter or bit *A* is held at an angle of 20° in holder *B* and is secured by screw *C* with short wrench *D*.

To turn cast iron the end of the cutter is shaped like a right diamond-point tool as at *E* and for cutting-in, facing, or squaring, the tool is shaped like a left diamond-point tool as at *F*, and used as *G*.

18. Holders and cutters for steel or wrought iron, Fig. 18. —Cutter or bit *A* is held at an angle of 20° in holder *B* and is secured by binder *C*.

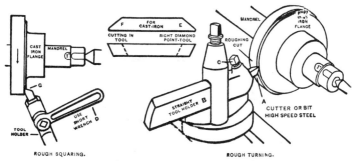

FIG. 17.—HOLDERS AND CUTTERS FOR CAST IRON.

To turn steel or wrought iron the cutter is shaped like a right diamond-point tool with 25° to 30° rake as at *D*. To square steel or wrought iron the cutter is shaped like a right side tool with 25° to 30° rake as at *E*, and used as at *F*. See Method of Squaring, p. 301.

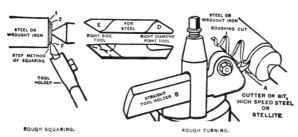

FIG. 18.—HOLDERS AND CUTTERS FOR STEEL OR WROUGHT IRON.

Fig. 20 shows at *D, E*, and *F* some uses to which straight holders and cutters may be put; and Fig. 21 shows some uses of the right and left bent or offset holders and cutters *G* and *H*.

19. Useful forms of cutters. — For convenience, one should
have a number of cutters ground to suitable cutting angles
for different materials and for different operations, as *I, J,
K, L, M, N,* Fig. 19.

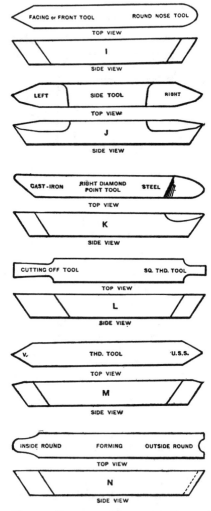

Fig. 19. — Useful Forms of High-speed Steel Cutters.

STOCK
C
D
E
F

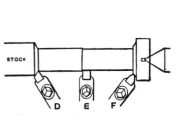

Fig. 20. — Operations with Holders and Cutters.

STOCK
C
RIGHT BENT
LEFT BENT
G
H

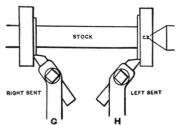

Fig. 21. — Squaring or Facing with Offset Tools.

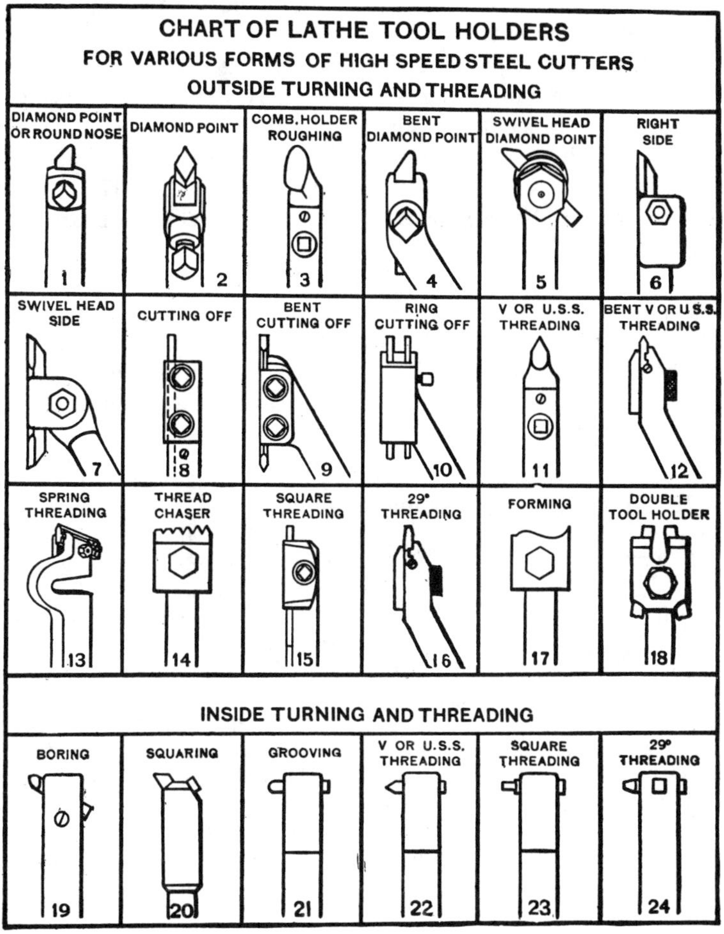

CHART OF LATHE TOOL HOLDERS
FOR VARIOUS FORMS OF HIGH SPEED STEEL CUTTERS
OUTSIDE TURNING AND THREADING
DIAMOND POINT OR ROUND NOSE
1
DIAMOND POINT
2
COMB. HOLDER ROUGHING
3
BENT DIAMOND POINT
4
SWIVEL HEAD DIAMOND POINT
5
RIGHT SIDE
6
SWIVEL HEAD SIDE
7
CUTTING OFF
8
BENT CUTTING OFF
9
RING CUTTING OFF
10
V OR U.S.S. THREADING
11
BENT V OR U.S.S. THREADING
12
SPRING THREADING
13
THREAD CHASER
14
SQUARE THREADING
15
29° THREADING
16
FORMING
17
DOUBLE TOOL HOLDER
18
INSIDE TURNING AND THREADING
BORING
19
SQUARING
20
GROOVING
21
V OR U.S.S. THREADING
22
SQUARE THREADING
23
29° THREADING
24

Fig. 22.

20. High-speed steel in general is made by alloying iron with tungsten and chromium or molybdenum and chromium. In some brands other ingredients are used such as vanadium, manganese, etc. The different kinds are hardened by heating from 1800° F. to 2200° F. and cooling in air blast or oil. High-speed steel has a much greater cutting efficiency than carbon steel. See *Principles of Machine Work*. It is largely used for cutters in tool holders, for engine and turret lathe work, screw machines, planers and shapers.

Bars and disks, annealed, are obtainable for making twist drills and milling cutters.

21. Stellite, a superior high-speed cutting metal, is an alloy of cobalt, chromium, tungsten and molybdenum in varying proportions. Since it contains no iron it is not a steel.

It is cast in bars and has a scleroscopic hardness of about 70. The cutters are used in tool holders, see Figs. 17, 18, or welded or brazed to a machine-steel shank to eliminate breakage. See *A*, Fig. 9.

Round-nose cutters are preferred, and sharp points avoided.

Owing to its brittleness, stellite should not be used for planer or sharper work but for lathe work where the cut is continuous.

The cutter should not project more than one-half inch beyond holder, should fit well on the bottom and be tightened only enough to give it a firm hold.

TURNING STEEL

22. To make plain machine handle, Fig. 23.

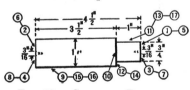

FIG. 23. — SCHEDULE DRAWING.

Specifications: Material, machine steel, $\frac{1}{16}''$ to $\frac{1}{8}''$ large; weight, 1 lb. 5. oz.

Hardness, 15 to 16 (scleroscope).

Carbon-steel cutting tools. See *Exception*, p. 205.

Lubricant, dry or lard oil.

While a lubricant may be used in all machining operations on steel or wrought iron, it is best to omit the lubricant when squaring (side tool) or turning (diamond-point tool) small work.

Time: Study drawing and schedule in advance, 5 min. — Oil lathe, 4 min.— Center and Square, 14 min.— Rough and finish turn, square shoulder and file, 27 min.— Clean lathe, 3 min.— Total. 53 min.

SCHEDULE OF OPERATIONS, MACHINES AND TOOLS

OPERATIONS.	MACHINES, SPEEDS, FEEDS.	TOOLS.
File stem off ends.	Vise.	Bastard or 2d-cut file.
Center to $\frac{3}{16}''$, **(1)**, **(2)**, hand or machine method. See that live center is nearly true and dead center in approximate alinement. See p. 116.	Speed lathe or centering machine. Engine lathe, 12″ to 16″.	$\frac{3}{32}''$ drill, 60° countersink, rule, lard oil.
Oil bearings.	. . . ., . .	Machine oil.
Rough square (3), reverse work on centers and repeat on **(4)**, by step method to $4\frac{1}{2}'' + \frac{1}{64}''$. See p. 301.	2d speed, or 25 F.P.M. Hand feed.	Dog, side tool 35° rake, calipers, rule.
Recenter to $\frac{3}{16}''$, **(5)**, **(6)**.	Speed lathe, 3d speed.	60° countersink, rule, lard oil.
Finish square (grind tool), to $4\frac{1}{2}''$, **(7)**, **(8)**.	Engine lathe, 2d or 3d speed or 50 F.P.M. Hand feed.	Side tool 35° rake, calipers, rule.
Rough turn to $1'' + \frac{1}{64}''$, **(9)**, one cut. Turn half way, reverse and turn other half.	1st or 2d speed, or 25 F.P.M. Medium power feed — 80 to 1″.	Diamond-point tool 35° rake, calipers, rule.
Draw line 1″ from one end **(10)**, for length of reduced part.	Vise.	Copper sulphate, scriber, rule.
Rough turn to $\frac{3}{4}'' + \frac{1}{64}''$, **(11)**, and within $\frac{1}{64}''$ of line **(10)**, two or three cuts.	Engine lathe, 1st or 2d speed, or 25 F.P.M. Medium power feed — 80 to 1″.	Dog, copper under set screw.
Rough square shoulder **(12)**. See p. 132.	2d or 3d speed, or 25 F.P.M. Hand feed.	Side tool 35° rake, rule.
Finish turn (grind tool), to $\frac{3}{4}''$, **(13)**, one cut.	3d speed, or 50 F.P.M. Fine power feed — 140 to 1″.	Diamond-point tool 35° rake, calipers, rule.
Finish square (grind tool) shoulder to length **(14)**.	. . . .	Side tool 35° rake, rule.
Finish turn to 1″, **(15)**, one cut. Turn half way, reverse and turn other half.	. . . .	Diamond-point tool 35° rake, calipers, rule.
File to remove tool marks (16), **(17)**.	4th speed, or 175 F.P.M., or speed lathe, 3d or 4th speed.	8″ or 10″ mill-bastard file, file card.
Stamp name on large end.	Vise, copper jaws.	$\frac{1}{8}''$ steel letters, hammer.

Attention. — Turn to dimensions with no allowances for filing or fitting unless reduced part **(13)** is to be a driving fit.

Note. — This handle may be nurled or formed to suit conditions. See pp. 437, 635. 636

23. Double holder. — Fig. 24 shows a "home-made" double tool holder A, facing both sides of blank B in one operation.

FIG. 24. — FACING GEAR BLANK WITH DOUBLE HOLDER AND CUTTERS.

Cutters C, C' are adjusted to cut as desired, then clamped by screws D, D'.

For duplicate pieces, clamp lathe carriage and use index pointer on mandrel press to locate blank on mandrel.

24. Two forged tools for facing. — Two bent tools A

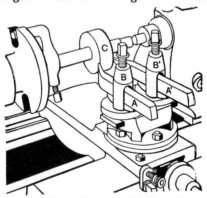

FIG. 25. — FACING GEAR BLANK WITH TWO FORGED TOOLS.

and A', Fig. 25, held in two tool-posts B and B' or by a bolt and strap, may be used to face the sides of gear blank C.

CUTTING–OFF TOOLS

25. Cutting-off tool. — Fig. 26 shows a cutting-off tool for all metals. *A* is its top face. For clearance it is made

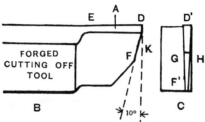

FIG. 26. — CUTTING-OFF TOOL FOR ALL METALS.

wider at the point, as shown. For steel or wrought iron a lubricant must be used, and the tool is often given front rake. For cast iron or brass the tool is used dry. The cutting speed is the same as for rough turning.

Cutting-off stock. — The tool is used close to the chuck jaws, as in Fig. 27, and should never be used to cut off stock

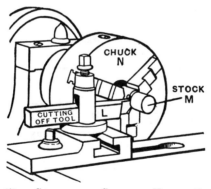

FIG. 27. — CUTTING-OFF STOCK IN ENGINE LATHE.

more than one diameter of stock away from chuck jaws lest it catch and break and also strain the chuck. Do not attempt to sever stock completely, as in Fig. 28, but finish cutting with a chisel or hack saw. Fig. 29 shows an offset cutting-off tool holder and cutter.

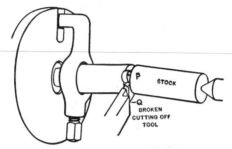

Fig. 28. — How to Break a Cutting-off Tool.

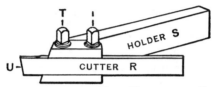

Fig. 29. — Cutting-off Tool Holder and Cutter.

26. To prepare two shaft blanks, one to turn and file running and driving fits (Fig. 30), and one to grind running and forcing fits (Fig. 31).

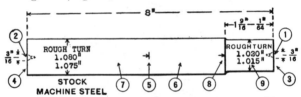

Fig. 30. — Preparing Shaft Blank for Turning and Filing Fits.
Schedule Drawing.

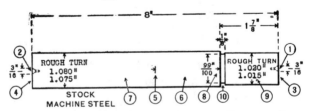

Fig. 31. — Preparing Shaft Blank for Grinding Fits.
Schedule Drawing.

Specifications: Material, machine steel $\frac{1}{16}''$ to $\frac{1}{8}''$ large; weight, 2 lb.
10 oz. each. Machine dry, or use lard oil.
For heat treatment see *Principles of Machine Work.*
Hardness, 15 to 16 (scleroscope).
True live center. Set dead center in approximate alinement, see p. 116.
High-speed steel or stellite cutting tools. See *Exception*, p. 205.
Time: Study drawing and schedule in advance, 5 min. — Oil lathe, 4 min.
— Center and square both shafts, 28 min. — Rough turn both shafts, 30
min. — (Center and square one shaft, 15 min. — Rough turn one shaft, 16
min.) Clean lathe, 3 min.

MULTIPLE SCHEDULE OF OPERATIONS, MACHINES AND TOOLS

Operations.	Machines, Speeds, Feeds.	Tools.
Center (1), (2), Fig. 30 and (1), (2), Fig. 31 to $\frac{3}{16}''$.	Centering machine. Drill speed, 1200 R.P.M. Countersink speed, 500 R.P.M.	$\frac{3}{32}''$ or No. 43 drill, 60° countersink, lard oil.
Rough square (3), (3), then (4), (4) to $8\frac{1}{64}''$.	Engine lathe, 12'' to 16''. 2d or 3d speed, or 50 F.P.M. Hand or power feed.	Dog, side tool or holder and cutter, 35° rake, calipers, rule.
Recenter to $\frac{3}{16}''$ and finish square (3), (3), then (4), (4).	3d or 4th speed, or 80 F.P.M. Hand or power feed.	
Mark lines approximately equidistant from ends, (5), (5). Rough turn (6), (6). Reverse and turn (7), (7). One or two cuts.	2d or 3d speed, or 60 F.P.M. Medium power feed 80 to 1''.	Two dogs with copper, diamond-point tool or holder and cutter, 35° rake, micrometer.
Mark lines at (8), (8).		Copper sulphate, rule, scriber.
Rough turn (9), (9) to (8), (8). One cut.	1st or 2d speed, or 60 F.P.M. Medium power feed —80 to 1''.	
Cut recess to $\frac{99}{100}''$ diameter, (10).	1st or 2d speed, or 30 F.P.M. Hand feed.	Cutting-off tool corners rounded, lard oil, calipers, rule.
To grind fits, see pp. 716, 720. To turn and file fits, see pp. 316, 317.		

27. To turn and file 1″ drive fit, (10), (11), Fig. 32.

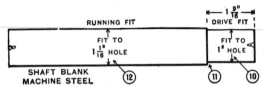

FIG. 32. — SCHEDULE DRAWING.

Specifications: Material, shaft blank machine steel, rough turned 1.015″ diameter. Hardness, 15 to 16 (scleroscope). Machine dry, or use lard oil. High-speed steel or stellite cutting tools. See *Exception*, p. **2**05.

Time: Study drawing and schedule in advance, 3 min. — Oil lathe, 4 min. — True live center. Set lathe straight, rough and finish turn, file and fit, 25 min. (With blank prepared.) — Clean lathe, 3 min. — Total, 35 min. (Preparation of blank, 31 min. extra.)

SCHEDULE OF OPERATIONS, MACHINES AND TOOLS

OPERATIONS.	MACHINES, SPEEDS, FEEDS.	TOOLS.
To MAKE FIT BY TRIAL AND CORRECTION, (10).		
True live center. Set dead center in accurate aline-ment to turn straight using this shaft or trial piece the same length, pp. **116, 117.**	Engine lathe, 12″ to 16″. 2d or 3d speed, or 70 F.P.M. Fine power feed — 140 to 1″.	Center truing tool. Dog, copper, dia-mond-point tool, or holder and cutter, 35° rake, micrometer.
Push mandrel into reamed hole in work (flange or gear). Set cali-pers to large end close to work. Take several light cuts about $\frac{1}{8}″$ in length until calipers fit work slightly harder than on mandrel.	2d or 3d speed, or 70 F.P.M. Fine power feed — 140 to 1″.	1″ mandrel, dia-mond-point tool, or holder and cut-ter, 35° rake, cali-pers, oil-stone.
Hold work in vise. Oil shaft and press into large end of hole. When it enters from $\frac{1}{64}″$ to $\frac{1}{32}″$ with hand pressure, continue cut to shoulder.	Vise.	Copper jaws.
Square shoulder, (11).	3d or 4th speed, or 70 F.P.M. Hand feed.	Side tool, rule.
File small portion at end. Clean, oil, and press into hole. Continue filing sparingly and testing until shaft will enter hole one-third length of fit. Keyways are cut and key fitted before shaft is pressed or driven to shoulder.	Engine lathe, 4th speed, or speed lathe, 1st or 2d speed, or 175 F.P.M.	8″ or 10″ mill bas-tard file, file card, lard oil.

Attention. — Hand pressure means grasping dog with both hands, and pressing shaft hard into hole with a right rotation. A left rotation will remove it. After each trial, the brightness of the surface and testing with calipers will indicate where to file. This is often called a wringing fit.

28. To make 1″ drive fit by allowance, Fig. 32.

Turn to $\begin{cases} 1.003''.................. \\ 1.004'' \ .... \end{cases}$	..................|..................	|..................
and	....	Micrometer.
File to $\begin{cases} 1.00125''.................. \\ 1.00100''.................. \end{cases}$	....	\}..................

See Belt-power Forcing Press, p. 220

29. To turn and file $1\frac{1}{16}''$ running fit by allowance, (12), Fig. 32.

Specifications: Material, shaft-blank, machine steel, rough-turned 1.078″ diameter. Hardness 15 to 16 (scleroscope).

High-speed steel or stellite cutting tools. See *Exception*, p. 205.

Time: Study drawing and schedule in advance, 3 min. — Oil lathe, 4 min. — True live center. Set lathe straight, rough and finish turn, file and fit, 42 min. (With blank prepared.) — Clean lathe, 3 min. — Total, 52 min. (Preparation of blank, 31 min. extra.)

SCHEDULE OF OPERATIONS, MACHINES AND TOOLS

OPERATIONS.	MACHINES, SPEEDS, FEEDS.	TOOLS.
True live center. Set dead center in accurate aline-ment to turn straight using this shaft or trial piece the same length, pp. 116, 117.	Engine lathe 12″ to 16″. 3d or 4th speed, or 70 F.P.M. Fine power feed — 140 to 1″.	Dog, copper, dia-mond-point tool, or holder and cutter, 35° rake, micrometer.
Grind and oilstone tool. Take light cuts at end about ¼ ″ in length until work measures $\begin{cases} 1.0650'' \\ 1.0645'' \end{cases}$ Take cut from end to end by re-versing work, one cut.	3d or 4th speed, or 70 F. P. M. Fine power feed—140 to 1″.	Diamond-point tool, or holder and cutter 35° rake, oilstone.
File over entire surface to erase tool marks and until work measures $\begin{cases} 1.0620'' \\ 1.0615'' \end{cases}$, (12).	Engine lathe 4th speed, or speed lathe, 2d or 3d speed, or 175 F. P. M.	8″ or 10″ mill bas-tard file, file card, micrometer.

THREADING OR SCREW CUTTING

30. Forms of threads. — There are four common forms of threads: the Sharp V, Fig. 33; the United States Standard, Fig. 34; the Square, see p. 510, and the Acme Standard or

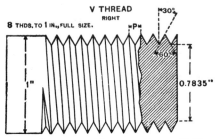

FIG. 33. — SECTIONAL VIEW OF SHARP V THREAD

29° thread, see p. 521. The Whitworth (English) Standard thread, Fig. 55, is very little used in the United States. It is standard in Great Britain for coarse pitches. The British Association Standard thread is standard for fine pitches. It

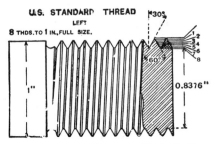

FIG. 34. —SECTIONAL VIEW OF U. S. S. THREAD.

is similar in form to the Whitworth, but the angle is $47\frac{1}{2}$ degrees. The International and French Standard threads, used with the Metric system in some foreign countries, are based on the same formulas as the United States Standard thread. See pp. 337, 351.

31. Right and left screw threads. — A right screw, Fig. 33, enters its nut when rotated to the right (clockwise). A left screw, Fig. 34, is the reverse. Screws are supposed to be right-threaded unless designated left.

32. Uses of different threads. — A right Sharp V or United States Standard thread is used to fasten parts together. A left thread is also used to fasten parts together, but only where a rotary motion would loosen a right thread as a nut on the near side of a wagon axle. The Square and 29° threads, both right and left, are used to transmit motion. See pp. **510–532**.

33. Method of threading screws and nuts. — Screws for fine machine parts are threaded in a lathe. Bolts, studs, and screws are threaded with dies by power and by hand. See Dies, *Principles of Machine Work*.

34. Lead screws, taps, and worm screws may be milled with a thread milling machine. See Thread Milling Machine, pp. **1047–104**9.

35. Small nuts are threaded with a tap by hand or power;

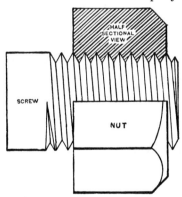

FIG. 35. — SCREW AND NUT.

large nuts are usually threaded in a lathe. Fig. 35 shows a screw and section of nut.

36. Rolled threads, produced by rolling the material between moving dies, are used on stove bolts, carriage bolts, some machine screws, etc.

37. Single and multiple threads. — Ordinary screws are single threaded, but for special purposes screws are double threaded, triple threaded, etc. Examine the end of the screw;

if only the end of one thread and one groove is found, it is single threaded, etc. See Multiple Threads, p. **531**.

38. **The pitch of a thread** is the distance along the axis of the work from the center of one thread to the center of the next, as P, Fig. 33.

39. **The lead of a thread** is the distance the screw advances in one revolution. In a single thread, the pitch is equal to the lead; in a double thread, the pitch is one-half the lead, etc.

40. **Threads or pitches per inch** signify the number of threads contained in one inch in length. For example, for $\frac{1}{10}''$ pitch we have 10 threads per 1$''$ (often called 10 P.)

41. **The diameter of threaded work** is measured over the tops of the threads, as 1$''$, Fig. 33.

42. **The root or bottom diameter** is measured at the root or bottom of groove, as .7835$''$, Fig. 33.

43. **Thread calipers,** Fig. 36, may be used to test the diameter of Sharp V, United States Standard or 29° thread

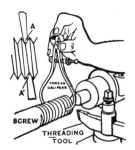

FIG. 36. — CALIPERING SCREW
WITH THREAD CALIPERS.

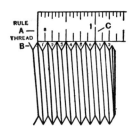

FIG. 37. — COUNTING THREADS
WITH STEEL RULE.

screws. As the fit is on the sides only the points are filed to fit the thread gage and are left truncated as at A and A' to avoid touching the bottom. The calipers are set to size by means of a thread gage, tap, or screw.

For accurate work, the screw is cut nearly to size of calipers and the final test is made by trying screw in the nut or work.

44. **To count threads.** — Place rule A on thread B, Fig. 37, count grooves in one inch. To count Square and 29°

threads, place the end of rule against the right edge of one of the threads and count the spaces to the right. For double threads, triple threads, etc., count all the spaces and divide by 2, 3, etc., respectively.

A screw-pitch gage A, Fig. 38, is used to determine number of threads to one inch on a screw or in a nut. Handle A

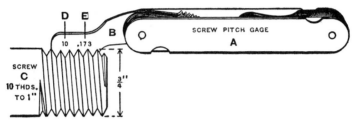

FIG. 38. — COUNTING THREADS WITH SCREW-PITCH GAGE.

contains many blades notched with all the ordinary threads to one inch. To determine number of threads to one inch, as at C, select by trial a blade that will match thread, and the number at D on blade, as 10, gives number of threads to one inch on screw. Decimal E is the double depth of a Sharp V thread of this pitch.

45. The Sharp V thread, Fig. 39. — The single depth for a 1″ pitch thread is .866″, double depth 1.732″. For pitch P, depth = .866 P = D. Double depth = 1.732 P.

$$P = \text{pitch} = \frac{1}{\text{No. of threads per inch}};$$

Root diameter = Outside diameter − double depth.

Formula: — Root diameter = Outside diameter −

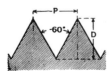

FIG. 39. — SECTION SHOWING PITCH AND DEPTH OF SHARP V THREAD.

$$\frac{1.732}{\text{No. of threads per inch}}.$$

Example. — To find root diameter of a screw 1″ diameter, 8 threads per 1″.

Solution. — $1 - \dfrac{1.732}{8} = .7835''$,

46. Table of Sharp V-thread screws.

Diameter of Screw.	No. Threads per Inch.	Diameter of Screw.	No. Threads per Inch.	Diameter of Screw.	No. Threads per Inch.
$\frac{1}{4}$	20	$1\frac{3}{8}$	6	$3\frac{1}{4}$	$3\frac{1}{2}$
$\frac{5}{16}$	18	$1\frac{1}{2}$	6	$3\frac{3}{8}$	$3\frac{1}{2}$
$\frac{3}{8}$	16	$1\frac{5}{8}$	5	$3\frac{1}{2}$	$3\frac{1}{2}$
$\frac{7}{16}$	14	$1\frac{3}{4}$	5	$3\frac{5}{8}$	$3\frac{1}{2}$
$\frac{1}{2}$	12	$1\frac{7}{8}$	$4\frac{1}{2}$	$3\frac{3}{4}$	3
$\frac{9}{16}$	12	2	$4\frac{1}{2}$	$3\frac{7}{8}$	3
$\frac{5}{8}$	11	$2\frac{1}{8}$	$4\frac{1}{2}$	4	3
$\frac{11}{16}$	11	$2\frac{1}{4}$	$4\frac{1}{2}$	$4\frac{1}{4}$	$2\frac{7}{8}$
$\frac{3}{4}$	10	$2\frac{3}{8}$	$4\frac{1}{2}$	$4\frac{1}{2}$	$2\frac{3}{4}$
$\frac{13}{16}$	10	$2\frac{1}{2}$	4	$4\frac{3}{4}$	$2\frac{5}{8}$
$\frac{7}{8}$	9	$2\frac{5}{8}$	4	5	$2\frac{1}{2}$
$\frac{15}{16}$	9	$2\frac{3}{4}$	4	$5\frac{1}{4}$	$2\frac{1}{2}$
1	8	$2\frac{7}{8}$	4	$5\frac{1}{2}$	$2\frac{3}{8}$
$1\frac{1}{8}$	7	3	$3\frac{1}{2}$	$5\frac{3}{4}$	$2\frac{3}{8}$
$1\frac{1}{4}$	7	3	$3\frac{1}{2}$	6	$2\frac{1}{4}$

For Diameter of Tap Drills for Sharp V Threads, see *Principles of Machine Work*.

47. The United States Standard thread, Fig. 40, has its top and bottom truncated by $\frac{1}{8}$ of the depth, shown by divisions 1, 2, 3, etc., Fig. 34. The single depth of a 1″ pitch is .6495, double depth is 1.299″. For pitch P, depth $= .6495 P = D$. Double depth $= 1.299 P$.

$$P = \text{pitch} = \frac{1}{\text{No. of threads per inch}};$$

Root diameter $=$ Outside diameter $-$ double depth.

Formula. — Root diameter $=$ Outside diameter $-$

$$\frac{1.299}{\text{No. of threads per inch}}.$$

$$F = \text{Flat} = \frac{P}{8}.$$

Example. — To find root diameter of a screw 1″ diameter, 8 threads per 1″.

Solution. —
$$1 - \frac{1.299}{8} = .8376''.$$

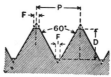

FIG. 40. — SECTION SHOWING PITCH AND DEPTH OF U. S. S. THREAD.

48. Table of United States Standard thread screws.

DIAM-ETER OF SCREW.	NO. THREADS PER INCH.	DIAM-ETER OF SCREW.	NO. THREADS PER INCH.	DIAM-ETER OF SCREW	NO. THREADS PER INCH.
$\frac{1}{4}$	20	$1\frac{5}{8}$	$5\frac{1}{2}$	$3\frac{3}{8}$	$3\frac{1}{4}$
$\frac{5}{16}$	18	$1\frac{3}{4}$	5	$3\frac{1}{2}$	$3\frac{1}{4}$
$\frac{3}{8}$	16	$1\frac{7}{8}$	5	$3\frac{5}{8}$	$3\frac{1}{4}$
$\frac{7}{16}$	14	2	$4\frac{1}{2}$	$3\frac{3}{4}$	3
$\frac{1}{2}$	13	$2\frac{1}{8}$	$4\frac{1}{2}$	$3\frac{7}{8}$	3
$\frac{9}{16}$	12	$2\frac{1}{4}$	$4\frac{1}{2}$	4	3
$\frac{5}{8}$	11	$2\frac{3}{8}$	4	$4\frac{1}{4}$	$2\frac{7}{8}$
$\frac{3}{4}$	10	$2\frac{1}{2}$	4	$4\frac{1}{2}$	$2\frac{3}{4}$
$\frac{7}{8}$	9	$2\frac{5}{8}$	4	$4\frac{3}{4}$	$2\frac{5}{8}$
1	8	$2\frac{3}{4}$	4	5	$2\frac{1}{2}$
$1\frac{1}{8}$	7	$2\frac{7}{8}$	$3\frac{1}{2}$	$5\frac{1}{4}$	$2\frac{1}{2}$
$1\frac{1}{4}$	7	3	$3\frac{1}{2}$	$5\frac{1}{2}$	$2\frac{3}{8}$
$1\frac{3}{8}$	6	$3\frac{1}{8}$	$3\frac{1}{2}$	$5\frac{3}{4}$	$2\frac{1}{4}$
$1\frac{1}{2}$	6	$3\frac{1}{4}$	$3\frac{1}{2}$	6	$2\frac{1}{4}$

For Diameter of Tap Drills for U. S. S. Thread, see *Principles of Machine Work.*

49. The forged threading tool, Fig. 41, is forged, hardened, and then tempered to a light straw color. The clearance at *FDE* is 15°, and the cutting edges are *GH* and *JK.*

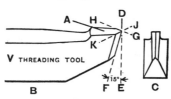

FIG. 41. — SHAPE OF SHARP V-THREADING TOOL.

The top is ground first and then the front faces are ground to fit center gage and the top is then set at height of center, as shown in Figs. 42, 43 at L, M, N and S.

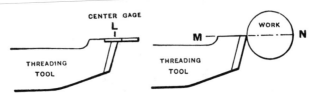

FIG. 42. — SETTING THREADING TOOL WITH CENTER GAGE AND TO HEIGHT OF CENTERS.

50. To set threading tool at height of center and at right angles to work. Fig. 43.

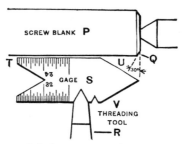

FIG. 43. — SETTING U. S. STANDARD OR SHARP V-THREADING TOOL AT RIGHT ANGLES WITH WORK.

SCHEDULE OF OPERATIONS

1. Turn work P to diameter.
2. Chamfer end Q depth of proposed thread.
3. Clamp tool R lightly with point 2″ from tool post.
4. Adjust point to height of dead center (MN, Fig. 42).
5. Hold gage S as shown. Adjust cross feed.
6. Rap tool around until edge TU is parallel to work and about $\frac{1}{32}$″ from it.
7. Clamp tool firmly.

51. Threading taper work. — Preferably use a taper attachment to cut a thread of correct pitch, as on pipe tap A, Fig. 44, as footstock "set over" will produce a thread slightly finer.

Set thread tool 90° to the axis of work as at *B*, not as at *C*. Thread tool *D* with gage *E* against shank as at *F* is correct.

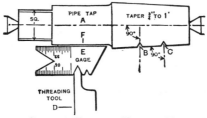

FIG. 44. — SETTING TOOL TO THREAD TAPER WORK.

52. A special threading tool that may be used straight or bent to thread to a shoulder, is shown in Fig. 45.

FIG. 45. — THREADING TOOL TO CUT TO SHOULDER.

53. United States Standard threading tool. — The United States Standard thread is cut with a V tool, *A*, Fig. 46,

FIG. 46. — SHAPE OF U. S. S. THREADING TOOL BEFORE TRUNCATING.

FIG. 47. — SHAPE OF U. S. S. THREADING TOOL.

truncated at point *B*, Fig. 47, $\frac{1}{8}$ depth of thread, which varies for every pitch. It is ground to fit notch *A* in U. S. S. thread gage, Fig. 48, then truncated at point to fit notch which corresponds with the threads to be cut, 8, Fig. 48. The tool is set the same as a Sharp V-thread tool.

Information. — The United States Standard form of thread, see table, p. 323, has been adopted by the United States government for its standard; the American Society of Mechanical Engineers, for the A. S. M. E. Standard; the Society of Automotive Engineers, for the S. A. E. Standard (U. S. F.), see pp. 354–356, and for nearly all manufacturers and railroads. The Sharp V thread cuts deeper than the U. S. S. thread and reduces both the tensile and torsional strength of the screw; it is also more easily damaged. The U. S. S. thread is more scientific and can be more readily made standard and interchangeable. Tap and die manufacturers have standardized U. S. S. threads and only furnish V threads on special orders and at higher prices.

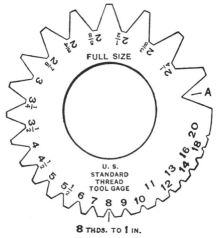

FULL SIZE

U. S.
STANDARD
THREAD
TOOL GAGE

8 THDS. TO 1 IN.

FIG. 48. — GAGE FOR U. S. S. THREAD TOOL.

54. The theory of screw cutting in the engine lathe. — To thread a screw in a lathe, the threading tool is moved along the bed a positive and uniform amount for each revolution of the lathe spindle. This motion is obtained by means of a train of change gears which connect the lathe spindle to the lead screw, and by half nuts in the apron which connect the lead screw to the carriage.

On most lathes the first change gear is on a separate shaft called the stud. On some lathes this stud is geared to rotate at the same speed as lathe spindle, and on others at a different ratio. Ordinary screws may be cut with simple gearing — two change gears as in Fig. 50. See also Compound Gearing, p. 333.

55. To calculate simple gearing with spindle gear 1 to stud gear 1:

$$\frac{\text{Lead screw threads per inch} \times \text{constant}}{\text{Threads per inch to be cut} \times \text{constant}}$$

$$= \frac{\text{teeth in gear on stud}}{\text{teeth in gear on lead screw}}.$$

The *constant* may be the common difference in number of teeth between the consecutive change gears, and this or any multiplier may be used to obtain available gears.

56. *Example.* — To cut 13 threads to 1″. Lead screw 5 threads to 1″; speed spindle same as stud; constant, 5.

Solution. —
$$\frac{5 \times 5}{13 \times 5} = \frac{25}{65} \quad \begin{matrix}\text{(gear on stud.)} \\ \text{(gear on lead screw).}\end{matrix}$$

Attention. — For stud ratios other than 1 to 1, multiply threads per inch of lead screw by ratio of the stud gear to the spindle gear and proceed as before.

57. *Example.* — To cut 13 threads to 1″ spindle gear 1 to stud gear 2, Fig. 50. Lead screw 8 threads per inch; constant, 6; speed of stud is one-half speed of spindle.

Solution. —
$$\frac{8 \times 2 \times 6}{13 \times 6} = \frac{96}{78} \quad \begin{matrix}\text{(gear on stud.)} \\ \text{(gear on lead screw.)}\end{matrix}$$

58. To prepare screw and nut blanks and to practice screw cutting, Fig. 49.

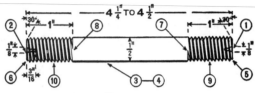

FIG. 49. — SCHEDULE DRAWING.

Specifications: Material, machine steel $\frac{1}{16}''$ large; weight, **6** oz. Hardness, 12 to 14 (scleroscope). Carbon-steel cutting tools.

If high-speed steel, or stellite cutting tools are used the speed may be increased. See *Exception*, p. **2**05.

Time: Study drawing and schedule in advance, 5 min. — Oil lathe, 4 min. — Prepare screw blank, 15 min. — Tap and square nut, 8 min. — "Set up" lathe for screw cutting, 6 min. — Thread one end, 7 min. — Thread the other end, 6 min. — Remove gears and clean lathe, 5 min. — Total, 56 min.

SCHEDULE OF OPERATIONS, MACHINES AND TOOLS

OPERATIONS.	MACHINES, SPEEDS, FEEDS.	TOOLS.
File ends flat.	Vise.	10″ or 12″ bastard file.
Center (1), (2).	Centering machine, drill 1700 R.P.M., countersink 600 R.P.M.	$\frac{1}{16}''$ drill, 60° countersink, rule, lard oil.
Omit squaring.		
See that live center is nearly true and dead center in approximate alinement. See p. **1**16.	Engine lathe, 12″ to 16″.	
Rough turn to $\frac{1}{2}'' + \frac{1}{64}''$, (**3**), one cut. Turn half way, reverse, and turn other half.	3d speed, or 30 F.P.M. Medium power feed — 80 to 1″.	Dog, diamond-point tool 35° rake, calipers, rule.
Finish turn to $\frac{1}{2}''$, (**4**), one cut. Turn half way, reverse and turn other half.	3d or 4th speed, or 50 F.P.M. Fine power feed — 140 to 1″.	

Chamfer ends to 30° approximately (**5**), (**6**). Fig. 43 and p. **3**41.	. . . .	Side tool 35° rake, center gage.
Draw lines 1″ from each end for length of thread (**7**), (**8**).	Vise, copper jaws.	Copper sulphate, scriber, rule.
Tap nut blank. Hole in ½″ nut blank is punched $\frac{13}{32}$″. See *Principles of Machine Work.*	Vise.	½ × 13 U. S. S. tap, tap wrench, lard oil.
Square both sides of nut to $\frac{7}{16}$″, p. **3**40.	Engine lathe, 3d speed, or 50 F.P.M. Hand feed.	Nut mandrel, side tool 35° rake, calipers, rule.
Thread (9) and (10) to fit nut. See pp. **3**18–**3**27, **3**29–**3**32. When through threading, remove gears, empty drip pan and clean lathe.	Arrange gears for 13 threads. 1st speed, or 20 F.P.M.	13 pitch U. S. S. thread tool, U. S. S. thread gage, center gage, clamp nut, lard oil.

Attention. — A student should learn screw cutting by threading blank ends until he can stop cut at same place every time without breaking point of tool and ruining work; also, until he can cut a smooth thread and make good fit in nut.

59. Description of screw-cutting mechanism, Fig. 50.

A — Gear on spindle.	*L* — Stud.
B — Spindle.	*M*— Radial arm.
C — Gear driven by one or both idle gears.	*N* — Bolt to clamp *M* so that *F* drives through *KH* to *G*.
D & *D′* — Idlers for reversing.	
E — Stud driven by *C*.	*P* — Reversing lever; shifts bracket for cutting right or left threads.
F — Stud gear.	
G — Lead screw.	
H — Lead-screw gear.	*Q* — Bracket carrying *D* and *D′*.
K — Idler gear, loose on stud.	

Exception. — On lathes that do not have reversing gears *D* and *D′*, use two idlers to cut a left thread.

60. To set up lathe for threading or screw cutting. Fig. 50.

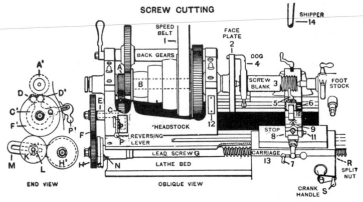

FIG. 50. — LATHE SET UP FOR THREADING U. S. S. OR SHARP V
THREAD — SIMPLE GEARING — TWO GEARS AND AN IDLER GEAR.

SCHEDULE OF OPERATIONS AND TOOLS
CHANGE OF GEARS, THREADING TOOL, CENTER GAGE, OIL BOX (TIN)

I. Place belt on step 1, screw face plate 2 hard against shoulder.

II. On blank 3, fasten dog 4 and mark with chalk the place in face plate where dog is inserted. (Always return dog to marked slot or tool will not resume its cut.)

III. Grind tool 5 to form of thread desired, U. S. S. or Sharp V, fasten tool in post 6, and set with gage, see pp. 140, 321–326.

IV. Feed tool in with handle 7 until it touches work, put on thread stop 8 with screw 9 in slide 10. Bring stop 8 against shoulder of 9 and clamp with screw 11.

V. Set footstock to allow $\frac{1}{2}''$ travel of tool beyond work.

VI. From index plate obtain gears for 13 threads and arrange as FF' and HH'.

VII. Select gear K for idler. Preferably a gear nearly the size of one of the change gears.

VIII. Place stud gear F on stud E, lead-screw gear H on G.

IX. Place gear K on stud L, in radial arm M, oil stud.

X. Adjust mesh of H and K and clamp in position.

XI. Swing radial arm M, to mesh K with F and clamp with bolt N.

XII. Be certain that friction feed is out before throwing in screw feed.

XIII. Connect lead screw to carriage 13 by handle S operating split nut R. Place tin box under tool to catch chips and drippings.

61. To operate lathe to cut the thread.

SCHEDULE OF OPERATIONS

I. Push shipper 14 toward headstock, adjust stop screw 9 and move tool 5 to trace a light line; stop lathe before tool reaches end of thread.

II. Finish length of thread by pulling belt by hand, the first time only.

III. Move tool out from work. Run carriage back. Adjust thread stop screw 9 to take cut.

IV. Count threads, Fig. 51.

Terminate cut by power by easing tool out when about ⅓ revolution from end of cut, or point of tool will snap off.

V. Stop lathe and reverse immediately.

VI. During return, lubricate work freely with lard oil and adjust stop screw 9 for next cut.

VII. Feed tool inward.

VIII. Start lathe forward and repeat.

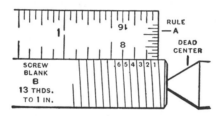

FIG. 51. — COUNTING THE TRACE OF A THREAD.

Attention. — The cutting speed for threading is about one-half to two-thirds that used for turning. The feed must be sufficient to allow tool to cut, for if tool is allowed to travel in groove without cutting, it will burnish and harden sides of thread so that on next cut the tool will be likely to dig in and tear the thread.

Oilstone top face of tool just before taking last few light cuts.

62. Number of cuts for U. S. S. thread, 13 P. — Take 8 cuts of .005″ each, 2 cuts of .002″ each, then 1 cut of .001″; clean, oil, and test in nut, and repeat until screw fits nut. Depth of thread when outside diameter is not reduced = .0499″.

63. Number of cuts for Sharp V thread, 13 P. — Take 10 cuts .005″ each, 2 cuts .002″ each, then 1 cut of .001″; clean,

oil, and test in nut, and repeat until screw fits nut. Depth
of thread when reduced $\frac{1}{64}''$ in diameter $= .0588''$.

64. To fit thread to nut. See p. 320. — Take light cuts as
the thread approaches size, and after each cut clean and oil
thread and try nut on. If a close fit is required, cut thread
until nut will go on easily with a wrench. This smooths down
the burr, after which the nut may go on with the fingers.
For a hard fit, force nut on with a wrench. For a loose fit,
cut the thread until the nut will go on with the fingers. After
thread is fitted, chamfer the end to the depth and angle of
the thread and file tops of thread slightly to remove burr.

65. To reset threading tool to resume cut. — If the tool
is dull and thread is only partly cut, remove the tool, regrind
and reset. If the end that receives the dog is cylindrical,
loosen the dog, and rotate the work until tool fits the groove.
Refasten the dog and feed tool away from work, run the lathe
forward a few revolutions, by hand, to take up back-lash; now
notice if tool and groove match; if they do not, repeat operation. If end of work is square or hexagonal, as a bolt head,
and driven by a clamp dog, disconnect lead screw from lathe
spindle by reversing lever in headstock, or by removing stud
gear, then adjust work.

66. To cut left threads. — Arrange gears as in cutting
right threads, with the exception that the lead screw must
rotate in opposite direction which is accomplished by the
reversing gears D, D', Fig. 50, or by an extra idler. At
beginning of thread, cut a groove in which to start tool, and
begin at left and cut to right.

67. To thread to shoulder or to terminate coarse thread. —
Cut groove or drill hole of diameter and depth equal to depth
and width of thread. See pp. 510, 511, 521.

68. Fractional threads. — A fractional thread is one whose
threads per inch are expressed by a mixed number, as $11\frac{1}{2}$
threads to $1''$; or a fraction, as $\frac{3}{4}$ of a thread to $1''$ ($1\frac{1}{3}''$ P).

To count fractional threads. — Take any number of threads that will match with even inches on the rule, then divide the number of threads by the number of inches.

69. To calculate simple gearing to cut fractional threads. — Proceed as for whole threads.

Example. — To cut $11\frac{1}{2}$ threads per inch (1″ pipe tap). Lead screw 5 threads per inch; constant, 4; speed of stud same as speed of spindle.

$$\frac{5 \times 4}{11\frac{1}{2} \times 4} = \frac{20 \text{ (gear on stud.)}}{46 \text{ (gear on lead screw.)}}$$

70. Compound gearing. — To cut fine or coarse threads that are not obtainable with simple gearing, compound the gearing, using 4 gears, Fig. 52.

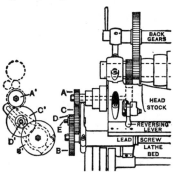

Fig. 52. — Compound Gearing for Threading — Four Gears.

To compound on some lathes, introduce between regular stud gear A and lead screw gear B two gears of different diameters, as C and D, which are feather keyed on a sleeve that runs freely on intermediate stud E, gear A driving gear C, and gear D driving gear B. The arrangement of gears is shown in end view at A', C', D', and B'.

71. To calculate compound gearing. — Factor first term into two fractions and treat each separately.

Example. — To cut 60 threads per inch.

Lead screw 6 threads per inch; speed of stud same as speed

of spindle; constant, select multiple of 5, the common difference between gears.

$$Solution. - \frac{6}{60} = \frac{2 \times 3}{5 \times 12}. \quad \frac{2 \times 20}{5 \times 20} = \frac{40}{100}. \quad \frac{3 \times 10}{12 \times 10} = \frac{30}{120}.$$

Gear on stud A, 40; gear on lead screw B, 120; first gear on sleeve C, 100; second gear on sleeve D, 30.

Attention. — If more convenient, exchange drivers A and C or driven C and B.

72. To calculate compound gearing for fractional threads, proceed as for whole threads.

Example. — To cut $2\frac{1}{4}$ threads per inch = $\frac{9}{4}$.

Lead screw 2 threads per inch; speed of stud same as that of spindle; constant, multiple of 5.

$$Solution. - \qquad \frac{2}{\frac{9}{4}} = \frac{8}{9} = \frac{2 \times 4}{3 \times 3}.$$

$$\left. \begin{array}{l} 2 \times 25 = 50 \\ 3 \times 25 = 75 \end{array} \right\} \text{one pair of gears.}$$

$$\left. \begin{array}{l} 4 \times 15 = 60 \\ 3 \times 15 = 45 \end{array} \right\} \text{other pair of gears.}$$

Attention. — It often happens that one or more gears have to be made or bought. If the threads per inch are expressed decimally, as 2.833 threads per inch, proceed as before, selecting such multiple as will give available gears, using the nearest whole tooth in case of resulting fractional teeth.

73. Result of gearing calculations may be checked as follows:

For stud ratio one: Threads on screw to be cut × teeth of stud gear = threads on lead screw × teeth of lead-screw gear.

For stud ratios other than one and for compound gearing: Threads on screw to be cut × teeth of all drivers in succession = threads on lead screw × teeth of all followers in succession.

Example. — p. **327** 13 × 25 = 5 × 65 = 325.

Example. — p. **327** 13 × 1 × 96 = 8 × 2 × 78 = 1248.

Example. — p. **334** 60 × 1 × 40 × 30 = 6 × 1 × 100
$$\times \; 120 = 72{,}000.$$

74. To calculate gearing for a given lead. — First change to threads per inch by dividing one by the lead of screw to be cut, and proceed as before.

Example. — To cut screw $\frac{3}{8}''$ lead.

Lead screw 2 threads per inch; speed of spindle same as speed of stud; constant, multiple of 5.

Solution. —
$$\frac{1}{\frac{3}{8}} = \frac{3}{2} \text{ (threads per inch)}.$$
$$\frac{2}{\frac{3}{8}} = \frac{4 \times 10}{3 \times 10} = \frac{40}{30}.$$

75. Compound gearing, ratio 2 to 1, is provided on some lathes on an extra adjustable stud. In such cases, gears are selected as in simple gearing for one-half or twice the desired pitch, and the 2 to 1 compound arranged to double the pitch or reduce it one-half.

76. To calculate gearing for metric screw threads with an English lead screw. — Gear up lathe as for cutting a Sharp V screw of the same number of threads per inch and use translating gears. One centimeter equals approximately $\frac{50}{127}$ of an inch. Provide lathe with pair of translating gears of 50 and 127 teeth. Arrange lathe as in compound gearing, Fig. 52.

Example. — To cut 13 threads to the centimeter.

Lead screw 5 threads per inch; speed spindle same as stud; constant, multiple of 5.

Solution. —
$$\frac{5 \times 5}{13 \times 5} = \frac{25 \text{ (gear on stud)}}{65 \text{ (gear on lead screw)}}.$$

Place translating gears on feathered sleeve, meshing 50 with gear 65 and 127 with gear 25. A metric lead screw may be used for cutting English threads.

77. Threading long screws. "Catching the thread" (threading without backing belt). — A method to save time in cutting long screws by quick return of tool carriage after each cut by hand.

1. If thread is same as lead screw or a multiple of it, throw half-nut out at end of each cut, return carriage by hand, throw in half-nut and tool will resume its cut.

2. If screw to be cut and lead screw are odd, or odd and even, move carriage any whole number of inches.

3. If both are even, any number of half-inches.

4. If screw is fractional, as $4\frac{1}{2}$ threads to 1″, move carriage number of inches equal to denominator or multiple of denominator. To obtain this, clamp steel rule to bed or use thread indicator, Fig. 53, which is included in the equipment of new lathes.

78. Thread indicator. — Example: To cut long screw 5 threads to 1″, using thread indicator.

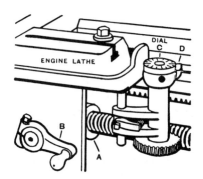

Fig. 53. — THREAD INDICATOR.

SCHEDULE OF OPERATIONS

I. Set up lathe to cut 5 threads to 1″ and adjust thread indicator to lead screw.

II. Start lathe and take one cut.

III. At end of cut withdraw tool and, without stopping lathe, throw out half-nut.

IV. Move carriage back by hand until tool is beyond end of work, then throw in half-nut A by handle B when any long line on dial C is at zero line D and take another cut.

V. Repeat this process for each cut until threading is completed.

Attention. — If thread indicator is used for all screw cutting, a backing belt is unnecessary. The second pulley on countershaft may be used for another range of forward speeds.

79. Whitworth (English) Standard thread.   Figs. 54, 55. — The tops and the bottoms of the 55° threads are rounded, $\frac{1}{6}$ of pitch, as shown by divisions 1, 2, 3, etc.

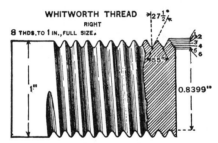

FIG. 54. — SECTIONAL VIEW OF WHITWORTH THREAD.

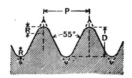

FIG. 55. — SECTION SHOWING PITCH AND DEPTH OF WHITWORTH THREAD

The single depth of a 1″ pitch is equal to .64033, and the double depth 1.28066. For pitch P, depth = .64033 $P = D$. Double depth = 1.28066 $P = 2 D$.

$$P = \text{pitch} = \frac{1}{\text{No. of threads to 1 inch}}.$$

R = radius = .1373 P.

Root diameter = Outside diameter − double depth.

Formula. — Root diameter = Outside diameter —

$$\frac{1.28066}{\text{No. of threads to 1 inch}}.$$

Example. — To find root diameter of a screw 1″ diameter, 8 threads to 1″.

Solution. —　　　$1 - \dfrac{1.28066}{8} = .8399.$

80. Table of Whitworth (English) Standard threads.

DIAMETER OF SCREW.	No. THREADS PER INCH.	DIAMETER OF SCREW.	No. THREADS PER INCH.	DIAMETER OF SCREW.	No. THREADS PER INCH.
$\frac{1}{4}$	20	$1\frac{3}{8}$	6	$3\frac{1}{4}$	$3\frac{1}{4}$
$\frac{5}{16}$	18	$1\frac{1}{2}$	6	$3\frac{3}{8}$	$3\frac{1}{4}$
$\frac{3}{8}$	16	$1\frac{5}{8}$	5	$3\frac{1}{2}$	$3\frac{1}{4}$
$\frac{7}{16}$	14	$1\frac{3}{4}$	5	$3\frac{5}{8}$	$3\frac{1}{4}$
$\frac{1}{2}$	12	$1\frac{7}{8}$	$4\frac{1}{2}$	$3\frac{3}{4}$	3
$\frac{9}{16}$	12	2	$4\frac{1}{2}$	$3\frac{7}{8}$	3
$\frac{5}{8}$	11	$2\frac{1}{8}$	$4\frac{1}{2}$	4	3
$\frac{11}{16}$	11	$2\frac{1}{4}$	4	$4\frac{1}{4}$	$2\frac{7}{8}$
$\frac{3}{4}$	10	$2\frac{3}{8}$	4	$4\frac{1}{2}$	$2\frac{7}{8}$
$\frac{13}{16}$	10	$2\frac{1}{2}$	4	$4\frac{3}{4}$	$2\frac{3}{4}$
$\frac{7}{8}$	9	$2\frac{5}{8}$	4	5	$2\frac{3}{4}$
$\frac{15}{16}$	9	$2\frac{3}{4}$	$3\frac{1}{2}$	$5\frac{1}{4}$	$2\frac{5}{8}$
1	8	$2\frac{7}{8}$	$3\frac{1}{2}$	$5\frac{1}{2}$	$2\frac{5}{8}$
$1\frac{1}{8}$	7	3	$3\frac{1}{4}$	$5\frac{3}{4}$	$2\frac{1}{2}$
$1\frac{1}{4}$	7	$3\frac{1}{8}$	$3\frac{1}{2}$	6	$2\frac{1}{2}$

WHITWORTH
THREADING
TOOL

FIG. 56. — SHAPE OF WHITWORTH THREADING TOOL.

81. The Whitworth threading tool, Fig. 56. — A tool of different size and shape is required for each pitch. It is made similarly to a formed cutter by milling. Grind on top face *A*.

BOLT AND NUT MAKING

82. The bolt and nut-making operations that follow apply to all work of this class, as bolts, studs, nuts, and screws.

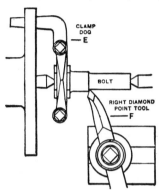

FIG. 57. — TURNING BODY OF BOLT TO HEAD.

To turn up to head of bolt, slant tool to left as at *F*, Fig. 57, and clamp tool firmly in tool post.

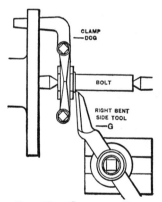

FIG. 58. — SQUARING BOLT
UNDER HEAD.

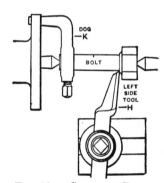

FIG. 59. — SQUARING BOLT
UNDER HEAD.

To square under head, use right bent side tool *G*, Fig. 58, or left side tool *H*, Fig. 59.   To drive from square or hexagonal heads use clamp or square dog *E*, Fig. 57, and to drive **from stem** use lathe dog *K*, Fig. 59.

83. Nut mandrels. — Nuts or similar pieces having tapped or threaded holes are screwed on a threaded mandrel and rough and finish squared to thickness. In Fig. 60 nut B is screwed on mandrel A against equalizing collar C, squared with side tool D, reversed and squared to thickness. It is then placed on a milling mandrel and the flats milled in a milling machine. After this it is replaced on regular nut mandrel, and side tool E set at 45°, as near as can be determined by the eye, and the edge chamfered as at F to about

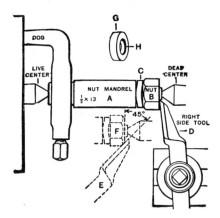

Fig. 60. — Squaring and Chamfering Nut.

$\frac{1}{32}''$ on edge of flats for $\frac{1}{2}''$ nuts and more or less for larger and smaller sizes, to give them a neat appearance.

Instead of equalizing collar C, a plain collar G is often used, with one inside edge rounded to fit over filleted shoulder of mandrel. For ordinary work, nut mandrels are often recessed at shoulder and used without collars. Threaded mandrels, for work threaded while held in a chuck, as a face plate for a lathe or chuck, need no collar, as the work is faced true with hole while held in chuck, and this true face is screwed against the shoulder on the mandrel. See Nut Mandrel, p. 1208.

84. Chamfering bolt heads, nuts, and screws. — A clamp nut (spring or split nut), Fig. 61, is used to protect the thread and prevent dog from bruising it.

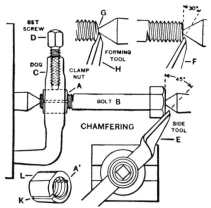

FIG. 61. — CHAMFERING AND FORMING BOLT HEAD AND POINT.

SCHEDULE OF OPERATIONS AND TOOLS

Bolt Heads.

1. Place clamp nut *A* on bolt *B* and fasten dog *C* with set screw *D*.

2. Set side tool *E* at angle of 45°, estimated.

3. Chamfer bolt head to remove corners and give neat appearance.

Nuts.

Mount nut on mandrel and chamfer same as bolts. See *E, F*, Fig. 60.

Screws.

Set side tool *F* at angle of 30° and chamfer to depth of thread.

To round end of screw *G*, use side tool *H* shaped into a forming tool.

85. To make a clamp nut, A′, Fig. 61.

1. Drill, tap, square and turn piece of round carbon steel to size.

2. Slit at *K* and mill or file flat *L* to receive set screw of dog.

3. Harden and draw to a spring temper.

Attention. — To make an improvised clamp nut, slit a nut at a corner with a hack saw.

A clamp nut and dog may be used to set or remove studs.

86. To make a finished bolt $\frac{1}{2}''$ diameter. Fig. 62.

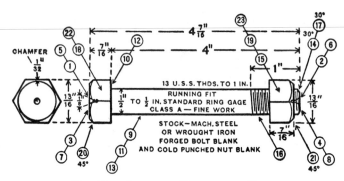

FIG. 62. — SCHEDULE DRAWING OF BOLT.

Specifications: Material, machine-steel forging, $\frac{1}{16}''$ large; weight, 10 oz. Machine-steel or wrought-iron hexagonal nut blank $\frac{1}{16}''$ large.

For heat treatment see *Principles of Machine Work*.

Hardness, 14 to 17 (scleroscope).

True live center, set dead center in approximate alinement, see page 116.

Carbon-steel cutting tools. See *Exception*, p. 205.

Time: Study drawing and schedule in advance, 10 min. — Oil lathe, 4 min. — Prepare blank bolt, 45 min. — Tap and square nut, 8 min. — "Set up" for thread, 6 min. — Cut thread, 7 min. — Mill bolt and nut, 10 min. — Chamfer, file and polish bolt head and nut, and stamp, 10 min. — Clean lathe, 3 min. — Total, 1 h. 43 min.

SCHEDULE OF OPERATIONS, MACHINES AND TOOLS

OPERATIONS.	MACHINES, SPEEDS, FEEDS.	TOOLS.
Straighten, see pp. 124, 125, and file ends flat.	Straightening press. Vise.	Chalk, file.
Center, machine method, $\frac{1}{8}''$, (1), (2), straighten.	Centering machine drill, 1700 R.P.M.; countersink, 600 R.P.M.; straightening press.	$\frac{1}{16}''$ drill, 60° countersink, lard oil. Chalk.
Mount on centers. Rough square to $4\frac{7}{16}'' + \frac{1}{64}''$, (3), (4). Take as little as possible off, (3). See step method of squaring, p. 301. Recenter to $\frac{1}{8}''$, (5), (6).	Engine lathe, 12″ to 16″. 3d speed, or 30 F.P.M. Hand feed. Speed lathe, drill 4th speed, countersink 3d speed.	Regular and clamp dogs, side tool, 35° rake, calipers, rule.
Regrind and oil stone tool. Finish square to $4\frac{7}{16}''$, (7), (8).	3d or 4th speed, or 50 F.P.M.	Side tool, 35° rake, calipers, rule.

Rough turn to $\frac{1}{2}''+\frac{1}{32}''$ (9) one or two cuts.	Engine lathe, 2d or 3d speed, or 30 F.P.M. Medium power feed — 80 to 1''.	Clamp dog, diamond-point tool, 35° rake, calipers, rule.
Rough square under head $\frac{7}{16}''+\frac{1}{64}''$, (10).	2d or 3d speed, or 30 F.P.M. Hand feed.	Left side or bent right side tool, calipers, rule.
Set dead center in accurate alinement to turn straight using this bolt or a trial piece the same length. See pp. 116, 117.	4th speed, or 50 F.P.M. Fine power feed — 140 to 1''.	Clamp dog, diamond-point tool, 35° rake, 1'' micrometer.
Rough turn $\frac{1}{2}''+\frac{1}{64}''$, (9) one cut.	3d speed, or 30 F.P.M. medium power feed — 80 to 1''.	Clamp dog, diamond-point tool, 35° rake, calipers, rule.
Regrind and oilstone tool. Finish turn to fit gage with allowance for filing, using calipers, mandrel and gage to obtain size; or measure with 1'' micrometer and allow .003'' for filing. (11). One cut.	3d or 4th speed, or 50 F.P.M. Fine power feed — 140 to 1''.	Diamond-point tool, 35° rake, calipers, $\frac{1}{2}''$ mandrel (or 1'' micrometer), $\frac{1}{2}''$ ring gage.
Regrind and oilstone tool and finish square under head, (12).	3d speed, or 50 F.P.M. Hand feed.	Left side or bent right side tool, 35° rake, calipers, rule.
File to fit gage, running fit. Oil work with machine or lard oil when testing in gage. (13.)	Engine lathe, 4th speed, or speed lathe 3d or 4th speed, or 175 F.P.M.	8'' or 10'' mill bastard file, calipers, $\frac{1}{2}''$ standard ring gage, oil, 1'' micrometer.
Chamfer point to 30°, (14).	Engine lathe, 3d speed, or 30 F.P.M. Hand feed.	Side tool 35° rake, center gage.
Tap $\frac{1}{2}''$ nut blank by hand, square to thickness, (15).	Vise. Engine lathe 3d speed, or 50 F.P.M. Hand feed.	$\frac{1}{2}'' \times 13$ U.S.S. tap, tap wrench, oil, nut mandrel, calipers, rule.
Draw line 1'' from end for length of thread (16).	Vise copper jaws.	Copper sulphate, scriber, rule.
Grind threading tool to fit gage, and thread bolt to fit nut. Make close fit, (16), see pp. 326, 331, rechamfer point, (17).	Engine lathe, 1st speed, or 20 F.P.M. Arrange for 13 threads.	Clamp dog, 13 pitch U.S.S. thread tool, center gage, thread calipers, rule, oil, drip (tin) pan.
File burr off top of thread. Remove gears, empty drip pan and clean lathe.	4th speed.	8'' to 10'' mill bastard file.
Mill head and nut to size $\frac{13}{16}''+$ .003'' for filing and polishing, (18), (19). See pp. 1020, 1026.	Milling machine. 3d or 4th speed. Back gears in, or 50 F.P.M. Medium power feed.	Heading mills, index head and chuck, milling machine nut mandrel, oil, 1'' micrometer.
Chamfer head and nut to 45°, (20), (21), pp. 340, 341.	Engine lathe, 3d speed, or 50 F.P.M.	$\frac{1}{2}'' \times 13$ clamp nut, and nut mandrel, rule, side tool.
File and polish milled sides of bolt head and nut only. (22), (23).	Vise.	8'' or 10'' hand-smooth file, nut mandrel, copper jaws, 90 emery cloth, lard oil.

87. To make pair of spring bolts, duplicate process, Fig. 63.

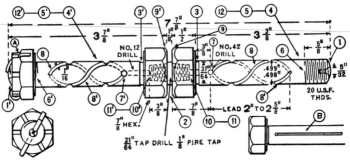

Fig. 63. — Schedule Drawing.

Specifications: Material, $\frac{7}{8}''$ hexagonal "special" steel $\frac{1}{8}''$ long; weight, 1 lb. 9 oz. Hardness, 15 to 16 (scleroscope).

For heat treatment see *Principles of Machine Work.*

High-speed steel, or stellite cutting tools.

Time: Study drawing and schedule in advance, 5 min. — Oil lathe, 4 min. — Center and square, 7 min. — Rough and finish turn, file and fit, 41 min. — Thread, 25 min. — Drill No. 42 hole and mill or cut spiral grooves, 34 min. — Cut off bolts and drill and tap for grease cups, 35 min. — Clean lathe, 3 min. — Total, 2 h. 34 min.

SCHEDULE OF OPERATIONS, MACHINES AND TOOLS

Operations.	Machines, Speeds, Feeds.	Tools.
Center. See that live center is nearly true and dead center in approximate alinement.	Engine lathe, 12″ to 16″.	
Rough square $7\frac{7}{8}'' + \frac{1}{64}''$ (1), (1′).	3d or 4th speed, or 50 F.P.M. Hand feed.	Dog, holder and cutter 35° rake, calipers, rule.
Recenter. **Omit finish square.**		
Draw lines for groove in center of work (2), and grooves for length of bodies (3), (3′).	Vise, copper jaws.	Copper sulphate, rule, scriber or half-round file.
Cut grooves to form heads at (2) and (3), (3′), to $\frac{31}{64}''$ diameter.	Engine lathe, 2d or 3d speed, or 40 F.P.M. Hand feed.	$\frac{1}{8}''$ cutting-off tool, calipers, rule, oil.
Rough turn to $\frac{1}{2}'' + \frac{1}{64}''$ (4), (4′), three or four cuts.	2d or 3d speed, or 50 F.P.M. Medium power feed — 80 to 1″.	Holder and cutter 35° rake, calipers, rule.

True live center. Set dead center in accurate alinement. Finish turn (5), (5') to .499″ + .003″.	Engine lathe. 3d or 4th speed or 70 F.P.M. Fine power feed — 140 to 1″.	1″ micrometer.
File to limit (5), (5') or to fit ½″ ring gage.	4th speed or speed lathe, 3d or 4th speed or 175 F.P.M.	8″ or 10″ mill bastard file, 1″ micrometer or ½″ ring gage.
Draw lines for length of thread (6) (6').	Vise, copper jaws.	Copper sulphate, rule, scriber.
Form ends (1), (1'), to length.	3d speed, or 70 F.P.M. Hand feed.	Forming tool and oil, or graver.
Thread ends to fit ½″ × 20 U. S. F. nut, (6), (6'), five cuts .005″ each, two cuts .003″ each, three cuts .001″ each.	Arrange for 20 threads. 1st or 2d speed, or 25 F.P.M.	20 pitch U. S. F. thread tool, center gage ½″× 20 U.S.F. clamp nut.
Drill grease holes through the body of bolts, (7), (7').	Speed lathe, 4th speed or 1400 R.P.M.	V center, drill chuck, No. 42 twist drill, oil.
Mill or cut spiral oil grooves (8), (8'), lead one turn in 2″ to 2⅝″.	Milling machine, 2d or 3d speed or engine lathe, 1st speed, back gears in.	₁₆″ convex cutter, milling machine centers, or ₁₆″ grooving tool, oil.
Mount in chuck. Cut off bolts (2).	.... 2d or 3d speed, or 40 F.P.M. Hand feed.	Universal chuck. ⅛″ cutting-off tool, oil.
Chamfer heads to 45° and ₃₂″ wide, (9), (9').	3d speed, or 70 F.P.M. Hand feed.	Holder and cutter 35° rake, or right side tool.
Drill holes into ends of bolts for grease cups and grease holes (10), (10').	4th speed, or 450 R.P.M. Hand feed.	Centering tool, drill chuck, ₆₄″, and No. 12 twist drills, oil.
Tap holes for grease cups (11), (11'). Start tap in lathe, pull belt downward and follow with dead center, remove work, finish tap in vise.	Engine lathe, vise, copper jaws.	⅛″ pipe tap, tap wrench, oil. See *Principles of Machine Work.*
Case-harden bodies only, or all over.	Gas furnace, 1325° F. to 1350° F.	Tongs, cyanide of potassium.

Exception. — If convenient to use a universal grinding machine, rough turn bodies of bolts to .499″ + .012″ (4), (4') and then after bolts are case-hardened mount on threaded center and

Grind to limit (12), (12'). See p. 715.	Universal grinding machine.	Grinding dog, grinding wheel 60 *K*, vitrified, 1″ micrometer.

Information. — If castle nuts are to be used, holes are drilled and cotter pins inserted and the points bent back as shown at *A*, Fig. 63.

Attention. — If desired, the spiral oil grooves may be omitted and a straight oil groove chipped, on one side only of each bolt, as at *B*, Fig. 63.

MAKING A TENSILE TEST SPECIMEN

88. To make tensile test specimen of steel, wrought iron, brass or bronze for testing these materials, Fig. 64.

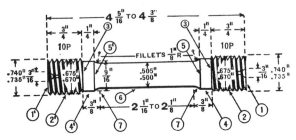

FIG. 64. — SCHEDULE DRAWING OF TENSILE TEST SPECIMEN.

Specifications: Material, machine steel $\frac{1}{16}''$ to $\frac{1}{8}''$ large; weight, 11 oz. Hardness, 15 to 18 (scleroscope).

For heat treatment see *Principles of Machine Work.*

High-speed steel, or stellite cutting tools.

Time: Study drawing and schedule in advance, 5 min. — Oil lathes, 5 min. — Prepare specimen, 35 min. — Thread, 30 min. — File and polish, 10 min. — Clean lathes, 5 min. — Total, 1 h. 30 min.

SCHEDULE OF OPERATIONS, MACHINES AND TOOLS

OPERATIONS.	MACHINES, SPEEDS, FEEDS.	TOOLS.
File ends flat. **Center.** See that live center is nearly true, and dead center in approximate alinement. If ends of work are nearly flat and length is within the limit, all squaring may be omitted, otherwise	Engine lathe, 12″ to 16″.	
Rough square to limit (1), (1').	3d speed, or 50 F.P.M. Hand feed.	Dog, holder and cutter 35° **rake,** calipers, rule.
Recenter. **Omit finish square.** **Rough turn** to $\frac{3}{4}'' + \frac{1}{64}''$, (**2**), (**2'**), one cut. Turn half way, reverse and turn other half.	2d or 3d speed, or 50 F.P.M. Medium power feed — 80 to 1″.	Holder and **cutter** 35° rake, calipers, rule.

True live center. Set dead center in accurate alinement using this piece.		
Draw lines ¾″ from each end for length of thread (**3**), (**3′**).	Lathe or vise.	Copper sulphate, scriber, rule.
Rough turn to $\frac{9}{16}$″ (**4**), (**4′**), one or two cuts.	Engine lathe, 2d or 3d speed, or 50 F.P.M. Medium power feed — 80 to 1″.	Calipers, rule.
Draw lines 1″ from each end to give length of body (**5**), (**5′**).	Lathe or vise.	Copper sulphate, scriber, rule.
Rough turn to $\frac{33}{64}$″ (**6**), one or two cuts.	2d or 3d speed, or 50 F.P.M. Medium power feed — 80 to 1″.	Calipers, rule.
Finish turn to $\begin{array}{c}.740'',\\.735'',\end{array}$ (**2**), (**2′**), one cut.	3d or 4th speed, or 70 F.P.M. Medium power feed — 80 to 1″.	1″ micrometer.
Finish turn to .505″ + .002″ for filing (**6**), two or three cuts. Adjust forming tool with thread stop and turn body and both fillets (**7**), (**7′**), with same setting.	2d or 3d speed, or 50 F.P.M. Fine power feed — 140 to 1″. Hand feed for fillets.	Forming tool 35° rake, 1″ micrometer, apply oil with brush or can.
Chamfer ends to 30°.	3d speed, or 70 F.P.M. Hand feed.	Right side tool, center gage.
Thread to limit (**2**), (**2′**), thread micrometer reading, $\begin{array}{c}.675''\\.670''\end{array}$ **or** to fit a nut or a gage, 9 cuts .005″ each, 5 cuts .003″ each, and 3 to 5 cuts .001″ each. See Adjustable Limit Snap Thread Gage, p. 225	Arrange for 10 threads. 1st or 2d speed, or 25 F.P.M.	10 pitch U. S. S. thread tool, U. S. S. limit thread gage, or thread micrometer (see p. **121**4), ¾″ × 10 clamp nut, lard oil.
File to remove tool marks and to limit (**6**).	Speed lathe, 175 F.P.M.	¾″ × 10 clamp nut, 8″ or 10″ mill-bastard or mill-smooth file.
Polish (**6**).	Highest speed.	120 emery cloth, polishing clamps, oil.

Attention. — If the feed is fine and the finishing cut smooth, filing may be omitted as polishing alone will remove the tool marks.

Important. — The body must be smooth and tangent with fillets as any shoulder or groove will make the specimen useless.

MAKING A STUD

89. To make a stud, Fig. 65.

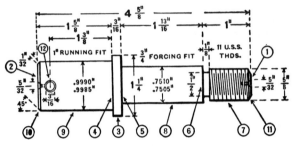

Fig. 65. — Schedule Drawing.

Specifications: Material, machine steel $\frac{1}{16}''$ to $\frac{1}{8}''$ large; weight, 2 lbs. Hardness, 14 to 16 (scleroscope).

For heat treatment see *Principles of Machine Work.*

High-speed steel, or stellite cutting tools.

Time: Study drawing and schedule in advance, 5 min. — Oil lathe, 4 min. — Square, rough turn, file and fit, 1 h. 10 min. — Thread and drill hole, 18 min. — Clean lathe, 3 min. — Total 1 h. 40 min.

SCHEDULE OF OPERATIONS, MACHINES AND TOOLS

Operations.	Machines, Speeds, Feeds.	Tools.
Center. See that live center is nearly true, and dead center in approximate alinement.	Engine lathe, 12″ to 16″.	
Rough square $4\frac{5}{8}''$, (1), (2).	2d or 3d speed, or 50 F.P.M. Hand feed.	Dog, holder and cutter 35° rake, calipers, rule.
Recenter. **Omit finish square.**		
Rough turn to $1\frac{1}{4}''$, (3), one or two cuts. Turn half way, reverse and turn other half.	Medium power feed — 80 to 1″.	Holder and cutter 35° rake, calipers, rule.
Draw lines (4), (5), (6), for lengths of running and forcing fits and thread.	Vise, copper jaws.	Copper sulphate, rule, scriber.
Cut grooves at lines (4), (5), .01″ smaller than diameters (8), (9), and $\frac{1}{2}''$ at (6), the root diameter of thread.	Engine lathe, 1st or 2d speed, or 30 F.P.M.	$\frac{1}{8}''$ cutting off tool, calipers, rule, oil.

Rough turn (7), (8), (9), $\frac{1}{64}''$ large, one or two cuts terminating in grooves.	2d or 3d speed, or 50 F.P.M. Medium power feed — 80 to 1''.	Calipers, rule.
True live center. Set dead center in accurate alinement.		
Finish turn (7) to $\frac{5}{8}''$, (8) to .7510'' + .003'', (9) to .9990'' + .003''.	3d or 4th speed. Fine power feed — 140 to 1''.	Calipers, 1'' micrometer.
Chamfer ends (10), 45°; (11) 30°.	Hand feed.	Center gage.
File (9) to a running fit in work or gage, or to limit, and (8), to a light forcing fit in work, or to limit.	4th speed.	8'' or 10'' mill bastard file, file card 1'' micrometer.
Thread (7) **to fit nut.** Take ten cuts .005'' each, three cuts .002'' each. Then take one cut .001''. Clean thread and test, and repeat cuts of .001'', if necessary. Depth of thread .059''.	Arrange for 11 threads. 1st or 2d speed, or 25 F.P.M.	11 pitch U. S. S. thread tool and gage, center gage, $\frac{5}{8}'' \times 11$ U. S. S. nut, oil.
Rechamfer to 30° (11).	3d speed, or 70 F.P.M.	Center gage.
Drill and slightly countersink cotter pin hole (12).	Vise, copper jaws. Speed lathe, 3d speed, or 600 R.P.M.	Scriber, rule, center punch, hammer, $\frac{3}{16}''$ twist drill, drill chuck, countersink, V center, oil.
Stamp name on large end.	Vise, copper jaws.	$\frac{1}{8}''$ steel letters, hammer.

Exception. — Grooves at (4) and (5) may be cut .01'' under size, and diameters (8) and (9) rough turned .008'' over size and ground to size.

Information. — There are two methods of making a shoulder on studs, bolts, shafts, etc. One is to cut a groove and terminate all cuts in this groove. The other is to turn close to a desired point, then square the shoulder by the "step method."

90. Formulas of bolt heads and nuts. — While finished heads and nuts (U. S. S.) are often made $\frac{1}{16}''$ smaller than the rough, it is best to make both the *same* size and to use the *same* wrench.

The short diameter or width across flats = $1\frac{1}{2} \times$ diameter of bolt + $\frac{1}{8}''$ for rough size and + $\frac{1}{16}''$ for finish size.

The long diameter or distance across corners of square head or nut = short diameter × 1.414.

The long diameter of hexagonal head or nut = short diameter × 1.155.

Thickness of nut = diameter of bolt.

Thickness of head = ½ short diameter of head.

Table of United States Standard bolt heads and nuts

Diam. of Screw.	Diam. in Decimals.	Threads per Inch.	Root Diam. of Screw.	Width of Flats.	Across Flats of Sq. and Hex. Nuts or Heads.	Across Corners of Hex. Nuts or Heads.	Across Corners of Sq. Nuts or Heads.	Thickness of Nuts.	Thickness of Heads.
	A	B	C	D	E	F	G	H	J
$\frac{1}{4}$	.25	20	.185	.0056	$\frac{1}{2}$	$\frac{37}{64}$	.707	$\frac{1}{4}$	$\frac{1}{4}$
$\frac{5}{16}$	.3125	18	.2403	.0069	$\frac{19}{32}$	$\frac{11}{16}$	.840	$\frac{5}{16}$	$\frac{19}{64}$
$\frac{3}{8}$	.375	16	.2996	.0078	$\frac{11}{16}$	$\frac{51}{64}$	.972	$\frac{3}{8}$	$\frac{11}{32}$
$\frac{7}{16}$	.4375	14	.3447	.0089	$\frac{25}{32}$	$\frac{9}{10}$	1.104	$\frac{7}{16}$	$\frac{25}{64}$
$\frac{1}{2}$	.5	13	.4001	.0096	$\frac{7}{8}$	1	1.237	$\frac{1}{2}$	$\frac{7}{16}$
$\frac{9}{16}$	.5625	12	.4542	.0104	$\frac{31}{32}$	$1\frac{1}{8}$	1.369	$\frac{9}{16}$	$\frac{31}{64}$
$\frac{5}{8}$	.625	11	.5069	.0114	$1\frac{1}{16}$	$1\frac{7}{32}$	1.502	$\frac{5}{8}$	$\frac{17}{32}$
$\frac{3}{4}$	.75	10	.6201	.0125	$1\frac{1}{4}$	$1\frac{7}{16}$	1.767	$\frac{3}{4}$	$\frac{5}{8}$
$\frac{7}{8}$	.875	9	.7307	.0139	$1\frac{7}{16}$	$1\frac{21}{32}$	2.032	$\frac{7}{8}$	$\frac{23}{32}$
1	1.	8	.8376	.0156	$1\frac{5}{8}$	$1\frac{7}{8}$	2.297	1	$\frac{13}{16}$
$1\frac{1}{8}$	1.125	7	.9394	.0179	$1\frac{13}{16}$	$2\frac{3}{32}$	2.562	$1\frac{1}{8}$	$\frac{29}{32}$
$1\frac{1}{4}$	1.25	7	1.0644	.0179	2	$2\frac{5}{16}$	2.828	$1\frac{1}{4}$	1
$1\frac{3}{8}$	1.375	6	1.1585	.0208	$2\frac{3}{16}$	$2\frac{17}{32}$	3.093	$1\frac{3}{8}$	$1\frac{3}{32}$
$1\frac{1}{2}$	1.5	6	1.2835	.0208	$2\frac{3}{8}$	$2\frac{3}{4}$	3.358	$1\frac{1}{2}$	$1\frac{3}{16}$
$1\frac{5}{8}$	1.625	$5\frac{1}{2}$	1.3888	.0227	$2\frac{9}{16}$	$2\frac{31}{32}$	3.623	$1\frac{5}{8}$	$1\frac{9}{32}$
$1\frac{3}{4}$	1.75	5	1.4902	.0250	$2\frac{3}{4}$	$3\frac{3}{16}$	3.888	$1\frac{3}{4}$	$1\frac{3}{8}$
$1\frac{7}{8}$	1.875	5	1.6152	.0250	$2\frac{15}{16}$	$3\frac{13}{32}$	4.153	$1\frac{7}{8}$	$1\frac{15}{32}$
2	2.	$4\frac{1}{2}$	1.7113	.0278	$3\frac{1}{8}$	$3\frac{5}{8}$	4.418	2	$1\frac{9}{16}$
$2\frac{1}{4}$	2.25	$4\frac{1}{2}$	1.9613	.0278	$3\frac{1}{2}$	$4\frac{1}{16}$	4.949	$2\frac{1}{4}$	$1\frac{3}{4}$
$2\frac{1}{2}$	2.5	4	2.1752	.0313	$3\frac{7}{8}$	$4\frac{1}{2}$	5.479	$2\frac{1}{2}$	$1\frac{15}{16}$
$2\frac{3}{4}$	2.75	4	2.4252	.0313	$4\frac{1}{4}$	$4\frac{29}{32}$	6.009	$2\frac{3}{4}$	$2\frac{1}{8}$
3	3.	$3\frac{1}{2}$	2.6288	.0357	$4\frac{5}{8}$	$5\frac{3}{8}$	6.539	3	$2\frac{5}{16}$

Attention. — A bolt is usually threaded a distance equal to twice the body diameter.

91. International and French Standard threads.

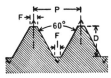

Fig. 66.

Diameter and pitch in Metric Measure.

$$\text{Formula} \begin{cases} p = \text{pitch.} \\ d = \text{depth} = p \times .6495. \\ f = \text{flat} = \dfrac{p}{8}. \end{cases}$$

92. International Standard thread.

Diameter in Millimeters.	Pitch in Millimeters.	Diam. at Root of Thread, m/m.	Diameter in Millimeters.	Pitch in Millimeters.	Diam. at Root of Thread, m/m.	Diameter in Millimeters.	Pitch in Millimeters.	Dam. at Root of Thread, m/m.
6	1.0	4.70	20	2.5	16.75	48	5.0	41.51
7	1.0	5.70	22	2.5	18.75	52	5.0	45.51
8	1.25	6.38	24	3.0	20.10	56	5.5	48.86
9	1.25	7.58	27	3.0	23.10	60	5.5	52.86
10	1.5	8.05	30	3.5	25.45	64	6.0	56.21
11	1.5	9.05	33	3.5	28.45	68	6.0	60.21
12	1.75	9.73	36	4.0	30.80	72	6.5	63.56
14	2.0	11.40	39	4.0	33.80	76	6.5	67.56
16	2.0	13.40	42	4.5	36.15	80	7.0	70.91
18	2.5	14.75	45	4.5	39.15	......	......	....

93. French Standard thread.

Diameter in Millimeters.	Pitch in Millimeters.	Diam. at Root of Thread, m/m.	Diameter in Millimeters.	Pitch in Millimeters.	Diam. at Root of Thread, m/m.	Diameter in Millimeters.	Pitch in Millimeters.	Diam. at Root of Thread, m/m.
3	0.5	2.35	16	2.0	13.40	36	4.0	30.80
4	0.75	3.03	18	2.5	14.75	38	4.0	32.80
5	0.75	4.03	20	2.5	16.75	40	4.0	34.80
6	1.0	4.70	22	2.5	18.75	42	4.5	36.15
7	1.0	5.70	24	3.0	20.10	44	4.5	38.15
8	1.0	6.70	26	3.0	22.10	46	4.5	40.15
9	1.0	7.70	28	3.0	24.10	48	5.0	41.51
10	1.5	8.05	30	3.5	25.45	50	5.0	43.51
12	1.5	10 05	32	3.5	27.45	......	......	....
14	2 0	11.40	34	3.5	29.45	......	......	....

INDEXING IN ENGINE LATHE

94. To index in engine lathe, Fig. 67. — To file round work square or hexagonal, or to drill diametrically through a shaft, equidistant lines may be drawn on the work to facilitate the operations.

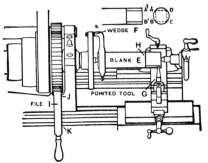

FIG. 67. — INDEXING IN THE LATHE.

SCHEDULE OF OPERATIONS

To divide tap shank or bolt head circumference into four equal parts, A, B, C, D, Fig. 67. Select engine lathe with headstock gear divisible by 4, as $72 \div 4 = 18$. Count headstock gear and mark divisions with chalk. Mount blank E on centers with wedge F between dog and face plate to prevent backlash. Use pointed tool G to mark line H required distance from end.

Place file I against under side of chalked tooth; rotate lathe until file touches headstock at J.

With the left hand press the handle downward until file touches the bed at K and hold it in this position.

With the right hand operating cross feed, move tool to lightly touch work, then change the right hand to long. feed handle and move the carriage to make a line with the tool. For the other lines repeat at the other chalked teeth on gear. Two lines are also shown at A', B'.

95. To make 14″ engine lathe live center, Fig. 68.

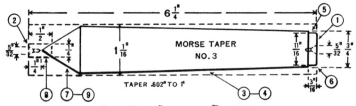

FIG. 68. — SCHEDULE DRAWING.

Specifications: Material, machine steel $\frac{1}{16}''$ to $\frac{1}{8}''$ large; weight, 1 lb. 9 oz. Hardness, 15 to 18 (scleroscope).
High-speed steel, or stellite cutting tools.
Time: Study drawing and schedule in advance, 5 min. — Oil lathe, 4 min. — Make center, 53 min. — Clean lathe, 3 min. — Total, 1 h. 5 min.

SCHEDULE OF OPERATIONS, MACHINES AND TOOLS

OPERATIONS.	MACHINES, SPEEDS, FEEDS.	TOOLS.
Center. See that live center is nearly true and dead center in approximate alinement.	Engine lathe, 12″ to 16″.	
Rough square to $6\frac{1}{4}''$, (**1**), (**2**).	3d or 4th speed, or 50 F.P.M. Hand or power feed.	Dog, holder and cutter 35° rake, calipers, rule.
Recenter. **Omit finish square.** **Turn taper shank** .602″ to 1′, (**3**).		
Set over footstock to .1568″ or $\frac{5}{32}''$, or use taper attachment.	. . . .	Rule, dividers.
Rough turn taper (**3**), to $1\frac{3}{16}''$ at small end, one or two cuts. Terminate cuts close to dog.	2d or 3d speed, or 40 F.P.M. Medium power feed — 80 to 1″.	Holder and cutter 35° rake, calipers, rule.
Take a light trial cut about .004″ or .005″, (**4**). To complete this taper, see Schedule of Operations, p. 2₂₉.	3d or 4th speed, or 70 F.P.M. Fine power feed — 140 to 1″.	M o r s e taper-ring gage No. 3, chalk or Prussian blue.
Smooth turn reduced part to $\frac{11}{16}''$, (**5**), one cut. Round corner (**6**).	3d or 4th speed, or 70 F.P.M. Fine power feed — 140 to 1″.	Holder and cutter, graver.
Reverse work and set tool at 30°, approximately, with work.	. . . .	Center gage.
Rough turn point of center (**7**), to leave stem (**8**), as shown, seven or eight cuts.	2d or 3d speed, or 50 F.P.M. Medium power feed — 80 to 1″, or hand feed.	
To finish point (**9**), place center in live spindle. See Truing centers, p. 1₁4.	2d or 3d speed, or 50 F.P.M. Hand feed.	Center truing tool, side or cutting-off tool, center gage.
Stamp name on (**1**).	Vise, copper jaws.	$\frac{1}{8}''$ steel letters, hammer.

Information. — Live centers are usually machine steel. Dead centers are carbon steel fitted to footstock spindle and of a length that when spindle is run back nearly as far as it will go, the center will be forced out. The conical point is hardened and tempered to a straw color and often ground. If a center is made of annealed carbon steel, the cutting speeds may have to be reduced.

AUTOMOBILE SCREWS AND NUTS

96. The Society of Automotive Engineers' standard screws and nuts.

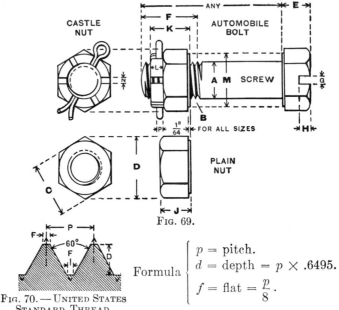

Fig. 69.

Fig. 70. — United States Standard Thread.

$$\text{Formula} \begin{cases} p = \text{pitch.} \\ d = \text{depth} = p \times .6495. \\ f = \text{flat} = \dfrac{p}{8}. \end{cases}$$

Diameter of screw = nominal diameter − .001″.

Thread = U. S. S. in form but with finer pitches.   Taps and dies are marked U.S.F.

Heads and nuts are semi-finished but smaller than U. S. S. Screws soft.   Plain nuts soft.   Castle nuts case-hardened. Nuts should be a good fit on screw— without perceptible shake.

The tap is from .002″ to .003″ larger than standard at the top of thread to give the screw clearance in the nut.   Material for screws and nuts, machine steel; tensile strength, 100,000 lbs. per square inch; elastic limit, 60,000 lbs. per square inch.

Threaded portions of screw should be one and one-half times the body diameter.

Attention. — The castle nut is used where a positive locking system is desired.

Important. — It is best to use U.S.S. threads on soft material such as aluminum and cast iron; and also on brass and bronze if subjected to great strains.

Table of Society of Automotive Engineers' Standard Screws and Nuts

Description	Symbol	Values
Diameter of Cotter Pin.	P	
Width of Slot in Castle and Diameter of Cotter Pin Hole.	N	
Diameter of Facing under Head and Nut, also Diameter of Castle.	M	
Height of Castle, also Depth of Slots.	L	
Thickness of Castle Nuts.	K	
Thickness of Plain Nut.	J	
Depth of Slot in Head.	H	
Width of Slot in Head.	G	
Length of Thread.	F	
Thickness of Head.	E	
Across Corners of Head and Nut.	D	.505 .577 .649 .756 .866 1.010 1.082 1.155 1.227 1.443 1.660 2.518 2.808 3.100 3.875
Across Flats of Head and Nut.	C	
Sizes of Body Drills.		
Sizes of Tap Drills.		No. 5 G P
Threads per Inch.	B	28 24 24 20 20 18 18 16 16 14 14 12 12 12 12
Nominal Diameter of Screw.	A	

97. Lock washers are used to hold plain automobile nuts, as in Fig. 71. The washer is cut open at A, and bent up at B and down at B'. When nut is screwed down hard these projections cut into both nut and seat and prevent nut becoming loose.

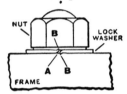

Fig. 71.—Fastening a Plain Nut with Lock Washer.

98. Constants for finding diameter at bottom of U. S. S., U. S. F. and V threads.

Threads per Inch	U. S. S. and U. S. F. Constant	V Thread Constant	Threads per Inch	U. S. S. and U. S. F. Constant	V Thread Constant
3	.43301	.57733	18	.07217	.09623
3½	.37115	.49487	20	.06495	.08660
4	.32476	.43301	22	.05905	.07873
4½	.28868	.38490	24	.05413	.07217
5	.25981	.34641	26	.04996	.06662
5½	.23619	.31492	28	.04639	.06186
6	.21651	.28868	30	.04330	.05773
7	.18558	.24744	32	.04059	.05413
8	.16238	.21651	36	.03608	.04811
9	.14434	.19245	40	.03248	.04330
10	.12990	.17321	44	.02952	.03936
11	.11809	.15746	48	.02706	.03608
12	.10825	.14434	50	.02598	.03464
13	.09993	.13323	56	.02320	.03093
14	.09279	.12372	60	.02165	.0288
16	.08119	.10825	64	.02030	.02706

$$Formula. - \begin{cases} C & = \text{Constant for number of threads per inch.} \\ D & = \text{Outside diameter.} \\ D_1 & = \text{Diameter at bottom of thread.} \\ D_1 & = D - C. \end{cases}$$

Example. — Given outside diameter of U. S. F. screw thread, ½ inch, 20 threads per inch, find diameter at bottom of thread. Expressed in decimals $D = .500''$ constant for 20 threads, U. S. F., $C = .06495$; diameter at bottom of thread $D_1 = .500'' - .06495 = .43605''$ or practically $\frac{7}{16}''$.

ADVANCED MACHINE WORK

SECTION 4

LATHE WORK

Chucks. Face Plates. Chucking. Reaming. Machining Alloys, etc. Mandrels or Arbors. Turning Flanges. Turning Pulleys. Polishing Lathe Work. Curve Turning and Forming. Making Formed Machine Handles. Polishing, Buffing and Lacquering.

CHUCKS

1. The term chuck has a double meaning. — First, it is the device used for holding work, drills or other tools. Second, it is the act of securing work in a holding device. See Chucking. Chucks are indispensable to a large class of work.

2. Attaching chucks to machine spindles. — Drill and other small chucks are attached by a double-ended taper arbor or shank, one end fitting the taper hole in chuck and the other the taper hole in spindle. Lathe chucks are usually attached by a threaded back plate. See p. 508.

3. Chuck jaws. — Four general kinds, Fig. 1:

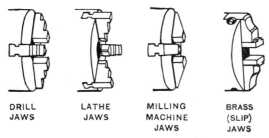

| DRILL JAWS | LATHE JAWS | MILLING MACHINE JAWS | BRASS (SLIP) JAWS |

Fig. 1. — Chuck Jaws. Four Kinds.

Drill jaws, for holding drills, rods, and similar pieces, also for holding hollow work by the inside.

Lathe jaws, for lathe work of large diameter.

Milling-machine jaws, used on milling machines.

Brass (slip) jaws, for brass work.

401

Information. — The jaws of some chucks may be reversed and used either as drill or lathe jaws.

4. Classes of chucks. — Drill, independent, universal, combination, and draw-in chucks.

5. A drill chuck is used to hold drills and small work.

6. An independent chuck, Fig. 2, is one in which each jaw is moved independently with wrench. Chuck *A* consists of disk *B* screwed to spindle of headstock *C*. Lathe jaws *D*, stepped to suit different diameters of work, slide in slots in disk *B* and are moved by screws *E, E,* operated by a special wrench.

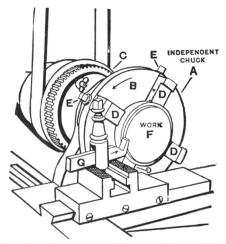

FIG. 2. — INDEPENDENT CHUCK. FACING DISK.

Important. — Concentric circles are marked on the face of some independent and combination chucks to facilitate setting jaws and work. See p. 1212.

Attention. — Independent chucks are better adapted for rough work than universal chucks.

7. To true up and hold work in an independent chuck. — Grip work tightly. Run lathe at a moderately high speed, rest hand on carriage and hold a piece of chalk to just touch work. Stop lathe, loosen jaw or jaws opposite part

marked by chalk and set others in. Erase chalk mark and
test again, continuing until work runs true, then set all jaws
up hard. Fig. 2 shows also the operation of facing work *F*
with cutting-in tool *G*.

8. A universal chuck, Fig. 3, is one whose jaws move
to and from the center simultaneously and concentrically.
Chuck *A* is screwed to spindle of headstock *B*. This is known
as a geared scroll chuck, and is made with either drill jaws, as
shown, or lathe jaws. It consists of shell *C*, three bevel
pinions *E* in mesh with an annular bevel gear, upon whose
face is a scroll which engages jaws *D*. This chuck should be
used for smooth work.

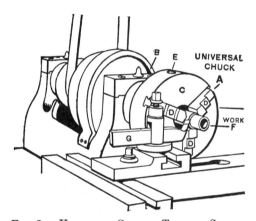

FIG. 3. — UNIVERSAL CHUCK. TURNING SLEEVE.

9. To true up and hold work in a universal chuck. —
Place work in chuck, "set up" jaws by one pinion, run lathe
and use chalk as before. If not true enough, loosen and
turn work about one-quarter of a revolution; tighten pinion
and test again; when right, tighten pinions hard.

Fig. 3 also shows how a bushing is made from bar *F*.
The bar is squared, then drilled and reamed by chucking
method (see pp. **4**09, **4**17), turned with tool *G*, and cut off with
cutting-off tool.

10. A combination chuck, Fig. 4, is a chuck in which the jaws may be moved independently or simultaneously. When moved simultaneously, jaws may be set either concentrically or eccentrically. Chuck *A* consists of shell *B* and jaws *C* moved by screws *D*. These screws mesh with thread on back of jaws, and carry pinions which can be placed in mesh with an annular bevel gear controlled by device on back of chuck. When in mesh, chuck is universal; out of mesh, each jaw can be moved independently. In Fig. 4 the

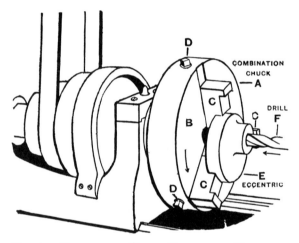

Fig. 4. — Combination Chuck Holding an Eccentric.

jaws are set eccentrically with annular out of mesh, then annular is thrown in mesh and the jaws are controlled as in a universal chuck. To make the chuck universally concentric, adjust each jaw to a circle on face of chuck and throw in annular.

11. Special chucks can be made or ordered from a manufacturer. For some classes of work, jaws of special shape may be home-made to fit a regular chuck.

12. Face plate jaws are obtainable which may be bolted to a face plate and used as a chuck.

13. Draw-in (spring) chucks, or collets, Fig. 5, are used on toolmakers' and watchmakers' lathes and also on some engine and turret lathes to hold bars or rods, as *BB'*. The rod is passed through the spindle and accurately held by the chuck, and from the rod small screws, studs, bolts, etc., may be conveniently made without preliminary cutting off, centering, squaring, etc.

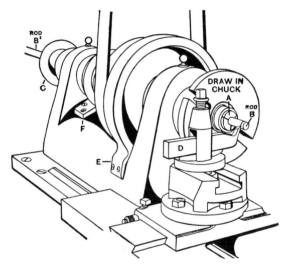

Fig. 5. — Turning Rod Held in Draw-in Chuck.

The steep taper on the chuck fits the conical hole in end of spindle. The chuck is slitted a short distance at three equidistant points, and rotating wheel *C*, which operates a hollow shaft that passes through the spindle and makes a threaded connection to end of chuck, draws the chuck into its seat to grip the bar or rod.

14. Care of chucks. — All chucks, and especially universal chucks, are short-lived for accurate work unless used intelligently and properly cared for; they should be cleaned and oiled frequently.

15. Mounting and removing chucks. — Clean and oil thread of both chuck and spindle, remove live center and plug hole with clean waste. Hold chuck square against nose of spindle with right hand or arm and rotate lathe with left hand until the chuck comes against shoulder on spindle. Small chucks may be loosened by grasping one jaw with a monkey wrench and striking the handle a sharp blow with the hand; large chucks, by placing a block of wood between a jaw and the bed of lathe and rotating lathe backward (with back gears in) by hand. Arbor or shank chucks are inserted and removed the same as lathe centers.

Attention. — To avoid springing work held in a chuck, the jaws should be forced against the solid parts, if convenient, as the arms of a pulley.

In some classes of light work, it is often necessary to loosen the jaws slightly before taking a finishing cut either when turning work held in a chuck or when boring or reaming.

FACE PLATES

16. To hold work on face plate. — Some work can be clamped to a large face plate and machined more accurately and conveniently than if held in a chuck. The work *B* is clamped to face plate *A*, Fig. 6, by clamps *C* and *C'* and bolts *D* and *D'*.

If a finished surface is to be clamped against a face plate or other finished surface, insert a sheet of paper between to prevent slipping.

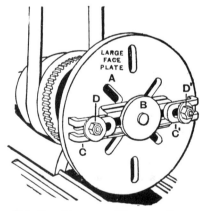

Fig. 6. — Clamping Work to Face Plate.

17. To clamp an engine crank to face plate, Fig. 7 — To face plate A, crank B is bolted by bolts C and C' and clamp

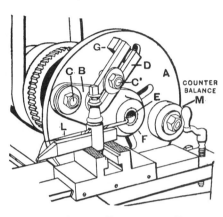

D, in order to bore out hole E and turn and face hub F. Before clamping to face plate, plane the crank on its face and line out the holes, as at H and I, Fig. 8. Describe circles of the required diameter around the cored holes the proper distance apart for the crank throw. To provide centers for circles, drive pieces of wood into holes to form

FIG. 7. — TESTING LOCATION OF ENGINE CRANK ON FACE PLATE.

bridges, as at J and J'. Turn down the corners of a piece of tin, and drive it into the center of the bridge, as at K and K'. Rotate lathe by hand and move crank by rapping until circle is true to axis of rotation when tested with scriber L, Fig. 7, then clamp crank hard to face plate.

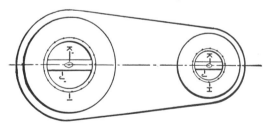

FIG. 8. — LAYING OUT HOLES IN ENGINE CRANK.

18. A counter weight to balance work is bolted to face plate at M, Fig. 7, in order to balance the eccentrically placed work and insure smooth running and accuracy.

19. To hold work with an angle plate. — Angle plate A, Fig. 9, is a useful fixture for various machine tools. It is planed all over with the faces at right angles (90°). It is bolted to face plate B, and pillow block C is clamped to inside surface by bolts D and D' and clamps E and E'. F is a counterbalance. Before boring, the pillow block has had its base planed and the cap fitted and screwed on. A circle of proper diameter is described around the cored hole and center punched. The angle plate and work must be adjusted until this circle runs true, after which hole G may be bored and reamed.

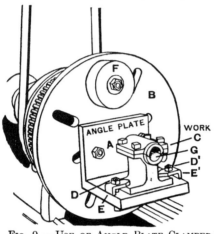

Fig. 9. — Use of Angle Plate Clamped to Face Plate.

CHUCKING

20. A method of drilling and reaming. — In chucking, the drill is stationary, while the work rotates. In drilling, the drill rotates and the work is stationary. Boring *is the enlarging* of a hole with a boring tool, see pp. 504–506, or boring bar, see pp. 610–619. Chucking is used where it would be impracticable to drill.

21. Chucking solid work with regular two-lipped twist drill in engine lathe, Figs. 10, 11, 12.

Specifications: Centering tool, twist drill $\frac{1}{100}''$ to $\frac{1}{64}''$ small, drill holder, chucking reamer .005'' small, hand reamer standard size, reamer wrench. Set dead center in approximate alinement.

Speed for chucking: Use the same cutting speed that would be used in rough turning work equal in diameter to the drill or reamer.

SCHEDULE OF OPERATIONS

1. Mount work A in chuck B, Fig. 10. 2. True up work, see pp. 402, 403.	3. To start drill in work A, cut true cavity as large as diameter of drill but at a slightly more acute angle than point of drill,

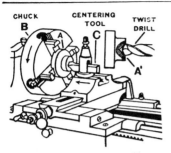

FIG. 10. — CUTTING A CONICAL CAVITY AXIALLY TRUE TO START TWIST DRILL.

as at A' with centering tool C, Fig. 10.

4. Insert taper shank drill D in drill holder E, Fig. 11.

5. Place point of drill in cavity and holder on dead center with handle resting on carriage.

6. Feed drill to work by operating footstock spindle with right hand and pressing holder on dead center with left hand.

7. Power ream with chucking reamer by same method as in drilling, see pp. 417, 418.

FIG. 11. — CHUCKING WITH TAPER SHANK TWIST DRILL.

8. To hand ream hole, see pp. 415–417. The hole may be drilled with a straight shank drill using holder, as in A, Fig. 12.

If a drill holder is not obtainable, a dog may be fastened to the drill, a tool placed upside down in tool post of lathe and pressed against the dog to keep drill on center, as in B, Fig. 12.

If the carriage is heavy, it may be helped along by left hand.

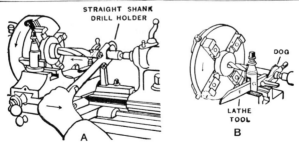

FIG. 12. — CHUCKING WITH STRAIGHT SHANK TWIST DRILL.

Warning. — Holder, drill, or reamer must be held firmly on dead center, for if allowed to slip off, work and tool will be ruined.

Information. — If a centering tool is not obtainable, cut a small true cavity in work with a side tool. Feed drill to the work and at the same time press butt end of lathe tool, as at *F*, Fig. 11, against side of drill and operate cross feed until drill runs true and cuts its full size. By this method, an expert can start a drill true without the cavity, but a student will need practice.

Important. — Reamed holes have a very slight taper and the end the reamers enter is always the larger; therefore drill and ream work from the side into which the shaft is to be fitted.

Attention. — Solid work, especially steel, is chucked with a two-lip twist drill. Three and four-groove drills are used for cored work (castings) or to follow a two-groove drill. Smooth holes may be made with a drill $\frac{1}{100}''$ to $\frac{1}{64}''$ small and a hand reamer; but better results are obtained by also using a fluted chucking reamer .005'' small before the hand reamer. See pp. 417, 418.

22. To chuck cored gear castings with four- or three-groove twist drill, Fig. 13. — Bevel cored hole in casting *A* to a slightly

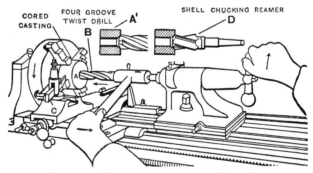

Fig. 13. — Chucking Cored Gear Casting With Four-Groove Twist Drill.

less angle than point of drill *B*, as at *A′*, and to the full diameter of drill with side tool *C*. This true and acute bevel starts the drill true and the drill cuts to its full size before its end reaches the irregular cored hole.

Drill hole $\frac{1}{100}''$ to $\frac{1}{64}''$ small, power ream .005'' small and hand ream to size. Cored holes over 1'' diameter may be drilled with a shell drill or shell chucking reamer, as at D, Fig. 13.

Attention. — For very accurate work, a hole should be bored, see pp. **5**04, **5**05, to about .005'' small, and then hand reamed, see p. **4**16.

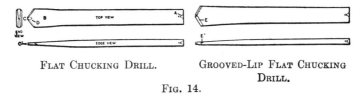

FLAT CHUCKING DRILL. GROOVED-LIP FLAT CHUCKING DRILL.

FIG. 14.

23. Flat chucking drills, Figs. 14–17, are often used for chucking in an engine lathe, for either cored holes or solid work. Large countersink A, Fig. 14, provides a firm bearing on dead center. End B is $\frac{1}{64}''$ smaller in diameter than the chucking reamer. Point CC' is central and either thinned or grooved on

FIG. 15. — TWISTED-LIP FLAT CHUCKING DRILL.

both sides, as at D. To give the cutting lips some rake, grooves may be ground above them, one of which is shown at EE', Fig. 14. A better way is to twist the lips as at F and G, Fig. 15.

FIG. 16. — FLAT CHUCKING REAMER.

24. Flat chucking reamers, Fig. 16. — The cutting edges are AB and CD. Head E is from .005'' to .010'' under size to allow for hand reaming.

25. Chucking with a flat drill and chucking reamer in an engine lathe, Fig. 17.

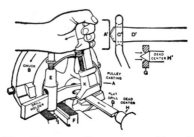

FIG. 17.—CHUCKING PULLEY WITH FLAT DRILL.

SCHEDULE OF OPERATIONS AND TOOLS

Set dead center in approximate alinement.
Drill, $\frac{1}{64}''$ to $\frac{1}{32}''$ small. Chucking reamer, $\frac{1}{100}''$ small.
Drill rest. Monkey wrench. Hand reamer, standard size.

1. Mount pulley A in independent chuck B.
2. True up by inside of rim.
3. Clamp drill holder C in post E to tool block F.
4. Set holder with dead center H' exactly in middle of slot as at G.
5. Set holder near work as at C', A'.
6. Place drill approximately central in holder, as C', D', with dead center H' in other end.
7. Place point of drill central against hub A'.
8. Pull wrench K forward to pinch drill in slot.
9. Start lathe at speed for twist drills.
10. Feed spindle until drill cuts half the depth of its point.

11. Remove wrench. Feed drill rapidly.
12. Hold drill back on dead center with left hand, when point breaks through.
13. Stop lathe when through.
14. Place flat chucking reamer in slot as in operation 4.
15. Hold with wrench until reamer is started. When through stop lathe. To ream hole, see pp. 414, 416, 417.

Attention. — If slot in holder is not at height of dead center, the drill or reamer will cut large and may spoil work. If drill moves sidewise when starting, replace wrench and repeat. Drill must cut true before reaching full diameter. For large holes or cored holes, use two or three flat drills of increasing diameters.

REAMING

26. **Reamers** are used for sizing, smoothing, and standardizing straight and taper holes. See Taper Reamers, pp. **4**18, **4**19, **6**19, **6**20.

There are two general classes: chucking or roughing reamers, used in a machine, and finishing reamers, used by hand or power. The usual amount for the finishing reamers to cut is .005″ to $\frac{1}{64}$″ for cast iron, and .005″ to .010″ for steel and brass.

27. Irregularly spaced teeth. — To prevent chattering, reamer teeth or blades are spaced progressively wider as in Fig. 18, *A* to 1 to the right, and from 1 to *A* to the left.

The clearance *G* is given the teeth or lands to relieve the cutting edge. The point of a hand reamer is slightly tapered a distance equal to its diameter, to enter the hole. The shank, *H,* Fig. 20, is ground .001″ small, to prevent binding in hole.

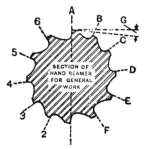

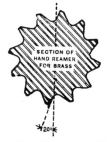

FIG. 18. — TEETH OF HAND
REAMER IRREGULARLY
SPACED.

FIG. 19. — TEETH OF HAND
REAMER WITH NEGATIVE
RAKE.

28. To ream brass, the face of teeth is inclined 20° back of radial (negative rake) to prevent chattering, as in Fig. 19. See Broaching, pp. **5**46–**5**48.

Attention. — Cast iron and brass are reamed dry; steel and wrought iron with oil.

Warning. — To ream thin work in a vise use a reaming jig. See p. **4**15.

Caution. — To ream work in a vise *without* a jig and also without chattering and spoiling the work, the length of hole should be about twice the diameter of the reamer. See pp. **4**15, **4**16.

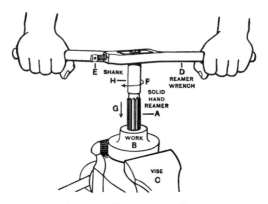

Fig. 20. — Reaming in Vise.

29. Hand reaming work held in vise, Fig. 20.

SCHEDULE OF OPERATIONS AND TOOLS

Hole in flanges casting drilled or drilled and reamed .005$''$ to $\frac{1}{64}''$ small.

1. Chuck work *B* in engine lathe then fasten firmly in vise *C*.	3. Rotate wrench rapidly in direction of arrow *F* and at same
2. Place adjustable reamer wrench *D* on square end of reamer *A* and fasten with thumb screw *E*.	time press downward in direction of arrow *G*, continue rotating and pressing downward until reamer passes clear through the casting.

Warning. — Hand reamers should never be rotated backward as it quickly destroys their cutting edges, and they should be used vertically when the nature of the work will permit.

Note. — Before placing reamer in hole, see that there are no burrs on shank *H*, which would be likely to scratch the reamed surface as the reamer is passed through the hole.

30. To hand ream thin work with reaming jig, as cutter disk, Fig. 21.

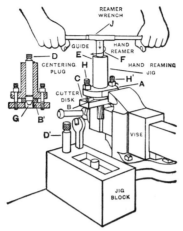

FIG. 21. — HAND REAMING THIN WORK HELD IN REAMING JIG.

Specifications: Hole in cutter disk or other thin work drilled by the chucking method $\frac{1}{100}''$ or $\frac{1}{64}''$ small, or, preferably, drilled and then bored and power reamed .005″ small, see pp. **4**08, **5**01–**5**06, **4**17.

SCHEDULE OF OPERATIONS

Hold jig base A firmly in vise, as in Fig. 21.

Place cutter disk or thin work B between clamp plate C and jig base A.

Insert centering or alining plug D in reamer guide E, where reamer F is shown, with end G, which is .005″ to $\frac{1}{64}''$ small, in hole in cutter disk B, as shown in detail at B'.

Tighten clamping bolts H, H' equally on both sides, first with thumb and finger then with the wrench, and remove plug D.

Place hand reamer F in reamer guide E, and with wrench J, revolve reamer to the right as shown by arrow, and press down lightly as the reamer is rotated, see pp. **4**13, **4**14.

Lubricant: If the work is steel, use plenty of lard oil; if cast iron, ream dry.

Important. — To turn disk, oil mandrel and if the disk is thin be careful and do not press mandrel in too hard, and do not remove mandrel from disk until all turning and squaring is completed.

Attention.—Clean jig and reamer before returning to tool room.

31. Adjustable reamers. — A, Fig. 23, may be adjusted to compensate for wear and to ream special-size holes.

32. Reaming stand *B*, Fig. 22, is more convenient than a vise. It consists of an independent chuck *C*, supported by column *D*. The work *E* is firmly gripped in the jaws *F* by operating the handle *G*.

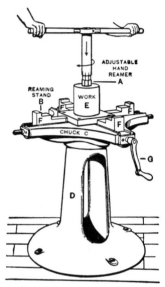

FIG. 22. — HAND REAMING IN REAMING STAND.

33. To ream work by hand in the lathe with a hand reamer, Fig. 23.

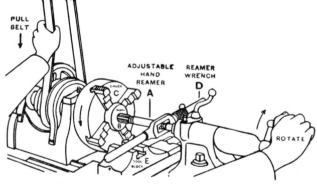

FIG. 23. — REAMING IN LATHE BY HAND.

SCHEDULE OF OPERATIONS AND TOOLS

Hole in blank drilled or bored .005″ to $\frac{1}{64}$″ small.
Set dead center in approximate alinement.

1. Place reamer *A* in drilled or bored hole in work *B* held in chuck *C*.	2. Place arm of wrench *D* against tool block *E*.
	3. Pull belt downward and follow with dead center.

Warning. — Carefully follow reamer with dead center to prevent spoiling reamer and work.

On thick work, start hand reamer in above manner in lathe, and then take both work and reamer to reaming stand or vise for finishing. See p. 414.

34. Fluted chucking reamer, Fig. 24, is obtainable in standard sizes or .005″ small, to be followed by standard hand reamer. This class of reamers has its points always beveled, and some points are slightly tapered the same as on hand reamers. Chucking reamers are also obtainable with taper shanks to fit drill sockets.

35. To ream in lathe by power with fluted chucking reamer, Fig. 24.

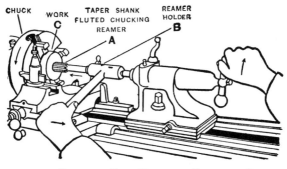

Fig. 24. — Reaming Gear Blank in Lathe by Power.

SCHEDULE OF OPERATIONS AND TOOLS

Hole in casting drilled $\frac{1}{100}''$ to $\frac{1}{64}''$ small.
Set dead center in approximate alinement.

1. Insert taper shank reamer *A*, Fig. 24, in reamer holder *B*.	**Speed for power reaming.** — Use about two-thirds the speed recommended for twist drills of same diameter.
2. Place reamer *A* in drilled hole in gear casting *C*, and holder *B* on dead center with handle resting on carriage.	*Warning.* — To prevent reamer *A*, Fig. 24, slipping off dead center, press holder hard against dead center. Neglect of this often ruins drill and work.
3. Feed reamer to work by operating footstock handle with right hand and press holder hard against center with left hand, as shown.	To hand ream, see pp. **4**14–**4**17.

36. Rose chucking reamer, Fig. 25, is made with either straight or helical flutes, with cutting teeth on end only. It will not produce as smooth a hole as a hand or fluted chucking reamer and is not used when a smooth hole is required. See also p. **4**17. It is obtainable in standard sizes and also with taper shanks to fit drill holders and sockets.

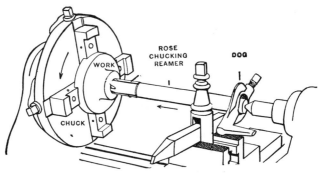

FIG. 25. — REAMING FLANGE CASTING BY POWER IN LATHE.

37. Broach reamers for small taper pins. — Collars, *A,* and similar work are often fastened to shafts by drilling, reaming, *B,* and pinning, *C,* Fig. 26.

Broach reamers may be supplied with a handle, as at E, and used by hand to enlarge holes and to remove burrs.   They are obtainable in tapers of $\frac{1}{8}''$ and $\frac{3}{16}''$ per foot, and in sizes from No. 1 to No. 70 (drill and wire gage sizes). Taper pin fluted reamers and taper pins having a taper of $\frac{1}{4}''$ per foot are used for pinning or doweling, and are obtainable in sizes from $\frac{1}{8}''$ to $1\frac{1}{4}''$.   Each reamer overlaps the next smaller size.

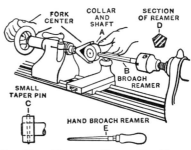

Fig. 26. — REAMING TAPER PIN HOLES WITH BROACH REAMER IN SPEED LATHE.

Attention. — To avoid breaking taper reamers, feed slowly, oil freely, and withdraw frequently to remove chips.

MACHINING ALLOYS, ETC.

38. To machine bronze. — Phosphor, Tobin, and other bronzes are tougher than ordinary brass, and are machined with cutting tools similar to those used for steel and wrought iron, and lubricated with lard oil.

39. To machine copper. — Copper is machined with tools similar to those used for steel but preferably with more rake as a keen edge is desirable, and with the point slightly rounded.   Lubricate with milk, soda water, or soap mixture. Use speed nearly as fast as for brass.

40. To machine aluminum. — Aluminum, owing to its light and ductile nature, is machined with tools having acute cutting angles, more rake than is used for steel, and at a moderately fast cutting speed.   See Lubricants for Cutting Tools, pp. 148–150.   A very high polish may be obtained on a cotton buffing wheel.

41. To machine babbitt and lead. — Babbitt and lead are machined dry and with keen tools.

42. To machine vulcanite or hard rubber or fiber. — Rubber and fiber are machined dry, at a moderately fast speed, with cutting tools similar to those used for steel.   They are finished and polished the same as steel, for a very high finish buff on a cotton buffing wheel charged with tripoli or rottenstone.

43. To machine rawhide, use tools similar to those used for steel, but preferably with more rake and with a cutting speed about the same as for brass, and machine dry. For gear blanks, the several layers are confined between riveted plates, and should be shellacked as soon as machined to prevent swelling. Rawhide is milled and drilled dry, and the chips removed by compressed air or a fan.

MANDRELS OR ARBORS

44. A mandrel, often called an arbor, is pressed, driven, or threaded into work to provide centers so that it may be machined.

An arbor is a shaft used to carry a cutting tool, as a milling machine arbor, a saw arbor, etc.

Four classes of mandrels are used: solid, expanding, built-up and gang. See pp. 4:21, 4:22, 6:13 and Nut Mandrel, pp. 340, 12:08.

45. Standard solid mandrel, AB, Fig. 27, is made of tool

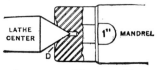

SECTION OF MANDREL END.

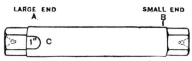

FIG. 27. — MANDREL OR ARBOR.

steel, hardened and ground. See Table of Mandrels, p. **12**04. The size is stamped at large end, as at C. A portion of each end is reduced and flattened to receive the driving dog. The ends are recessed as at D, Fig. 27, around the countersinks to protect them from injury.

Important. — For accurate work, mandrels should be tested to see if they run true. See **12**11.

46. Expanding mandrel. — Fig. 28 consists of shaft A having uniformly tapered slots as at B. Sleeve C guides tapered jaws D (see D') while they are being adjusted to the work by the sliding of the shaft A.

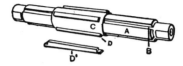

FIG. 28. — EXPANDING MANDREL.

47. Bridges in hollow castings A, Fig. 29, are often cast across the ends as at B, to provide for center holes.

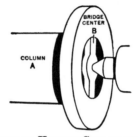

FIG. 29. — BRIDGE IN HOLLOW CASTING FOR CENTER HOLE.

48. Revolving dead center for pipe turning. — To square or turn small pipe, mount upon ordinary lathe centers. For large pipe or cored work, a special large dead center is needed,

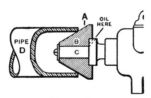

FIG. 30. — REVOLVING DEAD CENTER FOR PIPE TURNING.

preferably one that will revolve as at A, Fig. 30, in which cone B revolves upon shank C. A solid live center or chuck may be used at other end.

49. Built-up mandrels. — To face the ends of small engine cylinders or similar work, a built-up mandrel is used. To face the ends of a piece of pipe or cored work, make a mandrel by drilling and tapping three or four holes around each end of a stiff shaft and inserting adjusting screws to bear against the inside of work.

For large shells, tubes, etc., castings called " spiders " are used consisting of arms projecting from a hub, each arm supplied with a screw to bear against the work, and the hub supplied with set screws to fasten spider to mandrel.

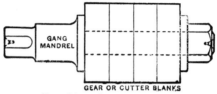

FIG. 31. — GANG MANDREL.

Gang mandrels are used to hold several gears or cutter blanks at once and are made with shoulder and reduced part to receive

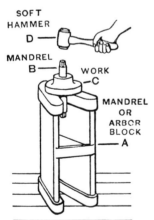

FIG. 32. — DRIVING MANDREL INTO FLANGE.

blanks which are fastened by collar and nut on end of mandrel, as in Fig. 31.

A special mandrel may be made of any suitable piece of stock by centering, turning, and filing to fit the hole.

50. Soft hammers for driving mandrels into work are made of lead or rawhide. A steel hammer should never be used without protecting work with a block of wood, copper, or lead.

Attention. — Molds are obtainable for molding lead hammers.

51. Mandrel or arbor block. — Fig. 32 shows the relative position of block *A*, mandrel *B*, work *C*, and soft hammer *D* when driving a mandrel into the work.

52. Mandrel or arbor press, *A*, Fig. 33, is used to press mandrel *B* into work *C* by forcing spindle *D* against the mandrel with handle *E*. Pointer *F* may be set to adjust mandrel for duplicate pieces. The press may be bolted to a lathe as at *G*, or mounted on a stand. It may be used also for forcing fits.

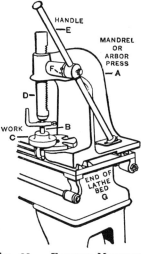

53. Center mandrel for bottom holes. — Mandrels made to fit the hole in the live spindle of a lathe are used to hold caps, oil cups, bottom or blind nuts and similar work that must be machined. The projecting end is turned either to a slight taper to receive work with a plain hole, or threaded to hold threaded pieces.

FIG. 33. — FORCING MANDREL INTO FLANGE WITH MANDREL PRESS.

54. Studs for driving large work instead of dog. — Work of large diameter, as a large pulley, may be driven by its arms with one or more studs fastened to the face plate of lathe.

Attention. — Oil mandrel before pressing or driving it into work. Take light cuts when using small mandrels to avoid springing of mandrel and chattering, producing irregular tool marks.

Warning. — To avoid breaking the spokes of a pulley or similar piece of work when driving a mandrel in or out at mandrel or arbor block or pressing a mandrel in or out with mandrel or arbor press, place a collar under hub of pulley so that the pressure will come on the hub and not on the rim of the pulley.

TURNING FLANGES

55. To clamp carriage to face large work in engine lathe. —
Feed tool to its cut by hand, then clamp carriage with binding screw and use power cross feed.

In the absence of power cross feed and clamp, throw feed
belt off or feed gears out of mesh. Run tool close to work,
tighten long. feed friction knob, revolve feed shaft by hand
until tool takes required cut and then feed tool inward or
outward with hand cross feed.

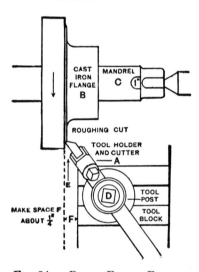

Fig. 34. — Rough Facing Flange.

56. To rough turn face of cast-iron flange, Fig. 34. —
Use cutting-in tool A for roughing flange B on mandrel C.
Fasten tool in post D, and clamp carriage. Feed in direction
of arrow E to outer edge of fillet.

57. To finish and scrape face of cast-iron flange to prepare for polishing, Fig. 35. — Take light finishing cuts with

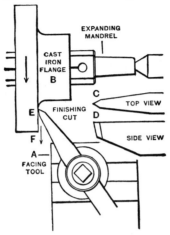

FIG. 35. — FINISH FACING FLANGE.

facing tool *A* ground to 60°, as at *C* and *D*, at medium high speed. Grind and oilstone top face of tool. Set it to drag a little as at *E* and feed in direction of arrow *F*. Use scraper

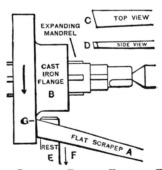

FIG. 36. — SCRAPING RADIAL FACE OF FLANGE.

A, Fig. 36, with cutting edge *CD* ground straight, or slightly convex, at medium speed. Hold scraper on rest *E* and move in direction of *F* with point dragging a little as at *G*. A piece of leather placed under scraper will often prevent chattering

58. To scrape inside rounds or fillets to the desired curvature. First shape with a round-nose lathe tool. Then use a

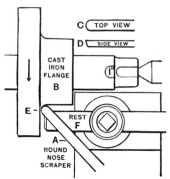

round-nose hand tool, *A*, Fig. 37, to prepare filleted corner of flange *B* for polishing. *C* and *D* show point and clearance of tool. To avoid chattering, curvature of end *E* should be greater than that of fillet for clearance and the tool should be held firmly on rest. A round or half-round file is sometimes used to smooth a fillet.

FIG. 37. — SCRAPING FILLET IN FLANGE.

Attention. — Cylindrical surfaces are more easily prepared for polishing than radial and curved surfaces. On small work a light finishing cut is taken with a fine feed, and on very large work a light finishing cut is taken with a broad-nose tool and a coarse feed, and the surfaces filed to erase the tool marks.

Warning. — Do not rub greasy fingers on cast iron before or during filing or scraping as the file or scraper will slip, glaze and scratch a greasy cast iron surface.

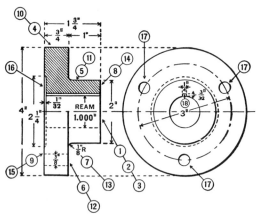

FIG. 38. — SCHEDULE DRAWING OF MAKING A CAST-IRON FLANGE.

59. To make a cast-iron flange finished all over, Fig. 38.

Specifications: Material, iron casting $\frac{1}{8}''$ large; weight 4 lb. 4 oz. Hardness, 26 to 28 (scleroscope).

Set dead center in approximate alinement.

High-speed steel or stellite cutting tools.

Time: Study drawing and schedule in advance, 5 min. — Oil machines, 7 min. — Chuck, 25 min. — Rough turn, 25 min. — Finish, 25 min. — Polish, 20 min. — Drill, 10 min. — Cut keyway, 8 min. — Clean machines, 12 min. — Total, 2 h. 17 min.

SCHEDULE OF OPERATIONS, MACHINES AND TOOLS

OPERATIONS.	MACHINES, FEEDS, SPEEDS.	TOOLS.
Snag casting. Mount in chuck and adjust until hub runs true. Drill and power (rough) ream hole (1), (2). See pp. 408–412, 417.	Engine lathe, 12″ to 16′. 3d speed, or 200 R.P.M. 1st or 2d speed, or 115 R.P.M. Hand feed.	File, independent, chuck, chalk, centering tool, $\frac{63}{64}''$ twist drill and holder, power fluted chucking reamer .005″ small.
Hand (finish) ream hole by starting reamer in lathe, pull belt by hand, complete at reaming stand or vise, (3). See pp. 416, 414, Figs. 23, 22.	Reaming stand or vise.	1″ standard hand reamer, reamer wrench.
True live center. See p. 114. Oil mandrel and press into hole in flange in direction hole was reamed. See pp. 422, 423; mount on centers.	Mandrel press.	1″ mandrel, dog.
Rough turn $\frac{1}{64}''$ oversize, (4), (5).	Engine lathe, 3d or 4th speed. Back gears in, or 35 F.P.M. Hand or medium power feed — 80 to 1″.	Holder and cutter or diamond-point tool 15° rake, calipers, rule.
Rough face (6); feed inward; rough fillet, (7); rough face, (8); reverse work on centers, and rough face (9) to $\frac{3}{4}'' + \frac{1}{32}''$.	3d or 4th speed. Back gears in or 35 F.P.M. Hand or medium power feed — 80 to 1″.	Holder and cutter 15° rake or round-nose tool. Two rules to measure length of hub, or inside calipers.
Finish turn (10), (11). See *Attention*, p. 428.	2d or 3d speed (back gears out) or 125 F.P.M. Fine power feed — 140 to 1″.	Holder and cutter or diamond-point tool 15° rake, calipers, rule.
File (10), (11).	Engine lathe, 2d or 3d speed, or 175 F.P.M.	8″ or 10″ mill bastard file, file card.
Finish face (12) with cuts .002″ to .003″ deep and scrape. See p. 425.	2d or 3d speed, or 125 F.P.M.	Facing tool, flat scraper.

Finish and scrape fillet, (13). See p. 426.	2d or 3d speed, or 125 F.P.M.	Round - nose tool, round-nose scraper.
Finish face and scrape, (14).	2d or 3d speed, or 125 F.P.M.	Facing tool, flat scraper.
Finish (do not scrape or polish), (15) (see *Attention*).	1st or 2d speed, or 60 F.P.M.	Facing tool, calipers, rule.
Cut recess $\frac{1}{32}''$ deep. Set side tool at 45°, feed outward, two or three cuts, (16).	1st or 2d speed, or 50 F.P.M. Hand or medium power feed — 80 to 1″.	Side tool, calipers, rule.
Polish, (10), (11), (12), (13), (14). See pp. 432, 433.	Speed lathe, highest speed.	Polishing stick, 60, 90,120, emery cloth, oil.
Jig drill, and ream holes, (17). See p. 604.	2-spindle drilling machine. Drill 600 R. P. M. Power reamer, 400 R.P.M. Hand feed.	Box jig, $\frac{63}{64}''$ drill, $\frac{1}{2}''$ power reamer.
Cut keyway central, (18). See p. 926.	Planer and vise or key seating machine.	Keyway tool holder and cutter, $\frac{1}{4}''$ wide.
Stamp name or initials in recess.		Steel name stamp, hammer.

Attention. — Finishing cuts 10 and 15 may be omitted until flange is keyed to its shaft. The hole may be chucked with a $\frac{63}{64}''$ twist drill or, if cored, with a three or four-groove drill, and hand reamed (1), (3). The hand tools may be carbon steel.

Note. — Bushings, collars and step pulleys or any other work mounted on a mandrel, with one, two, or more diameters, are machined in the same general manner as the flange in Fig. 38.

TURNING PULLEYS

60. The taper or crowning on the face of pulleys ranges from $\frac{1}{4}''$ per foot for wide pulleys to $\frac{3}{4}''$ per foot for narrow pulleys. Tapered-face pulleys are made commercially in a pulley lathe with two cutting tools operating at same time, — one on the front, the other on the back of machine, or in turret lathes with special tools.

The face of the pulley is often crowned in a forming lathe or ground in special grinding machines from the rough casting. The hubs of pulleys used in combination, as tight and loose pulley mechanism, are squared.

Pulley *A*, Fig. 39, is turned in an engine lathe. The rim is faced as at *B*. The face is tapered as at *C*. The rim may be chamfered as at *D*, or rounded with a right-and-left forming tool as at *E, E'*.

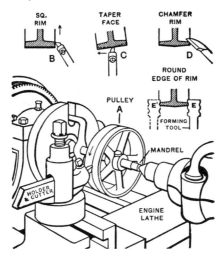

FIG. 39. — PULLEY TURNING IN ENGINE LATHE.

The set-over of footstock for ordinary tapers is forward, **as at** *A*, Fig. 40, and for pulleys it is backward as at *B*.

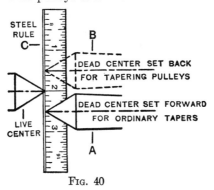

FIG. 40

Information. — Omit straight turning on large pulleys and pulleys for rough work. Rub chalk on face, find middle with rule, turn taper from rough casting, then round edges of rim with forming tool (see *E, E'*, Fig. 39) and file face.

61. To make a pulley 5″ in diameter. Fig. 41.

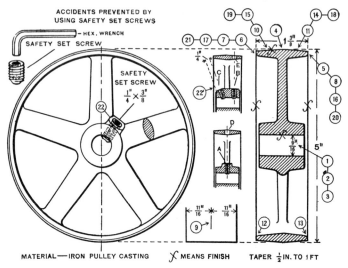

MATERIAL—IRON PULLEY CASTING   𝒳 MEANS FINISH   TAPER ½ IN. TO 1 FT

FIG. 41. — SCHEDULE DRAWING.

Specifications: Material, pulley casting with surfaces to be finished $\frac{1}{16}$″ large; weight, 2 lb. 5 oz. Hardness, 24 to 30 (scleroscope).
Set dead center in approximate alinement.
High-speed steel or stellite cutting tools.
Time: Study drawing and schedule in advance, 5 min. — Oil machines, 5 min. — Chuck, 20 min. — Turn, square, and taper, 40 min. — File and polish, 15 min. — Drill and tap, 10 min. — Stamp, snag and paint, 8 min. — Clean machines, 7 min. — Total, 1 h. 50 min.

SCHEDULE OF OPERATIONS, MACHINES AND TOOLS

Operations.	Machines, Speeds, Feeds.	Tools.
Snag casting. Mount in chuck and adjust until edge and inside of rim run true. Drill and rough ream hole, (1), (2). See pp. 408, 409, 417.	Engine lathe, 12″ to 16″. 3d speed, or 400 R.P.M. Hand feed.	File, chuck, chalk, centering tool, $\frac{35}{64}$″ twist drill, holder or dog, chucking reamer .005″ small. $\frac{9}{16}$″ standard hand reamer, reamer wrench.
Finish ream hole by hand at reaming stand or vise, pp. 415, 416, 414, (3).	Reaming stand or vise.	$\frac{9}{16}$″ standard mandrel.
Oil mandrel and press into hole. See pp. 422, 423.	Mandrel press.	Center truing tool,
True live center. See p. 114.	3d speed.	center gage.

Mount pulley on centers. Rough turn to 5″, (4), one or two cuts. See A, Fig. 39.	Engine lathe, 3d or 4th speed, back gears in, or 50 F.P.M. Medium power feed — 80 to 1″.	Dog, holder and cutter or diamond, point tool, 15° rake calipers, rule.
Rough square edge of rim (5), reverse pulley on centers and rough square (6) to $1\frac{3}{8}″ + \frac{1}{64}″$.* See B, Fig. 39.	3d or 4th speed, back gears in, or 50 F. P. M. Hand or medium power feed — 80 to 1″.	Holder and cutter or round-n o s e t o o l calipers, rule.
Finish square edge (7), reverse pulley on centers and finish square (8) to $1\frac{3}{8}″$ one cut each. See B, Fig. 39.		
Coat a small portion of face with copper sulphate or chalk and make line midway between edges, (9).	....	Copper sulphate or chalk, divider-calipers or dividers, rule.
Regrind tool.		
Set footstock back to turn taper of $\frac{1}{2}″$ per foot ($\frac{7}{64}″$ for $5\frac{1}{4}″$ between centers) and set tool to touch line (9), see C, Fig. 39, and feed toward headstock. Reverse pulley, reset tool and repeat, (10), (11).	3d or 4th speed, back gears in, or 50 F.P.M. Fine power feed — 140 to 1″.	Rule, dividers, holder and cutter or diamond-point tool.
Chamfer at 45° inside corners of rim, see D, Fig. 39, or round with forming tool, (12), (13).	2d speed, back gears in, or 25 F.P.M., hand feed.	Side tool, 15° rake.
File face, (14), (15). Scrape edges, (16), (17).	Engine lathe, 1st or 2d speed, or 175 F.P.M. Speed lathe, highest speed.	8″ or 10″ mill bastard file, graver.
Polish face and edges, (18), (19), (20), (21), See pp. 432, 433.		60, 90, 120 emery cloth, polishing stick, oil.
Drill and tap hole for set screw, insert $\frac{1}{4}″ \times \frac{1}{2}″$ round-point set screw.	Vertical drilling machine, 3d speed, hand feed. Tilting vise or fixture.	$\frac{3}{16}″$ extension drill, $\frac{1}{4}″ \times 20$ U.S.S., pulley tap, wrench, lard oil.
Stamp name or initial on end of hub.	....	Steel name stamp, hammer.
Snag and paint unfinished parts.	....	Special black paint.

* *Information.* — Take an equal amount of material from each edge of rim so that the arms will be approximately central.

Important. — Filing, scraping and polishing are omitted on large pulleys and rough work.

Attention. — Power chucking reamer may be omitted, as a good hole may be obtained with a drill and hand reamer.

When possible, place arms of pulley opposite chuck jaws. The hand tools may be carbon steel.

62. To locate set screws, Fig. 41. — If the boss is central as at *A*, drill body hole *D* in rim through which to tap hole *A*, and use a cup or round-pointed set screw. Where the boss is flat as at *B* and placed at one side, drill a body hole through rim *E*. If a body hole through rim is objectionable, hold pulley in a tilting vise or fixture and drill and tap hole at angle, as at *C*. In large pulleys use two set screws or key.

POLISHING LATHE WORK

63. Use a speed lathe for polishing but never an engine lathe when it can be avoided, as the gritty spatterings injure its bearings. Make exception only to work machined in a chuck of an engine lathe which must be polished before it is removed.

An effective polishing speed for emery, alundum, carborundum or aloxite cloth is between 5000 and 6000 F.P.M.; that is, surface speed of about a mile a minute (5280 feet).

It is not always possible to obtain this high speed, as speed lathes are usually belted for a great variety of work, as drilling, hand turning, etc., and their highest speed will not give the maximum polishing speed for small work.

On work that is unbalanced, it is not desirable, and often dangerous, to polish at the maximum speed, as it will shake the lathe and the work may fly off the centers. It is best to use as high a speed as the nature of the work will permit, since the polishing can then be done with less labor and with less tendency to destroy the truth of the work.

64. Order of applying different numbers of emery cloth. — If the work is carefully finished, limited application of 60 and 90 will produce an effective polish; and if the work is finished extra fine, 90 will be sufficient. If a more brilliant polish is desired, use 120 and flour. For large work not given a fine finish it may be necessary to begin with 46 or 54.

Apply 60 emery first with hard pressure until all tool marks and scratches are removed and the pores in the metal have nearly disappeared. Then apply 90 emery with lighter pressure until all evidence of 60 is removed. If on applying

90, tool marks, deep scratches, or large pores appear, return at once to 60 emery.   Follow this method in applying successive grades of emery.

Use lard oil on the emery cloth or work, distributing it with the fingers.

65. To polish flange, Fig. 42. — Oil speed lathe and shift belt to highest speed.   Press mandrel into flange A, which has been carefully finished by turning, filing, and scraping.   Fasten dog on mandrel.   Mount and adjust on centers.   Wrap a strip of No. 60 emery cloth B around end of wedge-shaped soft-pine stick C.   Drop a little oil on emery cloth or flange and distribute with fingers.   Start lathe and pivot stick on rest D which should be clamped from $2''$ to $2\frac{1}{2}''$ from the work.

Keep emery moving back and forth slowly along the work so that the marks will cross and recross each other, to avoid cutting grooves.   Grades 60 and 90 will produce a good polish on a flange.

To polish radial face of flange, clamp rest D parallel to face and from $2''$ to $2\frac{1}{2}''$ distant.   Apply emery cloth in same order

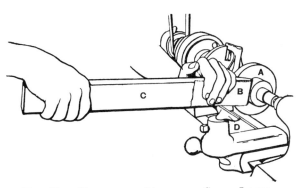

FIG. 42. — POLISHING A FLANGE IN SPEED LATHE.

as on hub.   Keep emery moving slowly, advance and recede in short strokes toward the center to avoid cutting grooves, then recede in the same order.   To polish the fillet, move emery cloth back and forth in short strokes following the curve.

66. To polish a shaft with wood polishing clamp, AA′
Fig. 43, hinged by leather at end. — Oil speed lathe and shift
belt to highest speed.   Place dog on shaft with copper under
set screw, and mount shaft on centers.   Drop a little oil on
emery cloth or shaft and distribute with the fingers.   **Wrap**

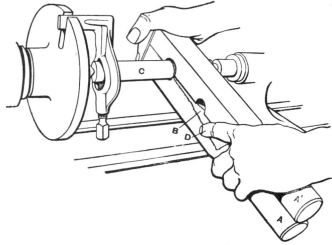

FIG. 43. — POLISHING A SHAFT IN SPEED LATHE.

emery cloth *B* around shaft *C* in one fold only, and hold end
as at *D*, to prevent it from winding around shaft.   Apply
pressure with the left hand, and with the right move clamp
back and forth in short strokes along shaft so that the marks
will cross and recross each other.   Move clamp continuously
when in contact with revolving work, otherwise emery will cut
grooves.   Grades 60 and 90 will produce a good polish on a
shaft.

67. To polish brass and copper, use a finer emery cloth than
for steel or iron, as the material is softer.   Start with a No. 90,
continue with No. 120 flour, and crocus cloth.   If a more
brilliant luster is desired, the work may be buffed.

For Polishing and Buffing with Wheels and Belts, and
Lacquering, see p. 440.

Attention. — Use oil sparingly to avoid excessive spattering. Do not allow the emery cloth to slip off the edge of work as it will round the corners. Adjust mandrel, or work freely on centers, and occasionally loosen and oil dead center, as the heat generated will expand the work or mandrel and burn off the dead center.

To lay lines uniformly after work is polished, moisten a piece of worn emery cloth of fine grade with oil (or oil the work, distributing it with the fingers) and move the emery cloth slowly along the work.

Important. — If holes are to be drilled in a surface that is to be polished, polish the surface first and drill the holes afterward.

Warning. — To avoid excessive vibration of lathe when polishing which may cause the work to fly off the centers, use a light malleable iron dog.

CURVE TURNING AND FORMING

68. Curve turning. — Small outside and inside rounds (convex and concave surfaces), ogees and other irregular curves may be rough formed, as in Fig. 44, with one hand operating the cross feed and the other the long. feed. The tool

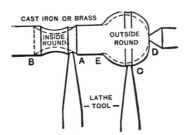

Fig. 44. — Curve Turning, Operating Feeds by Hand

for the inside round is moved, approximately to the correct curve, repeatedly from *A* to *B*; for the outside round, from *C* to *D* and from *C* to *E*. By moving one feed more rapidly than the other, as the curve may require, good results may be obtained. If the work is slender or the curve large, it is usually finished with hand tools in the speed lathe (see Templet, p. 437). If the curve is small and work stiff, forming tools may be used.

69. Forming tools for engine lathe work are used for

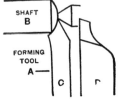

forming duplicate pieces. They may be made by milling or filing. If much stock has to be removed, rough form with a suitable lathe tool. Forming tool *A*, Fig. 45, is rounding the end of shaft, after which it may be filed and polished.

Forming tools are also used in turret

Fig. 45.—Forming with lathes, screw machines and planers.
Forged Tool.

70. Forming cutters and holders. —
Forming cutters, to fit thread tool holders, are obtainable. In Fig. 46, forming cutter *A*, held in holder *B*, is forming wheel *C*.

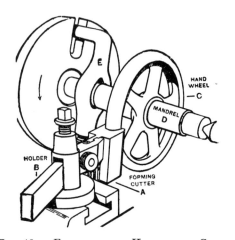

Fig. 46. — Forming with Holder and Cutter.

The manufacturing method of machining hand wheel rims is to mill them with a vertical milling machine or form them in a special machine which carries a pointed tool mounted on a turn table around the curve. See pp. 127, 309, 1265–1275.

MAKING FORMED MACHINE HANDLES

71. Single handles are made as outlined in Figs. 47, 48, 49. When manufactured in quantities, they are drop-forged to shape or made from bar stock with a forming tool in

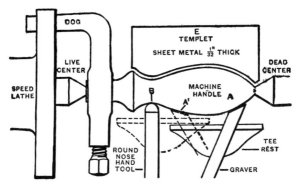

FIG. 47. — CURVE TURNING WITH HAND TOOLS. TESTING WITH TEMPLET.

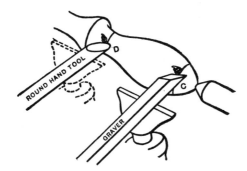

FIG. 48. — FINISHING FORMED HANDLE WITH HAND TOOLS. OVERHEAD TURNING.

a turret lathe. By either of the latter methods, they require but a small amount of hand finishing.

72. Templets of sheet brass or steel, E, Fig. 47, are used for duplicate work, serving as a pattern or guide to uniform production.

They are lined out from specifications on the drawing, and are rough cut by chipping, or with shears, and finished by filing. Templets are used as guides when rough turning or planing, and for careful tests when finishing.

73. To make a formed machine handle to templet, Fig. 49.

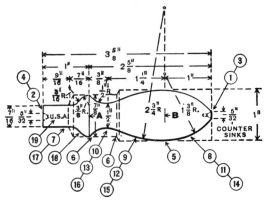

Fig. 49. — Schedule Drawing of Machine Handle.

Specifications: Material, machine steel $\frac{1}{16}$" to $\frac{1}{8}$" large; weight, 15 oz. Hardness, 15 to 16 (scleroscope).

True live center. Set dead center in approximate alinement.

Lard oil may or may not be used in squaring and turning steel or wrought iron. Carbon-steel or stellite cutting tools.

Time: Study drawing and schedule in advance, 5 min. — Oil lathe, 6 min. — Center and square, 10 min. — Rough turn, 10 min. — Rough form, 10 min. — Finish with hand tools, 30 min. — File and polish, 14 min. — Clean lathes, 5 min. — Total, 1 h. 30 min.

SCHEDULE OF OPERATIONS, MACHINES AND TOOLS.

Operations.	Machines, Speeds, Feeds.	Tools.
File ends flat.	Vise.	File.
Center, (1), (2).	Centering machine. Drill, 1250 R.P.M. Countersink, 700 R.P.M.	$\frac{3}{32}$" drill, 60° countersink, lard oil.
Rough square, (3), (4) to $3\frac{5}{8}$". See Step Method of Rough Squaring, p. 301.	Engine lathe, 12 to 16". 2d speed, or 35 F.P.M. Hand feed.	Dog, side tool, 35° rake, calipers, rule.

Recenter. Omit finish square.

Rough turn (5), $\frac{1}{64}''$ large. One cut. Turn one-half of length, reverse and turn the other half.	1st or 2d speed, or 35 F.P.M. Medium power feed — 80 to 1''.	Diamond-point tool, or holder and cutter, 35° rake, calipers, rule.
Rough turn (6), $\frac{1}{64}''$ large. One cut. Turn (6'), to about $1\frac{1}{2}''$ from end.		
Hold in vise and mark lines 1'' from each end at (A) and (B).	Vise.	Copper sulphate, scriber, rule.
Rough turn (7), $\frac{1}{64}''$ large. Two or three cuts.	3d speed, or 35 F.P.M. Medium power feed — 80 to 1''.	Calipers, rule.
Rough turn convex parts (8), (9), and concave part (10), several cuts. Operate both feeds in combination, by hand, and test with templet, as at E, Fig. 47.	2d or 3d speed, or 50 F.P.M.	Copper under set screw of dog, diamond-point tool, or holder and cutter, 35° rake, templet.
Rough turn with graver, (11), (12). See A and A', Fig. 47. Test often with templet.	Speed lathe, 8 to 12'', 2d speed, or 150 or F.P.M.	Graver.
Rough turn, (13). See B, Fig. 47. Test with templet and calipers at diameters $\frac{1}{2}''$ and 1''.	....	Round-nose hand tool, templet, calipers, rule.
Finish turn, (14), (15). See C, Fig. 48.	3d speed, or 225 F.P.M.	Graver, templet, calipers, rule.
Finish turn, (16). See D, Fig. 48.	....	Round-hand tool, templet, calipers, rule.
Finish turn, (17).	Engine lathe, 12 to 16'', 3d speed, or 50 F.P.M. Fine power feed — 140 to 1''.	Diamond-point tool, or holder and cutter 35° rake, calipers, rule.
Rough and finish turn, (18). Copy the reverse curve (the ogee) as accurately as can be determined by the eye, or make a templet from specifications or curve on drawing. See p. 437.	Speed lathe, 3d speed, or 225 F.P.M.	Graver, round-nose hand tool. Use two pieces of copper under No. 2 clamp dog. •
File, (18) and (16)...........	....	5'' to 8'' half-round smooth file
Polish, (18).	Speed lathe, highest speed.	60, 90, 120 and flour emery cloth and stick, oil.

File, (14) and (15).	3d speed, or 175 F.P.M.	8″ to 10″ mill bastard file.
Polish, (14) and (15). Stamp name or initials, (19).	Vise.	Steel name stamp, hammer, copper jaws.
(File off burr.)		

Attention. — Part 17 is usually fitted to a hand wheel, or other machine part, by a forcing or threaded fit; if so, operation 17 is omitted to allow the desired fit to be made later.

Note. — The machine handle may be finished with graver and round-nose hand tool used horizontally, as in Fig. 47, or with graver and round hand tool, by overhead turning, as in Fig. 48.

POLISHING, BUFFING, AND LACQUERING

74. Polishing and buffing with wheels and belts. — The term polishing is applied to the production of surfaces of the ordinary degree of finish; and buffing, to the production of brilliant lustre. These methods are employed upon castings and forgings which require lustre rather than accuracy; and on work of irregular outline, as automobile and bicycle parts, electrical instruments, silverware and nickel-plated work. The polishing is done on felt wheels and canvas belts to which emery is glued.

Buffing is done on cloth wheels of cotton or muslin charged with flour of emery, crocus, tripoli, rouge, etc., according to grade of lustre desired. The work is held with the hands and brought against the revolving wheel or belt, moved and turned, changing from a coarse to a finer wheel and so on, and then continued on the buffing wheels.

75. Lacquering is coating highly polished useful and ornamental articles of silver, brass, bronze, etc., with a liquid called lacquer to prevent them from tarnishing or oxidizing. Some lacquers are transparent, others are colored, according to the shade desired upon the work. To lacquer, clean work by wiping or, preferably, by dipping in a solution of potash and apply lacquer with a brush. Some work is dipped in a tank of lacquer. See *Principles of Machine Work.*

ADVANCED MACHINE WORK

SECTION 5

LATHE WORK

Inside Calipers and Inside Micrometers. Boring and Inside Threading.
Square Threads. Acme Standard or 29° Threads. Multiple Threads.
Brass Finishing. Alinement Drilling and Tapping.
Drilling, Tapping and Hand Threading In
Speed Lathe. Broaching Holes,
Keyways and Slots.

INSIDE CALIPERS AND INSIDE MICROMETERS

1. To set inside calipers. Fig. 1. — Hold rule *A* perpendicularly against carriage *B*. Place calipers *C* with point at *D*, and adjust nut *E* until other point coincides with middle of line *F*.

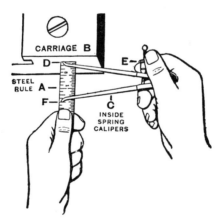

Fig. 1. — Setting Inside Calipers.

Another method is to set calipers to a standard ring gage, or to a hole of the desired diameter in any piece of work.

2. To measure diameters of holes with inside calipers. — Work to be measured may be held in any position on the bench,

501

in vise or chuck. To measure work G, Fig. 2, in chuck H, set calipers K to size and insert point M in the lower side of hole, and steady with finger while a gentle effort is made to

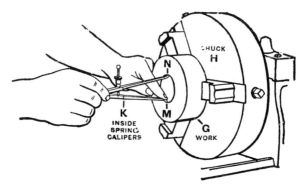

FIG. 2. — MEASURING WITH INSIDE CALIPERS.

insert point N, pivoting calipers on point M by raising and lowering the outer end; also move point N to right and left to locate maximum diameter.

3. To adjust the tool to bore a hole to diameter to which inside calipers are set.— Take trial cuts and test frequently with calipers. See pp. **5**04, **5**05.

4. Small inside micrometer calipers, Figs. 3, 4 and 5, are obtainable in two sizes, one measuring from two-tenths

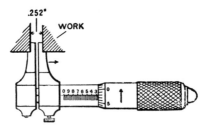

FIG. 3. — MEASURING WITH 1″ INSIDE MICROMETER.

of an inch to one inch, and the other from one inch to two inches.

Except that the barrels are figured from right to left, they are similar to outside micrometers having 40 threads to the inch.

The reading of the one-inch micrometer in Fig. 3 equals $10 \times .025 = .250'' + .002 = .252''$ or $\frac{1}{4}'' + .002''$.

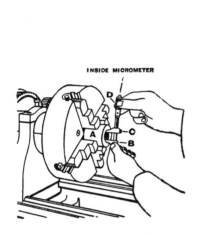

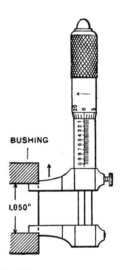

FIG. 4. — MEASURING BORE OF BUSHING.

FIG. 5.— MEASURING WITH 2″ INSIDE MICROMETER.

In Fig. 4 a two-inch micrometer is shown as used to measure the hole in bushing A.

Solid jaw B is placed against the lower wall of hole.

Sliding jaw C is moved against the upper wall of hole by turning thimble D to right. The hole is bored out with a boring tool (see pp. **5**04, **5**05) until the two-inch micrometer reads $2 \times .025'' = .050''$.

$1'' + .050'' = 1.050''$ diameter of bore of bushing, as in Fig. 5.

5. Large inside micrometer caliper. — Fig. 6 consists of barrel A graduated into 40 divisions to the inch, thimble B graduated into 25 divisions, attached to a screw having 40 threads to the inch, passing through a nut in end of barrel A. The screw has a movement of half an inch. Measuring point C is fixed to the thimble, but measuring point or rod D is held in chuck E and clamped by nut F and is removable in

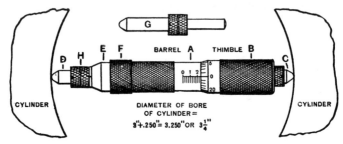

FIG. 6. — MEASURING BORE OF CYLINDER WITH LARGE INSIDE MICROMETER.

order to insert extension rods, a number being supplied varying in length by half an inch ($\frac{1}{2}''$), one being shown at G. Adjustment for wear on rods is provided by adjusting nuts H. Fig. 6 shows how the micrometer is used to measure the bore of a cylinder. The net length of micrometer is $3''$ and the reading is $3.200'' + 2 \times .025 = 3.250''$.

BORING AND INSIDE THREADING

6. Boring tools. — Fig. 7 shows a forged boring or inside turning tool. Cutting edges A and B must be shaped with accuracy. The point is rounded slightly in order to make it

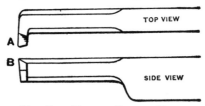

FIG. 7. — FORGED BORING TOOL.

cut smoothly, and, also, not dull quickly; if rounded too much, the tool will spring away from the cut, or chatter. See Charts, pp. 127, 309.

7. To set and use boring tool in lathe. — Fig. 8.

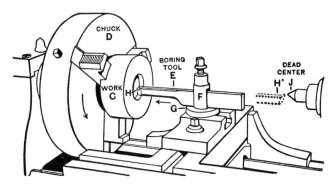

FIG. 8. — BORING IN ENGINE LATHE.

SCHEDULE OF OPERATIONS

1. True up cored work C in chuck D and face front end.

2. Set tool E height of dead center J as at H', reverse tool and post F into position parallel to ways of lathe and clamp tightly on G.

3. Run tool in length of hole to see that shank clears walls, also chalk top of tool to show length of hole.

4. Rough bore hole and caliper frequently.

5. Take two or three light finishing cuts to leave hole smooth and true. See pp. 501–503.

Speed for boring. — Use same cutting speed for boring or inside turning as for outside turning, see pp. 145–148.

Attention. — A method sometimes used to bore a straight and smooth hole is to take a light finishing cut inward, then reverse the feed and let the tool cut outward.

Note. — While a boring tool will cut satisfactorily if set at *height* of centers, as H', Fig. 8, still the tool will cut better if set *below* the centers, the amount increasing with the diameter of the hole, as is inversely true with outside turning tools. See pp. 133, 134.

This, however, is not always possible, especially in small holes as the size of tool will not allow sufficient clearance and will cause it to ride on the wall of the hole, which must be avoided.

8. Squaring of an inside shoulder with the tool K, Fig. 9. L is a section taken at MN. The rounded point of a boring tool leaves a fillet at the termination of the cut and if a square shoulder is desired an inside squaring tool is used to remove the fillet and square the shoulder by cutting to or from the center.

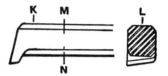

FIG. 9. — INSIDE SQUARING TOOL.

9. Boring holders and cutters are used in the same manner as a forged tool. *A*, Fig. 10, shows holder and double-end cutter rough boring a cored hole, *B*, in work *C*. For inside squaring a special cap is supplied which holds the cutter at an angle of 45° as at *D*. See No. 20, Chart, p. **309**.

FIG. 10. — BORING AND INSIDE SQUARING WITH HOLDERS AND CUTTERS.

FIG. 11 — BORING AND FACING WITH BENT HOLDER AND CUTTER.

At *E*, Fig. 11, a right-bent outside holder and cutter is shown used as a boring tool and for squaring and facing as at *F*.

For boring long holes it is more practical to use drills, boring bars, boring heads, etc.

10. Inside threading tools, United States Standard or Sharp V threads. — Fig. 12 shows point of inside threading tool. It

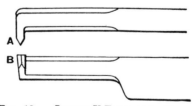

FIG. 12. — INSIDE V-THREADING TOOL.

may be ground as at *A* and *B* for United States Standard or Sharp V threads. It is similar to a boring tool. See No. 34, Chart, p. **127**.

The method of setting an inside United States Standard or Sharp V-threading tool at right angles to the work is shown

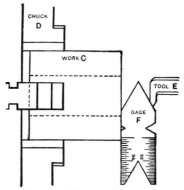

FIG. 13. — SETTING UNITED STATES STANDARD OR
SHARP V-THREADING TOOL WITH CENTER GAGE.

in Fig. 13. Work *C* is held in chuck *D* and is bored to size and end rough and finish squared. Tool *E* is then set to gage *F*.

11. To cut an inside thread in lathe. — Fig. 14.

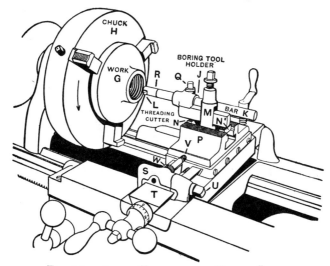

FIG. 14. — INSIDE THREADING IN ENGINE LATHE.

SCHEDULE OF OPERATIONS

1. True up work G in chuck H, square end and bore hole about $\frac{1}{64}''$ larger than root diameter of screw that it is to fit. See Table of Tap Drills, *Principles of Machine Work*, and Threading, p. 318.

2. Assemble holder J, cutter bar K, threading tool L, which should be ground to fit thread gage, and clamp in tool post M supported by blocks N and N' in tool block P.

Adjust tool to height of center, set by gage F, Fig. 13, and clamp bar and cutter firmly by screw Q and cap R.

3. The cutter may be removed, reground and reset to resume its cut by means of cap R.

4. Clamp thread stop S to slide T by screw U and adjust feed of tool by rotating nurled head W which is fast on screw V.

5. Rough thread with cuts from .003'' to .004'' nearly to size. Finish thread with cuts from .001''

to .002'' until thread fits screw, as follows:

Set outside calipers to the outside of the thread, then transfer setting to inside calipers or to a wire filed to fit outside calipers, and pointed at each end more acute than the thread. Cut thread slightly smaller than inside calipers or wire, then test it with the screw.

If work is to fit a lathe spindle or other work that cannot be removed, take chuck and work from lathe, clean, oil, and try on the screw. If it does not fit take another light cut, and so on until desired fit is obtained.

Attention.—If work is cast iron, thread dry; if steel or wrought iron, use lard oil.

Use oil for all materials when fitting to screw or work may seize screw and have to be split off, thereby destroying work and possibly the screw.

12. To finish tap the back plate of a chuck, Fig. 15. — When cutting an accurate thread such as that in a back plate, A, of a chuck or face plate of a lathe, it is best, if a suitable tap is available, to cut about three-quarters of a full thread with a threading tool, then finish tap as in Fig. 15. Clean and oil thread and tap and place tap wrench B on tap C, and mount on dead center with tap in thread. Start tap carefully so that it will follow the thread already cut and not split and destroy it. Pull belt D by hand and follow tap with dead center with handle E. To back out tap, unclamp footstock and run lathe backward by hand or by power. Then square up end F and bore out about two threads, as at G, to permit screwing plate to shoulder on nose of spindle or on a mandrel to be machined.

13. Interrupted thread tap, Fig. 15. — Instead of using adjustable tap *C*, with regular thread, preferably, use an

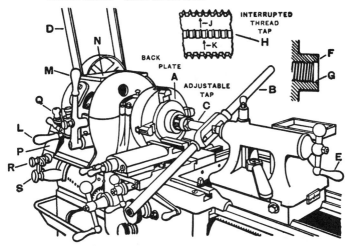

FIG. 15. — FINISHING THREAD WITH A TAP.

interrupted thread tap, as at *H*, which is obtainable. This tap requires less power to drive it. Every other tooth is cut away. The teeth of each land follow in the spaces of the land preceding as shown by arrows *J* and *K*.

14. Engine lathe, gear headstock, Fig. 15, shows an engine lathe equipped with a gear speed change located in the head in place of the cone pulley. The different speeds and positions of the levers *L* and *M* to obtain them are given in a table at *N*.

Attention. — Some all-gear headstock engine lathes have a variable speed countershaft to give a still greater variety of speeds.

This lathe is also equipped with a rapid change-gear mechanism for feeds and threads. In a table at *P* are given the different threads, and positions of the levers *Q*, *R*, and

S, to obtain them. The feed is usually two to four times threads per inch.

15. To cut right inside thread to shoulder. — Fig. 16.

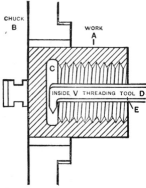

FIG. 16. — CUTTING INSIDE UNITED STATES STANDARD OR SHARP
V THREAD TO A SHOULDER.

SCHEDULE OF OPERATIONS

1. Work *A*, held in chuck *B*, is bored to correct diameter and depth.
2. Cut groove *C* to full diameter of thread with inside forming tool.

3. Mark line *E* with chalk to indicate when tool *D* reaches groove.
4. Cut thread to desired depth.

16. To cut a left inside thread to a shoulder, or any portion of the hole less than its entire length, cut groove as at *C*, Fig. 16, from which start tool outward, and make a mark at *E* to know when to adjust the tool forward into groove preparatory to starting to cut outward.

SQUARE THREADS

17. Square threads, Fig. 17, right or left, are used for screws to transmit motion, as the cross feed and lead screws of an engine lathe, valve stems, presses, rock drill feed screws, etc. They cannot be cut successfully with dies, or milled with a thread milling machine. See Acme Standard or 29° Threads, pp. **5**21-**5**31.

The thickness of thread and width of space are each nominally one-half the pitch. The depth is one-half pitch plus the clearance. The fit is on the sides of the thread with clearance top and bottom. A larger clearance is advisable for large diameters and coarse pitches.

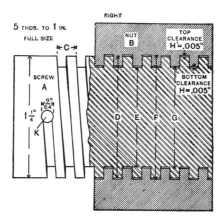

Fig. 17. — Sectional View of Square Thread Screw and Nut.

NAMES OF PRINCIPAL PARTS OF SQUARE THREAD

C. — Pitch.
D. — Diameter standard.
E. — Root diameter (which is also root diameter of tap).
F. — Diameter of bottom of nut (which is also diameter of tap).

G. — Bore of nut.
H and *H′*, clearances; same for all pitches. *H* made by cutting the thread .005″ deeper than ½ *P*. *H′* made by making tap .01″ larger in diameter than *D*.

18. To obtain parts of Square thread. —
Width of tool for screw thread =

$$\frac{1}{\text{No. of threads per inch}} \div 2.$$

Width of tool for tap thread =

$$\frac{1}{\text{No. of threads per inch}} \div 2 -$$

.0005″ for each linear inch of nut for shrinkage of tap.

Diameter of screw = any

Diameter of tap   = diameter of screw + .010″

Diameter of screw or tap at root of thread =
diameter of screw −

$$\frac{1}{\text{No. threads per inch}} + .010''.$$

Diameter to bore nut = diameter of screw −

$$\frac{1}{\text{No. of threads per inch}}.$$

Clearance = .005″, top and bottom of thread.

19. Table of Square threads. — While any pitch may be assumed, it is best, when it will answer the purpose, to use whole numbers of threads per inch as near as possible to three-quarters of the United States Standard thread. See p. 323.

Diameter.	Threads Per Inch.	Diameter.	Threads Per Inch.
½″	10	1″	6
⅝″	9	1¼″	5
¾″	8	1½″	4
⅞″	7	2″	3

20. Square threading tool, Fig. 18.

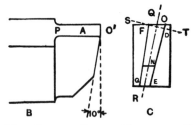

Fig. 18. — Outside Square Threading Tool.

A, top face, *B*, side view partially rotated, and *C*, end view. *DE* and *FG* inclined to line *QR* to give clearance.

21. To find inclination of thread and to file tool.

SCHEDULE OF OPERATIONS

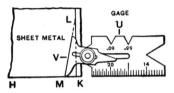

FIG. 19. — MAKING A TEMPLET AND SETTING GAGE TO ANGLE OF INCLINATION FOR SQUARE THREADING TOOL.

Inclination. Fig. 19.

1. File edge *HK*, true.
2. Draw *KL* at 90° to *HK*.
3. Make *KL* equal to circumference at root of thread.
4. Make *MK* equal to pitch.

5. File to *LM* which gives inclination of thread and use as a templet to test tool or omit filing and set angle gage *U*, Fig. 19, to angle *LM* as at *V*.

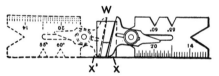

FIG. 20. — TESTING INCLINATION OF SQUARE THREADING TOOL WITH ANGLE GAGE.

To File Tool. Fig. 20.

1. File bottom of tool flat, and end square as in Fig. 18. File sides with an 8″ or 10″ hand smooth file and 8″ dead smooth file to give inclination, clearance, and size.

2. Test tool with gage *U* as at *X* and *X'*, Fig. 20, or with sheet metal templet *V*, as in Fig. 19.

3. Measure with micrometer calipers.

4. Harden and temper to a straw color.

5. Grind on end and a little on top.

Do not grind on sides.

Attention. — The amount of inclination varies with different pitches and diameters but with a generous side clearance one tool will do for several diameters. The tool is parallel from *O'* to *P*, Fig. 18.

For very coarse pitches tool is narrower at P than at O'. For fine pitches top face A is ground horizontal; for coarse threads it is ground on line ST at right angles to line QR.

Note. — The tool may be forged and ground to the proper angle with a universal tool grinder, see *Principles of Machine Work*, then hardened and tempered, reground and the filing omitted.

22. An outside Square threading holder and cutter is shown at A and B, Fig. 21. Holder C supports cutter D. Clamp E and bolt F fasten cutter to holder and hold different widths of

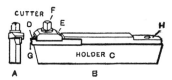

Fig. 21. — Outside Square Threading Holder and Cutter.

cutters. Cutter is ground upon end G only. Roughing and finishing cutters are used for coarse threads. End H is for left threads.

23. Method of setting outside Square threading tool. — Fasten tool A, Fig. 22, in tool-post lightly and adjust to height of dead center. Mount screw blank B on centers.

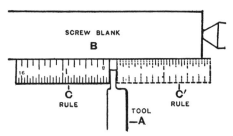

Fig. 22. — Setting Square Threading Tool with Steel Rule.

Place steel rule C against blank parallel to axis and rap tool until parallel to end of rule, testing from both sides as at C and C'. Fasten tool firmly.

24. An inside Square threading tool, Fig. 23, is used for cutting inside threads. The blade AB is inclined, shaped and sized to suit thread to be cut in the same manner as the outside Square threading tool, Fig. 18, p. **5**12. Inside Square threads

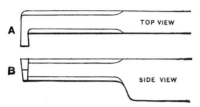

FIG. 23. — INSIDE SQUARE THREADING TOOL.

are cut also by inserting properly shaped cutters in boring tool holders. See No. 23, Chart, p. **3**09.

25. Method of setting inside Square threading tool. — Nut blank C, Fig. 24, is held in chuck D. End E is faced and hole bored to size. Place tool G in tool-post and adjust to height of dead center (see No. 2, p. **5**05) and fasten lightly

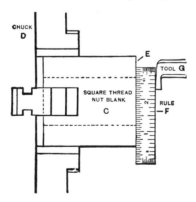

FIG. 24. — SETTING INSIDE SQUARE THREADING TOOL.

in approximate position. Hold rule F against end E and rap tool until blade of tool is parallel with rule. Then clamp firmly.

26. Roughing tool. — For Square threads, five pitch or coarser, use a roughing tool .010″ less in width, O, Fig. 18, than finishing tool.

27. Square thread tap 1¼″ diameter, 5 threads to 1″, Fig. 25, is used to tap both loose and fixed nuts. Loose nuts are usually rough threaded in the engine lathe with an inside Square thread roughing tool, Fig. 23, then finish threaded with one or more taps.

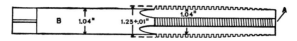

FIG. 25. — SQUARE THREAD TAP, 1¼ × 5.

For fixed nuts such as parts of machine frames, one or two roughing taps are used, followed with the finishing tap.

They are made without a leader as at A, Fig. 25, or with plain or threaded leaders, A, Fig. 33. See Alinement Drilling and Tapping, p. **5**38.

The diameter is made one-hundredth of an inch larger than the screw for clearance. The root diameter is the same as the screw. See pp. **5**11, **5**12.

Shank B is made about one-hundredth of an inch smaller than the bore of the nut.

These taps cannot be obtained commercially, but must be specially made.

28. To cut a Square thread screw, Fig. 26.

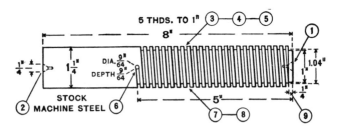

FIG. 26. — SCHEDULE DRAWING OF SQUARE THREAD SCREW

Specifications: Material, machine steel $\frac{7}{16}''$ large; weight, 3 lb. 8 oz.
For heat treatment see *Principles of Machine Work.*
Hardness, 14 to 16 (scleroscope).
True live center. Set dead center in approximate alinement.
High-speed steel or stellite cutting tools.
Time: Study drawing and schedule in advance, 5 min. —
Oil lathe, 4 min. — Prepare screw blank, 1 h. 5 min. —
Rough and finish thread and fit to nut, 1 h. 10 min. — (All tools
furnished.) —
Clean lathe, 3 min. — Total, 2 h. 27 min.

SCHEDULE OF OPERATIONS, MACHINES AND TOOLS

OPERATIONS.	MACHINES, SPEEDS, FEEDS.	TOOLS.
Center.	Centering machine.	$\frac{3}{32}''$ drill, 60° counter-sink, lard oil.
Rough square, (1), (2).	Engine lathe, 12″ to 16″. 2d or 3d speed, or 50 F.P.M. Hand or power feed.	Dog, side tool or holder and cutter 35° rake, calipers, rule.
Recenter.	Speed lathe, drill, 4th speed; countersink 3d speed.	
Finish square, (1), (2).	Engine lathe, 3d or 4th speed or 80 F.P.M. Hand or power feed.	
Rough turn $\frac{1}{64}''$ large, (3), one cut.	1st or 2d speed, or 60 F.P.M. Medium power feed — 80 to 1″.	Diamond-point tool or holder and cutter, 35° rake, calipers, rule.
Set dead center in accurate alinement to turn straight using this shaft or a trial piece the same length. See pp. 116, 117.	3d or 4th speed or 80 F.P.M. Fine power feed —140 to 1″.	Dog, copper, diamond-point tool or holder and cutter, 35° rake, micrometer.
Finish turn 1.25″ + .003″, (4).	3d or 4th speed, or 80 F.P.M. Fine power feed — 140 to 1″.	Copper under set screw of dog, diamond-point tool or holder and cutter, 35° rake, micrometer.
File 1.25″ + .001″, (5).	4th or 5th speed, or 175 F.P.M.	8″ or 10″ mill bastard file.
Polish 1.25″, (5).	Speed lathe, highest speed.	60 and 90 emery cloth, lard oil, polishing clamps.
Or rough turn .01″ large, and grind, after threading.	See Cylindrical Grinding Machine, p. 701.	

OPERATIONS.	MACHINES, SPEEDS, FEEDS.	TOOLS.
Drill hole, (6).	Speed lathe, drill chuck, 3d or 4th speed, or 1000 R.P.M.	Center punch, $\frac{9}{64}''$ straight shank twist drill, V center, depth gage, lard oil.
Grind square thread roughing tool.		
Set tool (see p. **5**14) and thread stop, arrange lathe for 5 threads. Pull belt downward to take up back-lash, loosen set screw of dog and adjust shaft until tool terminates in hole, (6). Tighten set screw. Rough thread to 1.04″ + .01″, (7), twenty cuts .005″ each. Depth of thread .100″.	Engine lathe. 1st speed, or 25 F.P.M.	Forged Square thread roughing tool, width .090″. See Fig. 18, or use holder and cutter, see Fig. 21, calipers, rule, lubricate freely with lard oil.
Set finishing tool to cut on both sides of groove by loosening dog, adjusting shaft and testing cut at end thread. Tighten set screw. Finish thread to 1.04″, (8), twenty cuts .005″, two cuts .002″, one cut .001″. Depth of thread .105″.	1st speed, or 25 F.P.M.	Forged finishing tool, width .100″ + .002″ for fit, 1″ micrometer, file, harden and temper and grind or use holder and cutter, calipers, rule, lard oil.
Turn off thick end thread, (9), and smooth thread with file.	2d or 3d speed, or 50 F.P.M. Hand feed.	Diamond-point and side tools, or holder and cutter, 8″ or 10″ mill bastard file.
File top of threads to remove burr.	4th or 5th speed, or 175 F.P.M.	8″ or 10″ mill bastard file.
File sides of thread slightly if needed to make the fit easier. (Preferably with file and speed reversed.)	1st speed, or 10 F.P.M.	5″ or 6″ warding bastard file, two safe edges.

Attention. — Terminate each cut as follows: stop lathe when tool is $\frac{1}{2}$ or $\frac{1}{3}$ revolution from hole, then carefully pull belt to continue cut almost to hole and end the cut by moving tail of dog in slot of face plate.

Note. — Roughing tool may be removed, ground, and reset if necessary, but not the finishing tool.

29. To make a Square thread nut, Fig. 27.

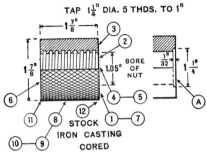

Fig. 27. — Schedule Drawing of Square Thread Nut.

Specifications: Preparing nut blank. Rough threading. Tapping.
Material, iron casting, cored; weight, 1 lb. 6 oz.
Hardness, 29 to 31 (scleroscope).
High-speed steel or stellite cutting tools.
Time: Study drawing and schedule in advance, 5 min. —
Oil lathe, 4 min. — Bore, thread, and tap nut, 40 min. —
Square, turn, and nurl nut, 28 min. — (All tools furnished.)
Clean lathe, 3 min. — Total, 1 h. 20 min.

SCHEDULE OF OPERATIONS, MACHINES AND TOOLS

Operations.	Machines, Speeds, Feeds.	Tools.
Mount in chuck, true up and clamp hard in chuck.	Engine lathe, 12″ to 16″. 3d speed, or 200 R.P.M.	Independent chuck, chalk.
Rough square, (1), one or two cuts. Feed inward.	2d or 3d speed, or 40 F.P.M. Hand or power feed.	Round-nose tool, or holder and cutter, 15° rake.
Rough bore hole to about 1.03″, (2), two or three cuts.	1st or 2d speed, or 40 F.P.M. Medium power feed — 80 to 1″.	Boring tool, see p. **5**04. Inside calipers, rule.
Finish bore hole, (2), two or three cuts.	3d speed, or 60 F.P.M. fine power feed — 140 to 1″.	
Or omit boring, bevel corner of hole and drill to size.	2d or 3d speed, or 60 F.P.M.	3 or 4-groove high-speed steel twist drill (1.05″). See p. **4**10 and *Principles of Machine Work*.

SCHEDULE OF OPERATIONS, MACHINES AND TOOLS
Continued

Operations.	Machines, Speeds, Feeds.	Tools.
Set inside Square thread tool (see Fig. 24), and cut recess for improvised gage $\frac{1}{32}''$ $\times$ $1\frac{1}{4}''$, (3). See (A), Fig. 27. Arrange lathe for 5 threads with thread stop reversed.	1st speed or 30 F.P.M. Hand feed.	Forged Square thread tool, width .090″, or holder and cutter, inside calipers, rule.
Rough thread to $1\frac{1}{4}''$ the diameter of gage A, (4), fifteen cuts, .006″ each, five cuts, .002″ each. Depth of thread, .100″.	1st speed, or 30 F.P.M.	Without oil.
Start tap in lathe, pull belt downward and follow with dead center. Be sure that tap follows thread or it may ream. Remove nut and tap to reaming stand or vise, and finish tapping, (5).	Reaming stand or vise and grooved wooden jaws.	$1\frac{1}{4}''\times5$ Square thread, tap and tap wrench, lubricate tap freely with lard oil.
Mount nut on nut mandrel and rough square, (6), one or two cuts.	2d or 3d speed, or 40 F.P.M.	$1\frac{1}{4}''\times5$ Square thread nut mandrel, dog, round-nose tool, or holder and cutter, calipers, rule.
Finish square, (6), one or two cuts.	3d or 4th speed, or 60 F.P.M.	Facing tool or holder and cutter.
Reverse nut and square to length to remove recess, (7), two or three cuts.	2d or 3d speed or 60 F.P.M.	Round-nose tool or holder and cutter, and facing tool, calipers, rule.
Rough turn, (8), one or two cuts.	1st or 2d speed, or 35 F.P.M. Medium power feed — 80 to 1″.	Diamond-point tool or holder and cutter, 15° rake, calipers, rule.
Finish turn, (8), one cut.	2d or 3d speed, or 50 F.P.M.	Diamond-point or round-nose tool, or holder and cutter, calipers, rule.
File, (9).	4th speed, or 175 F.P.M.	8″ or 10″ mill bastard file.
Nurl, two to four times, (10). See Machine Nurling, p. 636.	1st or 2d speed, or 35 F.P.M. Medium power feed — 80 to 1″.	Machine nurling tool, medium pitch, oil.
File corners slightly to remove burr, (11), (12).	4th speed or 175 F.P.M.	8″ or 10″ mill bastard file.

See *Attention* and *Note*, p. 521.

Attention. — In the absence of a tap an inside tool may be made one-half pitch + .001″ to finish the thread, but the thread will not be as smooth or the fit of screw and nut as good.

Note. — The nut in Fig. 27 is nurled for convenience in handling as a problem. For practical styles of nuts to transmit motion, see p. 538.

30. To fit screw to nut, Figs. 26 and 27.

SCHEDULE OF OPERATIONS

1. Ascertain if thread binds on top or bottom by testing with calipers and comparing with tap, Fig. 25; if so, file top of threads on screw or cut thread deeper.

2. Hold nut in grooved wooden jaws in vise. Oil screw, grasp dog with both hands and force in with hand pressure.

3. If necessary file sides of threads with a warding file.

4. After screw is fitted to nut, polish top of thread with 90 emery cloth, polishing clamps and lard oil.

Attention. — To file sides of a right thread, preferably, run lathe backward at a slow speed, reverse file and file toward footstock.

ACME STANDARD OR 29° THREADS

31. Acme Standard or 29° threads, Fig. 28, right or left, are used for screws to transmit motion, as on lead screws, feed

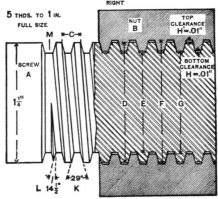

Fig. 28. — Sectional View of Acme Standard or 29° Thread Screw and Nut.

screws, elevating screws, valve stems, presses, rock drills, etc., and they are displacing Square threads for other purposes because they can be more readily cut in the lathe, and also successfully and rapidly cut with dies, with bolt cutters and turret lathes, and milled with thread milling machines.

The depth of the thread is equal to one-half the pitch plus .01″ for clearance. The fit is on the sides with clearance on top and bottom.

NAMES OF PRINCIPAL PARTS OF 29° THREAD

C — Pitch.
D — Diameter, standard.
E — Root diameter (which is also root diameter of tap).
F — Diameter at bottom of thread on nut (which is also diameter of tap).
G — Bore of nut.

H and H' — Clearances, same for all pitches. (H obtained by cutting thread .01″ deeper than $\frac{1}{2} P$; H', by making tap .02″ larger than diameter of screw.)
K — Included angle 29°.
L — Side angle 14½°.

32. To obtain parts of thread. —
Width of point of tool for screw or tap thread $=$

$$\frac{.3707}{\text{No. threads per inch}} - .0052.$$

Width of point of screw or nut thread $=$

$$\frac{.3707}{\text{No. threads per inch}}$$

Diameter of tap $=$ diameter of screw $+$ .020.
Diameter of screw $=$ any
Diameter of tap or screw at root of thread $=$ diameter of

$$\text{screw} - \left(\frac{1}{\text{No. threads per inch}} + .020'' \right).$$

Depth of thread $= \dfrac{1}{2 \times \text{No. threads per inch}} + .010.$

Diameter to bore nut $=$ diameter of screw $-$

$$\frac{1}{\text{No. threads per inch}}.$$

Clearance $=$ .01″ top and bottom of thread.

33. Table of Thread Parts for Acme Standard or 29° Thread.

Pitch.	No. of Threads per Inch.	Depth of Thread.	Width at Top of Thread.	Width at Bottom of Thr'd.	Space at Top of Thread.	Thickness at Root of Thread.
2	1/2	1.010	.7414	.7362	1.2586	1.2637
1 7/8	8/15	.9475	.6950	.6897	1.1799	1.1850
1 3/4	4/7	.8850	.6487	.6435	1.1012	1.1064
1 5/8	8/13	.8225	.6025	.5973	1.0226	1.0277
1 1/2	2/3	.7600	.5560	.5508	.9439	.9491
1 7/16	16/23	.7287	.5329	.5277	.9046	.9097
1 3/8	8/11	.6975	.5097	.5045	.8652	.8704
1 5/16	16/21	.6662	.4865	.4813	.8259	.8311
1 1/4	4/5	.635	.4633	.4581	.7866	.7918
1 3/16	16/19	.6037	.4402	.4350	.7472	.7525
1 1/8	8/9	.5725	.4170	.4118	.7079	.7131
1 1/16	16/17	.5412	.3938	.3886	.6686	.6739
1	1	.510	.3707	.3655	.6293	.6345
15/16	1 1/15	.4787	.3476	.3424	.5898	.5950
7/8	1 1/7	.4475	.3243	.3191	.5506	.5558
13/16	1 3/13	.4162	.3012	.2960	.5112	.5164
3/4	1 1/3	.385	.2780	.2728	.4720	.4772
11/16	1 5/11	.3537	.2548	.2496	.4327	.4379
2/3	1 1/2	.3433	.2471	.2419	.4194	.4246
5/8	1 3/5	.3225	.2316	.2264	.3934	.3986
9/16	1 7/9	.2912	.2085	.2033	.3539	.3591
1/2	2	.260	.1853	.1801	.3147	.3199
7/16	2 2/7	.2287	.1622	.1570	.2752	.2804
2/5	2 1/2	.210	.1482	.1430	.2518	.2570
3/8	2 2/3	.1975	1390	.1338	.2359	.2411
1/3	3	.1766	.1235	.1183	.2098	.2150
5/16	3 1/5	.1662	.1158	.1106	.1966	.2018
2/7	3 1/2	.1528	.1059	.1007	.1797	.1849
1/4	4	.1350	.0927	.0875	.1573	.1625
2/9	4 1/2	.1211	.0824	.0772	.1398	.1450
1/5	5	.110	.0741	.0689	.1259	.1311
3/16	5 1/3	.1037	.0695	.0643	.1179	.1232
1/6	6	.0933	.0617	.0565	.1049	.1101
1/7	7	.0814	.0530	.0478	.0899	.0951
1/8	8	.0725	.0463	.0411	.0787	.0839
1/9	9	.0655	.0413	.0361	.0699	.0751
1/10	10	.060	.0371	.0319	0629	.0681
1/16	16	.0412	.0232	.0180	.0392	.0444

34. Table of Acme Standard or 29° threads. — While any
pitch may be assumed, it is best to use whole numbers of
threads per inch as near as possible to three-quarters of the
United States Standard. See p. **3**23.

Diameter.	Threads Per Inch.	Diameter.	Threads Per Inch.
½″	10	1″	6
⅝″	9	1¼″	5
¾″	8	1½″	4
⅞″	7	2″	3

35. To file or grind Acme Standard or 29° threading tool, Fig.
29. — *A* is top face, *B* and *C* are side and end views. For
method of finding inclination of thread, see Square Threading
Tool, pp. **5**13, **5**14.

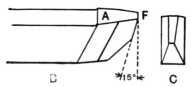

FIG. 29. — ACME STANDARD OR 29° THREADING TOOL.

SCHEDULE OF OPERATIONS

1. File bottom and end of tool.
2. File sides to fit gage *D* as at *E*, Fig. 30.
3. Harden and temper to a straw color.

4. Grind end until point *F*, Fig. 29, will fit notch in gage as at *G*, Fig. 30, the desired pitch,
5. Grind top face, *A*, Fig. 29, slightly.

Attention. — For fine pitches, use same tool for roughing and
finishing. For 5 pitch or coarser, rough with Square threading tool
.01″ narrower than point of 29° tool. In case of very coarse pitches,
cut a square groove, then with right and left side tool cut down sides
of thread, after which use finishing tool of desired shape. A compound rest is often used for coarse pitches.

Note.— The tool may be forged and ground to the proper angle with
a universal tool grinder (see *Principles of Machine Work*), then hardened and tempered, reground, and the filing omitted. The thread is also
cut with 29° threading holder and cutter. See No. 16, Chart, p. **3**09.

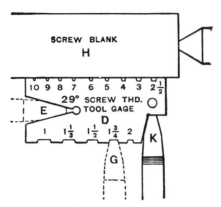

FIG. 30. — SETTING TOOL TO CUT ACME
STANDARD OR 29° THREAD.

36. Method of setting outside Acme Standard or 29° thread-ing tool. — Fasten tool K, Fig. 30, in tool post lightly and ad-just to height of dead center.   Mount screw blank H on centers. Place gage D against screw blank H parallel to axis and rap tool until angle of tool fits angle of gage.   Then fasten tool firmly.

37. An inside Acme Standard or 29° threading tool. — Fig. 31 is used for cutting inside threads.   The blade AB is in-clined, shaped and ground to suit thread to be cut in the same

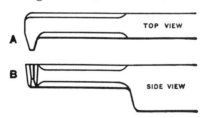

FIG. 31. — INSIDE ACME STANDARD OR 29°
THREADING TOOL.

manner as the outside Acme Standard or 29° threading tool, Fig. 29.

Inside 29° threads are cut also by inserting properly shaped cutters in boring tool holders.   See No. 24, Chart, p. 309.

38. Method of setting inside Acme Standard or 29° threading tool. — Nut blank *C*, Fig. 32, is held in chuck *D*. End *E* is faced and hole bored to size. Place tool *F* in tool post

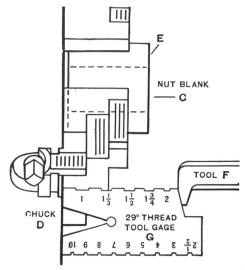

Fig. 32. — Setting Inside Acme Standard 29° Threading Tool.

and adjust to height of dead center (see No. 2, p. 505) and fasten lightly in approximate position. Hold gage *G* against face of chuck *D* and rap tool until angle of tool fits angle of gage. Then fasten tool securely.

39. Acme Standard or 29° thread tap $1\frac{1}{4}''$ in diameter, 5 threads to 1, Fig. 33, is used to tap both loose and fixed nuts.

Fig. 33.— Acme Standard or 29° Thread Tap $1\frac{1}{4} \times 5$.

Loose nuts are usually rough threaded in the engine lathe with an inside threading tool, Fig. 32, to about $\frac{3}{4}$ of a full thread, then finish threaded with a tap.

For fixed nuts, such as parts of machine frames, one or two roughing taps are used, followed by the finishing tap, preferably of type shown at H, Fig. 15. These taps are made with leaders as at A, Fig. 33, or without leaders, A, Fig. 25. See Alinement Drilling and Tapping, p. **538**. The diameter is made two-hundredths of an inch larger than the screw for clearance. See p. **522**. The root diameter of tap is the same as screw.

Leader A and shank B are two-thousandths of an inch smaller than bore of nut, and the leader may be used as a gage to test the bore of nut. These taps cannot be obtained commercially but must be specially made.

40. To cut Acme Standard or 29° thread screw, Fig. 34.

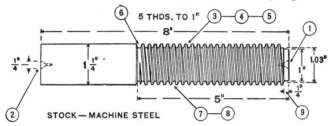

FIG. 34. — SCHEDULE DRAWING OF ACME STANDARD OR 29° THREAD SCREW.

Specifications: Preparing screw blank. Threading. Material, machine steel $\frac{1}{16}''$ large; weight, 3 lb. 8 oz.

For heat treatment see *Principles of Machine Work.*

Hardness, 14 to 16 (scleroscope).

True live center. Set dead center in approximate alinement.

High-speed steel or stellite cutting tools.

Time: Study drawing and schedule in advance, 5 min. — Oil lathe, 4 min. — Prepare screw blank, 65 min. — Rough and finish thread, and fit to nut, 1 h. 10 min. — (All tools furnished.) Clean lathe, 3 min. — Total, 2 h. 27 min.

SCHEDULE OF OPERATIONS, MACHINES AND TOOLS

OPERATIONS.	MACHINES, SPEEDS, FEEDS.	TOOLS.
Center.	Centering machine.	$\frac{3}{32}''$ drill, 60° countersink, lard oil.
Rough square, (1), (2).	Engine lathe, 12″ to 16″. 2d or 3d speed, or 50 F.P.M. Hand or power feed.	Dog, side tool or holder and cutter 35° rake, calipers, rule.

SCHEDULE OF OPERATIONS, MACHINES AND TOOLS
Continued

Operations.	Machines, Speeds, Feeds.	Tools.
Recenter.	Speed lathe, drill, 4th speed; countersink, 3d speed.	
Finish square, (1), (2).	Engine lathe, 3d or 4th speed, or 80 F.P.M. Hand or power feed.	
Rough turn $\frac{1}{64}''$ large, (3), one cut.	1st or 2d speed, or 60 F.P.M. Medium power feed — 80 to 1″.	Diamond-point tool or holder and cutter, 35° rake, calipers, rule.
Set dead center in accurate alinement to turn straight using this shaft or a trial piece the same length. See pp. 116, 117.	3d or 4th speed or 80 F.P.M. Fine power feed — 140 to 1″.	Dog, copper, diamond-point tool or holder and cutter, 35° rake, micrometer.
Finish turn 1.25″ + .003″, (4).	3d or 4th speed, or 80 F.P.M. Fine power feed — 140 to 1″.	Copper under set screw of dog, diamond-point tool or holder and cutter, 35° rake, micrometer.
File 1.25″ + .001″, (5).	4th or 5th speed, or 175 F.P.M.	8″ or 10″ mill bastard file.
Polish 1.25″, (5).	Speed lathe, highest speed.	60 and 90 emery cloth, lard oil, polishing clamps. 29° grooving tool, lard oil, calipers, rule.
Cut groove, (6), to root diameter of thread, 1.03″.	Engine lathe, 1st or 2d speed, or 25 F.P.M. Hand feed.	
Grind Square thread 29° roughing tool.		Forged Square thread (29° roughing tool) — width .060″, see Fig. 18, or holder and cutter, see Fig. 21.
Set tool (see p. 514) and thread stop, arrange lathe for 5 threads.		
Rough thread to 1.03″ + .02″, (7), twenty cuts .005″ each. Depth of thread, .100″.	1st speed, or 30 F.P.M.	Calipers, rule; lubricate freely with lard oil.
Set finishing tool to cut on both sides of groove by taking up back-lash, loosening dog, adjusting shaft and testing cut at end thread.		Forged 29° finishing tool, ground to fit angle and notch 5 on gage, Fig. 30, or use holder and cutter, calipers, rule, lard oil.
Finish thread, (8), twenty cuts of .005″ each, two cuts of .002″ each. Then take 1 cut .001″, clean, oil, and test and repeat until screw fits nut. Depth of thread .110″.	1st speed, or 30 F.P.M.	

SCHEDULE OF OPERATIONS, MACHINES AND TOOLS
Concluded

OPERATIONS.	MACHINES, SPEEDS, FEEDS.	TOOLS.
Turn off end of thread, (9), and smooth off thread with file.	2d or 3d speed, or 50 F.P.M. Hand feed.	Diamond-point and side tools, or holder and cutter, 8″ or 10″ mill bastard file.
File top of threads to remove burr.	4th or 5th speed, or 175 F.P.M.	8″ or 10″ mill bastard file.
File sides of thread slightly if needed to make the fit easier. (Preferably with file and speed reversed.)	1st speed, or 10 F.P.M.	5″ or 6″ warding bastard file, two safe edges.
Polish tops of threads.	Speed lathe, highest speed.	90 emery cloth, polishing clamps, lard oil.

Attention. — Terminate each cut as follows: stop lathe when tool is ½ or ⅓ of a revolution from groove, then carefully pull belt to continue cut to groove.

A hole instead of a groove is sometimes used in which to terminate the cut. See K, Fig. 17.

For fine pitches a groove, or hole, is sometimes omitted and a tapering termination used the same as United States Standard and Sharp V threads. See pp. 318, 319, 330, 331.

Warning. — Roughing tool may be removed, ground, and reset, if necessary, but it is best not to remove the finishing tool until thread is completed.

Important. — A 29° thread nut must be made first in order to fit the screw to it.

41. To make an Acme Standard or 29° thread nut, Fig. 35.

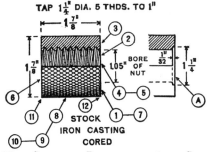

FIG. 35. — SCHEDULE DRAWING OF ACME STANDARD OR 29° THREAD NUT.

Specifications: Preparing nut blank. Rough threading. Tapping.
Material, iron casting, cored; weight, 1 lb. 5 oz.
Hardness, 29 to 31 (scleroscope).
High-speed steel or stellite cutting tools.
Time: Study drawing and schedule in advance, 5 min. —
Oil lathe, 4 min. — Bore, thread, and tap nut, 40 min. — Square,
turn, and nurl nut, 28 min. — (All tools furnished.)   Clean lathe, 3
min. — Total, 1 h. 20 min.

SCHEDULE OF OPERATIONS, MACHINES AND TOOLS

OPERATIONS.	MACHINES, SPEEDS, FEEDS.	TOOLS.
Mount in chuck, true up, and clamp hard in chuck.	Engine lathe 12″ to 16″. 3d speed, or 200 R.P.M.	Independent chuck, chalk.
Rough square, (1), one or two cuts. Feed inward.	2d or 3d speed, or 40 F.P.M. Hand or power feed.	Round-nose tool or holder and cutter, 15° rake.
Rough bore hole to about 1.03″, (2), two or three cuts.	1st or 2d speed, or 40 F.P.M. Medium power feed — 80 to 1″.	Boring tool, p. **5**04. Inside calipers, rule.
Finish bore hole 1.05″, (2), two or three cuts.	3d speed, or 60 F.P.M. Fine power feed — 140 to 1″.	
Or omit boring, bevel corner of hole and drill to size.	2d or 3d speed, or 60 F.P.M.	3 or 4-groove high speed steel twist drill (1.05″). See p. **4**10 and *Principles of Machine Work*.
Set inside 29° thread tool, Fig. 32, cut recess for improvised gage $\frac{1}{32}″ \times 1\frac{1}{4}″$, (3). See *A*, Fig. 35. Arrange lathe for 5 threads with thread stop reversed.	1st speed, or 30 F.P.M. Hand feed.	Forged 29° thread tool, or holder and cutter, inside calipers, rule.
Rough thread to $1\frac{1}{4}″$ the diameter of gage A, (4), fifteen cuts .006″ each, five cuts .002″ each. Depth of thread .100″.	1st speed, or 30 F.P.M.	Without oil.
Start tap in lathe, pull belt downward and follow with dead center. Be sure that tap follows thread or it may ream out thread. Remove nut and tap to reaming stand or vise, and finish tapping, (5).	Reaming stand, or vise and grooved wooden jaws.	$1\frac{1}{4}″ \times 5$, 29° thread tap and tap wrench, lubricate tap freely with lard oil.

SCHEDULE OF OPERATIONS, MACHINES AND TOOLS
Concluded

OPERATIONS.	MACHINES, SPEEDS, FEEDS.	TOOLS.
Mount nut on nut mandrel and rough square, (6), one or two cuts.	2d or 3d speed, or 40 F.P.M.	1¼″ × 5, 29° thread nut mandrel, dog, round-nose tool or holder and cutter, calipers, rule.
Finish square, (6), one or two cuts.	3d or 4th speed, or 60 F.P.M.	Facing tool, or holder and cutter.
Reverse nut and square to length to remove recess, (7), two or three cuts.	2d or 3d speed, or 60 F.P.M.	Round-nose tool, or holder and cutter and facing tool.
Rough turn, (8), one or two cuts.	1st or 2d speed, or 35 F.P.M. Medium power feed.	Diamond-point tool or holder and cutter, 15° rake, calipers, rule.
Finish turn, (8), one cut. File, (9).	2d or 3d speed, or 50 F.P.M. 4th speed, or 175 F.P.M.	Diamond-point or round-nose tool, or holder and cutter, calipers, rule. 8″ or 10″ mill bastard file.
Nurl, (10), two to four times. See Machine Nurling, p. 636.	1st or 2d speed, or 35 F.P.M. Medium power feed — 80 to 1″.	Machine nurling tool, medium pitch, lard oil.
File corners slightly to remove burr, (11), (12).	4th speed, or 175 F.P.M.	8″ or 10″ mill bastard file.

Attention. — In the absence of a tap the inside 29° thread tool may be used to finish the thread but the thread will not be as smooth or the fit of screw and nut as good.

Note. — The nut in Fig. 35 is nurled for convenience in handling as a problem. For practical styles of nuts to transmit motion, see p. 538.

MULTIPLE THREADS

42. Multiple-threaded screws, such as double and triple threads, etc., in Square, 29°, and other forms of threads, are used in cases where a quick lead is required, but a deep thread is not desirable.

43. To cut double Square thread, Fig. 36, 8 threads to 1″, pitch ⅛″.

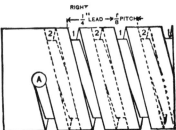

FIG. 36. — DOUBLE SQUARE THREAD SCREW.

SCHEDULE OF OPERATIONS

1. Gear lathe to cut 4 threads per inch. Fasten dog. Mount work on centers and with threading tool trace a line for groove 1.

2. At end of line center punch and drill finishing hole A to terminate groove 1. See pp. **516–518**.

3. Cut groove 1 to diameter.

4. If ratio of stud spindle and lathe spindle is 1 to 1, and stud gear has an even number of teeth, mark with chalk a tooth on stud gear and the corresponding space in idler gear. Then mark a tooth on stud gear diametrically opposite, which is determined by counting half the teeth in the stud gear.

Disengage idler gear from stud gear, rotate lathe spindle and bring gears into mesh as indicated by chalk marks on the teeth of the gears.

5. Now, with threading tool trace a line for groove 2 and on this line opposite A, center punch and drill finishing hole A' (not shown) to terminate groove 2.

6. Cut groove 2 to diameter.

Attention. — To cut a triple thread, the lathe spindle is rotated one-third of a revolution to cut the second thread, and another third to cut the third thread. If gear on stud is not divisible by number of threads to be cut, select change gears that have a stud gear that is.

Important. — Special face plates with multiple equidistant slots (index milled) are convenient for cutting multiple threads, as the tail of the dog can be more readily shifted than the gears after each thread is cut.

44. Width and inclination of tool for multiple threads. — The width or shape of tools for multiple threads is governed by pitch of the screw; and the inclination by the lead of the screw.

45. Multiple-thread taps are similar to those for square and 29° thread nuts. See pp. **516, 526, 527.** The Sharp V and 29° forms of this thread can be cut with dies.

BRASS FINISHING

46. To turn brass in the engine lathe. — Brass, also known as composition, ranging in hardness from soft yellow to hard brass or bronze, is generally turned with tools having little or no rake. High speeds and fine feeds are used and the tools must be kept sharp by frequent grinding to obtain the best results. See To Machine Bronze, p. 419. For Cutting Speed, see p. 145.

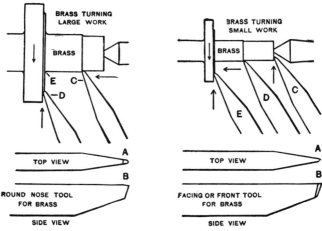

FIG. 37. — BRASS TURNING WITH ROUND-NOSE TOOL.

FIG. 38. — BRASS TURNING WITH FRONT TOOL.

47. The round-nose tool, as at *A* and *B*, Fig. 37, is generally used for large brass work. It may be used to turn the diameter as at *C*, the shoulder, as at *D*, and the round corner or fillet as at *E*, without change of position in the tool post.

48. Use front tool, Fig. 38, with point *A*, *B* ground at angle from 50° to 60° for squaring, turning, or facing, without change, as at *C*, *D*, and *E*. If the work is rigid, cut inward or outward; if slender, cut inward only   This tool may be ground to fit thread gage, and used as a threading tool for United States Standard or Sharp V threads.

Attention. — As bronze and brass will stretch, it is neces sary, when fitting a steel screw to a bronze or brass nut, to force the screw in hard at first to avoid a loose fit.

Reamed and drilled holes in brass or bronze are usually smaller than holes made with the same tools in cast iron on account of the resilient and adhesive quality of these materials.

Taps, dies, reamers, and files that are used on steel or iron may be used on brass, but it is best to have separate tools, for such tools never work as well on brass after they are used on steel or iron.

Brass in general is machined dry, but to thread rolled brass, as rod and tubing, with a die, it is usually necessary to use lard oil to prevent the chips from clinging to the threads of the die and stripping the threads on the work.

Monel metal is a natural nickel-copper alloy made directly from the ore, 67 per cent nickel, 28 per cent copper and 5 per cent of other metals.

It takes the same finish as pure nickel and is used for pump rods, valve stems, boat fittings, dairy and tannery machinery and other purposes where strength and non-corrodibility are required. It can be machined, forged, soldered, brazed and welded by oxy-acetylene and electric processes.

To machine, use tools with 25° to 30° rake, dry or with a lubricant of lard oil, borax and aquadag.

49. To make brass binding post, Fig. 39.

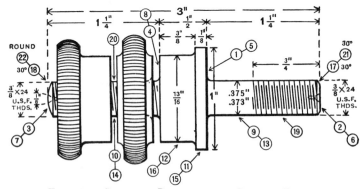

Fig. 39. — Schedule Drawing of a Binding Post.

Specifications: Material, composition or brass castings $\frac{1}{16}''$ large; weight, 10 oz. Hardness, 10 to 12 (scleroscope). True live center. Set dead center in approximate alinement. Carbon-steel cutting tools. **Time:** Study drawing and schedule in advance, 10 min. — Oil lathe, 3 min. — Make post, 38 min. — Chuck and turn nuts, 37 min. —Nurl nuts and thread post, 40 min. — Assemble, finish, polish, and lacquer, 34 min. — Clean lathes, 6 min. — Total, 2 h. 48 min.

SCHEDULE OF OPERATIONS, MACHINES AND TOOLS

Operation.	Machines, Speeds, Feeds.	Tools.
File ends flat and center to $\frac{1}{8}''$ diameter.	Vise, centering machine, drill, 3500 R.P.M., countersink, 1700 R.P.M.	File, $\frac{1}{16}''$ drill, 60° countersink.
Rough square, (1), (2), (3), (4). Take as little as possible off 1. See pp. 533, 534.	Engine lathe, 12″ to 16″, 4th or 5th speed, or 90 F.P.M. Hand feed.	Round-nose tool $\frac{1}{8}''$ wide, (1), (4), front tool, (2), (3), calipers, rule.
Recenter to $\frac{1}{8}''$ diameter.	Speed lathe, 3d or 4th speed.	60° countersink.
Finish square, (5), (6), (7), (8).	Engine lathe, 4th or 5th speed, or 150 F.P.M. Hand feed.	Round-nose tool, (5), (8), front tool, (6), (7), calipers, rule.
Rough turn diameters (9) and (10) to $\frac{3}{8}'' + \frac{1}{64}''$, (11) to $1'' + \frac{1}{64}''$, and (12) to $\frac{13}{16}'' + \frac{1}{64}''$, two cuts each.	4th or 5th speed, or 90 F. P. M. Fine power feed — 140 to 1″.	Copper under set screw of dog, round-nose tool, calipers, rule.
Finish turn, (13), (14), to fit $\frac{3}{8}''$ gage (easy fit) or use micrometer, tool finish, filing not necessary. Finish turn, (15), (16).	4th or 5th speed, or 150 F.P.M. Fine power feed — 140 to 1″.	Calipers, rule, $\frac{3}{8}''$ mandrel, $\frac{3}{8}''$ ring gage, or micrometer.
Chamfer ends 30° to depth of thread, (17), (18).	Hand feed.	Front tool.
Thread in engine lathe to fit nuts, (19), (20), or	2d or 3d speed, or 25 F.P.M. Arrange change gears for 24 threads.	U.S.F.-threading tool.

Thread with ⅜″ × 24 U.S.F. die, (19), (20). Start in lathe. Pull belt by hand, and finish in vise. See p. 542. If desired, post and nuts may be threaded ⅜″ × 16, United States Standard thread.	Speed lathe — 8″ to 12″. Vise.	3″ universal chuck, ⅜″ × 24 U.S.F. thread die and die stock, special vise jaws. See Automobile Screws, pp. 354, 355.
Rechamfer, (21).	Engine lathe, 4th or 5th speed, or 90 F.P.M.	Clamp nut, front tool.
Round end with forming tool, or	. . . .	Forming tool.
By hand tool, (22).	Speed lathe, 4th or 5th speed, or 500 F.P.M.	Planisher or graver.

50. To make brass nurled thumb nuts, Fig. 40.

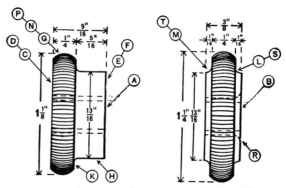

Fig. 40. — Schedule Drawing of Nurled Nuts.

SCHEDULE OF OPERATIONS, MACHINES AND TOOLS

Operations.	Machines, Speeds, Feeds.	Tools.
Chuck, drill, and tap nuts, (A), (B), see pp. 540–542.	Speed lathe, 4th or 5th speed, or 1500 R.P.M. Vise.	3″ universal chuck, planisher or graver. P, or 21/64″ drill, dog, ⅜″ × 24 U.S.F.-thread tap, tap wrench.
Screw on nut mandrel, mount on centers, rough square, (C). Finish square, (D).	Engine lathe, 4th or 5th speed, or 90 to 150 F.P.M. Hand feed.	Dog, ⅜″ × 24 U.S.F. thread nut mandrel, front tool.

OPERATIONS.	MACHINES, SPEEDS, FEEDS.	TOOLS.
Reverse on mandrel, rough square, (E). Finish square 'to thickness, (F).	....	Front tool, calipers, rule.
Rough turn to diameter, 1⅛", (G).	Engine lathe, 4th or 5th speed, or 90 to 150 F.P.M. Fine power feed — 140 to 1".	Round-nose tool, calipers, rule.
Rough and finish turn to diameter, (H). Rough and finish square to thickness, (K).	....	Round-nose tool, calipers, rule.
Square and turn nut (B) in same order as (A), except leave diameters (L) (M) $\frac{1}{64}$" large.		
Round nut (N) to fit concave single nurl. See p. 635.	Speed lathe, 4th or 5th speed, or 500 F.P.M.	Planisher.
Nurl (P), also round and nurl nut (B). See p. 635.	2d or 3d speed or 300 F.P.M.	Hand nurling tool.
Hold nut mandrel in chuck, screw on nut (B), and recess (R) to fit against 4 on binding post, Fig. 39.	4th or 5th speed, or 500 F.P.M.	Chuck, nut mandrel with thick collar, planisher or graver.
Screw nuts on binding post, mount on centers and hand turn, (S), (T), to correct diameter of post and nut (A).	4th or 5th speed, or 500 F.P.M.	Clamp nut, round-nose hand tool.
File.	3d or 4th speed, or 300 F.P.M.	5" half-round smooth file.
Stamp name or initials at end, (C).	Place on metal block.	Steel name stamp, hammer.
Polish and lacquer posts and nuts all over, except threads and nurling. See pp. 432, 440.	Highest speed, or 6000 F.P.M. See *Principles of Machine Work.*	120 and flour emery cloth, crocus cloth (an oxide of iron), rouge (a red iron peroxide), lacquer, brush.

Attention. — The cutting speeds in schedule are for soft brass or composition castings; but if castings are hard, reduce the cutting **speed.**

ALINEMENT DRILLING AND TAPPING

51. Fixed nuts, Fig. 41, which receive screws to transmit motion are made in various forms, as bushing nut *A*, which is threaded and forced into a bored hole and used to

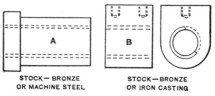

STOCK — BRONZE   STOCK — BRONZE
OR MACHINE STEEL   OR IRON CASTING

Fig. 41. — Bushing and Bracket Nuts.

receive vertical feed screw on a milling machine, or bracket nut *B*, used to receive cross-slide screw on a lathe, or bracket nut *C*, Fig. 42, for cross-slide screw on a milling machine.

Bracket nuts are drilled and tapped in alinement with scraped slide and bearing of screw, and the work may be done with a regular jig, or with a part of the machine itself used as an improvised jig, as knee *B*, Fig. 42.

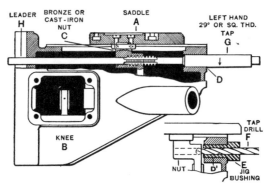

Fig. 42. — Alinement Drilling and Tapping, Milling Machine Cross-Feed Screw Nut. Hole in Knee Used as a Jig.

52. To drill and tap cross-feed screw nut in axial alinement in milling machine saddle and knee, Fig. 42. Improvised jig, 29° or Square thread tap.

SCHEDULE OF OPERATIONS

Place saddle *A* on knee *B*. Fit and bolt nut *C* to saddle and move close to hole *D*.

Clamp saddle and knee to angle plate at vertical drilling machine or to table of horizontal drilling machine and aline with spindle.

In hole *D′* insert jig bushing *E*, and drill nut with tap drill *F*.

Move saddle and nut away from hole *D*, and clamp saddle. Remove bushing, insert tap *G*, and tap hole by hand. Three or four taps of increasing diameters are used.

Attention. — Taps without leaders are used, but taps with leaders, either threaded or plain, as at *H*, are preferred for accuracy.

53. **To make bronze bushing, Fig. 43.**

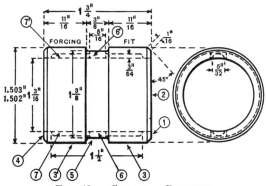

FIG. 43. — SCHEDULE DRAWING.

Specifications: Material, bronze casting $\frac{1}{16}''$ large, cored; weight, 10 oz.

Hardness, 11 to 14 (scleroscope).

High-speed steel, or stellite cutting tools.

Time: 1 h. 30 min.

SCHEDULE OF OPERATIONS, MACHINES AND TOOLS

OPERATIONS.	MACHINES, SPEEDS, FEEDS.	TOOLS.
Snag casting. Set dead center in approximate alinement. **Mount casting in chuck, true up and clamp firmly.**	Engine lathe, 12″ to 16″. 3d speed.	Independent chuck, chalk.

Rough and finish square end (**1**), two or three cuts.	4th or 5th speed, or 150 F.P.M. Hand feed.	Holder and cutter no rake, or front tool.
Rough bore hole to $1\frac{11}{64}''$, (**2**), two or three cuts. See p. **5**04.	Medium power feed — 80 to 1''.	Boring tool no rake, inside calipers, rule.
Finish bore hole to 1.1825'', (**2**), one cut.	Fine power feed — 140 to 1''.	2'' inside micrometer (or calipers and rule).
Finish ream hole by hand with hand reamer in lathe (**2**). See p. **4**16.	....	$1\frac{3}{16}''$ hand reamer, reamer wrench.
Oil mandrel and press lightly into hole in direction reamed.	Mandrel press.	$1\frac{3}{16}''$ mandrel, oil.
True live center. Set dead center in accurate alinement (see pp. **1**16, **1**17), using a shaft same length as mandrel.	Engine lathe.	
Rough turn to 1.503'' + .010'', (**3**), (**3**'), one or two cuts.	4th or 5th speed, or 150 F.P.M. Medium power feed — 80 to 1''.	Holder and cutter no rake, or round-nose tool, 2'' micrometer.
Finish turn to 1.503'' + .003'', (**3**), (**3**'), one cut.	Fine power feed — 140 to 1''.	
Rough and finish square to length (**4**), three or four cuts.	Engine lathe, 4th or 5th speed, or 150 F.P.M. Hand feed.	Calipers, rule.
Turn groove in center of bushing (**5**), two cuts.	Fine power feed — 140 to 1''.	Tool holder and round-nose cutter.
Chamfer ends (**1**), (**4**), to 45°.		
File to limit (**3**), (**3**').	4th speed.	8'' or 10'' mill-bastard file, file card, 2'' micrometer.
Drill oil holes (**6**), (**6**').	Vise, copper jaws. Speed lathe, 4th or 5th speed, or 1500 R.P.M.	Rule, scriber, center punch, hammer, $1\frac{5}{16}''$ straight fluted drill, V center.
Chip (or plane) oil grooves (**7**), (**7**').	Vise, copper jaws (shaper).	Bent center chisel, hammer (or inside grooving tool).
Remove burr from hole by running reamer through by hand.	....	Half-round smooth file.

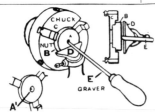

FIG. 44. — CUTTING A CONICAL CAVITY AXIALLY TRUE WITH GRAVER TO START TWIST DRILL.

DRILLING, TAPPING, AND HAND THREADING IN SPEED LATHE

54. To cut a conical cavity axially true, A, Fig. 44. — Hold nut B in chuck C. Set Tee rest D to bring edge of graver E at height of center. Hold graver down hard on Tee rest and at the same time press it into work. If graver is held firmly, it will produce a cavity axially true, as at A'. If the work is brass, use the graver or the corner of the planisher.

55. To chuck with twist drill in speed lathe, Fig. 45. — First make a conical cavity in work F true to axis of rotation, and approximately, to the angle and diameter of drill (see A', Fig. 44). Insert end of drill G in drill holder H, as at G' and H', with conical end in hole in floating center J. Fasten drill in holder by rotating nurled handle and tighten with pin K. Place point of drill in conical cavity in work F and floating

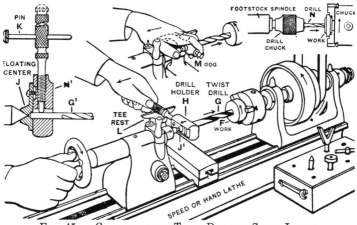

FIG. 45. — CHUCKING WITH TWIST DRILL IN SPEED LATHE.

center J' on center with holder H on Tee rest L. With left hand on holder, as shown, press floating center J' against dead center and with the right hand feed drill to cut.

In the absence of a drill holder, center end of drill and use a dog, as at M.

Warning. — To prevent floating center or drill from slipping off dead center when point breaks through the work, press hard against dead center. To neglect this often ruins drill and work.

Attention. — Reamers and taps are also held with this drill holder. The floating center provides a center hole in alinement with the point of drills, reamers, and taps that are not centered.

Information. — Small drills, as at N, Fig. 45, may be conveniently held for chucking by fitting a chuck to the footstock spindle of a speed lathe or engine lathe.

56. To tap work in speed lathe, Fig. 46. See Tapping, *Principles of Machine Work.* — Place tap wrench K on tap L and then insert in hole in work M, held in chuck N. Guide tap by dead center O, allowing handle of wrench to bear on rest P.

Rotate work by pulling belt Q, or rotate cone R, and feed footstock spindle to follow up tap with dead center.

Attention. — Carefully follow up tap with dead center, as the tap will break if it slips off.

Important. — As small taps break easily, it is best only to start tap in the lathe and then finish at vise.

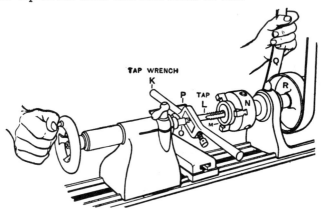

FIG. 46. — TAPPING IN A SPEED LATHE.

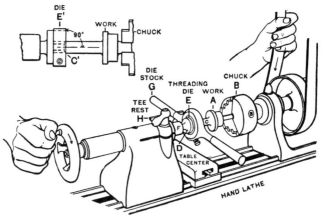

FIG. 47. — THREADING WORK IN SPEED LATHE WITH A DIE BY HAND.

57. To thread work held in chuck with die and stock by hand, Fig. 47. — Chamfer end of work A, held in chuck B, as at C', to assist starting die centrally.   Remove dead center, insert table center D, place muzzle side of die E against work (see E') and handle G on Tee rest H. Feed spindle F in until D presses against die stock. Rotate lathe by hand, following up die with D until thread is started. To terminate thread abruptly reverse die.

Attention. — The work must be adjusted in chuck to run true, otherwise die will cut a crooked thread.

Warning. — Chamfer ends of work as at C', Fig. 47, before mounting in chuck.

BROACHING HOLES, KEYWAYS AND SLOTS

58. Machine broaching is a process used in producing non-circular holes, such as square, hexagonal, rectangular and irregularly-shaped holes that have been drilled, cored or punched; also for cutting slots, keyways, etc. See Chart, Fig. 49. For large quantities of work, the broaching process is more economical than the production of such holes by milling, slotting, planing, filing, etc. One or more long cutters called broaches, as in Fig. 48, are drawn through a drilled or cored hole in work in a broaching machine, as in Fig. 51, or one or more short broaches are forced or pushed through in a press, as in Fig. 53. The broach is the same shape as the cross section of the desired hole. A typical square machine broach is shown in Fig. 48. Each tooth on the tapered part A is from .002″ to .005″ larger than the preceding tooth and gradually changes a round hole to a square hole. Part B is straight and full size. The broach is fastened in the broach holder C by key D. The threaded end E is screwed into the end of the driving screw on machine. Guide bushing F centers the broach.

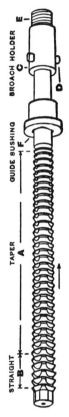

FIG. 48. — SQUARE MACHINE BROACH.

A sectional view of a broach producing a square hole is shown in Fig. 50. The teeth are often inclined and in opposite directions on the sides, and may be undercut, as at A, to give rake and

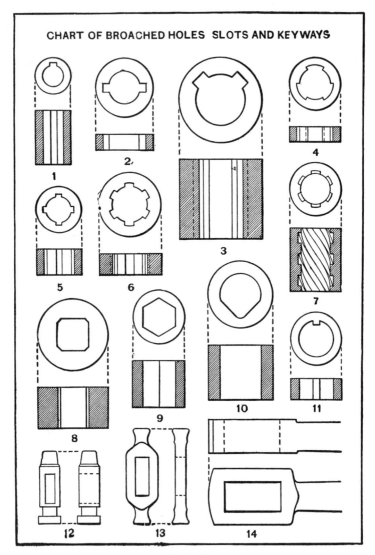

CHART OF BROACHED HOLES SLOTS AND KEYWAYS

FIG. 49.

produce a shearing cut.   Rounded or truncated corners should be used whenever possible as sharp corners are liable to break.

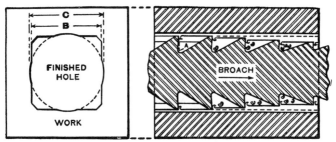

FIG. 50. — SHOWING HOW A BROACH CUTS.

A broaching machine (draw-cut principle) is shown in Fig. 51 broaching six keyways in yoke *A* of an automobile universal

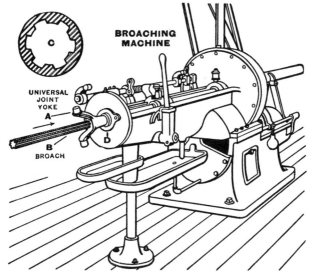

FIG. 51. — BROACHING SIX KEYWAYS IN YOKE FOR AUTOMOBILE UNIVERSAL JOINT.

joint with one stroke of broach *B*.   The finished hole is shown in section at *C*.   The work is held against bushing *D* by the pressure of the cut.

59. Preparation of holes for broaching. — Only a small amount of stock should remain in holes to be broached. Cored castings and drop forgings may be broached successfully. The amount of stock that the broach has to remove is reduced by drilling hole *C*, Fig. 50, larger than desired size of broached hole *B* which leaves sufficient bearing for all mechanical purposes.

Where keyways are to be broached in the work, the hole is prepared in the regular way by reaming.

60. Cutting speeds for broaching. — Nickel steel 18 F.P.M., machine steel and wrought iron 40 F.P.M., cast iron, malleable iron, and brass 108 F.P.M.

61. Lubricants for broaching. — For steel, wrought and malleable iron the broach and work are lubricated with lard oil or a good lubricating compound. For cast iron the sides of broach may be lubricated with lard oil or the broach may be flooded with a soap mixture. Brass and bronze are machined dry.

62. Hand broaching or drifting. — A plain broach or drift is used for bottom hole, as the socket or chuck screw wrench in

FIG. 52. — SOCKET WRENCH.

Fig. 52. The broach at *A* and *A*, Fig. 53, is made of carbon steel with shank to fit the spindle of mandrel press. It is hardened, tempered and ground a little smaller at *B* than at *C* for clearance, and sharpened by grinding end *D*. The hole is prepared by drilling slightly larger than the square and considerably deeper, as at *E* and *E'*, to receive the chips. The socket wrench *F* is held by holder *G* and the square broach is then pressed into the drilled hole as shown in Fig. 53. Small broaches of this type may be driven in with a hammer.

63. Improvised broach or drift. — For small bottom holes a broach or drift may be made by grinding a chisel to a desired shape and driving it into a drilled hole with a hammer.

64. Broaching circular holes, Fig. 54. — Broaching produces better holes in bronze and brass than reaming and is used on some classes of work.

Broach *A* is forced through bronze bushing *B*. Cored holes

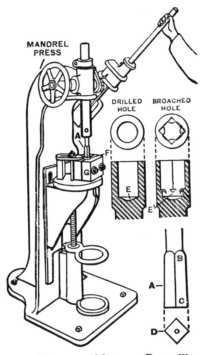

FIG. 53. — BROACHING WRENCH IN MANDREL PRESS WITH PLAIN BROACH.

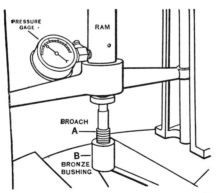

FIG. 54.—BROACHING HOLE IN BRONZE BUSHING. BELT POWER FORCING
PRESS.

require two or more broaches. Holes bored nearly to size require but one. Broaches are made 1/100″ oversize, then hardened and ground to size. For finishing, hole is .002″ undersize and leader on broach must fit hole. Broaches are also grooved spirally with a ratchet thread. Ratchet groove on broach A is cut 60°, making perpendicular tooth.

ADVANCED MACHINE WORK

SECTION 6

DRILLING JIGS   BORING BARS   ECCENTRIC TURNING

Drilling Jigs and Multiple-Spindle Drilling Machines. Radial Drilling Machines. Boring, Boring Bars and Boring Machines. Vertical Boring and Reaming. Taper Holes, Taper Reamers and Spindle Making. Magnetic Chucks. Steady and Follower Rests. Eccentric Turning. Nurling.

DRILLING JIGS, AND MULTIPLE-SPINDLE DRILLING MACHINES

1. The demand for machinery with interchangeable parts has compelled builders to design drilling, milling and planing jigs, gages, and templets to produce standard and duplicate parts accurately and economically.   See Precision Methods of Locating Holes for Jigs and Accurate Machine Parts, pp. **1248–1263**.

2. Drilling jigs are fixtures carefully made with hardened bushings to guide drills, reamers, etc., so that their operation shall be the same on each piece.   They may be divided into about five classes, and each class is best adapted to some particular kind of work.

I. Plate jigs are used for flanges, machine frames, etc. See pp. **602, 603**.

II. Solid jigs are used for work that can be readily clamped to jig body.   These jigs are preferred by some for general work.   See pp. **609, 610**.

III. Box jigs are used for general work.   The work is placed in a box, the hinged cover closed and fastened and the work held in place by binding screws.   See pp. **604, 606**.

IV. Rotary jigs are for work where the jig is too heavy to be easily turned over at the drilling machine.   The jig consists of a box mounted on trunnions to facilitate revolution.

V. Multiple jigs are for work that is to be index-drilled. The bushings are placed in a turret head.

3. An improvised jig or templet. — To drill and ream bolt holes *B*, *C*, and *D*, Fig. 1, equidistant in two or more flanges: First, lay out, drill, and ream one flange carefully. Next, clamp one drilled flange to another in the way they are

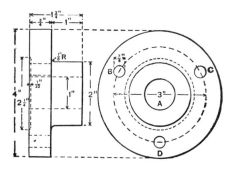

FIG. 1. — COUPLING FLANGE.

to fit, and insert a plug or mandrel through centers A to aline them. Drill and ream holes in the other flange. Other methods are to drill both flanges then ream together, or drill and ream in pairs.

4. Plate (flange) drilling jig, Fig. 2, is a cast-iron disk E supplied with a plug FG to aline flange and jig, a hardened

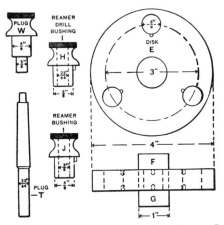

FIG. 2. — FLANGE DRILLING JIG. (PLATE JIG.)

steel drill bushing H and reamer bushing J. Plug T is for alining drill and jig and plug W to prevent relative movement of flange and jig.

5. To use flange jig, Fig. 3. — Place flange K on parallel pieces L and M, with jig N on top of flange; clamp lightly by strap P, block Q, and bolt R to table S. Aline jig and drill spindle with plug T, Fig. 2, moving work by rapping with a soft hammer until the plug will enter bushing exactly central; clamp firmly. Drill hole with reamer drill U. Substi-

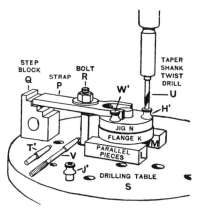

Fig. 3. — Drilling and Reaming with Flange Jig.

tute reamer bushing J' for drilling bushing H', and ream with fluted reamer V. Place plug W, Fig. 2, in first hole to keep jig and flange in alinement as at W', Fig. 3, while drilling and reaming second and third holes. Reverse jig to drill the second flange.

6. Drilling and tapping engine cylinder heads. — When two pieces of work are to be clamped together with cap screws or stud bolts, as a cylinder and cylinder head, two removable bushings are required for the jig; the first with body drill holes for the head, the second with tap drill holes for the cylinder. To tap the holes by hand, a tap bushing is sometimes used to guide the tap. See Automatic Tapping Attachment, Fig. 6.

7. Multiple-spindle drilling machines are used to save time in changing drills, reamers, and counterbores, in moving work from one machine to another, as is necessary with a one-spin-dle drilling machine.   One spindle holds a tap drill, another a body drill, another a counterbore, etc., all running, and the work is moved along the table from one spindle to another without stopping the machine.   See Figs. 4 and 8.

8. Box jigs are used to further increase the rapidity and accuracy of drilling, reaming, tapping, and counterboring.

The work is locked in a box provided with accurately machined bearings to rest on the drilling table.   The jigs are made heavy and are held in position by their own weight, thus saving the time that would be consumed in clamping and alining.

9. To drill and ream bolt holes in coupling flange with a box jig and a two-spindle high-speed drilling machine, Fig. 4.

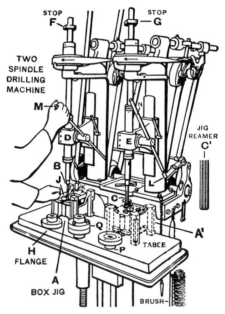

Fig. 4. — Drilling and Reaming Holes in Coupling Flange with Box Jig.

SCHEDULE OF OPERATIONS

$\frac{33}{44}''$ High-speed steel drill,
600 R.P.M.

Time, 6 min.

$\frac{3}{8}''$ High-speed steel reamer,
400 R.P.M.

---

1. Place box jig *A* on table. Insert drill *B* and reamer *C*, *C'* in spindles.

Locate heads *D* and *E* on column and clamp stops *F* and *G* on spindles to limit travel of drill and reamer as tested by passing drill and reamer through bushing into empty jig, or test drill and reamer by lines placed on outside of jig which indicate position of bushings.

Place flange *H* in jig with hub up, and secure with button *J* and screw *K*. Start machine by shipper *L*.

2. With left hand move jig to aline drill bushings with drill. Use lever feed *M* and drill three holes.

3. Turn jig over as shown, dotted, at *A'*.

4. Ream the holes, using lever feed *N*. *P* shows position of flange as it is reamed, and *Q* shows two flanges bolted together.

Attention. — Stop *G* must check reamer before it strikes drill bushing or reamer and bushing will be spoiled.

Important. — A reamer tapered at the end has a tendency to follow the drilled hole. Jig reamers are beveled at the point, as at *C'*, Fig. 4, to correct the error of the hole when used in a jig, see *F*, p. **820**.

Note. — Before placing jig on table, and before turning jig over, brush off table.

To prevent abrasion of bushings, drill, and reamer, apply a little oil with finger to upper part of drill and reamer.

10. Box jig for pieces to be drilled, reamed, and tapped in different directions, Fig. 5.

SCHEDULE OF PARTS

A — Box jig for pieces to be drilled, reamed, and tapped in different directions.

B — Duplicate of work held in jig *A*.

C — Cover held in place by thumb nut *D*.

E and *E'* — Two of the binding screws for adjusting work in jig.

F — Table.

G — Hole in piece *B* that is being drilled in the duplicate.

H — Drilling bushing.

K — Reaming bushing put in position after drilling.

L — Drill $\frac{1}{64}''$ small to allow for reaming.

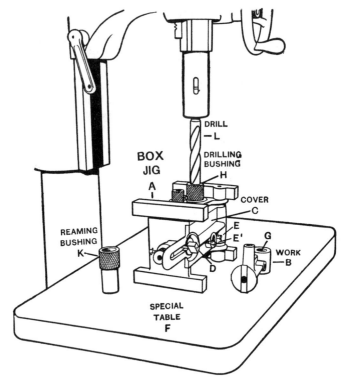

FIG. 5.—A Box Jig for Drilling, Reaming, and Tapping in Different Directions.

RADIAL DRILLING MACHINES

11. Radial drilling machines, commonly called radial drills, differ from vertical drilling machines in that the drill is moved to aline with the work, which is more convenient for large work, such as machine frames, that cannot be moved easily.

A plain radial drilling machine can be used only for vertical drilling, while a universal radial drilling machine may be used not only for vertical drilling but may be adjusted, also, to drill at almost any angle and used with hand or power feed.

12. Plain radial drilling machine, Fig. 6.

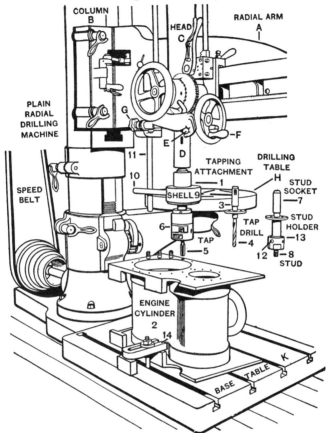

FIG. 6. — DRILLING, AUTOMATIC TAPPING, AND STUD SETTING.

SCHEDULE OF PARTS

A — Radial arm; may be swung around column by hand, and raised up or down by power.

B — Column.

C — Spindle head; may be moved back and forth on arm *A*.

D—Spindle; may be moved up and down in head *C* by hand.

E — Lever for quick movement of spindle *D*.

F — Hand wheel for slow movement of spindle.

G — Knob operating clutch throwing " in " or " out " power feed.

H—Drilling table for light work.

K — Base table for heavy work.

13. Automatic tapping attachment, friction drive is shown at 1, Fig. 6, fitted to the spindle of a radial drilling machine for drilling, tapping, and setting studs in an engine cylinder 2, without stopping or reversing the drill spindle. Socket 3 holding tap drill 4 is used to drill the hole, and is followed by tap 5 in tap socket 6, which in turn is replaced by stud socket 7 that sets stud 8 in place. Shell 9 holds reversing mechanism and is kept from revolving by rod 10 resting against rod 11. The attachment holds drill sockets and will drill holes in the regular way. To tap, press spindle *D* downward by hand lever *E* until the tap reaches the bottom of the hole as indicated by the slip of the friction drive, then raise handle *E*, which throws in the reversing mechanism and backs out the tap. After setting stud 8, raise handle *E* to back off holder 13, leaving stud 8 in cylinder 2. See *Principles of Machine Work*.

With some stud holders, stud 8 is released by rapping pin 12 lightly with a hammer before reversing the holder; on others, the release of stud is obtained by a stud nut which operates on the principle that the coarser threads of the holder, when reversed, will release the finer threads of the stud. The cylinder is clamped to base table *K* at 14.

Attention. — The studs may be set to project to any uniform height by using a gage block between stud holder and cylinder.

Note. — Cap screws, nuts, and slotted screws may be set in like manner by using special wrenches and screw drivers.

14. Jig vise. — In the absence of a regular jig, duplicate work of certain classes may be done with the aid of a jig vise as in Fig. 7. To use this vise lay out and drill one piece and use it as a gage by which to set the stop and jig plate. Vise *A* is heavy and rests without clamping on table *B* of drilling machine *C*. Work *D* is set against adjusting stop *E* and clamped by setting up sliding jaw *F* with lever *G*. Jig plate *H* carrying removable bushing *K* is then adjusted to the desired position for the hole and clamped to fixed **jaw**

of vise. A radial drilling machine being used, spindle L and drill M may be moved to suit the position of work. Cylindrical distance gage N, cut to its center, is sometimes used to set the center of the hole in the jig a given distance from the fixed jaw.

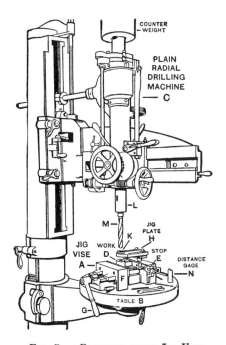

FIG. 7. — DRILLING WITH JIG VISE.

15. To drill and counterbore duplicate parts. — In Fig. 8, a multiple-spindle drilling machine is arranged to drill and spot face casting, as A, A' held by jig B. Insert drill C in spindle D and drill a hole until arrested by stop E. A smaller drill in spindle F finishes the hole as shown in section at F'. The jig is turned over and drill G used to drill the holes in lugs H and H' through bushings K and K'. Counterbore L is fitted to spindle *M* for spot facing lugs *H* and *H'* for the screw or bolt heads. Fixture *N* of the jig is used to clamp the rod of the other half of the strap while it is being drilled.

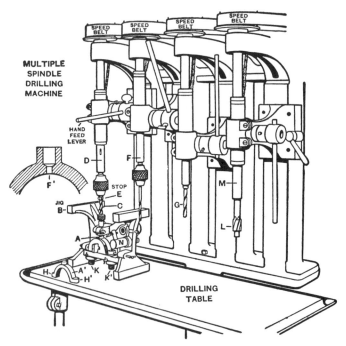

Fig. 8. — Drilling and Counterboring Duplicate Parts.

BORING, BORING BARS AND BORING MACHINES

16. To bore a cylinder or holes in frame of machine where it is not practicable to use drill or boring tool, a boring bar and cutter is used.

17. Three types of boring bars are the fixed cutter, Fig. 9, traveling head, Figs. 11 and 12, and sliding bar, Figs. 15, 17.

18. Fixed cutter type of boring bar boring cylinder, Figs. 9, 10.

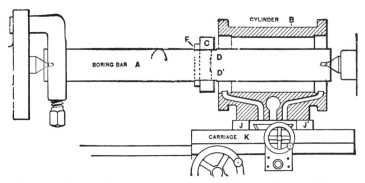

FIG. 9. — BORING CYLINDER WITH FIXED CUTTER BORING BAR IN
ENGINE LATHE.

SCHEDULE OF PARTS

Boring bar A is mounted on centers of an engine lathe to bore

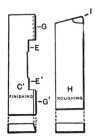

FIG. 10. — BORING BAR CUTTERS.

cylinder B. Double-end high-speed finishing cutter C, detail C', Fig. 10, is used. Bar is filed at D, D', Fig. 9, and cutter at E, E', Fig. 10, to set cutter true. Cutter is fastened by key F, turned on top and face, clearance filed as at G, G', Fig. 10, hardened and tempered. The bore is first roughed out with double-end cutters to within $\frac{1}{64}''$ or $\frac{1}{32}''$ of size or with a single-end roughing cutter H, Fig. 10, cutting point at I.

Single-end cutters of high-speed steel, Fig. 13, of round stock, for roughing and finishing, fastened by set screw and adjustable, are often used. The cylinder is set on parallels J, J' and clamped to carriage K.

A suitable speed and feed are used.

Information. — Cylinders are set true with bar by using inside calipers to test work and bar at either end, or by fastening piece of wire in head in place of one of cutters and revolving bar, testing cylinder at both ends and adjusting. When set true, cylinder is securely fastened. The screws may be slacked slightly for finishing cut.

19. Continuous finishing cut. — The machine should not be stopped during finishing cut as change in size of work by expansion and contraction caused by heat generated by cut, combined with spring of bar will make a ridge wherever cutter is stopped and a second cut may be necessary.

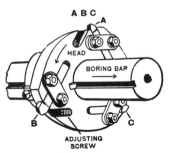

FIG. 11. — BORING HEAD WITH HIGH-SPEED STEEL CUTTERS.

20. Boring heads.—For holes too large for boring bar, boring heads of different sizes with adjustable multiple cutters A, B, C, Fig. 11, are used.

21. Traveling head type boring bar boring cylinder, Figs. 12, 13.

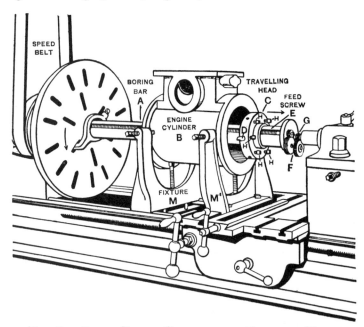

FIG. 12. — BORING ENGINE CYLINDER WITH TRAVELING HEAD BORING BAR IN ENGINE LATHE.

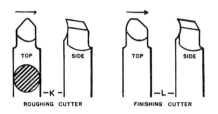

FIG. 13. — CUTTERS FOR TRAVELING-HEAD BORING BAR.

SCHEDULE OF PARTS

Boring bar A is mounted on centers to bore cylinder B.

Traveling head C carries cutters as at D. Head is fed along bar by feed screw E operated by gears F meshing with stationary gear G fast to dead center.

Several cutters may be used, held by set screws H. Roughing and finishing cutters are shown at K, L, Fig. 13.

Special fixtures M, M' for holding cylinder, are bolted to lathe bed.

See Heavy Duty Boring, Facing and Turning, p. **14**39.

22. Portable boring machines are obtainable for use when it is more convenient to take machine to work; and for reboring large cylinders, and work of that class, in place.

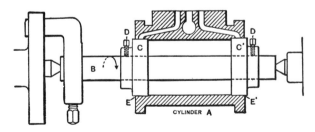

FIG. 14. — TO FACE AND TURN CYLINDER ENDS ON ''BUILT-UP'' MANDREL.

23. Radial facing or turning ends of cylinders. — After boring, small cylinders are sometimes placed on built-up mandrels mounted on lathe centers, as at *A*, Fig. 14, and the ends squared with lathe tools. Mandrel consists of shaft *B*, two large collars *C* and *C'* which are fastened by screws *D* and *D'*. Collars are turned slightly taper to fit counterbored portions *E* and *E'* of cylinder.

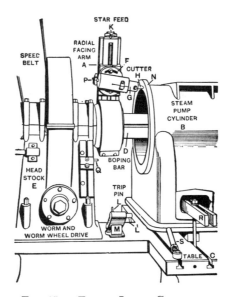

FIG. 15. — FACING LARGE CYLINDER.

A facing arm may be clamped upon an ordinary boring bar for facing cylinders.

A compound rest, or any slide that will carry a cutter, may be clamped to lathe face plate and operated by hand or star feed.

24. Radial facing arm with star feed, Fig. 15.

SCHEDULE OF OPERATIONS

Radial facing arm A is attached to nose of spindle of boring machine. Cylinder B clamped to table C is bored with sliding bar D, sliding in headstock spindle E. Tool block F carries tool holder G and cutter H.

Feed is obtained by star wheel K striking pin L on bracket M at each revolution of radial facing arm. To feed in opposite direction, pin L' is used. To turn outside of flange, as at N, tool block is set parallel to bar, and feed obtained by wheel P and pin Q. Cylinder is clamped as at R and S.

As a time saver, a similar radial facing arm is sometimes used on opposite end and both ends faced and cylinder bored simultaneously.

25. Drilling, reaming and counterboring in horizontal boring machines, Fig. 16. Bracket casting A, is faced and

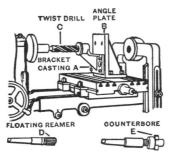

FIG. 16. — DRILLING, REAMING AND COUNTERBORING IN HORIZONTAL BORING MACHINE.

bolted to angle plate B. If casting A is solid, the hole is drilled with two-groove twist drill; if cored, with three- or four-groove twist drill C, reamed with floating reamer D, and faced to length with counterbore E. The floating reamer has side or floating movement and will produce a true hole.

26. Sliding bar type boring bar, Fig. 17.

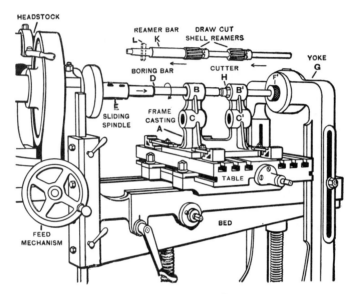

FIG. 17. — ALINEMENT AND PARALLEL BORING AND REAMING. HORIZONTAL BORING MACHINE.

SCHEDULE OF PARTS

Parallel Boring   Alinement Boring and Reaming

Frame casting A is bolted to table of horizontal boring machine. Cored holes B, B' and C, C', are bored and reamed parallel to each other.

Circles are described about holes. Boring bar D held in sliding spindle E, is supported in bushing F' in yoke G.

Holes are rough and finish bored with cutter H, and line-reamed with shell reamers supported on bar K. Bar is held by key L and drawn in direction of arrows.

Holes C, C', are bored and reamed exact distance from B and B' by raising bed required number of thousandths as indicated by dial graduated to read to one-thousandth of an inch.

See Horizontal Boring Machine, p. 1439.

VERTICAL BORING AND REAMING

27. Vertical drilling machines, radial drilling machines and vertical boring mills are used for vertical boring and reaming holes in alinement in frames of machines, small cylinders, etc. Holes are cored nearly to size then bored with boring bars and reamed with special or shell reamers. The boring bar has taper shank to fit sliding spindle of machine, and lower end is straight to run in guide bushing fitted to hole in table or base. Work may be held in jigs, fixtures, or clamped to table or base of machine.

28. Shell reamers are hollow reamer heads fitted to an arbor. One arbor can be used for a number of sizes. They are obtainable to 5″ diameter, fluted, rose and adjustable, with straight or spiral teeth, and are much used in chucking lathes, horizontal and vertical boring machines, boring mills, etc.

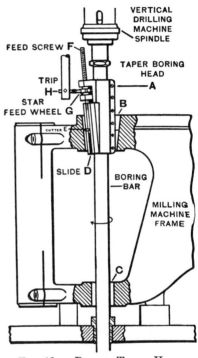

FIG. 18. — BORING TAPER HOLE.

29. Boring taper holes with boring bar, Fig. 18.

SCHEDULE OF PARTS

Taper hole in milling machine frame is bored with taper boring head *A*. Holes *B* and *C* are first bored straight. Hole *B* is then step bored. Slide *D* carries cutter *E*, which is fed downward by star-feed mechanism *F*, *G*, *H*.

30. Boring taper holes with boring bar in lathe. — Large cored work may be bored taper by clamping to lathe carriage or in steady rest, and bored with traveling head boring bar. A special center is clamped to face plate a calculated distance from line of centers. Bar is mounted on special live center and dead center. One cutter is used. In principle this method is similar to taper boring head.

31. Hand alinement reaming with shell reamer, Fig. 19.

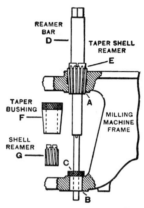

FIG. 19. — HAND ALINEMENT REAMING.

SCHEDULE OF OPERATIONS

1. — Bore taper hole A and straight hole B in alinement.

2. — Place bushing C in lower hole to aline reamer bar D.

3. — Ream taper hole A with reamer E then remove both C and E.

4. — Place taper bushing F in hole A and reamer G on bar, and ream hole B.

32. Power alinement reaming with jig, Fig. 20.

SCHEDULE OF OPERATIONS

1. — Place frame casting A in jig B and drill holes C and C'.

2. — Bring reamer bar D and shell reamer E down part way.

3. — Place shell reamer E' and its driving collar F in position and adjust bushings G and G' to keep reamer bar in alinement.

4. — Ream both holes at once.

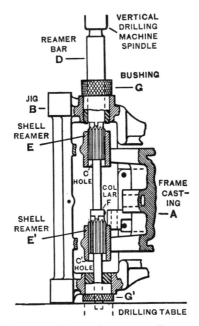

FIG. 20. — POWER ALINEMENT REAMING.

TAPER HOLES, TAPER REAMERS AND SPINDLE MAKING

33. Small taper holes are made by drilling a straight hole slightly smaller than diameter of small end, then reaming with roughing and finishing taper reamers.   Large taper holes are made by boring and reaming, or boring without reaming.

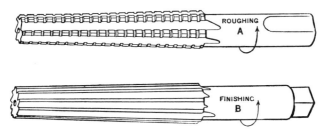

FIG. 21. — TAPER REAMERS.

34. Taper reamers *A* and *B*, Fig. 21, obtainable in Morse and Brown & Sharpe tapers and also in special tapers for spindles, sockets and collets. Fig. 22. is a stepped-roughing reamer for large holes.

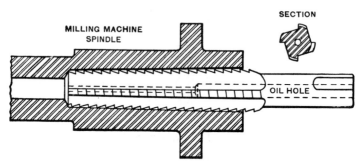

FIG. 22. — STEPPED ROUGHING TAPER REAMER.

35. Stepping work for taper holes. — Above 2″ at small end, taper holes are stepped with drill or boring tools to remove stock, as in Fig. 23, or bored with taper boring head.

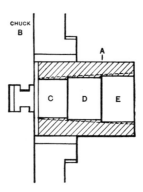

FIG. 23. —STEP-BORING FOR TAPER HOLES.

36. Boring taper holes. — Instead of stepping hole, it may be bored to required taper with taper attachment or compound rest and may or may not be reamed.

37. Taper plug and ring gages are used for standardizing taper holes and shanks, diameters and lengths. See Morse tapers, pp. **2**33, **2**34, Brown & Sharpe tapers, pp. **2**33, **2**35. Spindle A, Fig. 24, is reamed until line B of gage C is even with end.

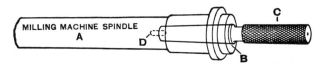

FIG. 24. — TAPER PLUG GAGE.

Shank 1, Fig. 25, of arbor 2, is filed or ground to fit gage 3 until end is even with end of gage at 4. The flat portion at 5 is the gage for milled tang. Line 6 gives reamer depth; line 7, plug depth. The gage is adjustable to compensate for wear.

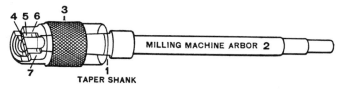

FIG. 25. — TAPER RING GAGE.

38. Stub mandrel, Fig. 26, is used for spindles and drill sockets which have to be turned after taper hole is reamed. The body is accurately ground to fit taper hole in spindle B.

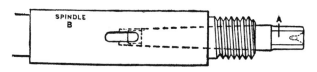

FIG. 26. — STUB MANDREL FOR TAPER WORK IN SPINDLE OR SOCKET.

39. To make small lathe spindle, Fig. 27.

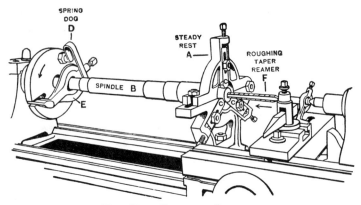

Fig. 27. — CHUCKING SPINDLE.

Specifications: Material, crucible steel spindle forging $\frac{1}{8}''$ to $\frac{1}{4}''$ large.

Machines and tools: Engine lathe, holder and cutter, calipers, rule, micrometer, steady rest, drills, roughing and finishing taper reamers, proof bar, test indicator, center scraper, stub mandrel, spring dog, threading tool, universal grinding machine.

SCHEDULE OF OPERATIONS

I. Preparing blank. Center, square to length, rough turn all diameters $\frac{1}{16}''$ large, turn and file spot on nose-end of spindle for steady rest.

II. Mounting in steady rest. Clamp steady rest A, Fig. 27, to lathe bed. Mount spindle B on centers and adjust jaws CCC to *spot*, fasten spring dog D to face plate and secure dog E (or tie dog to face plate with belt lacing) to hold spindle on live center. Move dead center away from work.

III. Drilling and reaming hole.

Drill hole in spindle B proper size, deeper than required depth of reamer hole, power ream with roughing taper reamer F. Hand ream taper hole carefully at vise with finishing reamer.

IV. Testing truth of hole with proof bar and indicator. Place tapered end of proof bar H in spindle and mount on centers, as in Fig. 28. Adjust feeler K of indicator to proof bar, rotate spindle by hand and determine error.

V. Correcting error. Remove spindle from lathe and scrape center hole on side where spindle is most eccentric, with center scraper. Test again. Repeat process until error is less than .001″.

VI. Mount opposite end of spindle in steady rest. Insert stub mandrel, as in Fig. 26, reverse spindle in lathe, turn and file spot on spindle near dead center for steady rest. Adjust steady rest to spot and secure work on live center.

VII. Drill hole through spindle. Drill hole through spindle to meet hole in opposite end. Center and fit plug to hole.

VIII. Test. Again use proof bar. Mount on centers, test and correct as in V.

IX. Finish bearings. Place stub mandrel again in taper hole. Mount on centers, turn and grind all bearings, and thread nose of spindle.

Attention. — Solid spindles are slotted at end of taper hole to allow use of center key to force out arbor or center.

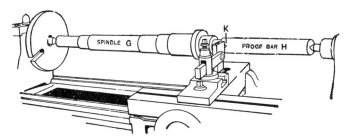

FIG. 28. — TESTING SPINDLE.

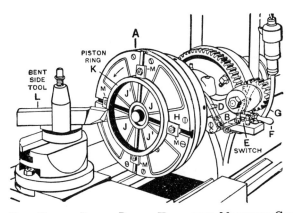

FIG. 29. — FACING PISTON RING. HELD WITH MAGNETIC CHUCK.

40. To make large spindles, first turn and fit all bearings, then mount in its own bearings and drill straight hole through it. Bore taper hole in end, either with taper attachment or compound rest, finish with hand reamer.

MAGNETIC CHUCKS

41. Magnetic chucks are used to hold accurately by magnetic attraction thin work to be finished, as iron or steel disks, rings and parallels. These chucks are best adapted for surface and cylindrical grinding machines, but may be applied to any machine tool. See pp. 802–804.

Note. — Magnetic chucks always require direct current.

42. Lathe magnetic chuck, Fig. 29, shows magnetic rotary chuck *A*, applied to an engine lathe. Projecting from back of chuck shell are two insulated brass, or copper, contact rings (not shown) to connect a direct current circuit with the coils of the electro-magnet system in the rotating chuck. Bracket *B* is fastened to headstock of lathe and carries at one end fiber collars *C* and brushes *D*, which bear against contact rings. Bracket *B* is also fastened to switch *E* which is opened and closed by handle *F*. The switch is connected at one end to brushes and at other end to supply wire *G*. The chuck is supplied with detachable face plate *H*. This face plate is made up of magnetic poles *JJ* and *J'J'*, adjacent segments being of opposite polarity. The spaces between the poles are filled up by non-magnetic metal to give a continuous even surface. The arrangement of these magnetic poles may be varied for different classes of work and to accommodate the exciting coils within the chuck.

In Fig. 29, edge of piston ring *K*, is being faced with tool *L*. The piston ring is turned inside and outside and finished on one side in usual way, then cut off a little oversize.

The finished side is held against magnetic chuck and handle *F* swung downward closing circuit, as in full lines, and the rough side faced smooth to size and perfectly parallel. To release ring, handle *F* is swung upward breaking circuit as dotted at *F'*. To avoid slipping under heavy cut, stops, *M, M, M, M,* are placed against ring.

43. Magnetic planer chuck. — Fig. 30 is used to hold work to be finished in iron or steel by magnetic contact. It is adapted to thin work that cannot be accurately held by other methods.

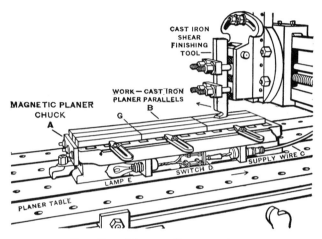

FIG. 30. — HOLDING PLANER WORK.

Magnetic chuck A is clamped to planer table and arranged to hold pieces B. The pieces are separated about $\frac{1}{16}''$ by non-magnetic material, as brass, wood or fiber, otherwise holding action will be weakened. The direct current is supplied by wire C, and switch D is closed to grip work and opened to release it. Lamp E indicates whether current is on or off.

44. To demagnetize work, Fig. 31. — Hardened steel and, to some extent, cast iron, coming in contact with magnetic chucks, becomes permanently magnetized. To demagnetize work, start demagnetizer A (1200 R.P.M.) and place work B on plates C, C'. Then vibrate work upward $1''$ to $2''$ from plates as shown dotted at B', and replace on plates. Repeat operation several times. The action of the demagnetizer is to magnetize the piece of steel alternately, in both directions. As the piece of steel is moved out of its influence, the steel is magnetized a little less each time the magnetism is reversed.

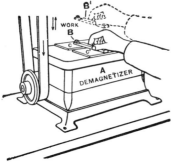

FIG. 31.—DEMAGNETIZING STEEL OR CAST IRON.

Attention. — An improvised demagnetizer may be made with a coil of wire using A. C. current at 60 cycles: The steel is placed in the coil and then moved slowly out. This will demagnetize a watch perfectly.

45. Electrical units. —

VOLT. — The unit of electromotive force (E.M.F.). The force required to send one ampere of current through one ohm of resistance.

AMPERE. — The unit of current. The current which will pass through one ohm resistance when impelled by one volt. A milli-ampere = one-thousandth of an ampere.

OHM. — The unit of resistance. The resistance offered to the passage of one ampere when impelled by one volt. The megohm = one million ohms.

WATT. — The unit of power:

Amperes × volts = watts.
(Amperes)^2 × ohms = watts.
(Volts)^2 ÷ ohms = watts.
746 watts = 1 horse power.
1000 watts = 1 kilo watt.

COULOMB. — The quantity of current which impelled by one volt would pass through one ohm in one second.

JOULE. — The unit of work. The work done by one watt in one second.

Information. — In ordering electric motors, it is absolutely necessary to state *voltage* and whether current is *direct* or *alternating*, and if alternating, whether *one, two* or *three phase* and *number of cycles.*

STEADY AND FOLLOWER RESTS

46. A steady rest, Fig. 32, is used to support a slender shaft to prevent vibration.

SCHEDULE OF PARTS

A — Rest; two parts, — base and top.	*F* — Bolt, hinges top to base.
B — Base.	*G* — Clamp, fastens top to base.
C — Top, hinged to base.	*H* — Sliding jaws adjusted to spot *J* on shaft.
D — Clamp strap.	
E — Bolt to clamp rest at any location, on ways of lathe.	$K_1K_2K_3$ — Adjusting screws.
	L — Nuts to clamp jaws in position.

47. To turn spot on shaft, then adjust jaws to that spot and turn shaft, Fig. 32.

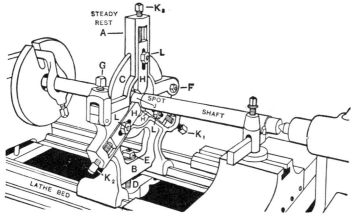

FIG. 32. — TURNING SLENDER SHAFT SUPPORTED BY A STEADY REST SET TO A TURNED SPOT.

SCHEDULE OF OPERATIONS
True live center   Set dead center in accurate alinement

1. Center and straighten shaft, rough square, recenter, finish square.

2. Turn spot *J* $\frac{1}{64}''$ to $\frac{1}{32}''$ large and wider than jaws *H*, central or nearer live center, and file smooth.

3. Move jaws *H* back to clear shaft and swing top *C* backward. Mount and clamp rest opposite spot. Adjust jaws *H* by screws: K_1 first, then K_2 to touch shaft, then clamp both jaws by nuts *L*.

Swing top *C* forward and clamp with screw *G*. Adjust third jaw *H* to touch shaft by screw K_3 and clamp jaw by nut *L* and oil spot *J*.

4. Turn one-half shaft, reverse shaft, adjust jaws to turned portion and turn rest of shaft, or spot the shaft in center and rough and finish both halves to spot by reversing shaft, then move rest along toward live center, readjust jaws and finish spot to size.

48. The cat head, *M*, Fig. 33, is used to hold slender shafts to be spotted as well as for steadying slender shafts to be turned without spotting. It is also used to hold square, hexagonal, or work of irregular section to be turned, as valve stems that have one part square and the other round.

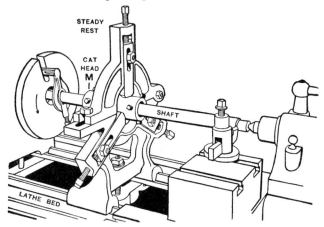

FIG. 33. — TURNING SLENDER SHAFT SUPPORTED BY A STEADY REST SET TO A CAT HEAD.

49. To true cat head on shaft. — Place it on shaft and true up accurately by adjusting screws. Test its truth by chalk, copper tool, or a test indicator. If the cat head is used in turning the second end of the shaft, it must be placed on the turned part and trued up as before. Usually the jaws of the steady rest are set directly on the shaft for the second half.

Attention. — Do not adjust jaws carelessly to cause the shaft to spring. If piece springs when outside skin is removed, straighten in a press, or if the shaft is slender, mount on centers and straighten. Use wooden jaws in steady rests for finished work.

50. Follower (Traveling) rest, *A*, Fig. 34, is used for turning shaft from end to end, and it is more convenient and produces more accurate work than a steady rest. The rest consists of two jaws and a frame bolted to carriage *B*. A spot of the desired diameter is turned at end *C* of shaft *D*, and the jaws

E and *E'* are adjusted to it. The tool must be slightly in advance of the jaws, which should be well lubricated where

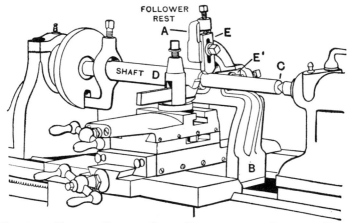

FIG. 34. — TURNING SLENDER SHAFT SUPPORTED BY A FOLLOWER REST.

they bear on the work. One-half of shaft is turned, then it is reversed and the second half turned. For more effective support, use bushings to suit different diameters of shafts in place of jaws. When cutting Square or 29° threads on slender pieces a follower rest is necessary to support the work.

ECCENTRIC TURNING

51. Eccentric turning. — Besides ordinary straight and taper turning, in which there is only a single axis and a single pair of centers, there is another class of turning, known as eccentric or offset turning, in which there is more than one axis and consequently more than one pair of centers. See Eccentric Shaft, Fig. 35, Engine Eccentric, Fig. 38, and crankshaft, Fig. 41.

Crankshafts are made with one or more cranks and may have solid eccentrics as well. Small crankshafts, in the rough, may be obtained drop-forged of machine steel, cut from steel slabs or in steel castings. The larger sizes come partly machined after being forged under the steam hammer.

52. To make eccentric shaft for engine lathe headstock, Figs. 35–37.

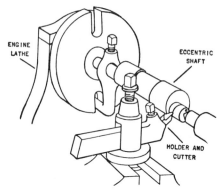

Fig. 35. — Turning an Eccentric Shaft in an Engine Lathe.

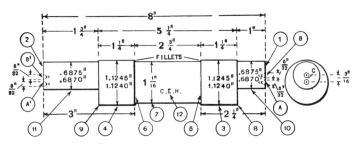

Fig. 35a. — Schedule Drawing.

Specifications: Material, machine steel $\frac{1}{16}''$ to $\frac{1}{8}''$ large; weight, 2 lb. 10 oz. Hardness, 15 to 16 (scleroscope).

High-speed steel, or stellite cutting tools.

Time: Study drawing and schedule in advance, 7 min. — Oil lathe, 4 min. — Centering, squaring, and rough turning blank, 30 min. — Laying out and drilling hole for driver, 6 min. — Grinding journals (**3**), (**4**), on concentric centers A, A', 35 min. — Aline jigs and make eccentric centers, 10 min. — Turning, squaring, filing, and fitting (**10**), (**11**), on eccentric centers B, B', 55 min. — Clean lathe, 3 min. — Total, 2 h. 30 min.

Exceptions: If jigs are not available and eccentric centers have to be layed out, add 8 min.

If a grinding machine is not available and the journals (**3**), (**4**), have to be turned, filed, and fitted, add 10 min.

SCHEDULE OF OPERATIONS, MACHINES AND TOOLS

OPERATIONS.	MACHINES, FEEDS, SPEEDS.	TOOLS.
Center to $\frac{5}{32}''$, (A), (A'), Fig. 35a.	Centering machine. Drill speed 1200 R.P.M., countersink speed 500 R.P.M.	$\frac{5}{64}''$ drill, or combination drill and countersink 60°. Lard oil.
See that live center is true and dead center in approximate alinement.	Engine lathe, 12″ to 16″.	
Rough square (1), (2).	2d or 3d speed or 50 F.P.M. Hand feed.	Dog, holder and cutter 35° rake, calipers, rule.
Recenter to $\frac{5}{32}''$, (A), (A'). **Finish square** to length, (1), (2).	3d or 4th speed, or 80 F.P.M. Hand or power feed.	
Rough turn to $\dfrac{1.145''}{1.140''}$ (3), (4) one cut. Turn halfway, reverse and turn other half.	2d or 3d speed, or 60 F.P.M. medium power feed — 80 to 1″.	Holder and cutter 35° rake, 2″ micrometer
Draw lines (5), (6).		Copper sulphate, rule, scriber.
Rough turn to $1\frac{1}{16}''$, (7), one cut.	Engine lathe, 3d speed, or 60 F.P.M. Medium power feed — 80 to 1″.	Holder and cutter 35° rake, calipers, rule.

Use drill jig for drilling end-driving hole, Fig. 35b.

Tools: $\frac{1}{4}''$ drill, drill chuck, drilling jig, sensitive drilling machine, lard oil.

Warning. — To avoid breaking drill when drilling driving hole or center holes in shaft, work must be held firmly on drilling table or center. The drill should be withdrawn two or three times to remove chips to prevent clogging, and to be lubricated.

Place jig *A*, Fig. 35b, on either end of eccentric shaft blank *B*, as at *C*, and tighten set screw *D*.

Place shaft blank vertically on table *E* of sensitive drilling machine.

Lubricate drill with lard oil and drill $\frac{1}{4}''$ hole $\frac{1}{4}''$ deep, or until stop collar *F* strikes jig. If a jig is not available,

Lay out and drill hole as in Fig. 35c.

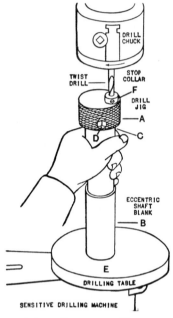

FIG. 35b. — DRILLING HOLE FOR GRINDING MACHINE END-DRIVING DOG WITH JIG.

FIG. 35c. — DRILLING HOLE FOR GRINDING MACHINE END-DRIVING DOG WITHOUT JIG.

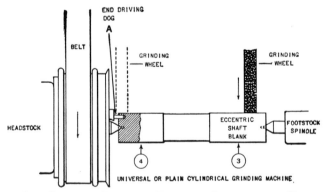

FIG. 35d. — GRINDING RUNNING FITS FOR JOURNALS OF ECCENTRIC SHAFT BLANK USING END–DRIVING DOG.

To grind journals, Fig. 35d. See pp. 716, 717.

OPERATIONS.	MACHINES, FEEDS, SPEEDS.	TOOLS.
Set grinding machine straight; limit .0002″ taper. Grind journals for running fit (3), (4), to limit $\frac{1.1245''}{1.1240''}$	See p. 712. Plain or universal grinding machine.	End-driving dog or pin, 2″ micrometer.

Exception. — If grinding machine is not available, turn, file and fit to $1\frac{1}{8}''$ reamed hole or ring gage; see pp. 231, 232; or measure with 2″ micrometer and allow .003″ for filing (1.125″ + .003″).

File to limit $\frac{1.1245''}{1.1240''}$ (3), (4).

To drill and countersink eccentric center with jigs, Fig. 35e.

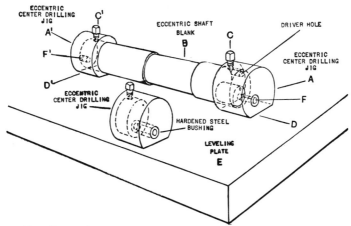

FIG. 35e. — ALINING DRILLING JIGS FOR ECCENTRIC CENTERS.

Tools: Two drilling jigs, leveling plate, combination drill and countersink, chuck and sensitive drilling machine (3d speed, or 700 R.P.M.).

Place jig *A*, Fig. 35e, on end of eccentric shaft blank *B*, and tighten set screw *C* on driver hole side. Place jig *A'* on other end of shaft *B*. On leveling plate *E*, aline flats *D,D'* of jigs which hold *F,F'* in alinement. Tighten set screw *C'*. Place hole *A* of jig *B*, Fig. 35f, on center *C*.

Drill and countersink end D of shaft E until stop collar F on countersink G strikes jig H, or until countersunk hole is $\frac{5}{32}''$ diameter. Reverse and drill and countersink end J.

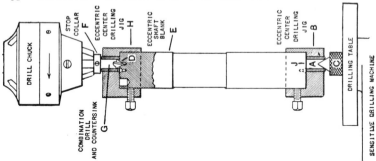

FIG. 35f. — DRILLING AND COUNTERSINKING WITH JIGS.

Limit of accuracy. — Eccentric centers located with jigs may give an accuracy of throw within $.002''$ to $.003''$. If greater accuracy is desired, test with indicator and correct throw with center hole scraper, Fig. 37.

Lay out eccentric centers. — If jigs for drilling the eccentric centers are not available, lay out eccentric centers B,B', Fig. 35a, $\frac{3}{16}''$ from center A,A', as at C, as follows: Clamp pointed tool in tool post at height of centers. Mount shaft on centers without dog, and make a short line at each end in same plane as at (**1**) and (**2**), Fig. 36. Draw a line on each end as at D, with pointed tool or center square and scriber. Slant pointed tool to the left with point $\frac{3}{16}''$ (half the required throw of eccentric) from lathe center, and draw lines at both ends intersecting radial lines as at E. Make center punch marks at these intersections, and drill and countersink $\frac{1}{64}''$ small.

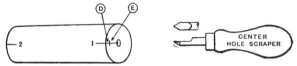

FIG. 36.—ECCENTRIC SHAFT BLANK.   FIG. 37.—CENTER HOLE SCRAPER.

Test and correct throw. — Mount in lathe on eccentric centers B, B', Fig. 35a, and measure throw between lathe tool and work with rule and inside calipers. Correct throw to $\frac{3}{16}''$

at both ends by scraping over countersinks with center hole scraper, Fig. 37. Test also with indicator, p. 1210, and make throw same at both ends. Countersink to $\frac{5}{32}''$, test and scrape again, if necessary. Limit of error permissible .001''

To turn and fit eccentrics, Fig. 35a.

OPERATIONS.	MACHINES, FEEDS, SPEEDS.	TOOLS.
Mark lines on concentric side of shaft at (8) and (9), for length of reduced parts. See Fig. 35a.	Vise, copper jaws.	Copper sulphate, rule, scriber.
Rough turn reduced parts to $\frac{11}{16}'' + \frac{1}{64}''$, (10), (11), five or six cuts each.	Engine lathe. 2d or 3d speed, or 50 F.P.M. Medium power feed — 80 to 1''.	Holder and cutter 35° rake, calipers, rule.
Rough square shoulders to within $\frac{1}{64}''$ of lines. True live center.	2d or 3d speed, or 50 F.P.M. Hand feed.	Holder and cutter 35° rake, rule.
Set dead center in accurate alinement to turn straight using trial piece the same length, pp. 116, 117.	Engine lathe 12'' to 16''. 3d or 4th speed, or 70 F.P.M. Fine power feed — 140 to 1''.	Dog, copper, holder and cutter 35° rake, micrometer.
Make wringing fit, Fig. 35a, (10), (11), as follows: **Finish turn** reduced parts, one cut (10), (11), with allowance for filing to fit $\frac{11}{16}''$ ring gage or reamed hole; or measure with 1'' micrometer and allow .003'' for filing (.6875'' + .003''). See pp. 232, 317.	3d speed, or 70 F.P. M. Fine power feed — 140 to 1''.	Holder and cutter 35° rake, calipers, $\frac{11}{16}''$ mandrel, $\frac{11}{16}''$ ring gage, 1'' micrometer.
Finish square shoulders (8), (9).	Hand feed.	Holder and cutter 35° rake.
File (10), (11), to fit ring gage or reamed hole, wringing fit; or File to limit $\frac{.6875''}{.6870''}$ **Stamp name** on surface (12).	Engine lathe, 4th speed, or speed lathe, 2d or 3d speed. Vise, copper jaws.	8'' or 10'' mill bastard file, file card, $\frac{11}{16}''$ ring gage, micrometer, lard oil. Stamp or letters.

Information. — Shoulders (8), (9) of eccentric shafts are usually squared to fit an exact space. See eccentric shaft in back gear mechanisms of engine lathes.

Note. — Eccentric shafts may also be machined by first making a special hollow chuck to fit the nose of the spindle and then making an eccentric sleeve, each end of which will fit the hollow chuck.

53. To turn engine eccentric, Fig. 38.

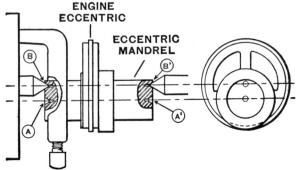

FIG. 38. — TURNING AN ENGINE ECCENTRIC.

SCHEDULE OF OPERATIONS

1. Make eccentric mandrel Fig. 38, of required throw by method, pp. **6**30, **6**31.

2. Chuck, ream, and push eccentric casting lightly onto mandrel. Mount in lathe and rap eccentric until it runs true. Take from lathe and press mandrel hard. On large eccentrics, set screws are used to fasten eccentric to mandrel. Do all facing operations and turning of hubs on regular centers AA', and all eccentric diameters on eccentric centers BB'.

Attention. — When eccentric centers lie outside shaft, eccentric may be turned on disk, carrying split stud with expanding screw, the whole being bolted to face plate of lathe in position to give desired throw. Some prefer this fixture to an eccentric mandrel.

54. Crankshaft turning, Fig. 41. — The centers or axes lie outside of the shaft, so that fixtures, arms or flanges, carrying the centers must be provided.

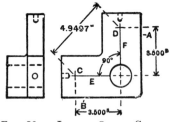

FIG. 39. — LAYING OUT A CENTER FIXTURE FOR A TWO-THROW 90° ENGINE CRANKSHAFT.

55. To lay out two-throw 90° crankshaft center fixtures, Fig. 39.

SCHEDULE OF OPERATIONS

1. In two fixture castings chuck and ream holes to next regular diameter larger than required diameter of ends of crankshaft, mount on mandrel and square arms and hub. Plane

edges AB, at right angles and an equal distance from hole.

2. Set pointed tool $3\frac{1}{2}''$ from point of dead center and describe arcs CD. If fixtures are not planed, points CD are obtained as follows:

Center punch point on C, in middle of casting on the arc. Set dividers 4.9497″ the hypothenuse of right triangle whose base is $3\frac{1}{2}''$, using the one-hundredth graduation of the rule, and with C as center intersect arc at D. Distance CD, may be obtained by the following rule: Find the square root of the sum of the squares of the two throws.

$Example:$ $\quad \sqrt{3\frac{1}{2}^2 + 3\frac{1}{2}^2} =$

4.9497, answer. Lines EF, on each fixture are used to aline cranks with fixtures when sides of fixtures are not 90°.

Attention. — If only one crankshaft is to be turned, the fixture may be drilled and countersunk in casting. For a number of crankshafts, chuck large holes and drive in hardened and ground steel plugs carrying large center holes.

To make these holes in correct location, clamp fixtures to face plate of lathe with marks C and D, in axis of rotation as tested with axis indicator (see Lathe Axis Indicator, **1213**), then drill and bore to size.

56. To aline center fixtures and lay out cranks, Fig. 40.

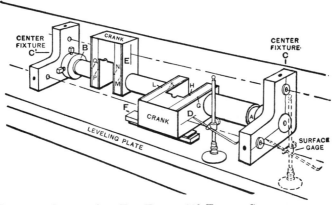

FIG. 40. — LAYING OUT TWO-THROW 90° ENGINE CRANKSHAFT AND ALINING CRANKS WITH CENTER FIXTURES.

SCHEDULE OF OPERATIONS

Adjust fixtures, center and square shaft in relation to crank webs.	Turn AB to fit center fixtures CC'.
	Mount fixtures and adjust

centers of cranks D and E to aline with centers of fixtures by wedges F and test with surface gage. Lay out crank web by lines and center punch at G, H, K, L, M, N, P, Q.

Mount in lathe on one pair of crank centers and revolve; regular centers are horizontally opposite. Test top of turned ends with surface gage.

57. To turn crankshaft, Fig. 41.

FIG. 41. — TURNING A TWO-THROW 90° ENGINE CRANKSHAFT. SCHEDULE DRAWING.

SCHEDULE OF OPERATIONS

Rough turn blank. Mount upon regular centers AA', Fig. 41. Rough square faces of webs 1, 2, 3, 4, and rough turn shaft 5, 6, and 7.

Change to crank centers BB'. Rough square and turn 8, 9, and 10.

Change to crank centers CC' and rough square and turn 11, 12, 13.

Counterbalance crank fixtures by using weight D, or preferably by adjustable weights EE'.

Use driver FF' with piece of leather to reduce jar.

Season crankshaft between roughing and finishing, if time will permit.

Finish square and turn in reverse order.

It is best to spot 7 and use steady rest for finishing. For slender crankshafts use jack or braces as shown by dotted lines. European tool post G is preferred to a single tool post. Mill or plane edges of webs.

Attention. — Crankshaft lathes are obtainable. Automobile and motorboat crankshafts may be rough turned, then ground, or ground direct from drop forgings or steel castings.

NURLING

58. Hand and machine nurling or milling tools, Figs. 42 and 43, are used to check or mill surfaces of nuts, screw heads, handles, knobs, etc., to increase the grip and facilitate rotation. These indentations are similar to those on the edges of silver and gold coins to detect the removal of metal and called milled edges.

59. Hand nurling. Fig. 42. — Thumb nuts and screw heads are often nurled with single nurl wheel *A* and holder *B*, Fig. 42. Oil work and nurl.

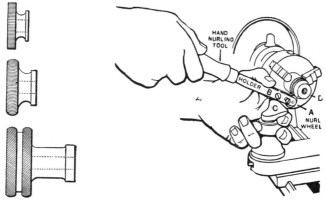

FIG. 42. — HAND NURLING IN SPEED LATHE.

SCHEDULE OF OPERATIONS

Speed for hand nurling brass, 300 F.P.M.

Speed for hand nurling steel, 200 F.P.M.

Place holder on rest *C* which should be firmly clamped, and place wheel under thumb nut *D*. Press down firmly with right hand, at same time steadying tool with the left, until desired effect is produced.

Information. — Hand nurls are obtainable in a variety of patterns for thumb nuts and screw heads, as shown in Fig. 42.

Attention. — Use fine nurls on steel.

60. Machine nurling, Fig. 43.

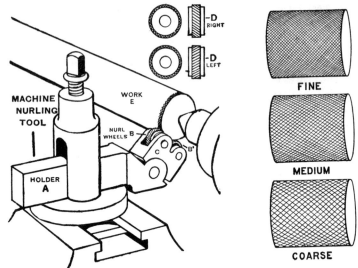

FINE

MEDIUM

COARSE

FIG. 43. — MACHINE NURLING IN ENGINE LATHE.

SCHEDULE OF OPERATIONS

Speed for machine nurling iron and steel, 40 F.P.M.

Feed for nurling, medium power feed, 80 to 1″.

Nurling tool, Fig. 43, consists of holder A and two hardened steel nurling wheels B, B', mounted in head C, connected with holder by rocking joint. Nurls have opposed helical teeth as at D, D'. Holder is fastened at right angles to work E, and adjusted to have both wheels bear equally on work.

Use power feed. Oil work and nurls. Start lathe. Force nurls with cross feed screw hard against work, preferably with half width of nurls. Stop lathe, and if pitch of checking on work is same as nurls, reduce pressure and throw in power long. feed to travel back and forth over surface until projections come nearly to a point.

Attention. — To prevent machine nurls from cutting double set of projections twice as fine as the nurls and spoiling the work, press nurls hard at start and move carriage slightly to left and right by hand until desired effect is produced, then partially relieve pressure before starting power feed, to avoid undue wear on centers and center holes. A slow but sure process is to press nurls hard on work and pull belt forward and backward by hand until nurling is reproduced all around work, then start machine and use power feed.

Note.— Nurls for machine work are obtainable for fine, medium, and coarse pitches as shown in Fig. 43. Medium is most used.

ADVANCED MACHINE WORK

SECTION 7

CYLINDRICAL GRINDING (EXTERNAL) INTERNAL GRINDING

Cylindrical Grinding. Grinding on Two Dead Centers. Universal Grinding Machines. Grinding Wheels. Problems in Cylindrical Grinding. Internal Grinding.

CYLINDRICAL GRINDING

1. Machine grinding is a scientific method of producing cylindrical, conical, and plane surfaces accurately, rapidly, and economically, with automatic grinding machinery; and for duplicating machine parts.

Almost any material, as hardened and soft steel, wrought iron, cast iron, brass, copper, aluminum, vulcanite, and wood fiber, may be ground accurately. Grinding machines are not designed to remove a large amount of stock but to produce accurate dimensions on work that has been rough turned.

GRINDING ON TWO DEAD CENTERS

2. The principle of grinding on two dead centers. — Ground work is more accurate than turned work. This accuracy is obtained by grinding the work on two dead centers as in Fig. 1, which eliminates the error caused by wear of spindle bearings. The wear of dead centers is readily cor-

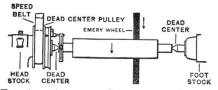

FIG. 1. — ACCURACY OBTAINED BY GRINDING ON TWO DEAD CENTERS.

rected by grinding the centers, but the wear of spindle bearings cannot be readily corrected.

Machine grinding is also used for finishing as a substitute for polishing as it is more economical than using files and emery cloth.

UNIVERSAL GRINDING MACHINES

3. Grinding machines may be divided into four general classes: Universal grinding machines for general work; plain grinding machines for outside work only; cutter grinding machines for grinding cutters and similar work; and surface grinding machines for surface grinding only.

Special grinding machines are designed for special purposes, as internal grinding machines, piston grinding machines, and crank grinding machines to rough as well as to finish automobile crank shafts without rough turning.

4. Universal grinding machine, Fig. 2, like the engine lathe, is designed for general work, as straight and taper grinding, both external and internal, facing sides of disks, grinding clearances on cutters, reamers, etc.

SCHEDULE OF PARTS

Countershaft Drive.	Belt Feed.
5. *Machine Parts.*	*N* — Water supply pipe.
A — Base, contains locker for small parts.	*P* — Belt to drive work.
B — Bed, contains feed and reversing, mechanism.	*Q* — Belt to drive grinding wheel.
C — Table.	*R* — Belt to drive feed cone.
C' — Swivel table.	*S* — Belt to drive pump.
D — Swivel headstock (90°), position adjustable.	
E and *E'* — Dead centers.	**6.** *Countershaft Parts.*
F and *F'*—Dead center pulleys; large or small may be used.	I. Line shaft.
G — Live spindle pulley.	II. Tight and loose pulley mechanism controlled by shipper.
H — Stop for live spindle pulley.	III. Belt to drive sleeve carrying feed cone.
J — Footstock, position adjustable.	IV. Belt to drive work drum.
K — Lever to spring dead center " in " or " out."	V. Work drum.
L — Grinding wheel, head mounted on swivel slide and adjustable " forward " or " backward; " it can also be fed by hand to grind steep tapers.	VI. Clutch operated by shipper brake.
	VII. Belt to drive grinding wheel drum.
	VIII. Grinding wheel drum.
	IX. Belt to drive dead center pulley *F* to give higher speed.
M — Grinding wheel.	X. Belt to drive live spindle pulley *G*.

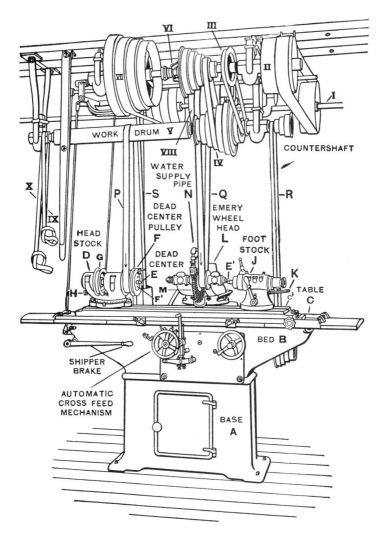

FIG. 2. — UNIVERSAL GRINDING MACHINE.

GRINDING WHEELS

7. Grinding wheels, Chart, Fig. 3, are made from various abrasives, — emery, corundum, alundum, aloxite, carborundum, crystolon, carbolite. The abrasive is mixed with a bond, the amount and composition of which makes a wheel hard or soft and determines its grade or degree of hardness.

They are molded, baked, turned, balanced, and tested for hardness and speed.

The number of a wheel is determined by the number of meshes per linear inch of the sieve through which the abrasive has passed, and for grinding wheels runs from No. 12 to No. 120. Lower numbers indicate coarser wheels and higher numbers finer wheels. See *Principles of Machine Work*.

8. Coarseness.	**Grade of Hardness.**
Coarseness determined by No. of abrasive.	**Hardness determined by** bond binding grains of abrasive together.

Processes.

Vitrified.	Tanite.
Silicate.	Vulcanite.
Elastic.	Celluloid.

Each process is best adapted to certain *classes* of grinding, yet each may be used for *general* grinding.

9. Vitrified wheels, reddish brown in appearance, are made by mixing the abrasive with a bond of clay, sand, spar, etc., they are then molded into shape, placed in ovens similar to pottery kilns and subjected to a prolonged and intense heat (3000° F.). These wheels are even in texture, open and porous. They are especially adapted to grinding hardened steel, as they will not glaze easily and are thus cool-cutting. Wheels made by this process are used for general work and for cylindrical and cutter grinding. See Table of Grades, p. 707.

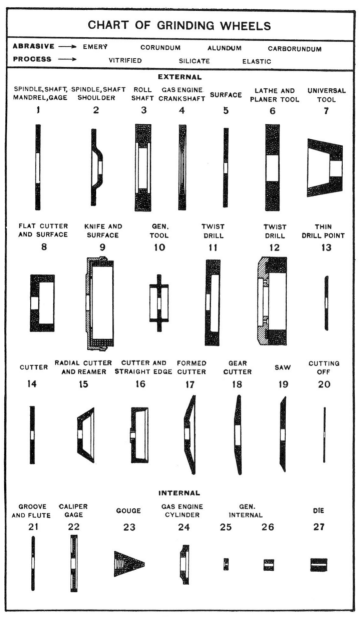

CHART OF GRINDING WHEELS

ABRASIVE ⟶ EMERY   CORUNDUM   ALUNDUM   CARBORUNDUM

PROCESS ⟶   VITRIFIED   SILICATE   ELASTIC

EXTERNAL

SPINDLE, SHAFT, MANDREL, GAGE	SPINDLE, SHAFT SHOULDER	ROLL SHAFT	GAS ENGINE CRANKSHAFT	SURFACE	LATHE AND PLANER TOOL	UNIVERSAL TOOL
1	2	3	4	5	6	7

FLAT CUTTER AND SURFACE	KNIFE AND SURFACE	GEN. TOOL	TWIST DRILL	TWIST DRILL	THIN DRILL POINT
8	9	10	11	12	13

CUTTER	RADIAL CUTTER AND REAMER	CUTTER AND STRAIGHT EDGE	FORMED CUTTER	GEAR CUTTER	SAW	CUTTING OFF
14	15	16	17	18	19	20

INTERNAL

GROOVE AND FLUTE	CALIPER GAGE	GOUGE	GAS ENGINE CYLINDER	GEN. INTERNAL		DIE
21	22	23	24	25	26	27

FIG. 3.

10. Silicate wheels, light gray in appearance, are made by mixing the abrasive with a bond of silicate of soda, and then subjecting to a low heat (300° to 400° F.). These wheels are made porous or dense, as desired, are even in hardness and have unusual strength. They are adapted for wet tool grinding and general purposes. All wheels above 30″ in diameter are made by this process. See Table of Grades, p. 707.

11. Elastic wheels, black in appearance, are made by mixing the abrasive with a bond of shellac, etc.; and then subjecting to a low heat (300° to 400° F.). These wheels are not brittle, and can be made as thin as $\frac{1}{16}$″. They are used for grinding arbors, cutters, reamers for saw gumming, and for cutting off small stock, such as thin steel tubing, wire, etc., also for grinding in narrow openings. See Table of Grades, p. 707.

12. Tanite wheels are made by mixing the abrasive with tanite (a chemical and mechanical transformation of leather) and then subjecting to heat under hydraulic pressure. They are used for general grinding in a similar manner to silicate wheels.

13. Vulcanite wheels are made by mixing the abrasive with a bond of rubber and sulphur, and subjecting to a low heat (225° F.). These wheels are not brittle and have a high factor of safety. They are used for special work.

14. Celluloid wheels, brown in appearance, are made by mixing the abrasive with celluloid; both are subjected to a low heat (250° F.).

15. Combination wheels are made by mixing several numbers of the abrasive together in the formation of the wheel. Three numbers are generally used for this purpose, as 24, 36, 46,—the coarsest grade is used for fast cutting and the finest grade is used for a smooth finish. The addition of a small amount of fine abrasive to a coarse wheel will produce wheel of great durability.

16. Table of Grades of Wheels Made by Different Processes

	VITRIFIED PROCESS.	SILICATE PROCESS.	ELASTIC PROCESS.
Very soft...	E	$\frac{3}{4}$	$\frac{3}{4}$ E
	F	1	1 E
	G	$1\frac{1}{4}$	$1\frac{1}{4}$ E
	H	$1\frac{1}{2}$	$1\frac{1}{2}$ E
Soft........	I	$1\frac{3}{4}$	$1\frac{3}{4}$ E
	J	2	2 E
	K	$2\frac{1}{4}$	$2\frac{1}{4}$ E
	L	$2\frac{1}{2}$	$2\frac{1}{2}$ E
Medium....	M	3	3 E
	N	$3\frac{1}{2}$	$3\frac{1}{2}$ E
	O	4	4 E
	P	$4\frac{1}{2}$	$4\frac{1}{2}$ E
Hard.......	Q	5	5 E
	R	$5\frac{1}{2}$	$5\frac{1}{2}$ E
	S	6	6 E
	T	$6\frac{1}{2}$	$6\frac{1}{2}$ E
Very hard..	U	7	7 E

Attention. — The grade of a grinding wheel is more important than the number of the abrasive. The number influences the grade to a slight extent, a fine wheel, as No. 90, grade M, is slightly harder than a coarse wheel same grade as No. 60, grade M.

17. Shapes of wheels. — Plain disk wheels are used for straight and taper work, and plain and cup wheels for surface grinding.

Cup, bevel, and disk wheels are used for grinding drills, cutters, and for grinding to shoulder. Small wheels are used for internal grinding. See Chart of Grinding Wheels, Fig. 3.

18. Selection of wheel depends on hardness of material and nature of work. In general, use a hard wheel for rough grinding and a medium or soft wheel for accurate grinding and fine finish. If a wheel is too hard, it will glaze quickly and heat the work; if too soft, it will wear rapidly and prevent accurate grinding. A coarse wheel is less liable to heat than a fine wheel, as the latter is apt to glaze. A perfect wheel is one with a bond that will release the grains before glazing takes place, and hard enough to take a series of cuts without changing its shape or diameter, as accurate grinding depends upon the sizing power of wheel.

19. To mount wheels. — Place rubber or blotting-paper washers between wheels and flanges, and clamp. Wheels should fit easily on arbors. Some are bushed with soft metal to fit standard shafting and will run fairly true. Wheels to be interchangeable without truing are provided with individual taper bushings. Small wheels for internal grinding have counterbore depressions for fillister head screws.

20. The periphery speed for wheels should be approximately from 5000 to 6500 feet per minute (F.P.M.). A higher speed may cause the wheel to break. Should a wheel heat or glaze, run it slower. If it is too soft, it can often be made to hold its size by using a higher speed.

Table of Speeds for Grinding Wheels

The table given below designates number of revolutions per minute for specified diameters of wheels, to cause them to run at the respective periphery rates of 4000, 5000, and 6000 feet per minute.

Diam. Wheel.	Revolutions per Minute for Surface Speed of 4000 Ft.	Revolutions per Minute for Surface Speed of 5000 Ft.	Revolutions per Minute for Surface Speed of 6000 Ft.
1 inch.	15,279	19,099	22,918
2 "	7,639	9,549	11,459
3 "	5,093	6,366	7,639
4 "	3,820	4,775	5,730
5 "	3,056	3,820	4,584
6 "	2,546	3,183	3,820
7 "	2,183	2,728	3,274
8 "	1,910	2,387	2,865
10 "	1,528	1,910	2,292
12 "	1,273	1,592	1,910
14 "	1,091	1,364	1,637
16 "	955	1,194	1,432
18 "	849	1,061	1,273
20 "	764	955	1,146
22 "	694	868	1,042
24 "	637	796	955
30 "	509	637	764
36 "	424	531	637

21. The periphery speed of work (F.P.M.) should be proportional to the grade and speed of wheel, also to the diameter of work. The speed varies from 15 to 60 feet per minute for different classes of work. The higher speed is best for cast iron and the lower for duplicate work.

22. Direction of rotation of wheel and work must be opposite at cutting point as in Fig. 4.

FIG. 4. — DIRECTION OF ROTATION OF WHEEL
AND WORK FOR EXTERNAL GRINDING.

23. Feed of table or traverse speed is in proportion to the width of wheel face and finish required. Use coarse feed for roughing. For large work with a heavy machine, use one-half to three-fourths width of wheel to each revolution of work; on light machines use one-third to one-half width of wheel. For a very fine and accurate finish, use one-fourth to one-third width of wheel.

24. Depth of cut. — For roughing take deep cuts, — .001″ to .004″ at each stroke; for finishing, light cuts, — .00025″ to .0005″ at each stroke. The sparks thrown off by the grinding wheel indicate the depth of cut: a large volume of sparks indicates a heavy cut, and a small volume a light cut.

25. Width of face of wheel.— A wide wheel with a coarse feed, removes stock rapidly, and is used where it can pass from one-fourth to one-half its width beyond the end of work or recess.

A narrow wheel is used to grind to a shoulder not recessed to produce an accurate diameter next to shoulder or with narrow recess. The face of a wheel may be narrowed by beveling a corner with a diamond tool.

26. To true grinding wheel.— Grinding wheels wear smooth or become glazed by use, so that they will not cut freely. A new wheel will not run true and must be *trued* before using.

To true *face* of wheel, mount diamond tool *A*, Fig. 5, in fixture *B*. Feed tool to touch revolving wheel with cross feed, then traverse tool by power or hand long. feed. To true

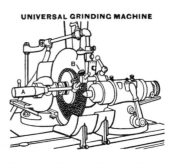

Fig. 5.— Truing Face of Wheel   Fig. 6. — Truing Side of Wheel
    with a Diamond Tool.   with a Diamond Tool.

side of wheel, clamp diamond tool *A*, Fig. 6, in slide of fixture *B* at right angles to wheel. Feed tool to wheel by hand long. feed until it touches revolving wheel, then operate tool by hand with cross feed.

If a wheel is soft, but little attention may be given it while roughing, as it will wear away fast enough to keep sharp, but it must be sharp and also true, to produce a fine finish.

Attention. — Grinding wheels may be trued dry, but the wear on the diamond tool is much less if plenty of water is used.

27. To true centers, Fig. 7. — Remove centers, clean holes with waste. Clean dead center and insert it in live center spindle. Belt machine to revolve live spindle, pull out pin *A*, swivel headstock to 30° and clamp. Set reversing dogs and grind center by trial and correction to fit center gage (60°) See p. 113. Replace dead center, clean and insert live center in its spindle and grind. Both centers are hardened and tempered. In a plain grinding machine, use a special attachment. Some small grinding machines permit swinging of table to grind centers.

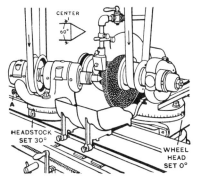

FIG. 7. — GRINDING THE CENTER OF UNIVERSAL GRINDING MACHINE.

28. Methods of driving work. — (Two dead centers.) Centered work is ground on two dead centers and driven by dead center pulley and balanced dog, as in Fig. 1. To

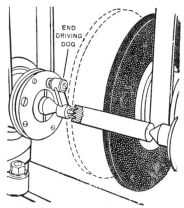

FIG. 8. — END DRIVING DOG FOR WORK OF ONE DIAMETER.

grind work of one diameter, use end driving dog, as in Fig. 8. (With revolving spindle.) For work held in chuck, as face and internal grinding, the spindle revolves.

Some classes of centered work may be rapidly handled and ground by using spindle and triangular live center drive, as in Fig. 9.

A triangular center punch is driven lightly into one center hole to make it fit center. The work is then mounted on centers and not moved until finished.

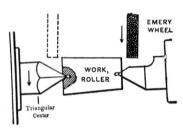

FIG. 9. — TRIANGULAR CENTER DRIVE FOR RAPID PRODUCTION.

29. To set swivel table to grind straight work. — Set both zero lines on headstock swivel (graduated in degrees) to coincide, then set both zero lines on swivel table (graduated scale in inches per foot and degrees) to coincide. To obtain accurate setting, take a few light cuts and caliper both ends with micrometer. If they differ, release clamp bolts and swing swivel table a little with adjusting screw at end, clamp bolts and repeat above process until desired accuracy is obtained.

Attention. — The graduations on headstock swivel and swivel table are helpful, but the coincidence of lines, even if located with a magnifying glass, give only approximate accuracy, and it is only by the process of trial and correction that accuracy is obtained.

30. Slight tapers are obtained by setting swivel table to approximate taper by scale and using a standard taper ring gage to determine exact taper, which is obtained by trial and correction in the same manner as a straight setting.

31. Steep tapers on work held in chuck or on headstock spindle are obtained by swiveling the headstock. For work mounted on centers, tapers are obtained by setting wheel slide, and for two abrupt tapers, outside or inside, by setting both headstock and wheel slide at proper angles.

32. Wet and dry grinding. — Wet grinding is rapid and accurate. An insufficient or fluctuating supply of water will cause a change of temperature and produce inaccurate work. If the nature of the work will not permit, or the machine is not arranged for wet grinding, good results may be obtained by grinding dry with very light cuts. Water guards for wet grinding are supplied. See *A*, Fig. 10.

33. Lubricants for grinding. — Use water, and to prevent rusting of machine, add enough sal soda to the water to show a deposit on machine. Machine oil is sometimes added to mixture.

Aluminum is ground with a lubricant of kerosene, or a mixture of kerosene and machine oil.

As most cutter grinders are not arranged for wet grinding, cutters and reamers are ground dry.

34. To prepare work to be ground. Allowance for grinding. — Short work that is not to be hardened is rough turned from .006″ to .010″ large. For large work and long, slender work that will spring and run out of true after turning, and for hardened steel that is liable to spring in hardening, it is good practice to allow from .020″ to .030″ to grind off. A large allowance means but one cut in the lathe, and it is more economical to grind with the modern grinding machine than to take a second cut in the lathe to reduce the allowance.

35. Rough and finish grinding. — In manufacturing machine parts in lots, it is good practice to rough grind all pieces to within .002″ or .003″ before finishing any. The wheel then need be trued only twice, — once in roughing and once for finishing.

36. The finish of work. — An ordinary shaft should be given a " commercial " finish; that is, it should be ground to a good smooth surface. A forcing fit need not be ground to such a fine finish, but close to size. Gages and fine work of this class must be ground both to a fine finish and to an exact size. A fine accurate finish is generally obtained with light cuts, slow feed, a true wheel, and a liberal supply of water.

37. Expansion of work while grinding. — Expansion is caused by the friction of the wheel. In grinding work dry, particularly long shafts, it sometimes is made to run out of true and this is due to its own internal strains or unequal expansion. A grinding operation shows error or truth; an *uneven* volume of sparks indicates error, and an *even* volume of sparks truth of work. To remedy this, flood work with water, reduce speed of work and feed one-fourth thousandth at each end of stroke until work is cylindrically true. When the work becomes round again, increase its speed and the depth of cuts. Spring back rests *B, B, B*, Fig. 17, are often used to remedy this error.

Attention. — A distinct difference in the sparks, however, may indicate an error as small as one-tenth to one-fourth thousandth of an inch or less, which may be ignored for some classes of work.

38. Seasoning work. — Long slender work is often laid away for a period of time between roughing and finishing, to relieve internal strains, when great accuracy is required. This is usually unnecessary for short or heavy work.

39. Care of machine and work. — All bearings should be kept well oiled. The cross slide should be cleaned, oiled, and if need be, adjusted to move smoothly. The centers and center holes in work must be kept absolutely true and clean and well oiled to produce accurate work. While most of the bearings on modern grinding machines are dust-proof, nevertheless the machines should be kept clean.

40. Measuring tools for grinding. — Use micrometer calipers and limit gages for general work. For very accurate work use a ten-thousandth micrometer or measuring machine.

Warning. — Push the shipper slowly and watch the grinding wheel as it slowly starts. Never stand in front of a wheel when it is starting, to avoid injury if the wheel should happen to break. Modern grinding wheels have a high factor of safety and seldom burst at the speed recommended, but any wheel may be broken by careless usage; accidentally moving the headstock, or footstock, or work, against the side of the wheel, especially a thin, vitrified wheel, revolving or stationary, may break it.

PROBLEMS IN CYLINDRICAL GRINDING

41. Adjustments and movements to operate universal grinding machine, Fig. 10.

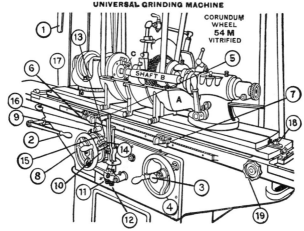

UNIVERSAL GRINDING MACHINE

CORUNDUM WHEEL 54 M VITRIFIED

FIG. 10. — GRINDING A RUNNING FIT.

SCHEDULE OF PARTS

1. Shipper to start and stop grinding wheel.
2. Shipper brake to start and stop work.
3. Knob to start or stop long. power feed work (table feed).
4. Long. feed hand wheel.
5. Lever to reverse long. feed by hand.

6 and 7. Dogs to reverse long. feed automatically.

8. Cross feed hand wheel.

Automatic Cross Feed.

9. Pawl.
10. Ratchet wheel (each tooth reduces work 1/4/1000″ in diameter and each graduation 1/1000″).

11 and 12. Adjusting screws which control movements of pawl 9 and depth of cut at each end of stroke or at each reversal of table.

13. Perpendicular latch.
14. Latch head.
15. Shield.
16. Horizontal latch.

Swivel Table.

17 and 18. Bolts to clamp swivel table.

19. Screw to adjust swivel table. To grind straight or taper, see Graduated Scale at end of swivel table.

42. To grind running fit on two dead centers, Fig. 11.

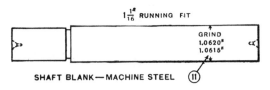

FIG. 11. — SCHEDULE DRAWING.

Specifications: Shaft blank, machine steel rough turned

Machines and tools: Universal grinding machine.
Grinding wheel 12″ × ¾″, No. 54, Grade M, vitrified, grinding dog.
Speed of wheel, 5000 F.P.M. Speed of work, 50 F.P.M.
Feed ⅓ width of wheel per revolution.
Oil bearings and two dead centers with machine oil.
Lubricant, solution of sal soda and water.
Time: Study drawing and schedule in advance, 10 min. — Oil machine and true wheel, 7 min. — Grind fit with machine "set up," 20 min. — Clean grinder, 3 min. — Total, 40 min. ("Set up machine to grind straight, 20 min. extra.)

SCHEDULE OF OPERATIONS

I. Preparatory Adjustments.
True wheel and centers, p. **7**09. Set zero lines on headstock and swivel table to grind approximately straight, p. **7**12. Arrange water guards. See Fig. 10. Unlock horizontal latch 16 from latch head 14, throw out pawl 9 and move grinding wheel back with hand cross-feed wheel 8, to allow space to mount work. Mount shaft B on centers with copper under set screw of grinding dog C. Start grinding wheel, work and feed, 1, 2, 3, and adjust table dogs 6, 7, to obtain length of stroke and avoid wheel striking dog or footstock.
Set automatic cross feed 11,

12, Fig. 10, to feed .0005″ to .001″ (2 to 4 teeth on ratchet), at each end of stroke or reversal of table.

II. Grind Trial Diameter (11), Fig. 11. (Grind out tool marks only.) Move revolving wheel to lightly cut revolving work with cross-feed hand wheel 8. Turn on water and throw in feed pawl 9 to mesh with ratchet wheel 10, and allow automatic cross feed to take three or four trial cuts whole length of work to grind out tool marks. Then throw out pawl 9 and, without moving cross feed, allow wheel to pass over work several times until cutting nearly dies out.

Stop work, feed and wheel, 2, 3, 1, with wheel at footstock end.

III. Correct Straightness. Measure work at both ends with 2″ micrometer, and if not straight, move swivel table (see p. 712). Take one or two cuts with automatic cross feed. Measure, and repeat until machine grinds straight.. If there is danger thereby of grinding shaft too small, use trial piece same length.

IV. Find Amount Oversize. Measure work with 2″ micrometer. Subtract 1.0640″ (1.0620″ + .002″ for finishing) from reading of micrometer. For example, reading of micrometer may be 1.0740″, then 1.0740″ − 1.0640″ = .0100″ for rough grinding. As each tooth in ratchet wheel 10 = ¼/1000″ in diameter of work and work is .010″ large, work is 40 teeth large.

V. Set Automatic Cross Feed. Raise perpendicular latch 13 in head 14, throw in pawl 9 and, without moving cross feed, move shield 15 to right or left until end of shield and pawl are 40 teeth apart, then drop latch 13 in ratchet wheel 10.

VI. Rough Grind (11), .0020″ Large, as follows: Start wheel, work and feed, 1, 2, 3, and rough grind automatically until shield 15 lifts pawl 9. Then lock latch head 14, with horizontal latch 16 (this locks coarse cross feed).

Throw out pawl 9 and allow wheel to pass over work several times until cutting nearly dies out. Stop work with wheel at footstock end with shipper brake 2, shut off water and measure, and from reading subtract 1.0620″ (1.0640″ − 1.0620″ = .0020″.)

VII. Finish Grind (11), Fig. 11, with pinch feed (fine cross feed) ¼/1000″ in diameter of work, as follows: Start work. Take two cuts ½/1000″ each (two pinches at each end of work). Stop work and measure (1.0630″). Take one cut ¼/1000″ (one pinch at footstock end of work) and measure and repeat until work measures 1.0620″.
Limit 1.0620″ to 1.0615″.

Attention. — This shaft is for running fit in 1 1/16″ hole. Hole limit 1.0625″ + or − .00025″. See Running Fits, pp. 214, 215.

Information. — The equipment for wet grinding consists of water tank, pump, piping and valve to supply water to work and water guards, channels in table and frame to carry water back to filter and tank.

Note. — When through grinding, unlock horizontal latch 16 from latch head 14, move grinding wheel back to allow space to mount next piece of work. Clean machine with waste.

Warning. — To avoid accident, start grinding wheel slowly.

43. To grind a phosphor bronze taper bushing, *A*, **Fig. 12.**
Bushing blank, phosphor bronze. Rough turned to enter taper hole within $\frac{3}{8}''$ of small end.

Corundum grinding wheel (plain), $12'' \times \frac{1}{2}''$, No. 54, Grade M, vitrified.

Speed of wheel, 6500 F.P.M. Speed of work, 75 F.P.M.
Feed, $\frac{1}{3}$ width of wheel.
True wheel and centers.

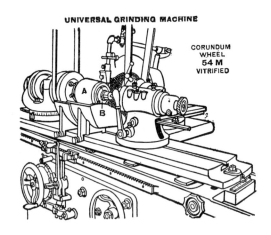

FIG. 12. — GRINDING A BRONZE TAPER BUSHING.

Set automatic feed to feed .001″ at both ends of stroke and to grind .002″ large, then use pinch feed to finish.

Place bushing *A* on a built-up taper mandrel *B* and mount on dead centers.

Set swivel table to grind taper. To obtain correct taper, take light cuts and test taper in taper hole in headstock or frame of machine, and adjust table until taper fits.

44. Adjustments and movements to operate universal and tool grinding machine, Fig. 13.

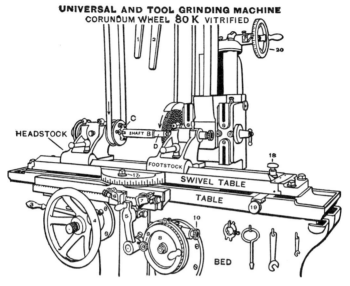

UNIVERSAL AND TOOL GRINDING MACHINE
CORUNDUM WHEEL 80 K VITRIFIED

FIG. 13. — GRINDING A FORCING FIT.

SCHEDULE OF PARTS

1. Shipper to start and stop grinding wheel.
2. Shipper brake to start and stop work.
3. Lever to start or stop long. power feed (table feed).
4. Long. feed hand wheel.
5. Lever to reverse long. feed by hand.
6 and 7. Dogs to reverse long. feed automatically.
8. Cross-feed hand wheel. Each graduation reduces work 1/1000″ in diameter.

Fine Cross Feed.

9. Thumb nut for controlling fine or coarse cross feed.

10. Thumb screw for operating fine cross feed.

Swivel Table.

17. Bolt for clamping swivel table.
18. Spring knob for quick adjustment of swivel table.
19. Thumb screw for fine adjustment of swivel table. To grind straight or taper, see Graduated Scale at end.

Grinding Wheels.

20. Hand wheel to elevate grinding wheel.

45. To grind forcing fit on two dead centers, Fig. 14.

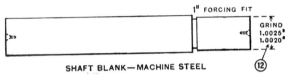

FIG. 14. — SCHEDULE DRAWING.

Specifications: Shaft blank, machine steel rough turned.

Machines and Tools: Universal grinding machine.
Grinding wheel $7'' \times \frac{1}{2}''$, No. 80, K, vitrified, grinding dog.
Speed of wheel, 5000 F.P.M.   Speed of work, 50 F.P.M.
Feed $\frac{1}{3}$ width of wheel per revolution.
Oil bearings and two dead centers with machine oil.   Grind dry.

Time: Study drawing and schedule in advance, 5 min. — Oil machine and true wheel, 7 min. — Grind fit with machine "set up," 15 min. — Clean grinder, 3 min. — Total, 30 min.   ("Set up" machine to grind straight, 10 min. extra.)

SCHEDULE OF OPERATIONS

I. Preparatory Adjustments. True wheel and centers, p. 709. Set zero lines on swivel table to grind approximately straight. See p. 712.   Loosen nut 9, Fig. 13, and move grinding wheel back with hand cross-feed wheel 8 to allow space to mount work.   Mount shaft B on centers with copper under set screw of grinding dog C.   Start grinding wheel, work and feed, 1, 2, 3, and adjust dogs 6, 7 to obtain length of stroke of part D and avoid wheel touching shoulder of work or footstock.

II. Grind Trial Diameter (12) Fig. 14. (Grind out tool marks only.)   Move **revolving** wheel to **lightly cut revolving work** with cross-feed hand wheel 8.   Tighten nut 9 (this locks coarse cross feed), and use fine cross-feed thumb screw 10.   Graduations on cross feed reduce work 1/1000″ in diameter.   Take two, three, or four trial cuts ($\frac{1}{2}$/1000″ each), to grind out tool marks, feeding at footstock end only, and allow wheel to pass over work several times until cutting nearly dies out, then stop table and work with grinding wheel at footstock end with shippers 3, 2.

III. Correct Straightness. Measure work at both ends with 2″ micrometer and, if not straight, move swivel table (see p. 712). Take one or two cuts with fine cross feed.   Measure and repeat until machine grinds straight. If there is danger thereby of grinding shaft too small, use trial piece same length.

IV. Find Amount Oversize. Measure work with 2″ micrometer. Subtract rough diameter of work 1.0025″ + .002″ for finishing from reading of micrometer. For example, reading of micrometer may be 1.012″, then 1.012″ − 1.0045″ = .0075″ for rough grinding.

V. Rough Grind (12), **.002″ Large,** as follows: Use $\frac{1}{2}$/1000″ cuts with thumb screw 10 at footstock end only, as follows: Take 12 cuts and measure (1.0060″); take 3 cuts and measure (1.0045″).

VI. Finish Grind (12), Fig. 14. Use $\frac{1}{4}$/1000″ cuts and feed at footstock end only, as follows: Take 5 cuts and measure (1.00325″); take one cut and measure, and repeat until work measures 1.0025″. Limit 1.0025″ to 1.0020.″.

Attention. — This shaft is to be forced into 1″ hole. Hole limit 1000″ + or −.00025″. See Forcing Fits, p. 217. Clean machine with brush and waste.

46. To grind a shoulder, Fig. 15. — Spindle, crucible steel. Grinding wheel (plain) 12″ × $\frac{1}{2}$″, No. 60, Grade K, vitrified.

Speed of wheel, 5000 F.P.M. Speed of work, 50 F.P.M.

Feed, $\frac{1}{3}$ width of wheel.

True wheel and centers. Set machine to grind straight.

Mount spindle blank A, recessed as at B for side clearance, on centers. Control depth of cut

FIG. 15. — GRINDING A SHOULDER.

by long. feed, and feed by cross feed. If shoulder is large use special wheel mounted on end of wheel spindle C and set swivel $\frac{1}{2}$° from 90° to give clearance.

47. Plain grinding machine for commercial grinding, Fig. 16. — The essential parts of this machine are similar to those of a universal grinding machine except that the headstock and wheel slide do not swivel, and the swivel table forms a water guard. It is designed for straight and taper outside work only. It is made strong and rigid to grind work rapidly.

Some plain grinding machines have a traveling grinding wheel instead of a traveling table.

48. To grind cast-iron roll, *A*, Fig. 16.

Roll blank, cast iron.   Rough turned .010″ to .020″ large.

Alundum grinding wheel (plain) 20″ × 2″, No. 24 Combination, Grade K, vitrified.

Speed of wheel, 5000 F.P.M.   Speed of work, 50 F.P.M.

Feed, ¾ width of wheel.

True wheel and centers.   Set machine to grind straight.

PLAIN GRINDING MACHINE

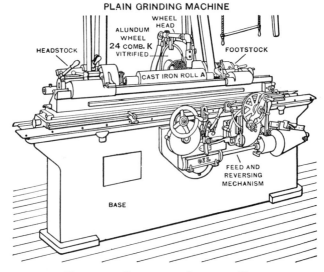

Fig. 16. — Grinding a Cast-iron Roll.

Set automatic feed to feed at both ends of stroke and to grind .002″ large, then use pinch feed to finish.

Mount roll *A* on dead centers.

Place universal back rests 6 to 10 diameters apart and grind to required diameter with water.

49. To grind slender shaft, *A*, Fig. 17.

Shaft blank, machine steel smooth turned .012″ large.

Alundum grinding wheel (plain) 12″ × ¾″, No. 60, Grade L, vitrified.

Speed of wheel, 6500 F.P.M.   Speed of work, 90 F.P.M. Feed, $\frac{3}{4}$ width of wheel.

True wheel and centers.   Set machine to grind straight.

Set automatic feed to feed at both ends of stroke and to grind .002″ large, then use pinch feed to finish.

Mount shaft A on dead centers.   Place three universal back rests B against shaft, 6 to 10 diameters apart, see Fig. 17, and grind to required diameter.

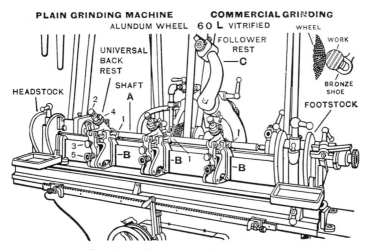

Fig. 17. — Grinding a Slender Shaft.

50. Back rests, plain and universal, are fixtures used to reduce vibration and permit a greater depth of cut on straight and taper work both large and small.

The plain rest has a single shoe of wood or soft metal and is used for small, short work.

The universal rests B, B, B can be delicately adjusted. Select bronze shoes 1, 1, 1, (see detail) the size of finished work, move screws 2 to maintain contact of shoe upon work while grinding trial piece, and adjust stop screw 3 to preserve diameter when finished.

Adjust screws 4 and 5 to regulate pressure of springs upon shoes and work.

Duplicate pieces are ground without disturbing adjustments except slightly for wear of shoe 2 and diameter 3.

Follower rest *C* is used on slender work that has been rough ground .001″ large. It cannot be used on taper work. Adjust with wheel and work in motion.

51. Automatic magnetic sizing grinder, Fig. 18.

The mechanical automatic feeding mechanism is electrically controlled and duplicates straight and taper work regardless of wear of grinding wheel.

A coarse feed for roughing automatically switches to a fine feed for finishing, and when work is finished to size the feed automatically stops.

52. To rough and finish grind spindles in an automatic magnetic sizing grinder, Fig. 18. Duplicate work.

Spindle blank, machine steel, rough turned 0.020″ large.

Corundum grinding wheel (plain) 18″ × 1½″, No. 40, Grade M, vitrified.

Speed of wheel, 5000 F.P.M. Speed of work, 50 F.P.M.

Feed, ¾ width of wheel.

True wheel and centers. Set machine to grind straight.

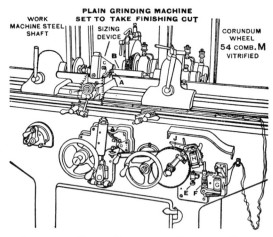

Fig. 18. — Automatic Sizing Device for Grinding Duplicate Work.

To set automatic sizing device, rough and finish grind one piece to required size with automatic mechanical feed, and use this as a master piece to set sizing device. Place diamond-point bearer *A*, on master piece *B* and adjust arm *C* by screw *D* to make an electrical contact. Then mount blank for next piece on centers. Adjust coarse feed for roughing by screw *E* to feed 001″ at each end of stroke, and adjust screw *F* to have magnet *G* throw out the coarse and throw in the fine feed one tooth (one-eighth thousandth) when piece is within .002″ of size.

Magnet *H* throws out latch *J* and stops feed when piece is ground to correct diameter

Attention. — The coarse and fine feeds and the amount allowed for finishing may be varied for different classes of work. Dry batteries or a plug in a lamp socket of a direct-current circuit will operate device.

53. To grind a straight and taper bearing on a spindle, Fig. 19.

Spindle blank, crucible steel unannealed. Rough turned .010″ to 020″ large.

Corundum grinding wheel (plain) 12″ × 1″, No. 54, Grade M, vitrified.

Speed of wheel, 5000 F.P.M.   Speed of work, 50 F.P.M.
Feed, ⅓ width of wheel.
True wheel and centers.

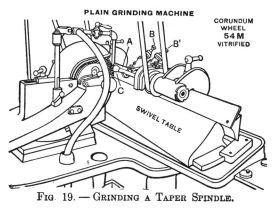

FIG. 19. — GRINDING A TAPER SPINDLE.

Set automatic feed to feed .001″ at both ends of stroke and to grind .002″ large, then use pinch feed to finish

Mount spindle *A* on dead centers.

Place two universal back rests *B* and *B′* against straight portions of work. After straight portions are ground, set swivel table to grind taper bearing *C*. To obtain correct taper, take light cuts, try taper in box and adjust table until taper fits. Use Prussian blue for marking in taper hole.

54. To grind taper collet, *A*, Fig. 20.

Collet blank, machine steel. Rough turned .010″ to .020″ large and recess at *B*.

Corundum grinding wheel (plain) 12″ × ½″, No. 54 Combination, Grade M, vitrified.

Speed of wheel, 5000 F.P.M.   Speed of work, 50 F.P.M.

Feed, ¾ width of wheel.

True wheel and centers.

UNIVERSAL GRINDING MACHINE
CORUNDUM WHEEL **54** COMB. **M** VITRIFIED

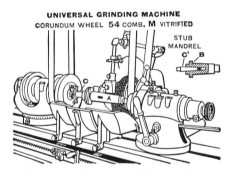

FIG. 20. — GRINDING A TAPER COLLET.

Set automatic feed to feed at both ends of stroke and to grind to .002″ large, then use pinch feed to finish.

Place stub mandrel *C*, *C′* in taper hole of collet *A*, mount on centers.

Set swivel table to grind taper. To obtain correct taper, take light cuts, try taper in gage or spindle hole and adjust table until taper fits. Use Prussian blue for marking in hole.

INTERNAL GRINDING

55. **Internal grinding** is the process of grinding straight or taper holes smooth and true, in gears, milling cutters, gages, spindles, cylinders, machine parts, etc., hardened, case-hardened or unhardened. Special internal grinding machines, and internal grinding attachments are used for universal and cutter grinding machines. Softer wheels are used than for external grinding as the contact is greater. Fig. 21 shows relation of work and wheel.

Fig. 21. — Direction of Rotation of Work and Wheel for Internal Grinding.

Vitrified grinding wheels for internal grinding range from $\frac{1}{4}''$ to $2\frac{1}{2}''$ in diameter, and from $\frac{1}{4}''$ to $\frac{3}{8}''$ face. They are made from Nos. 120, 90, 80, 70 and 60 of the different abrasives in grades J, K, $\frac{1}{4}$, $\frac{1}{2}$ and 1. Elastic wheels are used for brass.

Allowances for internal grinding should be much less than for external grinding as the operation is much slower on account of the small wheel and lack of rigidity.

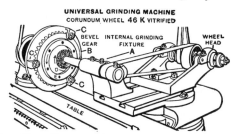

Fig. 22. — Grinding Hole in Automobile Gear.

56. **To set machine to grind straight hole,** select shaft about $1\frac{1}{4}''$ diameter and longer than hole to be ground and mount in chuck. Grind shaft and adjust swivel head until machine grinds perfectly straight, then arrange machine for internal grinding.

57. **To grind hole in case-hardened automobile bevel gear,** Fig. 22.

Specifications: Gear, case-hardened. Hole 0.010″ small.

Machine and tools: Universal grinding machine, grinding wheel (plain) 2″ × $\frac{3}{8}$″, No. 46, Grade K, vitrified. Plug gage.

Speed of wheel, 5000 F.P.M. Speed of work, 50 F.P.M.

Feed: Set automatic cross feed to feed .001″ at one end of stroke only and to grind .002″ small; then use pinch feed to finish or, after first cut, lock cross feed and use pinch feed for both roughing and finishing.

True wheel. Set machine to grind straight.

Arrange machine for internal grinding, clamp fixture A to wheel stand. Fasten gear B to face plate by clamps C and true up gear by indicator and soft hammer. Rough and finish grind hole D and face E before removing work from face plate. Test hole with plug gage or inside micrometer.

58. To grind hole in milling cutter, Fig. 23.

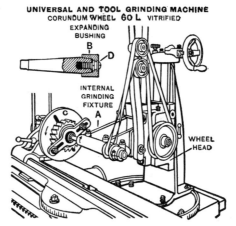

FIG. 23. — GRINDING HOLE IN MILLING CUTTER.

Specifications: Cutter hardened. Hole .008″ small.

Machine and tools: Universal and tool grinding machine, grinding wheel (plain) $\frac{3}{4}$″ × $\frac{1}{4}$″, No. 60, Grade L, vitrified. Plug gage.

Speed of wheel, 5000 F.P.M. Speed of work, 50 F.P.M. Feed, power long., automatic or hand cross feed. True wheel. Set machine to grind straight.

Arrange cutter-grinding machine for internal grinding. Set grinding wheel head height of centers by gage, and clamp fixture A to wheel head. Insert special chuck with expanding bushing B in spindle.

To set cutter true, clamp cutter C lightly to face plate, expand bushing with screw D, then clamp cutter firmly, loosen and remove bushing.

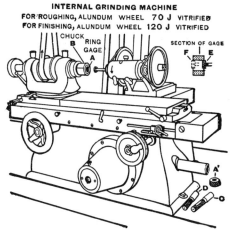

FIG. 24. — GRINDING STANDARD RING GAGE.

59. Special internal grinder, Fig. 24, is used for grinding holes as small as $\frac{1}{100}''$ in diameter with a range of speed from 10,000 to 100,000 R.P.M.

Small holes are ground with a steel plug charged with diamond dust or powdered bort.

To prevent the grinding wheel from following the curve of the hole caused by the hole being smaller at the middle than at the ends due to shrinkage in hardening, the table feed is controlled by a mechanism which retards and accelerates. It is progressively slower from near end to middle of gage and progressively faster from middle to far end, and inversely.

60. To grind 1″ standard ring gage, Fig. 24.

Specifications: Carbon steel gage blank, hardened.   Hole .005″ to .008″ small.

Machine and tools: Internal grinding machine, grinding wheel $\frac{5}{8}″ \times \frac{1}{4}″$, No. 70, Grade J, vitrified.   For very fine finishing, use No. 120, Grade J, vitrified.

Speed of wheel, 5000 F.P.M. (30,000 rev.).   Speed of work, 65 F.P.M. (250 rev.).

Feed, automatic.

True wheel.   Set machine to grind straight.

Mount gage A in special draw in chuck B.   Adjust to retard feed in center of gage.   Grind and test with limit gages C, D.   Allow .0004″ for lapping.

Attention. — Adjust stroke to reverse wheel, as at E and F to avoid grinding ends of hole tapering.

61. To grind taper hole in spindle, Fig. 25.

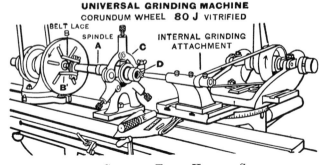

FIG. 25. — GRINDING TAPER HOLE IN SPINDLE.

Specifications: Crucible steel spindle blank.   Taper hole .010″ small.

Machine and tools: Universal grinding machine, grinding wheel (plain) $\frac{5}{8}″ \times \frac{3}{8}″$ No. 80, Grade J, vitrified.

Speed of the wheel, 5000 F.P.M.   Speed of work, 50 F.P.M.

Feed, automatic.

True wheel.

Depth of cut is obtained by hand cross feed.

Mount one end of spindle A on dead center with dog fastened to face plate by belt lacing BB' and the other end in steady rest C with a strip of cloth under jaws. Set swivel table to grind taper. To obtain correct taper take light cuts with grinding wheel D, try gage E in taper, and adjust table until taper fits. Use Prussian blue for marking.

Attention. — Set swivel table for slight tapers, and grinding wheel slide for steep tapers and large angles. Hold short work in a chuck.

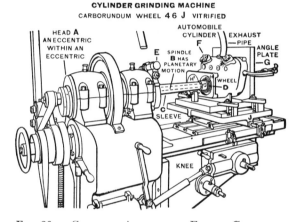

CYLINDER GRINDING MACHINE

CARBORUNDUM WHEEL **46 J** VITRIFIED

FIG. 26. — GRINDING AUTOMOBILE ENGINE CYLINDER.

62. To grind automobile cylinder, Fig. 26.

Use cylinder grinding machine to grind round, straight and smooth bores in gas-engine cylinders and other work held stationary, wet or dry.

Eccentric head A, is an eccentric within an eccentric and gives spindle B, which runs in eccentric extension sleeve C and wheel D, a planetary motion and also provides a feed by worm screw E for feeding wheel to its cut and to grind to correct diameter.

Specifications: Engine casting. Bore 0.010″ small.

Machine and tools: Cylinder grinding machine, grinding wheel (plain), $3\frac{1}{2}'' \times \frac{3}{4}''$ (offset), No. 46, Grade J, vitrified. Inside micrometer gage.

Speed of wheel, 5000 F.P.M.   Speed of work, 50 F.P.M.

True wheel.

Set and clamp rough-bored automobile cylinder F to angle plate G. Aline bore of cylinder H with spindle and grind, then move cross slide J and aline bore of cylinder H' and grind. Measure with inside micrometer or gage.

ADVANCED MACHINE WORK

SECTION 8

SURFACE GRINDING   CUTTER GRINDING

Surface Grinding. Surface Grinding Machines. Problems in Surface Grinding. Cutter Grinding. Adjusting Tooth Rest and Wheel for Obtaining Clearance. Problems in Cutter Grinding.

SURFACE GRINDING

1. Surface grinding machines are used to grind accurately flat surfaces of rough-planed or milled work hardened, case-hardened or unhardened.   See Grinding Wheels, pp. 704–710. High-power vertical manufacturing surface grinders are used to grind castings and drop forgings from the rough, rapidly and accurately.   Surface grinding is also used where accuracy is not required and takes the place of filing and polishing, saves time, files and emery cloth.   The work is either clamped to table, held in chuck or fixture, or held by magnetic chuck.

SURFACE GRINDING MACHINES

2. Surface grinding machine described, Fig. 1.

SCHEDULE OF PARTS

A — Base fitted with locker.	*G* — Grinding wheel head.
B — Table.	*H* — Handle to elevate wheel head *G*.
C — Hand long-feed handle.	
D — Hand cross-feed handle.	*J* — Exhaust pipe to carry away dust.
E — Automatic feed adjustment.	
	K — Magnetic chuck.
F and *F'* — Reversing table dogs.	*L* — Switch for turning electric current on or off.

PROBLEMS IN SURFACE GRINDING

3. To grind thin machine parts, Multiple grinding, Fig. 1.

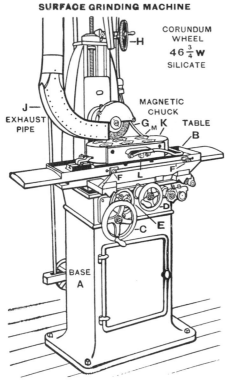

SURFACE GRINDING MACHINE

CORUNDUM WHEEL
46¾ W
SILICATE

H

J—
EXHAUST PIPE

MAGNETIC CHUCK

G
M K
TABLE
B

F L F
D
C E

BASE A

Fig. 1. — Grinding Thin Work Held by Magnetic Chuck on Surface Grinding Machine.

Specifications: Material, carbon steel rough planed 0.020″ large and hardened.

Machine and tools: Surface grinding machine, grinding wheel (plain) $9'' \times \frac{5}{8}''$, No. 46, Grade $\frac{3}{4}$ W, silicate.

Speed of wheel, 5000 F.P.M. Speed of work, 20 F.P.M.

Feed, hand or automatic.

True wheel.

Mount work M (six pieces at once), on magnetic chuck.

Finish grind one side, reverse and light grind. Remove work and measure with micrometer, replace and grind to size.

Attention. — Some work is rough ground on both sides, to relieve strains before finishing.

4. Magnetized master blocks. — Steel blocks hardened, ground and lapped square or at any desired angle and magnetized, are used to hold work on magnetic chucks to reproduce different angles accurately. Work may be held directly on block, or more firmly by placing the work between two blocks, Fig. 2, so that the magnetic circuit will pass through all three. See Magnetic Chucks, pp. 623–626.

5. To grind work square held with magnetized master blocks, Figs. 2, 3.

FIG. 2.—SQUARING SIDE OF WORK ON MAGNETIC CHUCK, USING MAGNETIZED MASTER BLOCKS.

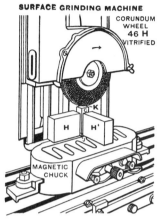

FIG. 3.—SQUARING END OF WORK ON MAGNETIC CHUCK, USING MAGNETIZED MASTER BLOCKS.

Specifications: Material, carbon steel planed 0.010″ large and hardened.

Machine and tools: Surface grinding machine, grinding wheel (plain) 9″ × $\frac{5}{8}$″, No. 46, Grade $\frac{3}{4}$ W, silicate.

Speed of wheel, 5000 F.P.M.   Speed of work, 20 F.P.M.

Feed, hand or automatic.

True wheel.

SCHEDULE OF OPERATIONS

I. Place master blocks A, A', on magnetic chuck, Fig. 2, with brass wire B (non-magnetic) between as packing block. Place work C between blocks, turn current on chuck to hold work and grind surface D.

II. Place surface D against A and grind E.

III. Place surface D against A and grind F.

IV. Place surface E against A and grind G.

V. *To square ends* place master blocks H, H', on magnetic chuck, Fig. 3, with work C on end separated from table by piece of brass J, and grind end K, reverse block and grind opposite end.

6. To grind angular work, Fig. 4.

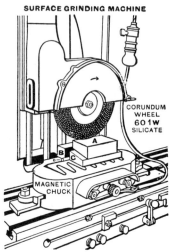

FIG. 4. — GRINDING ANGULAR WORK ON MAGNETIC CHUCK, USING
MAGNETIZED MASTER BLOCKS.

Specifications: Material, carbon steel rough planed 0.010″ large and hardened.

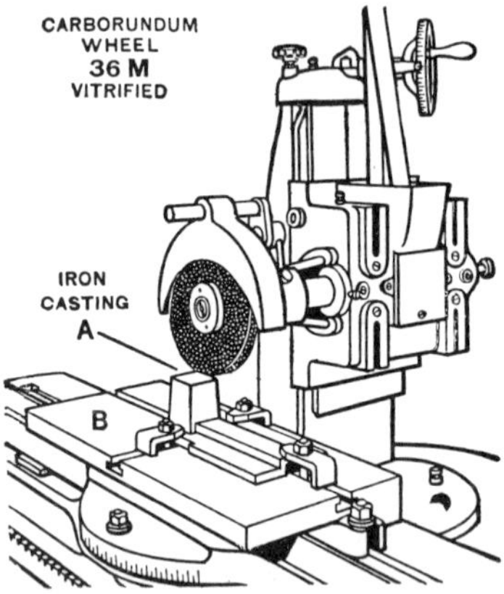

Machine and tools: Surface grinding machine, grinding wheel (plain) $9'' \times \frac{5}{8}''$, No. 60, Grade 1 W, silicate.

Speed of wheel, 5000 F.P.M.

Speed of work, 20 F.P.M.

Feed, hand or automatic.

True wheel.

Place square master block B and angular master block C on magnetic chuck with brass wire D between, in such a position that the magnetic circuit is completed through the two

FIG. 5. — SURFACE GRINDING WITH
CUTTER GRINDING MACHINE.

blocks and the work. Place work A on angular block and grind top surface.

7. To grind surface of casting, Fig. 5.

Specifications: Iron casting rough planed.

Machine and tools: Cutter grinder, clamps, crystolon or carborundum wheel (disk) 6″ × ⅜″, No. 36, Grade *M*, vitrified.

Speed of wheel, 5000 F.P.M. Speed of work, 50 F.P.M.

Feed, hand or automatic.

True wheel.

Clamp casting *A* to table plate *B*, adjust revolving wheel to touch work lightly and grind.

8. Surface grinding with planer type surface grinding machine, Fig. 6.

CARBORUNDUM WHEEL **36 M** VITRIFIED

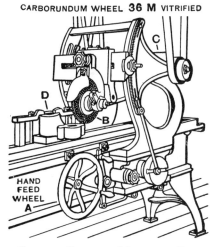

FIG. 6. — SURFACE GRINDING MACHINE. PLANER TYPE.

Specifications: Material, iron casting planed 0.010″ large, to be ground on both sides to produce a good finish.

Machine and tools: Surface grinding machine, planer type, grinding wheel (plain) 9″ × ⅝″, No. 36, Grade *M*, vitrified.

Speed of wheel, 5000 F.P.M. Speed of work, 20 F.P.M.

Feed, hand or automatic.

True wheel.

This machine is similar to planer with the addition of hand

long. feed wheel A.   Grinding wheel B is driven from drum C.
Work D is clamped to table.

9. **Surface grinding with motor-driven vertical spindle surface grinding machine,** Fig. 7.

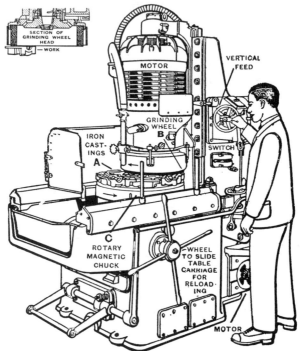

Fig. 7. — MOTOR-DRIVEN VERTICAL SPINDLE SURFACE GRINDING MACHINE.

Specifications: Cast-iron bearing in the rough $\frac{1}{32}''$ to $\frac{1}{16}''$ large.

Machine and tools: Motor-driven surface grinding machine, grinding wheel (ring) $18'' \times 5'' \times 1\frac{1}{2}''$ face, No. 20, Grade G, vitrified carbolite.

Speed of wheel 4200 F.P.M.   Speed of work 90 F.P.M.

Feed, hand or automatic.

True wheel.

This is a type of high-power vertical grinding machine used for grinding castings and forgings from the rough, as cast-iron bearings A, with ring wheel B. The grinding wheel has large contact and to grind work rapidly a coarse feed and a stream of water is used. Work is held on rotating magnetic chuck C by magnetic attraction. The work is prevented from slipping by a ring placed around the chuck and projecting $\frac{1}{8}$ of an inch or more above it.

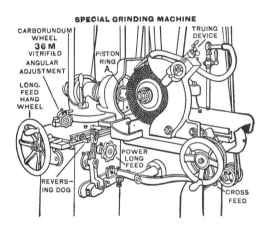

Fig. 8. — GRINDING PISTON RING.

10. To grind piston ring, Fig. 8.

Specifications: Material, cast-iron piston ring, turned and squared 0.010″ large.

Machine and tools: Special grinding machine, grinding wheel (plain) 9″ × $\frac{5}{8}$″, No. 36, Grade M, vitrified.

Speed of wheel, 5000 F.P.M. Speed of work, 100 F.P.M.

Feed, hand or automatic.

True wheel. Set machine to grind straight.

Mount ring A on magnetic rotary chuck and grind. Remove work and measure with micrometer. With a fine feed a smooth finish may be obtained.

CUTTER GRINDING

11. Cutter grinding machines are used to grind clearances on cutters, reamers and similar work. They are also used for light cylindrical and surface grinding and are a necessary supplement to milling machines.

12. Cutter grinding machines described, Fig. 9.

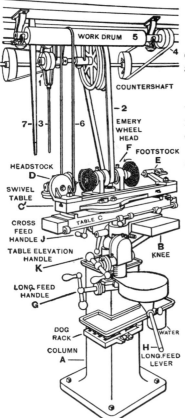

HEADSTOCK
D

SWIVEL TABLE
C'

CROSS FEED HANDLE J

TABLE ELEVATION HANDLE
K

LONG. FEED HANDLE
G

DOG RACK

COLUMN
A

FIG. 9. — UNIVERSAL CUTTER AND TOOL GRINDER.

SCHEDULE OF PARTS

A — Column.

B — Knee, can be swiveled completely around column and adjusted up or down.

C — Table. C' — Swivel table.

D — Universal headstock, has vertical and horizontal adjustment on head graduated in degrees.

E — Footstock has a compensating center to allow for expansion of work during grinding.

F — Grinding wheel head.

G — Long. feed handle.

H — Long. feed lever.

J — Cross-feed handle.

K — Table-elevation handle.

Countershaft.

1. Tight and loose pulley for driving grinding wheel mechanism.

2. Belt to drive grinding wheel.

3. Shipper to start and stop grinding wheel.

4. Tight and loose pulley mechanism for work.

5. Work drum.

6. Belt to drive work.

7. Shipper to start and stop work.

13. To grind centers, Fig. 10. — Adjust vertically to bring work centers same height as wheel spindle. Belt machine to revolve center. Remove live center, clean hole with waste, insert center. Swivel table to 30°. Use power feed and grind. Test with 60° center gage.

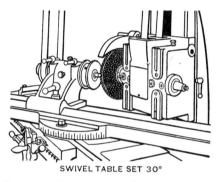

SWIVEL TABLE SET 30°

Fig. 10. — Grinding Center with Cutter Grinding Machine.

14. To grind straight. — Setting to zero lines at scale end is sufficiently accurate for ordinary cutter grinding, but for reamer grinding and to obtain accuracy, grind trial piece on centers and adjust until desired accuracy is obtained. See Setting Swivel Table for Straight Grinding, p. 712.

15. Grinding wheels for cutters should not be finer than 90 or coarser than 46, and of soft or medium grades I, J or K. Periphery speed of wheel should be about 5000 feet per minute.

16. Depth of cut and feed. — Take light cuts from .002″ to .004″ and move cutter rapidly across face of wheel, if necessary go around cutter two or three times rather than take a heavy cut which would draw the temper.

ADJUSTING TOOTH REST AND WHEEL FOR OBTAINING
CLEARANCE

17. Center height gage. — To set center of wheel spindle at

FIG. 11. — SETTING GRINDING WHEEL HEIGHT OF CENTERS.

height of work centers, raise wheel head or lower table until
centering lug A, Fig. 11, is same height as center height gage B.
Tooth rest A, Figs. 12, 13, supports tooth when grinding

FIG. 12. — SETTING TOOTH REST HEIGHT OF CENTER REST FASTENED
TO TABLE.

clearances on cutters and reamers, and is set at height of
work centers by center height gage B, Figs. 12, 13.

FIG. 13. — SETTING TOOTH REST HEIGHT OF CENTERS. REST FAS-
TENED TO WHEEL HEAD.

18. Cutter and reamer clearances. — Milling cutters under
3″ in diameter are given a tooth clearance of from 6° to 7°;
those 3″ and over, 4° to 5°, see Figs. 14, 15.

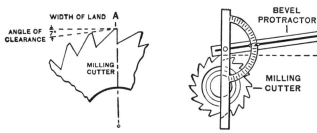

FIG. 14. — ANGLE OF TEETH AND   FIG. 15. — MEASURING CLEARANCE
WIDTH OF LAND OF MILLING   OF TEETH OF MILLING CUTTER.
CUTTER.

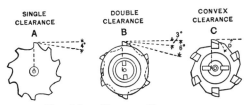

FIG. 16. — REAMER CLEARANCE.

Reamers are given 4° clearance, as at A, Fig. 16.

Reamers with thick teeth are given a double clearance; the first about 3° and the second 6°, as at B, to remove stock and leave a narrow land back of cutting edge. Some makers give reamers, C, a convex clearance, as at D.

19. Disk and Cup wheels. — Cutters and reamers are ground by two methods either the periphery of disk wheels or the side faces of cup wheels.

Disk Wheels give a slightly concave clearance, while **cup wheels** give a straight line clearance. The tendency seems to favor the latter method for wide lands.

20. Direction to revolve wheel. — Fig. 17, shows two methods (down and up) of grinding cutters and reamers

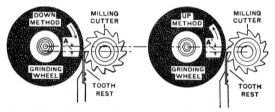

FIG. 17. — DISK WHEEL. RELATION OF TOOTH REST, MILLING CUTTER AND WHEEL TO GRIND TOOTH CLEARANCE.

There is a difference of opinion as to which method produces the keener cutting edge. The " down " method is safer as the tooth cannot be lifted by action of wheel.

21. To obtain clearances on cutter and reamer teeth when grinding with disk wheels, Fig. 17. — Set tooth rest at height of work centers with center height gage and center of wheel spindle (dial at zero) in same plane. Then raise wheel (or lower table on some grinders) amount for clearance, as at A, A' which varies with angle of clearance and diameter of wheel and is calculated by the following formula:

22. To calculate elevation for Disk Wheels:

Formula = Constant for 1° clearance (.0088″) $\times$ clearance angle $\times$ diameter of grinding wheel.

Example. — To find elevation of wheel center for 7° clearance angle, diameter of grinding wheel 6″.

Solution. — .0088″ × 7 × 6 = .3696″ or $\frac{3}{8}$″; amount to elevate wheel or to lower table.

Example. — To find elevation of emery wheel center for 4° clearance angle, diameter of grinding wheel 7″.

Solution. — .0088″ × 4 × 7 = .2464″ or $\frac{1}{4}$″; amount to elevate wheel or to lower table.

Attention. — The tooth rest may be lowered to obtain clearance instead of raising wheel or lowering table. Use formula as trial setting only. Test with bevel protractor, Fig. 15, and change elevation if necessary.

23. To obtain clearances on cutter and reamer teeth when grinding with Cup Wheels, Fig. 18. — Set wheel (or table) dial at zero. Fasten tooth rest to wheel head (stationary), Fig. 13, and adjust to center height gage. Then lower wheel head (or raise table) amount for clearance, as at A or A' which varies with angle of clearance and diameter of cutter or reamer.

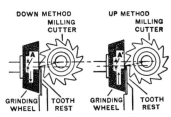

FIG. 18. — CUP WHEEL. RELATION OF TOOTH REST, MILLING CUTTER AND WHEEL TO GRIND TOOTH CLEARANCE.

24. To calculate distance to lower tooth rest for Cup Wheels.

Formula = Constant for 1° clearance (.0088″) × clearance angle × diameter of cutter or reamer.

Example. — 1″ hand reamer has angle clearance of 4°. Find amount to lower tooth rest.

Solution. — .0088″ × 4 × 1 = .0352″ or $\frac{1}{32}$″ +.

Example. — 1″ hand reamer has two clearance angles on each land, see p. 811, 1st of 3°, 2d of 6°. Find amount to lower tooth rest.

Solution for 3°. — .0088″ × 3 × 1 = .0264″ or $\frac{1}{32}$″ −.

Solution for 6°. — .0088″ × 6 × 1 = .0528″ or $\frac{3}{64}$″ +.

25. Indexing cutters and reamers. — To grind the teeth so that each tooth will be a counterpart of the others, the cutter or reamer is indexed from the teeth with the tooth rest. After the tooth rest and cutter are properly adjusted, a tooth is pressed lightly on tooth rest by hand and ground. To grind next tooth, first index by revolving cutter backward against spring tooth rest, then press the tooth to be ground lightly on the tooth rest and grind. Repeat process until all the teeth are ground.

PROBLEMS IN CUTTER GRINDING

26. To grind clearance on teeth of plain milling cutter. Fig. 19.

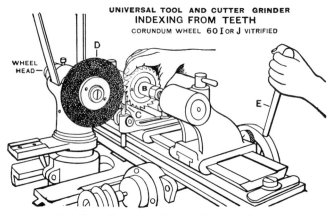

UNIVERSAL TOOL AND CUTTER GRINDER
INDEXING FROM TEETH
CORUNDUM WHEEL 60 I OR J VITRIFIED

Fig. 19. — Grinding Plain Milling Cutter.

Specifications: Plain milling cutter $2\frac{1}{2}'' \times \frac{1}{2}''$.

Machine and tools: Universal tool and cutter grinder, head and footstock, tooth rest and holder, grinding wheel (disk) $7'' \times \frac{1}{2}''$, No. 60, Grade I or J, vitrified. Bevel protractor. Speed of wheel, 5000 F. P.M. Hand feed (lever), $1''$ mandrel. Grind dry.

Time: 3 min. (once around) with machine " set up."

SCHEDULE OF OPERATIONS

Oil machine.

True wheel. Set machine to grind straight.

I. Hold cutter. Place cutter *A* on plain mandrel *B*, or gang mandrel.

II. Set tooth rest. Clamp tooth-rest holder to table and set tooth rest *C* at height of centers, as in Figs. 11, 12, with center height gage or surface gage, and elevate center of wheel *D* ⅜″ above centers for a 6″ wheel and a 7° clearance.

III. Grind clearance. Mount work on centers. Start grinding wheel. Press cutter lightly on tooth rest. Move cutter to touch revolving grinding wheel by hand cross-feed wheel and take a cut of from .002″ to .004″. With hand long. feed *E*, move cutter forward and backward quickly, to grind clearance.

IV. Test angle of clearance. With bevel protractor, test angle of clearance, as in Fig. 15; if necessary, change height of wheel to get desired clearance.

V. Index. Index cutter to next tooth by revolving it backward against spring tooth rest, then press the tooth to be ground lightly on the tooth rest and grind.

Repeat until all teeth are ground. Take cuts of from .002″ to .004″ each, and go around at least twice. Clean machine with brush and waste.

Warning. — Take several light cuts in preference to one heavy cut that may draw the temper.

Information. — There is no standard width of land for the teeth of milling cutters, but cutters not over 3″ in diameter are usually milled to give a land of about $\frac{1}{32}$″ in width, and ground to give a land of about $\frac{3}{64}$″.

Attention. — Plain cutters that are made in large lots, or

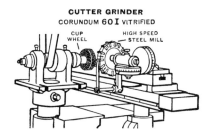

Fig. 20. — Grinding Inserted Tooth Milling Cutter.

when used in gang milling, are mounted on a mandrel and rotary-ground to diameter before grinding clearance.

27. To grind inserted tooth milling cutter, Fig. 20.

Specifications: 8″ inserted tooth milling cutter.

Machine and tools: Cutter grinder, universal head, tooth rest and holder, grinding wheel (cup) 5″, No. 60, Grade *I*, vitrified.

Speed of wheel, 5000 F.P.M.   Hand feed.

True wheel.   Set machine to grind straight.

For method of grinding clearance, see p. 814.

28. To grind radial teeth of side milling cutter, Fig. 21.

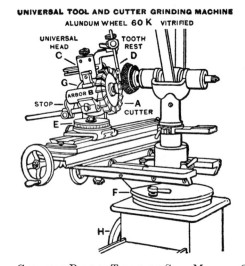

Fig. 21. — Grinding Radial Teeth of Side Milling Cutter.

Specifications: $3\frac{1}{2}$″ side milling cutter.

Machine and tools: Universal tool and cutter grinding machine, universal head, tooth rest and holder, grinding wheel (bevel cup) 6″ × $\frac{1}{2}$″, No. 60, Grade *K*, vitrified.   Bevel protractor.

Speed of wheel, 5000 F.P.M.   Hand feed.

True wheel.

Place cutter A on arbor B and mount in holder in universal head C which has both vertical and horizontal adjustments. Adjust tooth rest D at same height as center of cutter. Set horizontal swivel E about $\frac{1}{2}°$ away from 90°, knee F at 90° and vertical swivel G 4° from 90° to give required clearance. Raise wheel by hand wheel H to clear tooth below one being ground and move knee $\frac{1}{2}°$ from 90° to give wheel clearance. Arrange table dogs and grind teeth.

Attention. — Grind from 2° to 4° clearance on teeth of side and end mills, and from .001″ to .002″ concave by setting swivel head $\frac{1}{2}°$ away from 90°.

Information. — When the " lands " of radial teeth become

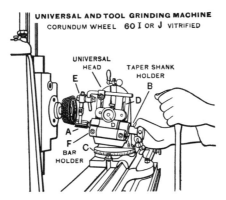

UNIVERSAL AND TOOL GRINDING MACHINE
CORUNDUM WHEEL   60 I OR J VITRIFIED

UNIVERSAL HEAD   TAPER SHANK HOLDER

FIG. 22. — GRINDING END MILL.

very wide from frequent grinding, the teeth may be " stocked " or ground by raising the swivel head about 30° and grinding the teeth to a larger angle leaving narrow "lands" to be ground to a smaller angle for clearance.

29. To grind end mill, Fig. 22.

Specifications: 1″ end mill.

Machine and tools: Universal and tool-grinding machine, universal head, taper shank holder, tooth rest and holder, grinding wheel (bevel cup) 4″ × $\frac{1}{2}$″, No. 60, Grade I or J, vitrified.

Speed of wheel, 5000 F.P.M.   Hand feed.
True wheel.

Place shank of mill A in taper shank holder B and mount in universal head.   Set horizontal swivel C about $\frac{1}{2}°$ away from 90°.   Set vertical swivel D at 3° for clearance.   Adjust tooth rest E to height of mill center, raise wheel to clear tooth below one being ground and set it $\frac{1}{2}°$ from 90° for wheel clearance.   Arrange table dogs and grind teeth with lever feed.

30. To grind angular cutter, Fig. 23.

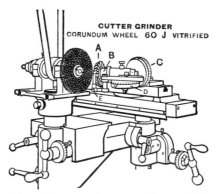

Fig. 23. — Grinding Angular Milling Cutter.

Specifications: 70° angular milling cutter.

Machine and tools: Cutter grinder, swivel head, tooth rest and holder, grinding wheel (disk) $6'' \times \frac{1}{2}''$, No. 60, Grade J, vitrified.

Speed of wheel, 5000 F.P.M.   Hand feed.
True wheel.

Place cutter A on arbor B and mount in universal head. Set vertical swivel C at 90° and horizontal swivel D at 70° Bolt tooth rest E to universal head and set at height of center of cutter.   Set cutter below center of wheel spindle for 6° clearance.   Set reversing stops and operate machine by lever feed.

Warning. — In grinding angular cutters and taper reamers, care should be taken to set the tooth rest the proper height, for any variation above or below center line of tooth will change angle of cutter or taper of reamer.

Attention. — The radial teeth of angular cutters are ground by the same process as the radial teeth of side and end mills, see pp. 816–818.

31. To grind clearance on teeth of spiral milling cutter, Fig. 24.

Specifications: $2\frac{1}{2}'' \times 3''$ spiral milling cutter.

Machine and tools: Cutter grinder, head and footstock, split tooth rest and holder, grinding wheel (disk) $7'' \times \frac{1}{2}''$, No. 60, Grade *I*, vitrified. Bevel protractor. Speed of wheel, 5000 F.P.M. Grind dry.

FIG. 24. — GRINDING SPIRAL MILLING CUTTER. INDEXING FROM TEETH.

Time: 4 min. (once around), with machine "set up."

SCHEDULE OF OPERATIONS

Oil machine.

True wheel. Set machine to grind straight. Set center of wheel at height of centers.

I. Hold cutter. Place cutter *A* on cutter bar *B* which is mounted in holder *C* shown in detail at *A'*, *B'*.

II. Set tooth rest. Clamp tooth-rest holder to wheel head and set tooth rest *D* at the same angle as spiral of mill and at height of centers, with center height gage *E* directly opposite the middle of wheel, so that flexible part *F* will clear the wheel when indexing cutter, then lower center of wheel $\frac{5}{16}''$ below centers for 6'' wheel and 6° clearance.

III. Grind clearance. Start grinding wheel and press mill *A* firmly against tooth rest *F*. Move cutter to touch revolving wheel lightly, by hand cross-feed wheel and with other hand, move cutter across face of wheel with a spiral motion, taking care to keep cutter teeth always against tooth rest. Move cutter back still pressing on tooth rest, and index to grind next tooth.

Clean machine with brush and waste.

32. To grind hand reamer *A* **with Cup Wheel** *B*, Fig. 25.
Specifications: 1″ solid hand reamer.

Machine and tools: Universal cutter grinder, head and footstock, tooth rest and holder, grinding wheel (cup) 4″ diameter, No. 60, Grade *K*, vitrified. Bevel protractor. Speed of wheel, 5000 F.P.M. Hand feed (lever). True wheel. Set machine to grind straight.

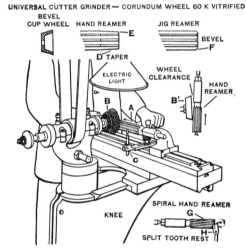

UNIVERSAL CUTTER GRINDER — CORUNDUM WHEEL 60 K VITRIFIED

FIG. 25. — GRINDING HAND REAMER WITH CUP WHEEL.

Set knee ½° away from center to give wheel *B* clearance, as at *B′*. Set tooth rest *C* at height of centers. Lower tooth rest to give 4° tooth clearance, see p. 813. Mount reamer on centers, adjust and clamp tooth rest.

Grind teeth straight and to required diameter 1.0004″. Swivel table to 1° and grind taper *D* on end, a distance equal to diameter of reamer or less. Bevel ends of teeth at *E*. For jig reamers, omit taper but grind bevel on end of teeth, as at *F*.

Attention. — To grind the teeth of a spiral hand reamer, as at *G*, holder for split tooth rest *H* is bolted to wheel head. See Fig. 13.

33. To grind hand reamer A **with Disk Wheel** B**,** Fig. 26.
Specifications: 1″ adjustable hand reamer.

Machine and tools: Universal grinding machine, head and footstock, tooth rest and holder, grinding wheel (disk) 7″ $\times$ ½″, No. 60, Grade I, vitrified.

Speed of wheel, 5000 F.P.M. Hand feed.

True wheel. Set machine to grind straight.

Set tooth rest C at height of centers. Mount reamer on centers, adjust and clamp tooth rest.

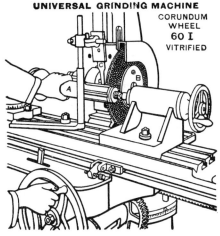

FIG. 26. — GRINDING HAND REAMER.

To obtain clearance, lower centers of wheel spindle below reamer center. Grind teeth straight and to standard diameter by trial cuts, then swivel table 1° and grind end of teeth taper, see D, Fig. 25, a distance from end equal to diameter of reamer or less.

34. To oilstone reamer for clearance and size, Fig. 27. — Hand reamers are often ground cylindrically true .0001″ to .0004″ oversize, then ground for clearance almost to edge and stoned to size, as follows: Hold reamer A in vise between copper jaws and stone each blade with oilstone B, used as in draw filing. Stone heel of tooth first then gradually approach to cutting edge. Ream hole and test with standard plug gage. Repeat process until hole fits gage.

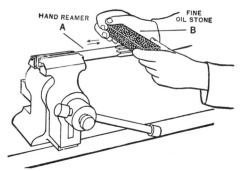

FIG. 27. — STONING REAMER TO SIZE.

35. To grind taper reamer *A*, Fig. 28.

Specifications: No. 3 Morse standard taper reamer.

Machine and tools: Cutter grinding machine, head and footstock, tooth rest and holder, grinding wheel (disk) 6″

FIG. 28. — GRINDING TAPER REAMER.

× ½″, No. 60, Grade *J*, vitrified. No. 3 Morse standard taper ring gage.

Speed of wheel, 5000 F.P.M. Hand feed.

True wheel.

Set center of wheel B at height of centers and machine to grind taper. Clamp tooth-rest holder to wheel head and set tooth rest C at height of centers. Mount reamer on centers, lower wheel head for clearance. Set dogs to limit stroke, press reamer on tooth rest and grind.

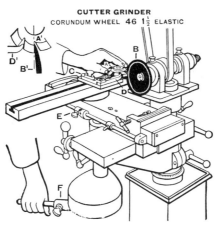

CUTTER GRINDER
CORUNDUM WHEEL 46 $1\frac{1}{2}$ ELASTIC

FIG. 29. — GRINDING GEAR CUTTER.

36. To grind gear cutter (formed cutter), Fig. 29.

Specifications: No. 4, 10 pitch gear cutter.

Machine and tools: Cutter grinder, gear cutter grinding attachment, grinding wheel (dish) $6''$, No. 46, Grade $1\frac{1}{2}$, elastic.

Speed of wheel, 5000 F.P.M. Hand feed.

True wheel.

Grind gear cutter A, A' with wheel B. Fasten gear cutter grinding attachment C to table and mount cutter upon it. Set knee at slight angle to axis of wheel spindle to bring edge of wheel in contact only, as at B'. Set cutting edge of wheel in same plane with center of cutter with special gage or straight edge, grind teeth radially. Set pawl tooth rest D and D' against back of tooth and adjust until tooth is radial, then clamp. Adjust stop E to limit travel so that wheel will just touch bottom of groove. To grind, press cutter lightly against tooth rest D and operate lever feed F. To index to next tooth, move cutter back with lever F, revolve cutter away from tooth rest D, then again press lightly against tooth rest D and grind. Take light cuts and go around twice, if necessary, rather than one heavy cut that may draw the temper.

Attention. — The teeth of formed cutters should be ground radially and equidistantly or they will not cut correct form.

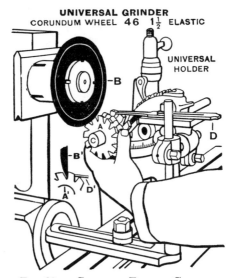

FIG. 30. — GRINDING FORMED CUTTER.

37. To grind formed cutter, or a hob, Fig. 30.

Specifications: $\frac{1}{2}''$ formed concave cutter.

Machine and tools: Cutter grinder, universal head, tooth rest and holder, grinding wheel (dish) 6'', No. 46, Grade $1\frac{1}{2}$, elastic.

Speed of wheel, 5000 F.P.M.   Hand feed.

True wheel.   Set machine to grind straight.

Mount cutter A on centers and wheel B on spindle, or mount work on bar C and hold in universal head. Clamp tooth rest D, D' to table and adjust to grind teeth radially.

38. To sharpen tap, Fig. 31.

Specifications: $\frac{3}{4}'' \times 10$ U. S. S. tap.

Machine and tools: Universal cutter grinder, universal head, tooth rest and holder, grinding wheel (round face), $6'' \times \frac{1}{2}''$, No. 80, Grade J, vitrified.

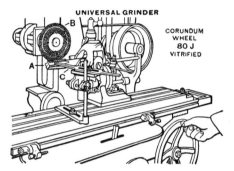

FIG. 31. — SHARPENING TAP.

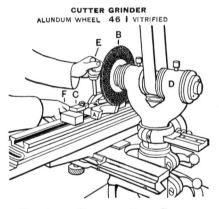

FIG. 32. — GRINDING FLAT CUTTER.

Speed of wheel, 5000 F.P.M. Hand feed.

True wheel. Set machine to grind straight.

Mount tap A in universal head or on centers. Use thin rounded-face wheel B. Clamp tooth rest C to table and adjust so that teeth will be ground radially. Use hand feed.

39. To grind flat cutter, Fig. 32.

Specifications: $1\frac{1}{2}''$ radius flat cutter.

Machine and tools: Cutter grinder, clamps, grinding wheel (disk) $6'' \times \frac{1}{2}''$, No. 46, Grade I, vitrified.

Speed of wheel, 5000 F.P.M. Hand feed.

True wheel.

Fasten cutter firmly to table by clamps C. Set wheel head D at right angles to table, lower wheel until it just touches work and grind with lever feed E and cross feed F.

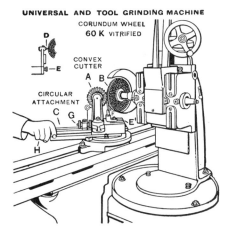

FIG. 33. — GRINDING CONVEX CUTTER.

40. To grind convex cutter, Fig. 33.

Specifications: Convex cutter.

Machine and tools: Cutter grinder, circular attachment, tooth rest and holder, grinding wheel (concave) $7'' \times \frac{1}{2}''$, No. 60, Grade K, vitrified.

Speed of wheel, 5000 F.P.M. Hand feed.

True and shape wheel.

Convex cutter A is ground with wheel B and circular attachment C which is for grinding convex and concave cutters. First true wheel to radius to be ground with carbon point D and holder E clamped in position at E'.

Mount cutter in position and set tooth rest F height of center of work. Locate wheel and cutter for clearance and bring center of radius to be ground directly over center of pivoting point of attachment. Grind teeth with light cuts, feed at G, and swing handle H.

CUTTER GRINDER

CORUNDUM WHEEL 46 $\frac{3}{4}$ W SILICATE

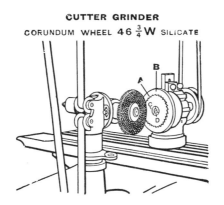

FIG. 34. — GRINDING SIDE OF SLITTING SAW.

41. To grind sides of slitting saw, Fig. 34.

Specifications: $6'' \times \frac{1}{8}''$ slitting saw.

Machine and tools: Cutter grinder, headstock, expansion bushing, grinding wheel (bevel disk) $8'' \times \frac{3}{8}''$, No. 46, Grade $\frac{3}{4}$ W, silicate. Micrometer.

Speed of wheel, 5000 F.P.M. Speed of work, 50 F.P.M. Automatic feed.

True wheel.

Mount slitting saw A on face chuck B and hold in place by expansion bushing C which is expanded by screw D, or use a magnetic rotary chuck. Swivel wheel head $\frac{1}{2}°$ to grind concave to give saw clearance. Set stops and grind.

Attention. — The sides of plain milling cutters are ground by the same process.

42. To grind caliper gage, Fig. 35.

Specifications: 6″ caliper gage.

FIG. 35. — GRINDING CALIPER GAGE.

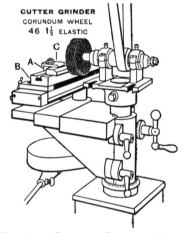

FIG. 36. — GRINDING STRAIGHT EDGE.

Machines and tools: Cutter grinder, vise and fixture, grinding wheel (double cup) $7'' \times \frac{3}{4}''$, No. 120, Grade G or H, vitrified. Inside micrometer gage.

Speed of wheel, 5000 F.P.M.   Hand feed.

True wheel.   Set machine to grind straight.

Grind gage *A*, with double-cup wheel *B*.   Set wheel-center height of center of work and grind both ends without moving in vise.   A small hole is drilled in corners of gage for wheel clearance.

43. To grind straight edge, Fig. 36.

Specifications: $12'' \times 1\frac{3}{8}''$ beveled straight edge.

Machine and tools: Cutter grinder, clamps, grinding wheel (cup) $6'' \times \frac{1}{2}''$, No. 46, Grade $1\frac{1}{2}$, elastic.

Speed of wheel, 5000 F.P.M.   Speed of work, 50 F.P.M. Hand feed.

True wheel.   Set machine to grind straight.

Mount straight edge *A* on table *B* and hold by clamps *C*, swivel knee $\frac{1}{2}°$ so that only edge of wheel will come in contact with work.   Grind to finish without removing work.

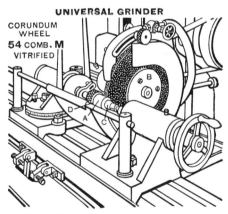

FIG. 37. — GRINDING JIG BUSHING.

44. To grind jig bushing, Fig. 37.

Specifications: $\frac{3}{4}''$ jig bushing.

Machine and tools: Universal grinder, grinding wheel (disk) $10'' \times \frac{1}{2}''$, No. 54 Comb., Grade *M*, vitrified. **Micrometer.**

Speed of wheel, 5000 F.P.M.   Speed of work, 50 F.P.M. Automatic feed.

True wheel.   Set machine to grind straight.

Bushing A is ground with wheel B.   Press mandrel C into bushing, mount on centers and grind as in plain cylindrical grinding.   A groove is cut in bushing at D for wheel clearance.

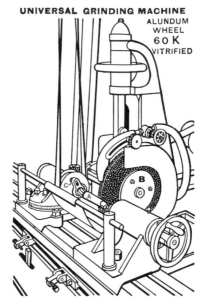

UNIVERSAL GRINDING MACHINE

ALUNDUM WHEEL 60 K VITRIFIED

Fig. 38. — Grinding Taper Shank of End Mill.

45. To grind taper shank of end mill, Fig. 38.

Specifications: 1″ end mill, Morse taper No. 2.

Machine and tools: Universal grinding machine, aloxite or alundum wheel (disk) 12″ × ½″, No. 60, Grade K, vitrified. Taper gage.   Prussian blue.

Speed of wheel, 5000 F.P.M.   Speed of work, 50 F.P.M. Automatic feed.

True wheel.

Shank of end mill A is ground with wheel B.   Mount on

centers with dog C. Set table to required taper and adjust table dogs.

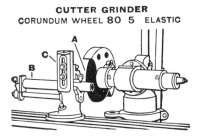

Fig. 39. — Cutting off Tubing.

46. To cut off tubing, Fig. 39.

Specifications: 1″ tubing.

Machine and tools: Cutter grinder, special holder, grinding wheel (thin $\frac{1}{8}$″) 7″ × $\frac{1}{16}$″, No. 80, Grade 5, elastic.

Speed of wheel, 5000 F.P.M.

True wheel A. Set machine to grind straight.

Mount tubing B in holder C. Start machine and feed by hand.

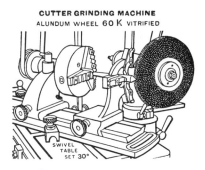

Fig. 40. — Grinding Hardened Lathe Center.

47. To grind hardened lathe center, Fig. 40.

Specifications: Hardened engine lathe center, Morse taper No. 2. Center gage.

Machines and tools: Cutter grinding machine, universal chuck, steady rest, grinding wheel (disk) $10'' \times \frac{1}{2}''$, No. 60, Grade K, vitrified.

Speed of wheel, 5000 F.P.M.   Speed of work, 50 F.P.M.

True wheel.

Set swivel table at 30°, mount center in universal chuck and steady rest, as shown, and use automatic or hand feed.   Test with center gage.

ADVANCED MACHINE WORK

SECTION 9

PLANING

Planers. Cutting Speeds and Feeds. Cutting Tools. Horizontal, Vertical and Angular Planing. Measuring Work. Holding Work, and Problems in Planing. Fixtures, Vises and Jigs. Shapers.

PLANERS

1. The planer bears the same relation to plane or flat surfaces that the lathe does to cylindrical surfaces. — The larger part of planer information and skill consists of setting up the work. The tools and cutting angles are similar to those of lathe tools.

2. The three general methods of holding work are the mechanical, magnetic and physical.

The mechanical method is holding work by clamps and bolts, fixtures, jigs or shoes, vises or chucks, and applies to the larger part of planer work. See pp. 913–917, 921–933.

The magnetic method is holding work by magnetic chucks. See pp. 624–626.

The physical method is holding work by cementing with cement, shellac or resin. See p. 920.

3. Types of planers. — There are two leading designs of planers, which are distinguished by their driving mechanisms: The spur-gear type and the spiral-gear type. The size of planers is designated by the size of the work that it will plane.

4. The spur-gear type planer table is driven by a spur gear which meshes with a rack on bottom of table, and with a pinion on the driving shaft on the outer end of which are tight and loose pulleys driven by open and crossed belts from countershaft.

5. The spiral-gear type planer table is driven by a spiral gear which meshes with a rack on bottom of table. The spiral-gear shaft is connected to driving shaft by bevel gears. The rest of the mechanism is similar to the spur-gear planer.

6. Planer, spur-gear drive, described, Fig. 1.

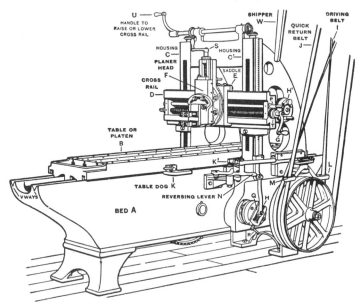

FIG. 1. — PLANER, SPUR-GEAR MECHANISM.

SCHEDULE OF PARTS

A — Bed.

B — Table operated by rack, gears, belts, slides in scraped *V* ways of bed, holds work and performs cutting movement.

CC' — Housings or posts.

D — Cross rail may be raised or lowered and clamped to housings.

E — Saddle moved along cross rail by hand or power.

F — Planer head.

G — Handle to operate horizontal feed screw, gives hand, horizontal or cross feed.

HH' — Power-feed mechanism. Friction feed disk, pawl and ratchet.

I — Driving belt (one speed), tight and loose pulley countershaft.

J — Quick return belt (return speed 3 to 1).

KK' — Table dogs, reverse motion of table and are located to give desired length of stroke.

LM — Belt shifters.

N — Reversing lever to reverse motion of table; operated by hand or dogs KK'.

P — Vertical and angular feed shaft turned with handle G, placed on end.

Q — Feed rod pin.

R — Feed adjusting handle. To obtain power horizontal feed, move pin Q off center to right or left, depending on whether cut is right or left, and drop pawl H in ratchet. The amount of cut depends on distance Q is removed off center.

S — Hand vertical feed. Some provided with dial to read to thousandths.

T — Feed gear. To obtain power vertical and angular feed place gear T on shaft P, adjust pin Q for desired number of teeth fed and drop in pawl H. May feed tool downward or upward.

U — Handle for raising and lowering cross rail.

W — Shipper.

---

7. Planer head and mechanism for cutting tool, Fig. 2.

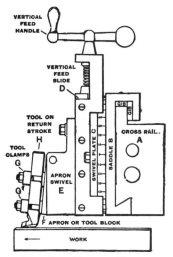

SCHEDULE OF PARTS

A — Cross rail.

B — Saddle.

C — Swivel plate.

D — Vertical feed slide.

E — Apron swivel or clapper box.

F — Apron or tool block, hinged to relieve tool on return stroke.

GG' — Tool clamps. (Used on large planers. Tool post used on small planers.)

H — Tool.

Fig. 2. — Planer Head.

See Inspection of a Planer, p. 1416.

8. Large planers are often equipped with two heads mounted upon the cross rail. Two other heads are mounted upon the housings so that work may be planed upon the top and both sides at the same time. Special designs are made for wide and extra-heavy work. Open side planers are obtainable for planing the ends of beams, frames and edges of steel or iron plates, and for other work that will not pass between the double housings or posts of an ordinary planer. The outer end may be supported upon an eye-beam which is mounted upon rollers running upon an improvised track.

CUTTING SPEEDS AND FEEDS

9. Cutting speed (F.P.M.). — Ordinary planers are belted to give one cutting speed only. Cast iron may be planed with carbon-steel tools at about 25 feet per minute, steel, wrought iron and bronze, at 17 feet per minute. When both iron and steel have to be planed on the same planer, as is usually the case, a cutting speed of about 20 feet is used. With high-speed tools, these speeds may be increased from 50 per cent to 100 per cent. For brass, composition and softer metals it is customary to use a planer speeded for cast iron. Variable-speed planers are obtainable which give changes of speed from 15 feet to 70 feet per minute. Variable-speed countershafts for a single-speed planer are obtainable.

10. Feed is obtained by a feed ratchet and pawl. The amount of feed 1 tooth on ratchet gives varies with the planer.

Cast iron may be rough planed with as deep a cut as the machine will carry. The finishing cut with a broad-nose tool should be light, from .005″ to .010″ in depth, with a feed nearly equal to the width of tool. Steel, wrought iron and aluminum are rough planed with a diamond-point or roughing tool in about the same manner as cast iron except that a little less depth of cut and a finer feed are used. For finish planing, use tools with narrow cutting edges, fine feed, 1 or 2 teeth, cut .005″ to .010″ in depth, with or without lubricant. See Table of Lubricants, p. 149.

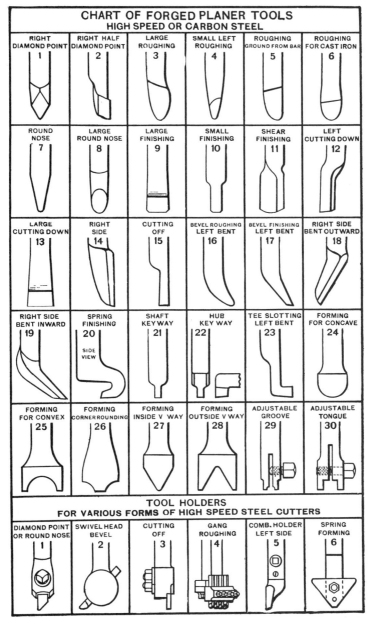

CHART OF FORGED PLANER TOOLS
HIGH SPEED OR CARBON STEEL

RIGHT DIAMOND POINT 1	RIGHT HALF DIAMOND POINT 2	LARGE ROUGHING 3	SMALL LEFT ROUGHING 4	ROUGHING GROUND FROM BAR 5	ROUGHING FOR CAST IRON 6
ROUND NOSE 7	LARGE ROUND NOSE 8	LARGE FINISHING 9	SMALL FINISHING 10	SHEAR FINISHING 11	LEFT CUTTING DOWN 12
LARGE CUTTING DOWN 13	RIGHT SIDE 14	CUTTING OFF 15	BEVEL ROUGHING LEFT BENT 16	BEVEL FINISHING LEFT BENT 17	RIGHT SIDE BENT OUTWARD 18
RIGHT SIDE BENT INWARD 19	SPRING FINISHING 20 SIDE VIEW	SHAFT KEY WAY 21	HUB KEY WAY 22	TEE SLOTTING LEFT BENT 23	FORMING FOR CONCAVE 24
FORMING FOR CONVEX 25	FORMING CORNER ROUNDING 26	FORMING INSIDE V WAY 27	FORMING OUTSIDE V WAY 28	ADJUSTABLE GROOVE 29	ADJUSTABLE TONGUE 30

TOOL HOLDERS
FOR VARIOUS FORMS OF HIGH SPEED STEEL CUTTERS

| DIAMOND POINT OR ROUND NOSE 1 | SWIVEL HEAD BEVEL 2 | CUTTING OFF 3 | GANG ROUGHING 4 | COMB. HOLDER LEFT SIDE 5 | SPRING FORMING 6 |

FIG. 3.

(905)

11. Planer cutting tools. — Chart of planer tools (forged, and tool holders and cutters) is shown in Fig. 3. They have the same rake for cutting different metals as lathe tools but less clearance (3° to 5°), Fig. 4, and are made heavier and stiffer.

Fig. 4. — LEFT DIAMOND-POINT PLANER TOOL.

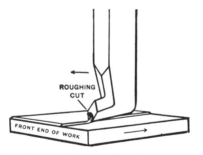

Fig. 5. — ROUGH PLANING CAST IRON WITH LEFT DIAMOND-POINT TOOL.

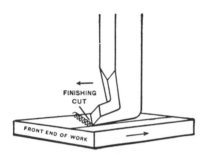

Fig. 6. — FINISHING STEEL OR WROUGHT IRON WITH LEFT DIAMOND-POINT TOOL.

HORIZONTAL, VERTICAL AND ANGULAR PLANING

12. Left diamond-point tool set to rough plane a horizontal surface on cast iron, Fig. 5. — Set tool perpendicular to table. Feed from 2 to 4 teeth.

13. Diamond-point tool to finish plane steel or wrought iron, Fig. 6. — Feed 1 or 2 teeth. Cut .005″ to .015″ in depth. With or without lubricant.

14. **Round-nose tool set to rough plane a horizontal surface on cast iron or brass, Fig. 7.**—Set tool perpendicular to table. Feed from 2 to 4 teeth.

15. Small square-nose tool to finish plane cast iron or brass, Fig. 8.— Feed from 4 to 8 teeth. Cut .005″ to .015″ in depth.

For a coarse feed, set cutting edge parallel to table, or line of feed.

For a fine feed, round bottom face slightly by grinding and oilstoning, and set corner *A* slightly lower than corner *B* to drag and give smoother cut. The feed must be less than the width of the tool.

16. Spring tool to finish plane cast iron, Fig. 9. — Set tool the same as in Fig. 8. Feed from 8 to 12 teeth ($\frac{3}{8}$″ to $\frac{1}{2}$″ wide). Cut about .003″ in depth.

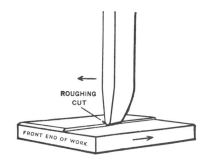

Fig. 7. — Rough Planing Cast Iron with Round-Nose Tool.

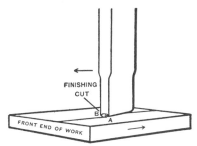

Fig. 8. — Finishing Cast Iron with Square-Nose Tool.

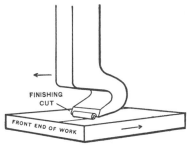

Fig. 9. — Finishing Cast Iron with Spring Tool.

17. Fine versus coarse feed for finishing cut. — If a surface is to be filed, scraped or fitted, finish nearly to size with a light cut and fine feed. When true surfaces only are required, use light cut and coarse feed.

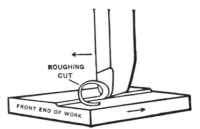

Fig. 10. — Rough Planing Steel or Wrought Iron with Roughing Tool.

18. Roughing tool to rough plane steel or wrought iron, Fig. 10. — Dry or with oil. Feed 1 to 3 teeth.

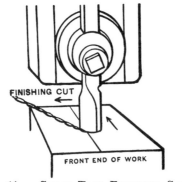

Fig. 11. — Shear Tool Finishing Steel or Wrought Iron.

19. Shear tool to finish plane steel or wrought iron, Fig. 11. — It is forged with a twisted blade. The edge is ground and oilstoned to an arc of a 3″ circle with clearance. Used with lard oil or soda water. Feed 1 tooth. Cut .003″ in depth. See Helical Milling Cutter, pp. **10**47, **10**48.

20. Side tool to finish plane vertical or angular surfaces, Fig. 12. — Bevel point A to produce a smooth surface. Feed downward by power or hand.

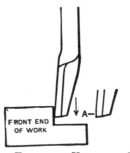

Fig. 12. — Finishing Vertical Surface with Side Tool.

21. Holders and cutters (high-speed steel), Figs. 13, 14, 15. — Used for horizontal, vertical and angular planing.

Fig. 13. — Horizontal Planing. Tool Holder and High-Speed Steel Cutter.

Fig. 14. — Vertical Planing.

Fig. 15. — Angular Planing.

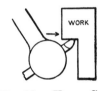

Fig. 16. — Horizontal Planing. Special Tool Holder.

Fig. 17. — Vertical Planing.

Fig. 18. — Under-Cut Planing.

22. Special holder and cutter (high-speed steel), Figs. 16, 17, 18. — Used for horizontal, vertical, angular and undercut planing.

23. To set planer in alinement. — *Approximate method.* Unclamp bolts back of rail. Lower rail nearly to work, then raise it about 1″ above work to take out back lash, leave handle up, and clamp.

Accurate method. Test table at both sides with micrometer, as in Fig. 19. If the two readings show a difference, loosen bolts at back of rail, adjust one of the bevel gears on horizontal shaft at top of posts and repeat the whole operation.

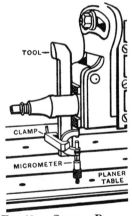

Fig. 19. — Setting Planer Cross Rail in Alinement.

Other methods of testing are: To lower cross rail on parallel, or test with surface gage and paper, or test with tool held in tool post and paper.

PLANER TABLE

FIG. 20. — REDRESSING PLANER TABLE WITH SPRING FINISHING TOOL.

24. Redress planer or shaper table with a spring tool, Fig. 20, with a light cut, .003″ to .005″ in depth, and a very coarse feed, when it has become bruised and uneven from careless handling of work, fixtures, hammering or wear.

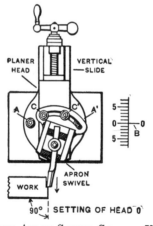

FIG. 21. — HEAD AND APRON SWIVEL SET FOR VERTICAL PLANING.

25. To set planer head and apron swivel to plane vertical surface, Fig. 21. — Release bolts A, A' and adjust head until zero lines coincide, as at B, then clamp. Release bolts C, C' and swing apron swivel away from surface to be planed in order that tool, as it lifts on the backward stroke, will move away from the vertical face of work. Work should never be planed vertically if it can be held and planed horizontally.

26. To set planer head and apron swivel for angular planing, Figs. 22, 23. — Release bolts and swing head to

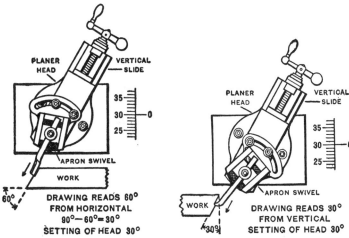

FIG. 22. — HEAD AND APRON SWIVEL SET FOR ANGULAR PLANING.

FIG. 23. — HEAD AND APRON SWIVEL SET FOR INSIDE BEVEL PLANING.

required angle and clamp; then swing apron swivel away from surface to be planed and clamp.

27. Planer head graduated to read from vertical. — On most American planers and shapers if drawing reads an angle of 60° from horizontal, the head is set to read 30°, as in Fig. 22. If drawing reads 30° from vertical, the head is set to read 30°, as in Fig. 23.

28. Planer head graduated to read from horizontal. — On planers and shapers where the zero line and 90° graduation coincide, when the slide is vertical, reverse above rule.

MEASURING WORK

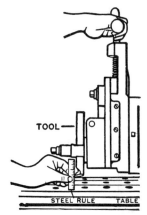

FIG. 24. — STEEL RULE TO SET PLANER TOOL.

29. Use steel rule to set planer tool, Fig. 24, to give results that will compare with those obtained with spring calipers in lathe work.

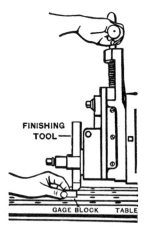

FIG. 25. — GAGE BLOCK TO SET PLANER TOOL.

30. Use gage block to set planer tool, Fig. 25, to give results that will compare with those obtained with microm-

eter in lathe work. To set tool, feed it down until the block
will slide between the table and tool
point, touching tool lightly. To
plane a piece .001″ thicker than the
gage, .001″ tissue paper is placed between gage and tool. Tissue paper of
suitable thickness is often used with
gage when there is danger of tool
pulling into the work by springing or
by looseness in head. See pp. 9₂8, 9₂9.

Set of gage blocks for use on planer
work is shown in Fig. 26.

FIG. 26. — GAGE BLOCKS.

HOLDING WORK, AND PROBLEMS IN PLANING

31. Planer parallels are often necessary to level work.
They are made of cast iron in
pairs of equal width, thickness
and length, Fig. 27.

**32. Iron clamps for planers and
shapers** are used in a great variety
of shapes also with bolts, blocking
and leveling jacks for clamping
work to fixture or to table. See
Figs. 28, 33.

FIG. 27. — PLANER PARALLELS.

**33. Step blocking and leveling
jacks** are convenient for blocking up the end of a clamp or to
level work. See Fig. 28, also Figs. 39, 41.

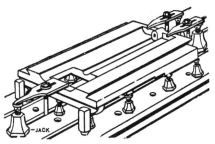

FIG. 28. — LEVELING WORK WITH PLANER JACKS.

34. Planer bolts of different lengths with square and Tee shaped heads, long threads and a variety of washers, are necessary. Studs and Tee nuts are a good substitute for bolts.

35. Screw pins and screw bunters, Fig. 29, are used in various ways to fasten planer work to table. The circular portion of the screw pin fits the hole in the table and the

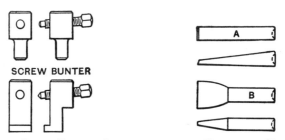

SCREW BUNTER

FIG. 29. — SCREW PIN AND SCREW   FIG. 30. — FINGERS OR TOE DOGS.
BUNTER.

squared portion is drilled and tapped, preferably at an angle of from 5° to 10°, to receive a set screw. The screw bunter fits the Tee slot.

36. Fingers or toe dogs, Fig. 30, are used with screw pins or bunters to fasten thin work to planer table.

FIG. 31. — METHODS OF CLAMPING WORK. GOOD AND BAD.
DIFFERENT KINDS OF PLANER BOLTS.

37. Good and bad methods of clamping or strapping work to table, Fig. 31. — Always place blocking *A* farther away from bolt *B* than work *C*, so that the most pressure will come on work.

38. Holding rectangular or oblong work by bedding on table with a tongue or rib fixture, *A*, bunters *B*, fingers *C* and thrust pin D, Fig. 32. — The tongue fixture is clamped in the long Tee slot in table. To be sure that work is properly

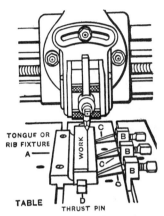

FIG. 32. — PLANING WORK BED-
DED TO TABLE.

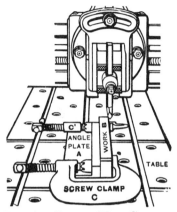

FIG. 33. — PLANING WORK CLAMPED
TO ANGLE PLATE.

bedded to table gently pound it to seat with lead hammer during and after tightening screws. Tightening screws too hard will lift the work. When taking heavy roughing cuts,

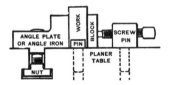

FIG. 34. — SCREW PIN FOR CLAMPING WORK AGAINST ANGLE PLATE.

apron swivel should be swung a little away from the direction of feed so that tool will not bear excessively on return stroke.

39. Angle plate (angle iron), A, Fig. 33, has planed surfaces at right angles (90°). To plane edges, hold work B by clamps C, C', or by block and screw pins, Fig. 34.

40. Tee fixture for holding work at right angle, A, Fig. 35, clamped to table by bolts B, B'. Its front face is at right angle to edge of table. The work C is held by clamps D and E. The vertical face F is planed with cutting-down tool G. Use fine feed.

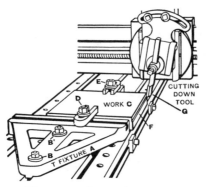

FIG. 35. — PLANING WORK AT A RIGHT ANGLE HELD BY TEE FIXTURE.

41. Adjustable tongue fixture. — In Fig. 36 is a tongue fixture in two parts, A and A', for holding thin work of any length. Work is held against stop B and fastened by screw stops C, C' and wedges D, D'.

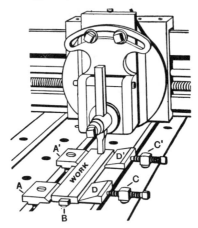

FIG. 36. — TO HOLD THIN WORK OF ANY LENGTH.

42. Beveling thin work. — Any thin work, as gib A, may be beveled with a bent side tool B, as in Fig. 37.

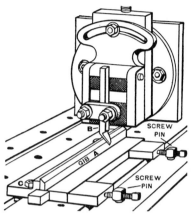

Fig. 37. — Planing Bevel with Bent Side Tool.

43. To plane taper parallels or wedges, a taper shoe is supplied which is adjusted to give any desired taper in inches per foot.

Feed at beginning or end of cut. — The feed movement of the tool should be arranged to take place at the beginning or at the end of the cut, but not during the cut. For general work either end will do. For accurately finishing broad surfaces it is best to feed just before beginning the cut (although the cutting stroke is longer) as the drag of the tool on the return stroke wears the tool less.

Direction of feed or cut. — In horizontal planing the tool may be fed either away from the operator by using a left tool, or toward the operator by using a right tool. The general practice is to feed away from the operator, as in Figs. 5, 6, 32, 33. In vertical and angular planing, the tool is fed downward as in Figs. 21, 22, 23.

Beveling end of castings. — At end of cut on iron castings, steel castings or brittle alloy castings, the pressure of the tool will break off the metal deeper than the planed surface. ˙ To avoid this, file or chip a bevel to depth of cut, as in Figs. 32, 33. Small work may be held and beveled at the vise, and large work, after it is clamped to the planer.

44.   To plane rectangular work clamped to planer table, Fig. 38.

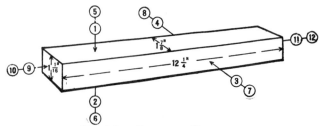

FIG. 38. — SCHEDULE DRAWING.

Specifications: Material, iron casting $\frac{1}{8}''$ large, free from visible defects; weight, 7 lbs. 14 oz.   Hardness, 23 to 35 (scleroscope). High-speed steel cutting tools.   Speed, 22 to 30 F.P.M.

Time: Study drawing and schedule in advance, 10 min. — Oil planer, 8 min. — " Set up " tongue and angle plate fixtures, 12 min. — Rough plane surfaces **(1)**, **(2)**, **(3)**, **(4)**, 30 min. — Finish plane surfaces **(5)**, **(6)**, **(7)**, **(8)**, 22 min. — Set Tee fixture, apron swivel and stroke for vertical planing, 8 min. — Rough and finish plane ends **(9)**, **(10)**, **(11)**, **(12)**, 22 min. — Clean planer, 8 min. — Total, 2 h.

SCHEDULE OF OPERATIONS, MACHINES, FIXTURES AND TOOLS

OPERATIONS.	MACHINES, FIXTURES, SPEEDS, FEEDS.	TOOLS.
Snag casting. Clean planer table. Set planer in alinement. See p. **9**09.	Machinist's vise. Planer, 16″ to 24″ cutting speed 25 F.P.M. With a high-speed planer this may be increased from 25% to 100%.	Hammer, cold chisel, 10″ or 12″ worn bastard file. Brush, waste.
Oil bearings. See p. **9**20. Horizontal planing.	. . . .	Machine oil.
To table and fixture, clamp work. See p. **9**14.	Tongue fixture, bunters or screw pins, fingers.	Lead hammer, planer wrench.
Adjust table dogs to length of stroke.	. . . .	Planer wrench.
Rough plane (Bevel end.) **(1)**, one cut $\frac{1}{16}''$ deep. See p. **9**17.	Feed 3 teeth or $\frac{1}{32}''$.	10″ or 12″ bastard file, diamond-point tool, or holder and cutter 15° rake, rule.

Rough plane (Bevel end.) (2), one cut. Set tool $1\frac{1}{16}''$ $+ \frac{1}{64}''$ from table.	....	....
To angle plate and table, clamp work. See p. 915.	Angle plate, screw pins and blocking or clamps.	.... •
Rough plane (Bevel end.) (3), with (1) against angle plate, one cut $\frac{1}{16}''$ deep.	....	....
Rough plane (Bevel end.) (4), with (1) against angle plate, one cut. Set tool $1\frac{7}{8}'' + \frac{1}{64}''$ from table.		
To table and fixture, clamp work.	Tongue fixture, bunters or screw pins, fingers.	....
Finish plane (File burr off edges.) (5), one cut $\frac{1}{128}''$ deep, estimated.	Feed 6 teeth or $\frac{1}{16}''$.	8'' or 10'' hand-smooth file, square-nose finishing tool, oil-stone.
Finish plane (6), one cut. Set tool $1\frac{1}{16}''$ from table.	....	Rule or gage block.
To angle plate and table, clamp work.	Angle plate, screw pins and blocking or clamps.	....
Finish plane (7), with (5) against angle plate, one cut $\frac{1}{128}''$ deep, estimated.	Feed 6 teeth or $\frac{1}{16}''$.	....
Finish plane (8), with (5) against angle plate, one cut. Set tool $1\frac{7}{8}''$ from table. (File burr off edges.)	....	Rule or gage block.
Vertical planing. To Tee fixture and table, clamp work.	Tee fixture, clamps, bolts, blocks, swing apron swivel. See Fig. 35.	....
Adjust table dogs to length of stroke.		
Rough plane (Bevel end.) (9), with (3) against Tee fixture, one cut $\frac{1}{16}''$ deep.	Feed 2 teeth, or $\frac{1}{32}''$.	Holder and cutter 15° rake, or cutting down tool.
Finish plane (10), one cut $\frac{1}{128}''$ deep. Reverse work.	Feed 1 tooth or $\frac{1}{64}''$.	Side tool with point beveled.
Rough plane (Bevel end.) (11), with (3) against Tee fixture, one cut. Set tool $12\frac{1}{4}'' + 1\frac{1}{8}''$ from (10).	....	Long rule.
Finish plane (12), one cut. Set tool $12\frac{1}{4}''$ from (10). (File burr off edges.)	....	....

Attention. — It is good practice to file the burr off the edges slightly with a hand-smooth file after planing each surface.

45. Oiling. — Oil every bearing of a planer once a day, and the V ways, loose pulleys, friction disk and sliding parts more often. The V ways are the most important and must be kept clean and well oiled.

46. Cleaning. — The chips should always be brushed toward side of table to avoid brushing them into the V ways which would be disastrous.

Warning. — Before starting a planer, see that the table dogs are fastened firmly and nothing is clamped to the table that will strike the housings or cross rail. Also that all wrenches, bolts, clamps and other pieces that are not fastened to the table are removed.

47. Holding very thin work by gluing or cementing. — Work too thin to be held by any mechanical means may be secured to resist the pressure of a light cut by gluing or cementing it to the planer table. Clean edges of work and table around work with gasoline or benzine and emery cloth. Apply melted shellac or rosin around the edges.

48. Spring of planer work is caused by severe and uneven clamping, also poor or unsuitable fixtures, bad blocking and shimming or wedging.

Another cause of springing is the removal of the outer surface of castings and forgings which relieves the internal strains. This is largely overcome by rough planing all surfaces to be finished before finishing any surface, then slightly relieving the pressure of the clamps and other holding devices for the finishing cuts.

Cold-rolled stock is liable to spring badly unless an equal amount of stock is planed off each surface. A casting or a piece that has been rough planed may be nominally flat, but to obtain very accurate results, select the straightest side. Rest this side on the table and, if necessary, place under corners or middle, strips of tissue paper to compensate for spring or distortion. Finish surface, then turn work over and finish side first selected. The edges of the work should be finished in same order. Take a light finishing cut about .003″.

49. To aline work on table with surface gage, use pieces of tin, brass or paper as shims and place under work directly under clamps. A triangular casting should have at least three bearing points and an oblong or square casting, four.

At *A*, Fig. 39, surface gage *B* is used to test the work as at *C*, *D*, *E* and *F*. Pinch bar *G* is used to lift casting to in-

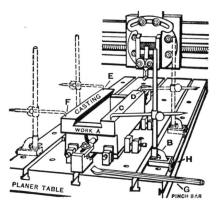

FIG. 39. — LEVELING AND ALINING WORK WITH SURFACE GAGE.

sert shims. To aline the work with the edges of the table, move guide *H* along edge and use straight point of scriber to test work. Screw pins and blocks clamp casting, and assist in alining.

50. To aline work in absence of surface gage, feed tool down at one end, insert paper between tool and work and adjust until tool pinches paper. Repeat on opposite end of work and at various points on top and sides.

FIXTURES, VISES AND JIGS

51. Planer jacks for leveling work are used to hold work that has ribs on its lower side, as casting *A*, Fig. 28.

52. To set tools for duplicate work with special gage or pattern, Fig. 40. — Bed work *A* to table and clamp to angle plate *B* by clamps *C* and *D*. To plane wide groove *E* accurately, use special gage *F* bolted to table and angle plate.

The top of gage is exact pattern of work, minus .001″.   For roughing, set tool G to cut .005″ large by a thickness gage; and for finishing, to cut .001″ large by tissue paper.   See pp. **9**12, **9**13.

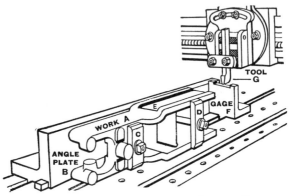

Fig. 40. — Special Gage or Pattern for Setting Planer Tool to Duplicate Work.

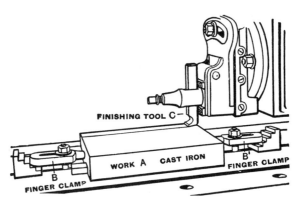

Fig. 41. — Holding Work with Finger Clamps.

53.   To hold work with finger clamps, Fig. 41, drill holes in ends of work A to receive finger ends of clamps B, B'.   To finish work, which is cast iron, use finishing tool C with a coarse feed.

54. Special planer clamps A, A', Fig. 42, are used for clamping work B to table.

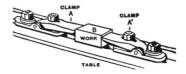

FIG. 42. — PLANER CLAMPS FOR HOLDING WORK. SPECIAL.

55. Planer jig or shoe to hold a number of pieces to be planed at once, A, Fig. 43. — The work consists of 6 castings, BB, etc., held with pinch pieces C and D and packing block E, adjusted by screws FF, etc.

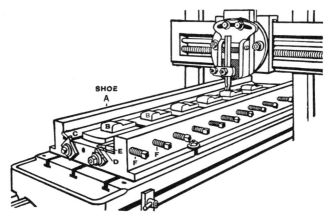

FIG. 43. — PLANER JIG OR SHOE FOR HOLDING WORK.

56. To plane Tee slot, Fig. 44. — First plane slot A, then bottom portion B, C with bent tools, one to a right angle, and other to a left. Any tool taking an under cut cannot be allowed to remain in slot during backward stroke.

57. Tool lifter D, Fig. 44, consists of metal plate with two lugs that receive pointed screws which engage large center punch marks or countersinks in the tool. During the cutting stroke, the lifter will drag, as at E, and lift on return, as at E'.

Some use a piece of leather in place of the tool lifter, to fasten apron so it cannot swing by wedge or brace between tool and apron swivel; and to lift tool by hand at end of each cut.

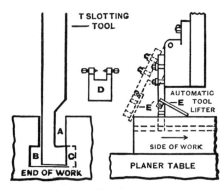

FIG. 44. — TEE SLOT PLANING.

Special planer tools and side heads for undercutting are obtainable.

58. To plane keyway in shaft A, Fig. 45, drill hole B in which to terminate cut as the tool must pass out of cut each

FIG. 45. — PLANING KEYWAY IN SHAFT.

time. Set tool central with square and rule. Use fine hand feed and oil.

The depth of keyway is measured at side.

59. To clamp shaft to cut keyway its whole length, hold as in Fig. 46. A special planer tool holder fitted with a keyway cutter is used as a spring tool.

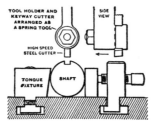

FIG. 46. — HOLDING LONG SHAFT TO PLANE KEYWAY WHOLE LENGTH.

60. To plane keyway in hub. — Use a tool holder and cutter or forged tool. Keyways may be cut by keyseating machine or by hand with chisel and file. See Motor-Driven Keyseater, p. 936.

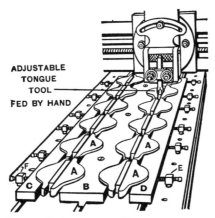

FIG. 47. — STRING FIXTURES FOR HOLDING VISE BASE PLATES.

61. String fixtures to hold several pieces A, A, etc., Fig. 47, **to be planed at once,** are economical and mechanical. Fixture B is clamped to table by screws. Work is clamped by rails C and D and screw pins E and F.

62. Planer vises (chucks) are obtainable with circular graduated bases or with square bases, as Vise, Fig. 48, and fastened to table by clamps. The sliding jaw is moved against work and nuts A, A', tightened, then is forced hard against work by screws B, B, B in block C which is kept from slipping by thrust strip D. Finally, pound work to its seat and tighten nuts A, A'. See Milling Machine Vises, pp. 1011–1016.

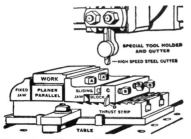

Fig. 48. — Planer Vise for Holding Work.

63. To plane keyway in flange, Fig. 49. — See Keyseater and Slotter, pp. 936, 937, and Broaching Keyways, pp. 543–548.

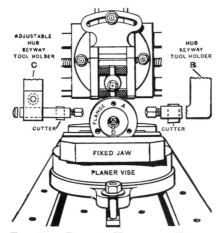

Fig. 49. — Planing Keyway in Flange.

Specifications: Material, cast-iron flange, turned, polished and drilled. Flange has 1″ hole. Plane keyway $\frac{1}{4}$″ wide, $\frac{3}{32}$″ deep.

Machine and tools: Planer vise, hub keyway tool holder and $\frac{1}{4}$″ cutter, rule, planer parallel. Plane dry.

Time: Study drawing and schedule in advance, 3 min. — Oil planer, 5 min. — "Set up" machine, 4 min. — Cut keyway, 8 min. — Clean machine, 2 min. — Total, 22 min.

SCHEDULE OF OPERATIONS, MACHINES AND TOOLS

I. Clean planer and vise and oil bearings.

II. Set vise. Fasten planer vise to table with jaws at right angle to line of travel and fixed jaw in front, as in Fig 49.

III. Clamp work. Clamp flange *A* lightly in center of vise as shown, with parallel underneath. Adjust flange until two holes are approximately parallel to fixed jaw by measurement, then clamp firmly with vise screws.

IV. Adjust cross rail. Raise or lower cross rail until work will pass under.

V. Set tool. Fasten hub keyway tool holder *B* in tool post so that tool post will not strike flange, and with face of cutter parallel to table or vise.

VI. Adjust length of stroke. Move saddle on cross rail to one side of flange, and adjust table dogs for length of stroke. Set feed rod clamp at dead center.

VII. Set cutter to plane keyway central. Move saddle back and set cutter central by placing a block against the side of hub and measuring to side of cutter with rule. Adjust until measurement is the same on both sides from block to cutter.

VIII. Plane keyway. Pull belt by hand to see if adjustments are right. Start machine, feed cutter down by fine hand vertical feed (.003″ to .004″ cuts) until keyway *D* is $\frac{3}{32}$″ deep, measuring at side of keyway.

64. To plane taper keyway in hub, Fig. 50. — Use taper parallels and clamp gear to these parallels.

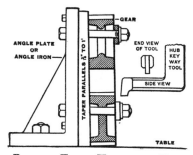

Fig. 50. — Planing Taper Keyway in Hub of Gear.

65. To plane lathe bed. — To obtain the most accurate results from large castings, the parts to be finished by planing should be rough planed and laid aside for a number of weeks to season to relieve internal strains. (Lathe beds are sometimes roughed out by gang mills, then finished by planing.) First finish bottom of the lathe bed A, then clamp, as in Fig. 51, and plane the top and V ways. Use templets of sheet steel, B and C, and test the inner and outer V's, as follows:

Plane one face on the same side of both inner V's to templet B. Tissue paper should be inserted between templet and

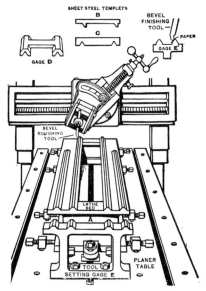

FIG. 51. — PLANING LATHE BED.

face of V at top and bottom to test the angle, so that angle of head may be corrected, if need be. Then plane second faces of the inner V's and test with the same templet. When finished both V's are tested with templet C.

Plane off apex of each V $\frac{1}{8}''$. The same method and type of templets are used for the outer V's. For the final test, use gage D. A tool setting gage, E, may be used to set tool. For each of the finishing cuts, the tool is adjusted until it pinches a piece of paper placed between the gage and tool, as at *E'*. The paper should be of sufficient thickness to allow for scraping and fitting. The horizontal parts and edges of the bed may also be planed to gage. See Scraping, *Principles of Machine Work*.

66. To plane head and footstock of lathe (two methods), Fig. 52. — For large lathes, the V grooves are planed and fitted to the V ways on the bed before the holes in headstock and footstock are bored. The headstock and footstock are then clamped to bed and bored with a special boring bar. For small lathes, use the reverse method. First bore holes in headstock *A* and footstock *B* and fit boxes in headstock. Next mount on a double arbor or mandrel *C* to hold them in alinement, clamp in three accurately V-grooved blocks *D, D, D*, which fit Tee slot and are bolted to table. Use braces *E* to steady the whole.

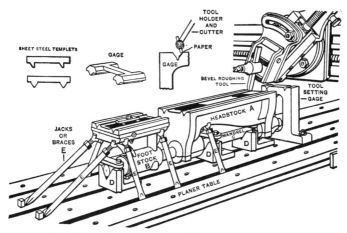

FIG. 52. — PLANING LATHE HEAD AND FOOTSTOCK.

To finish, clamp both lightly to avoid springing. First cut groove to the desired depth with a cutting-off tool in the center of each V groove, then plane the faces in the same order as the V's on the lathe bed, and test with the internal templets or gages.

SHAPERS

67. Shapers are particularly suitable for short work that can be held in a vise. The size is designated by size of work it will plane. A 14″ shaper will plane work 14″ × 14″ × 14″. **Two types of shapers** of leading designs are Geared Shapers and Crank Shapers.

68. The geared-shaper ram is driven by a spur gear meshing with a rack fastened to bottom of ram similar to a spur-geared planer. Suitable gearing, shafts, pulleys, belts, and countershaft, complete the driving mechanism for moving ram forward and backward.

69. The crank-shaper ram is driven by a direct or indirect crank movement which gives a quick return. Suitable gears, shafts, a stepped cone pulley and belt from countershaft, complete the driving mechanism for moving the ram forward and backward.

70. Feed mechanisms of various shapers consist of a feed disk and connections for obtaining one or more teeth of feed, measured with a pawl and ratchet in advance of each cutting stroke. See 902.

71. Length of stroke of shaper is positive and can be more readily set to terminate cuts close to a shoulder than that of a planer.

72. Shaper geared, described, Fig. 53.

SCHEDULE OF PARTS

A — Pillar or column.

B — Ram, slides in plane scraped bearings.

C — Shaper head bolted to ram. Mechanism similar to planer head. See p. 903.

D — Cross rail which may be raised and lowered on pillar by handle on opposite side.

E — Saddle.

F — Table which supports work and vise.

G — Handle to operate horizontal feed screw and give cross feed.

HH′ — Feed mechanism: Feed disk, pawl and ratchet. For power feed.

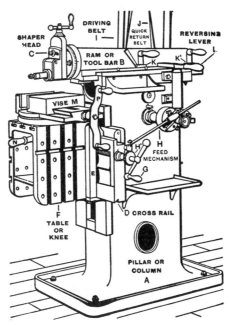

FIG. 53. — SHAPER, GEAR MECHANISM.

I — Driving belt for cutting stroke of ram (two speeds) from tight and loose pulley countershaft.

J — Belt from large pulley on countershaft gives a return speed about 3 to 1.

KK′ — Movable dogs reverse motion of ram and may be located to give desired length of stroke.

L — Reversing lever operating friction clutch which reverses ram and may be operated by hand or dogs *KK′*.

M — Vise.

N — Lever to clamp vise to table.

P — Handle to clamp work to vise.

73. Traveling head shaper or traverse shaper has a box-shaped bed supported by pillars at the ends and two tables on its side. The ram is mounted in a saddle which travels along the bed at a right angle to stroke of ram by hand or

power feed. It is designed for long work beyond the work
of a pillar shaper. Small work may be clamped to table or
held in vise, and large work on both tables, and further supported, if need be, by jacks and blocking resting on the
floor.

74. Draw-cut shaper used for heavy work. The apron
and tool are reversed. The tool cuts on the return stroke.

75. Swivel shaper vise (or chuck), Fig. 54. — To set vise
jaws at right angle to cutting stroke, remove taper pin, un-

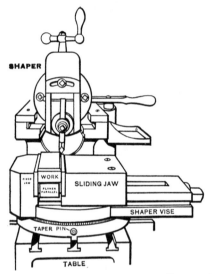

Fig. 54. — Planing Work Held in Shaper Vise.

clamp, swing vise and insert pin. As the base is graduated
into degrees of a circle, the jaws may be located and clamped
at any desired angle.

Attention. — To obtain very accurate results, a thin backing strip or a rod of small diameter is inserted between the sliding jaw of vise and work to insure even pressure on the fixed
jaw, so that the second cut will be exactly at right angles to
the first.

The pressure of the vise jaws has a tendency to lift the

work, and for finishing fine work strips of tissue paper may be inserted at each end between the planer parallel and work and the work lightly pounded to its seat with lead hammer until it pinches the paper.

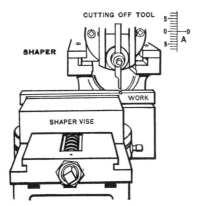

FIG. 55. — TO CUT OFF WORK HELD IN VISE.

76. Cutting off work held in a vise, Fig. 55. — Use cutting-off tool, the head as at A and apron swivel vertical. For deep cuts the apron must be clamped or the tool lifted out of groove on return stroke.

FIG. 56. — TAPER KEY HELD IN VISE WITH SWIVEL FIXTURE.

77. To plane taper work in a vise. — Use swivel fixture A, Fig. 56. Place it between the sliding jaw and taper key B. By matching the tapers, two taper keys may be held and planed without the fixture.

78. To plane rectangular work held in shaper vise, Fig. 57.

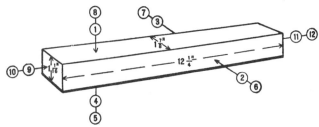

FIG. 57. — SCHEDULE DRAWING.

Specifications: Material, iron casting $\frac{1}{8}''$ large, free from visible defects; weight, 7 lbs. 14 oz. Hardness, 23 to 35 (scleroscope). High-speed steel cutting tools. Speed, 30 to 60 F.P.M.

Time: Study drawing and schedule in advance, 5 min. — Oil shaper, 3 min. — "Set up" machine, 13 min. — Plane rough surfaces (**1**), (**2**), (**3**), (**4**), 21 min. — Finish plane surfaces (**5**), (**6**), (**7**), (**8**), 11 min. — Set vise, apron swivel and stroke for vertical planing, 7 min. — Rough and finish plane ends (**9**), (**10**), (**11**), (**12**), 13 min. — Clean shaper, 5 min. — Total, 1 h. 18 min.

SCHEDULE OF OPERATIONS, MACHINES, FIXTURES AND TOOLS

OPERATIONS.	MACHINES, FIXTURES, SPEEDS, FEEDS.	TOOLS.
Snag casting.	Machinist's vise.	Hammer, cold chisel, 10″ or 12″ worn-bastard file.
Clean shaper and vise.	14″ shaper and shaper vise, cutting speed 50 F.P.M.	Brush, waste.
Oil bearings. Horizontal planing.	. . . .	Machine oil.
In vise, clamp work, see Fig. 54, with planer parallel underneath. Adjust dogs on ram to length of stroke.	Shaper vise, planer parallel.	Lead hammer, shaper wrench.
Rough plane (Bevel end.) (**1**), one cut $\frac{1}{16}''$ deep. See p. 917.	Feed 2 teeth or $\frac{1}{32}''$.	10″ or 12″ bastard file, holder and cutter or diamond-point tool 15° rake, 3″ rule.

In vise clamp work with planer parallel underneath. Rough plane (Bevel end.) (2), with (1) against fixed jaw, one cut $\frac{1}{16}''$ deep. In vise clamp work with planer parallel underneath. Rough plane (Bevel end.) (3), with (1) against fixed jaw, one cut. Set tool $1\frac{7}{8}'' + \frac{1}{64}''$ from parallel underneath. In vise clamp work with planer parallel underneath. Rough plane (Bevel end.) (4), with (2) against fixed jaw, one cut. Set tool $1\frac{1}{16}'' + \frac{1}{64}''$ from parallel underneath. In vise clamp work with planer parallel underneath. Finish plane (5), (6), (7), (8), in given order. For (5), (6), set tool to take cut $\frac{1}{128}''$, estimated; for (7), (8), set tool to dimensions by rule or gage block. (File burr off edges). Set vise at right angles to present position, reclamp. See Fig. 55. Vertical planing.		
	Feed 4 teeth or $\frac{1}{16}''$.	8'' or 10'' hand-smooth file, square-nose finishing tool, rule or gage block.
In vise clamp work with planer parallel underneath. Adjust dogs on ram to length of stroke.	Swing apron swivel. See Fig. 21.	
Rough plane (Bevel end.) (9), with (2) against fixed jaw, one cut $\frac{1}{16}''$ deep.	Hand feed, or if equipped with power feed, use 2 teeth or $\frac{1}{32}''$.	Holder and cutter 15° rake, or cutting down tool.
Finish plane (10), one cut $\frac{1}{128}''$ deep.	Hand feed or power feed, 1 tooth or $\frac{1}{64}''$.	Holder and cutter 15° rake, or side tool with point beveled.
Reverse work. Rough plane (11), with (2) against fixed jaw, one cut. Set tool $12\frac{1}{4}'' + \frac{1}{128}''$ from (10). Finish plane (12), one cut. Set tool $12\frac{1}{4}''$ from (10). (File burr off edges.)	. . . .	Long rule and short rule.
Stamp name on broad surface.	Vise, copper jaws.	$\frac{1}{8}''$ steel letters, hammer.

79. Motor-driven shaper, Fig. 58, is convenient and successful for all types of shaper work. This machine is shown planing the top of a machine leg. The table of shaper is removed and leg A is clamped to saddle B. The shaper is started by means of switch C and rheostat D. All adjustments of tool and feed are the same as those given on pp. **9**34, **9**35.

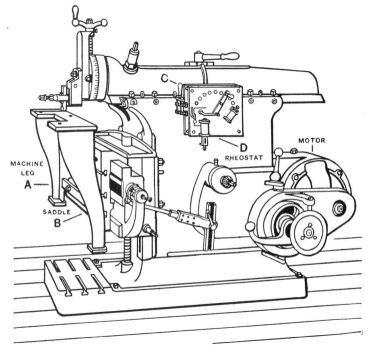

Fig. 58. — Planing Top of Machine Leg with Motor-Driven Shaper.

80. Keyseater and slotter, Fig. 59. — This machine will cut keyways in pulleys, flywheels, gears and cams. By the addition of fixtures, it will plane keys and take care of a large variety of internal and external slotter work. It is supplied with automatic feed, stop, and release. The table may be tilted for taper keyways and other taper work. In Fig. 59, cutter A is cutting keyway B in fly wheel C. The work is centered automatically by centering bushing D. The hub may be left rough. Keyways may be cut in several pieces by placing one above the other.

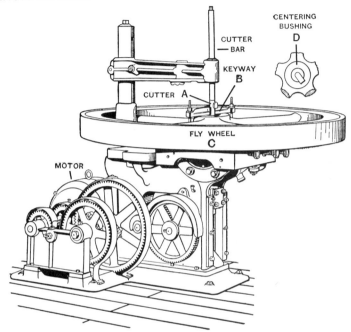

FIG. 59. — CUTTING KEYWAY IN FLYWHEEL WITH MOTOR–DRIVEN KEYSEATER.

Small keyways are usually cut with a saw-cutter outfit, as at *A*, Fig. 60.

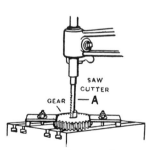

FIG. 60. — CUTTING SMALL KEYWAY IN GEAR.

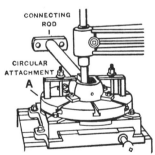

FIG. 61. — CUTTING SLOT IN CONNECTING ROD.

By the addition of a circular attachment, as at *A*, Fig. 61, the rectangular and circular slots in the ends of the connecting rods and other work of this nature, may be quickly and accurately machined. See Planing Keyways, pp. 926, 927; Broaching Keyways, pp. 543–548; Milling Keyways In Shafts, pp. 1014, 1015, 1018.

81. Slide rule for timing planer work.

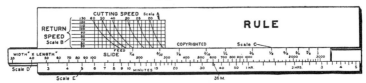

Fig. 62.—Slide Rule Set for Cutting Speed of 30 F.P.M., Return Speed 90 F.P.M., Feed $\frac{1}{16}''$.

The slide rule, Fig. 62, is used for calculating the time required to plane work at different cutting speeds, return speeds and feeds. It consists of a rule and slide with five scales. Scale *A* is the cutting speed, scale *B* is the return speed, scale *C* is the feed, scale *D* the area to be planed, and scale *E* the time required to do the planing.

To use the slide, first find the number of square inches to be planed; second, find the cutting and return speeds of the planer; third, decide on the feed to be used; fourth, locate the point of intersection of the cutting speed curve on scale *A* with the horizontal return speed line on scale *B*; fifth, move the slide until the feed figure on scale *C* is directly under this intersection; sixth, locate on scale *D* the number corresponding to the area of work to be planed and directly below on scale *E* will be found the time it will take to do the planing.

Example. — To plane a bench leveling plate 20 inches by 30 inches long.

Width × Length	20″ × 30″ = 600 square inches
Cutting speed	30 feet per minute
Return speed	90 feet per minute
Amount of feed	$\frac{1}{16}''$

Move the slide until the $\frac{1}{16}''$ on scale *C* comes directly under the intersection of 30 foot cutting speed on curve from scale *A*, and the 90 foot return speed line from scale *B*. Directly below the 600 on scale *D* will be found on scale *E*, the time required 36 minutes.

Attention. — In calculating the area to be planed, the over-travel of tool at front end of work, and tool clearance at back end, should be added to the length.

ADVANCED MACHINE WORK

SECTION 10

MILLING

Introduction. Milling Machines. Milling Cutters. Speeds and Feeds Measuring Work. Plain Milling. Jigs and Fixtures. Index Milling. Calculating Diameter of Blank to Mill Square or Hexagonal. Milling on Centers. Grooving Taps. Fluting Reamers. Milling Teeth of Milling Cutters. Circular Milling. Vertical Milling. Profiling. Rake Tooth Milling Cutters.

INTRODUCTION

1. Milling is a process of machining parts to dimensions, to regular and irregular shapes, to cut gears, to groove taps, reamers, drills, etc., with rotary cutters. Its greatest value is its power of duplication to produce interchangeable machine parts.

MILLING MACHINES

2. Milling machines may be classed as follows:

I. The column and knee type, Fig. 1, made both plain and universal, and used for general work.

II. The planer type, Fig. 26, for the heaviest gang and slab milling.

III. The manufacturing type, Fig. 29, for light plain milling.

IV. Vertical milling machines make use of the rapid end mill principle of cutting and are used for face milling, die sinking, routing, etc., see Fig. 64.

V. Profiling machines make use of the templet principle of milling irregular shapes, see Fig. 69.

VI. Special milling machines: Gear-cutting machines, bolt heading machines, thread milling machines, etc.

Speed changes are obtained by cone headstock (cone drive) or gear headstock (constant speed drive).

3. A plain milling machine, Fig. 27 (column and knee type), is used for various kinds of plain milling, as surface or face

1001

milling, keyways, slotting, etc., and with the addition of index centers, is used for index milling. See also, Fig. 28.

The table is provided with two movements — longitudinal, controlled by hand and power feeds; and transverse, by hand

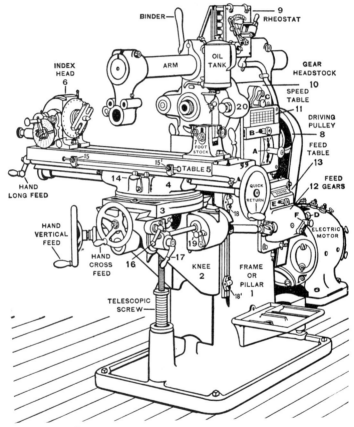

FIG. 1. — UNIVERSAL MILLING MACHINE.

feed and on some machines by power long. feed. The knee is adjustable on the column and may be set at different heights to suit the work, and may be fed upward to take vertical cuts by hand and on some machines by power.

4. A universal milling machine, Fig. 1 (column and knee type), with swivel table for spiral operations, is adapted to a great variety of work. It is used for all kinds of index, spiral and light plain milling and within its capacity can do the work of all the other classes of milling machines, planers and shapers, both regular and special.

5. Universal milling machine described, Fig. 1.

SCHEDULE OF PARTS

MOTOR DRIVE   CONSTANT BELT SPEED   GEAR HEADSTOCK
GEAR FEED

1. Frame.
2. Knee gibbed to frame.
3. Clamp bed.
4. Saddle pivots in clamp bed, is graduated in degrees.
5. Table gibbed to saddle, can be swiveled to any desired angle for milling spirals.
6. Index, spiral or dividing head (detachable) with mechanism for rapid, plain and differential indexing. Also connected by gears to feed screw to give rotary motion to mill work spirally.
7. Footstock (detachable).
8. Driving pulley driven by motor or belt from countershaft.
9. Rheostat, starting device for motor.
10. Gear headstock consists of spindle, back gears and mounted system of gearing, giving 16 spindle speeds, ranging in geometric progression from 15 to 376 revolutions per min.
11. Speed table gives spindle speeds, and position of the three levers A, B, C, to obtain them. To obtain a given speed, unlatch lever A and lower it. Move index lever B to required column

under speed table, then raise lever A to engage proper gears and latch it. Lever C gives fast or slow motion. Throw back gears in and out as in lathe.

12. Mounted system of feed gears similar to the headstock, driven by a chain.
13. Feed table gives position of three levers D, E, F, to obtain 12 feeds. To obtain a given feed, unlatch lever D and raise it. Move lever E to required column under feed table 13, then lower lever D to engage gears and latch. Lever F gives fast and slow motion.
14. Power long. feed lever.
15, 15'. Table dogs to stop power long. feed.
16. Power cross feed lever.
17. Power vertical feed lever.
18, 18'. Vertical feed dogs.
19. Lever to start, stop or reverse all feeds.
20. Knob to release spindle when desired to rotate it by hand. Dials on long. or horizontal, cross and vertical feeds are graduated in thousandths.

See Inspection of a Milling Machine, p. **14**10.

6. Milling-machine arbor A, Fig. 2, which is used for light work, is held in place by a tang and is removed with a rod. The cutter may be secured on arbor by wire key, collars and

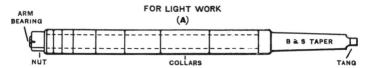

FIG. 2. — MILLING-MACHINE ARBOR FOR CUTTERS WITH PLAIN HOLES — LIGHT WORK.

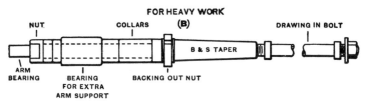

FIG. 3. — MILLING-MACHINE ARBOR FOR CUTTERS WITH PLAIN HOLES — HEAVY WORK.

nut, or the key may be omitted for light work. Arbor B, Fig. 3, is used for heavy work. It is held by drawing-in bolt, and backed out by nut.

FIG. 4. — SHELL-MILL ARBOR.

FIG. 5. — MILLING-MACHINE COLLETS FOR MILLS WITH TAPER SHANKS.

7. Shell-mill arbor, Fig. 4, is used to hold shell mills.

8. Milling-machine collets (C), (D), Fig. 5, are used to hold taper arbors and shanks, such as end mills.

MILLING CUTTERS

9. Milling cutters (classified in four types, see Chart, Fig. 10), are made of carbon and high-speed steel, hardened, tempered and ground. Large mills are made with cast-iron or machine-steel body and high-speed steel inserted teeth. Large mills are often nicked to break chips and reduce friction.

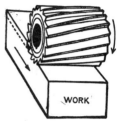

10. Face or plain milling cutter, Fig. 6, cuts surface *parallel* to its axis. Cut-ters over $\frac{3}{4}''$ in width have helical teeth; less than $\frac{3}{4}''$, have straight teeth. They

Fig. 6.—Face Milling Cutter.

are used for keyways, slots, cutting-off stock and similar work. Grind clearance on teeth 6° to 7° for cutters under 3″ in diameter, and 4° to 5° for those over 4″. Excessive clearance will cause vibration. See Coarse Tooth, Rake Tooth, and Helical Tooth Milling Cutters, pp. **1046–1048.**

11. Side or radial mills (shell-end mill), Fig. 7, cut surfaces *perpendicular* to axis. Grind clearance 2° to 4° and .001″ to .002″ hollow, or lower near center. Excessive clearance will cause vibration.

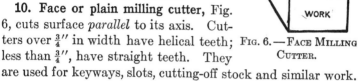

Fig. 7.—Shell-End or Side Mill.

12. Angular cutters, Fig. 8, cut surfaces *inclined* to axis, as 45°, 50°, 60°, 70° and 80°. Double-angle cutters have angles, as 40° and 12°, 48° and 12°, etc. Grind clearance same as face and side cutters.

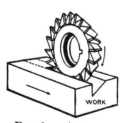

Fig. 8. — Angular Milling Cutter.

13. Formed milling cutter,

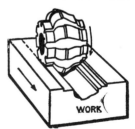

Fig. 9. — Formed Milling Cutter.

Fig. 9, cuts surfaces of *curved* or *irregular outline*. It is so relieved that clearance does not have to be considered. The

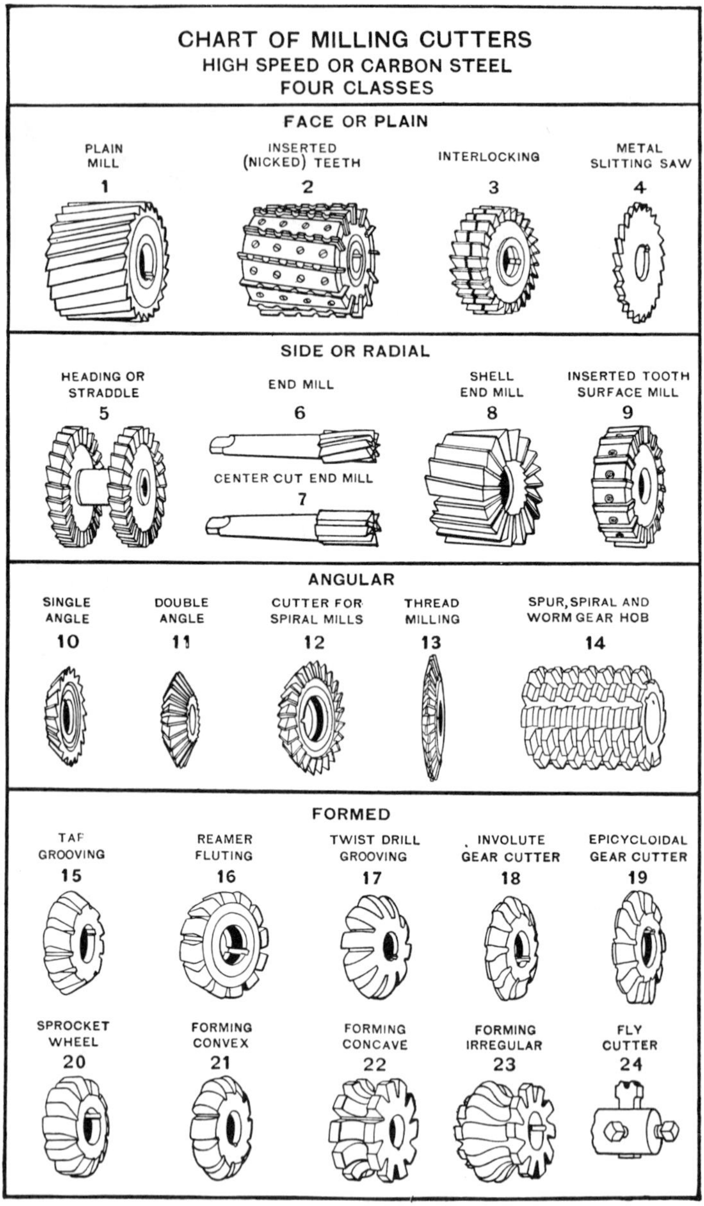

CHART OF MILLING CUTTERS
HIGH SPEED OR CARBON STEEL
FOUR CLASSES

FACE OR PLAIN

PLAIN MILL	INSERTED (NICKED) TEETH	INTERLOCKING	METAL SLITTING SAW
1	2	3	4

SIDE OR RADIAL

HEADING OR STRADDLE	END MILL	SHELL END MILL	INSERTED TOOTH SURFACE MILL
5	6	8	9
	CENTER CUT END MILL		
	7		

ANGULAR

SINGLE ANGLE	DOUBLE ANGLE	CUTTER FOR SPIRAL MILLS	THREAD MILLING	SPUR, SPIRAL AND WORM GEAR HOB
10	11	12	13	14

FORMED

TAP GROOVING	REAMER FLUTING	TWIST DRILL GROOVING	INVOLUTE GEAR CUTTER	EPICYCLOIDAL GEAR CUTTER
15	16	17	18	19
SPROCKET WHEEL	FORMING CONVEX	FORMING CONCAVE	FORMING IRREGULAR	FLY CUTTER
20	21	22	23	24

Fig. 10.

teeth must be ground radially and equidistantly or it will not mill its correct form.

14. **Form milling cutters** are made and sharpened by grinding the top of the teeth the same as ordinary cutters, but are largely superseded by formed cutters.

15. **Fly cutter,** as shown in Chart, No. 24, Fig. 10, is an *improvised cutter* that can be made to any desired form for any experimental operation, as plain, angular, or formed milling, also to cut the teeth of gears, as in Fig. 11. It is filed to shape with clearance back of cutting edge, then hardened, tempered and the face ground. It is securely held in the fly cutter arbor. As there is only one cutting edge, the feed must be very fine.

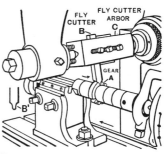

16. **A cotter mill** is often used to mill slots. See *C*, Fig. 42.

FIG. 11. — CUTTING GEAR TEETH WITH FLY CUTTER.

17. **Right and left cutters.** — Ordinary milling cutters can be operated both right and left, according to which side is placed on arbor first. Irregularly formed cutters are obtainable, right or left, to suit different milling machines. The cutting movement of a right cutter is the same as that of a regular twist drill.

18. **The correct and incorrect direction of rotation** for horizontal feed is shown in Figs. 12, 13. — To mill work by raising knee, that is vertical feed, work should be on side of mill moving downward to avoid tendency of mill to lift work. For a very thin cutter, set gib screws hard, or use a counterweight to hold table back because back lash will cause teeth to catch and break.

19. **Opposite cutting movements of different machines.** — Before placing a cutter on a milling-machine arbor, ascertain the position of headstock on the table, whether right or left, which differs on different makes of milling machines. See Figs. 40, 57. Also the direction of cutting movement and direction of rotation of cutter teeth.

20. Selection and mounting of cutters depends on nature of work, construction of machine and available cutters.

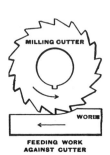

FIG. 12. — CORRECT CUTTING MOVEMENT. FOR HORIZONTAL FEED.

Use cutters as small as work will permit as they have a shorter cut and require less power. End mills or side mills are the fastest cutters as they have more teeth in contact at once. For some operations, as cutting gears, a certain type of formed cutter must be used. Before mounting a cutter, start machine and note direction of rotation.

FIG. 13. — INCORRECT CUTTING MOVEMENT. FOR HORIZONTAL FEED.

Attention. — When setting up machine, all parts that are fitted or clamped together must be wiped clean. Vise, fixtures and work must be held securely.

SPEEDS AND FEEDS

21. Cutting speeds for milling cutters depend on the hardness of the material, depth of cut and stability of the machine, therefore only approximate speeds are suggested.

Carbon steel cutters: Cast iron, 40 F.P.M.; machine steel, 40 F.P.M.; annealed carbon steel, 30 F.P.M.; brass or composition, 80 F.P.M.

High-speed steel cutters: Cast iron, 80 F.P.M.; machine steel, 80 F.P.M.; annealed carbon steel, 60 F.P.M.; brass or composition, 160 F.P.M.

For great accuracy, rough oversize and finish with a light cut, high speed and fine feed. A cutter may be safely rotated the same number of revolutions as lathe work equal in diameter to cutter. A slitting saw will stand twice the cutting speed of a large plane mill.

22. Feed for milling cutters is from .002″ to .250″ per cutter revolution and depends on diameter of cutter, kind of material, width and depth of cut, size of work, and whether light or heavy machine is used. Plain mills will stand a coarse feed; angular mills will not. A 3″ plain cutter at 40 F.P.M., feed .040″ per cutter revolution, will produce a finished surface.

As a large part of milling is done with one cut, the feed should be fine enough to produce a smooth surface. When two cuts are used, it is best to try a low speed and coarse feed for roughing, and a faster speed and finer feed for finishing.

Warning. — High speeds, *not coarse feeds*, ruin cutters. Watch action of cutter, belt and feed. Too fast a speed, or a dull cutter will make a squeaking, scraping noise. A dull cutter wears rapidly and produces a rough surface. As soon as a cutter shows dullness, grind it.

23. Lubricant for milling cutters. See Table, p. 149. — Cast iron and brass or composition are milled dry. A jet of air to remove chips will admit faster feed and prolong life of cutter. Lard oil is used on tenacious metals, such as steel and wrought iron. A stream of oil under pressure, will wash away chips, prevent clogging and recutting of chips. For ordinary work, drop oil on cutter, as in Fig. 50. While lard oil is best, milling compounds are obtainable which are less expensive, or a soda-water mixture may be made, as follows: ¼ lb. sal soda, ½ pt. lard oil, ½ pt. soft soap and water to make 10 qts. Boil one-half hour.

24. To mill aluminum. — Speed for high-speed steel milling cutters 110 F.P.M. Feed from .007″ to .250″ per cutter revolution. The speed and feed for milling aluminum may be from 25 per cent to 50 per cent higher than for brass.

Lubricant: equal parts of lard oil and kerosene, or soap, water and kerosene; or lard oil, 25 per cent, benzine, 75 per cent. Some prefer a compound made as follows: 3 gals. mineral lard oil, 5 lbs. sal soda, and 47 gals. of water.

End mills or, preferably, high-speed steel inserted-tooth face milling cutters, give better results than plain milling cutters as there is more chip room, and with a proper lubricant, clogging and recutting of the chips, which is the real trouble in milling aluminum, is avoided.

Example in milling aluminum. — $3\frac{1}{2}''$ high-speed steel inserted-tooth face milling cutter, as in Fig. 64: speed, 320 R.P.M.; feed, 4.8″ P.M., lubricated abundantly with a mixture, one part aqualene, 20 parts water, will produce a finely finished surface.

Information. — All machining operations, as turning, drilling, planing, milling, etc., on other metals, may be applied to aluminum by using a good lubricant. Pure aluminum does not machine as readily as some of the standard aluminum alloys. In general, the cutting speed should be higher and the feed lower than for brass.

25. To mill copper. — Speed 80 F.P.M. or a little lower than for brass. Feed .002″ to .125″ per cutter revolution. Lubricant same as for aluminum, or milk may be used.

26. Pickling and tumbling castings and forgings. — The hard scale and sand should be removed from castings before milling by tumbling in barrel or by pickling. Forgings should be pickled as the scale is destructive to cutters. See Pickling, *Principles of Machine Work*.

MEASURING WORK

27. To set cutter by trial to cut required depth. — Take short cuts and test depth with a rule, or line out work and take short cuts to line.

28. To set cutter by measurement, start machine and move work under cutter by hand long. feed; then raise knee by vertical feed shaft until work just touches revolving cutter (or take a short trial cut less than required depth and measure work with micrometer). Set dial on vertical feed at zero and move work away from cutter, then raise knee number of thousandths of an inch required. Move table slowly to cut by hand feed to avoid breaking cutter, then throw in power feed. At end of cut, trip feed by hand or automatically by table dog.

29. The limit in milling. — For plain milling, as bolt heads, nuts and similar work, a limit of from .002″ to .004″ is allow-

able. On fine machine parts, such as gun work, sewing machines, typewriters, electrical and scientific instruments, limits from .0005″ to .002″ are allowable. For close fits, mill as accurately as possible to save filing.

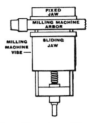

FIG. 14. — SETTING VISE JAWS PARALLEL TO ARBOR.

PLAIN MILLING

30. Methods of holding work in plain milling are similar to those for planing or drilling. There are three general methods: Holding in vise, clamping to table, or holding by jig or fixture.

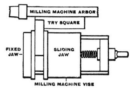

FIG. 15. — SETTING VISE JAWS AT RIGHT ANGLE TO ARBOR.

31. To set milling machine vise on table with jaws parallel to arbor, see Fig. 14, and at right angle to arbor, see Fig. 15. To set jaws at any other angle, use bevel protractor, or graduated circle at base of swivel vise, Fig. 17.

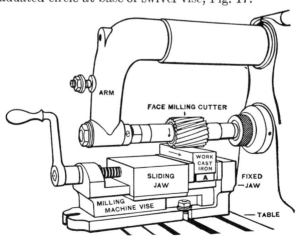

FIG. 16. — FACE MILLING.

32. To face mill work, Fig. 16. — Drive arbor lightly into spindle. Place cutter on arbor, clamp hard and adjust to arm. Bolt vise to table with jaws at right angle to arbor. Place work on planer parallel *A*, and clamp hard. As pressure of jaws will lift work, pound with soft hammer during and after clamping.

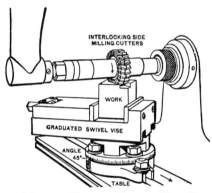

FIG. 17. — MILLING WORK IN SWIVEL VISE SET AT **45°**.

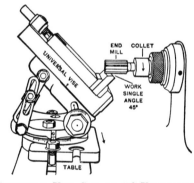

FIG. 18. — UNIVERSAL VISE SET AT 45° VERTICALLY TO SPINDLE.

33. Graduated swivel vise, Fig. 17, is used for milling at an angle as shown. The width of the slot in work is adjusted by inserting thin metal or paper washers between the interlocking side-milling cutters.

34. A universal vise can be swung to almost any position, at a right angle to plane of axis of spindle and 45° vertically, as in Fig. 18, or at an angle of 45° vertically and horizontally, as in Fig. 19.

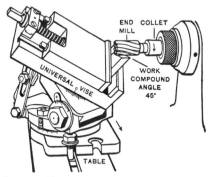

FIG. 19. — SETTING UNIVERSAL VISE AT 45° HORIZONTALLY AND VERTICALLY TO SPINDLE.

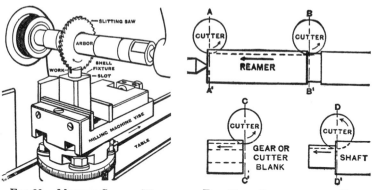

FIG. 20.—MILLING SLOT IN WORK HELD IN SHELL FIXTURE.

FIG. 21. — BEGINNING AND END- ING CUTS IN GROOVES, FLUTES, KEYWAYS AND SLOTS.

35. Shell fixture, Fig. 20, made square or hexagonal, and used to hold two, four or six-sided work without an index head. For shafts, spindles and screws, the hole is straight. For shanks of mills, drills and arbors, the hole is taper. A screw secures work in fixture, and a slot permits vise to clamp work.

36. In taking short trial cuts to obtain width of land in grooving taps, fluting reamers, milling teeth of cutters, etc., it is well to know that the milling cutter does not cut its full depth until the axis of the cutter has passed the end of the work as shown at AA', Fig. 21. To avoid a curved tooth or land, at end of cut, the feed must be continued by power or hand until the axis of the cutter has passed the end of the work as shown at BB' and CC', Fig. 21. To mill the bottom of a keyway or slot straight to a shoulder, do not stop the feed until the axis of cutter has passed the shoulder as shown at DD', Fig. 21.

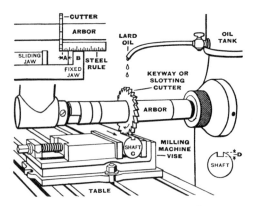

FIG. 22. — MILLING KEYWAY IN SHAFT.

Information. — To terminate a cut automatically at end of tooth or land, mount work on centers, raise table until work just clears cutter. Start machine and power feed. Sight axis of cutter and end of tooth or blank and when axis of cutter is nearing end of tooth or blank, clamp table dog. Begin cut as in AA', Fig. 21, and mill one groove and terminate cut with table dog. Then adjust dog to correct length of groove.

37. To mill keyway or slot in shaft. Fig. 22.

Specifications: Material, machine steel shaft cold rolled, turned or ground, $1\frac{1}{16}''$ diameter. Mill keyway $\frac{1}{4}''$ wide, $\frac{3}{32}''$ deep.

Machines and tools: Milling machine, milling machine vise, arbor $\frac{1}{4}''$ cutter, rule, planer parallel, lead hammer.

Speed, 1st or 2d speed or 60 F.P.M.

Feed, medium power feed, or $.010''$ per cutter revolution.

Lubricant, lard oil.

Time: 10 min. with machine "set up."

SCHEDULE OF OPERATIONS, MACHINES AND TOOLS

I. Clean and oil machine and vise.

II. Preparatory adjustments. Fasten milling machine vise to table with jaws at right angle to arbor. See Fig. 15.

Insert clean arbor in clean spindle and drive lightly with lead hammer. Place cutter on arbor without key, to cut in direction of rotation as shown, and clamp hard. Oil end of arbor and adjust arm to it.

III. Set cutter central. Raise table and move cross feed until distance A, from cutter to fixed jaw B, equals one-half diameter of shaft less one-half width of cutter.

Example: shaft $= 1\frac{1}{16}''$, cutter $= \frac{1}{4}''$. One-half diameter of shaft $= \frac{17}{32}''$, one-half width of cutter $= \frac{1}{8}''$. Thus distance $A = \frac{17}{32}'' - \frac{1}{8}'' = \frac{13}{32}''$.

Clamp cross feed.

IV. Clamp shaft in vise. Lower table. Place shaft C on parallel between vise jaws. Clamp hard and drive to seat with lead hammer. Draw line for length of keyway.

V. Take trial cuts. Start machine and take a few short trial cuts with hand feed until cutter mills its width.

VI. Raise table for depth of keyway. Move work away from cutter. Set vertical dial at zero and raise table required number of thousandths for depth D $(.094'')$.

Feed work to cutter by hand feed then throw in power feed and mill keyway with one cut. When nearing line, trip power feed by hand, and feed to line by hand feed.

Attention. — When milling two or more keyways, slots, grooves, etc., it is best to set and adjust table dog to trip feed automatically at desired length of cut.

Warning. — To avoid injury, keep clothing and fingers away from revolving cutters.

38. Metal slitting saw A, Fig. 23, is used to slit lathe foot-stock B, held by fixture C. Also to cut off pieces of work as in Fig. 24.

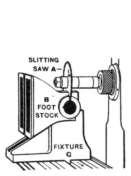

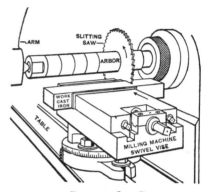

FIG. 23.—SLITTING LATHE   FIG. 24.—CUTTING OFF PIECES.
FOOTSTOCK.

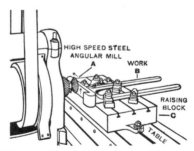

FIG. 25.—MILLING DOVETAIL. PLAIN MILLING MACHINE.

39. To mill dovetail, Fig. 25, use angular shell mill A, mounted on end of short arbor with hand vertical feed. If necessary, raising block C may be used to elevate work B.

40. Planer-type milling machine, Fig. 26, with elevating spindle, is designed for heavy, wide and long cuts. They are made with one or two horizontal heads and in addition one or two vertical heads. The work is held by bolts, clamps and screw pins, and should butt against pins or angle plate to prevent slipping.

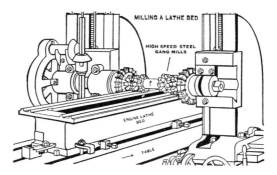

Fig. 26. — Planer-Type Milling Machine.

41. Gang mills are made by combining a number of mills of desired shape, as in Fig. 27, and enables work to be finished in one operation.

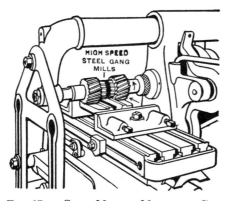

Fig. 27. — Gang Mills. Milling a Cap.

JIGS AND FIXTURES

42. Milling jigs and fixtures are used where a number of duplicate parts have to be milled. See Figs. 28, 29, 30.

43. Multiple milling. — Use milling fixture, Fig. 28, for rapid production of keyways. It is bolted to table and has

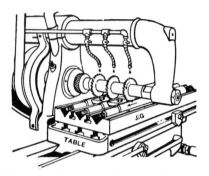

FIG. 28. — MULTIPLE MILLING. MILLING KEYWAYS IN THREE SHAFTS AT ONE TIME.

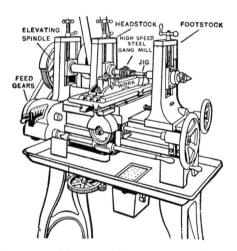

FIG. 29. — COMMERCIAL MILLING. MANUFACTURING MILLING MACHINE.

three V-grooves into which three shafts are clamped, and three keyways milled at a time.

44. Manufacturing type of milling machine, Fig. 29, adapted to accurate duplicate work which can be held in a vise or jig, and is largely used in gun, sewing machine, typewriter, and shoe machinery work.

45. A jig for milling duplicate machine parts, Fig. 30. — The mill has inserted high-speed steel cutters, held in place by taper pins.

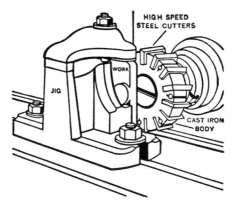

FIG. 30. — MILLING DUPLICATE WORK. INSERTED TOOTH MILL.

INDEX MILLING

46. Indexing is to divide work into any number of parts. There are three methods: Rapid, plain and differential.

Rapid Indexing

47. Rapid indexing is done by an index plate or disk fastened to the index head spindle which may have 24, 30 or 36 notches or holes to index work into any number of divisions that will divide these numbers evenly. It is preferred for work that requires only a small number of divisions, as bolt heads, nuts, grooving taps, etc.

48. Holding work for index milling is similar to holding work in an engine lathe. There are three general methods: Mounting on centers, on mandrel, or by securing in a chuck.

49. To mill bolt head and nut hexagonal with heading mills, bolt head milling machine. Fig. 31.

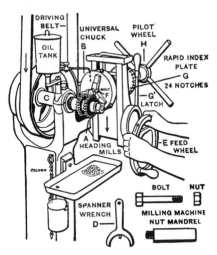

FIG. 31. — MILLING BOLT HEAD AND NUT HEXAGONAL.

Specifications: $\frac{1}{2}''$ bolt turned and threaded, with $\frac{7}{8}''$ hexagonal head and nut to be milled to $1\frac{3}{16}''$.

Machines and tools: Bolt head milling machine, arbor, index head and universal chuck, two high-speed steel heading mills, $1\frac{3}{16}''$ spacing collar, milling-machine nut mandrel, $1''$ micrometer.

Speed, 60 F.P.M., or 1st or 2d speed.

Feed, medium power feed, or .007$''$ per cutter revolution.

Lubricant, lard oil.

Time: for nut 4 min., bolt 5 min. with machine "set up."

SCHEDULE OF OPERATIONS, MACHINES AND TOOLS

I. Clean and oil machine and index head.

II. Preparatory adjustments. Place mills *A* on arbor without key, to cut in direction of rotation as shown, with $1\frac{3}{16}''$ spacing collar between them and clamp hard. Move index head until chuck jaws are $\frac{1}{4}''$ from mills and clamp with thumb screws on slide. Place latch G' in notch in index plate G.

III. Aline mills with chuck. Aline mills *A* with chuck *B*, by a nut mandrel (equal in diameter to space between mills) held in chuck.

Move spindle C endways by adjusting nuts on front box with spanner D, until mandrel moved by wheel E, will pass between mills.

IV. Test spacing of mills. Test spacing of mills by milling a trial piece, as a piece of shaft clamped in chuck. Mill two sides and measure with micrometer, reading should be .8125″ to .8155″. If measurement is not within the limit, adjustments may be made, as follows: To mill larger, place washers of thin metal or paper of the desired thickness between mills and spacing collar. To mill smaller, square off sides of spacing collar.

V. Mill nut hexagonal. Screw nut hard on a milling-machine nut mandrel held between copper jaws in a vise. Place mandrel in chuck B, with nut ⅜″ from the jaws and two sides of nut parallel to mills, and clamp hard. Start machine and feed nut downward slowly between mills by hand wheel E. Mill two sides, one cut. Move work upward until clear of mills. Withdraw latch G', index work four notches on index plate G by revolving with pilot wheel H. Mill remaining sides by the same method.

VI. Mill bolt head hexagonal. Clamp bolt by its body very hard in chuck in the same position as the nut, and mill with the same setting and by same process.

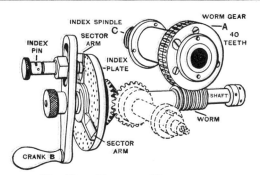

FIG. 32. — INDEXING MECHANISM.

Plain Indexing

50. Index head and footstock, Nos. 6, 7, Fig. 1, are used on plain or universal milling machines to do all kinds of index milling (see Spiral Milling). The index head spindle, Fig. 1, can be clamped at any angle with table, from 5° below horizontal to 10° beyond vertical. The footstock center, Fig. 1, can be adjusted at any angle in vertical plane.

51. Index mechanism, plain indexing. — To determine number of turns or any part of a turn to revolve crank for each division, divide 40 by number of divisions of work. Worm gear A, Fig. 32, has 40 teeth, and 40 turns of crank B will rotate spindle C one revolution. Where 40 divides without a remainder, adjust index crank pin to fit any hole in circle, and turn crank twice for twenty divisions, four times for ten divisions, five times for eight divisions, etc.

52. Use of index plate for obtaining fractional part of revolution of crank. The index plate contains concentric rows of equidistant holes, as in Fig. 33. The three index plates supplied with head have different numbers of holes

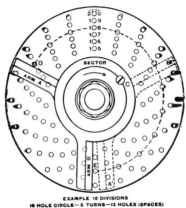

EXAMPLE 15 DIVISIONS
18 HOLE CIRCLE—2 TURNS—12 HOLES (SPACES)

Fig. 33. — Index Plate.

53. The sector, Fig. 33, consists of two radial arms A, B, that span required number of holes, and are clamped by screw C. This saves counting the holes after first time, and eliminates liability of error.

Example 1. — To divide work into 15 divisions.

Solution. — $\frac{40}{15} = 2\frac{2}{3}$, number of turns to move handle.

Select index plate with hole circle divisible by 3, as in Fig. 33 (18-hole circle), adjust crank to allow pin D to drop into a hole in this circle, move arm A against it. Release screw C and move arm B until it spans 12 spaces,

13 holes = $\frac{2}{3}$ of 18. Clamp screw C. Take first cut, withdraw index pin and secure with slight turn. Rotate crank 2 turns and 12 spaces, and drop crank pin in hole E. Then with finger, draw arm A around until it is against index pin, when sector arms will be in position shown by dotted arms A', B'. Take second cut.

Example 2. — To divide work into 48 divisions.

Solution. — $\frac{40}{48} = \frac{20}{24}$.

Select plate with a 24-hole circle set sector arms to span 21 holes, or 20 spaces, and move index pin that amount for each division, or use an 18-hole circle and 15 holes.

54. Graduated sector permits arm to be set to span any number of holes without counting by data given in table.

55. Adjustable crank allows index pin to drop into nearest hole to start cut readily at a desired point on work.

Warning. — There must be one more hole between sector arms than number obtained by calculation, or given on index table for holes in indexing mean spaces.

56. The index table supplied with machine gives data for plain and differential indexing.

Differential Indexing

57. Differential indexing is used to index prime, fractional and other numbers not obtainable by plain indexing.

58. Differential principle, Figs. 34, 35. — The index spindle is geared to index plate with gears supplied for spiral milling. Indexing is done on rotating index plate with mechanism arranged for plain indexing. With gears at a ratio of 1 to 1 and one idler to rotate index plate in direction of index crank, one division will be subtracted from regular indexing. With two idlers to rotate index plate opposite to index crank, one division will be added.

59. To obtain ratio of gearing, select a number near desired number of divisions, which can be indexed by plain method. The difference between these two numbers will represent gear for spindle. Divide selected number by 40 and quotient will represent gear for worm. Multiply each by a constant to give available gears. If selected number is larger than desired number, gearing must rotate plate with crank; if smaller, opposite to crank.

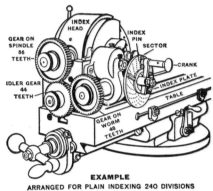

EXAMPLE
ARRANGED FOR PLAIN INDEXING 240 DIVISIONS
18 HOLE CIRCLE, 3 HOLES
DECREASED BY SIMPLE GEARING TO 233 DIVISIONS
(NO. OF SECTOR GRADUATION 32)

FIG. 34. — DIFFERENTIAL INDEXING WITH SIMPLE GEARING.

Example 1. — *Simple gearing*, Fig. 34. Desired number of divisions 233, selected number 240.

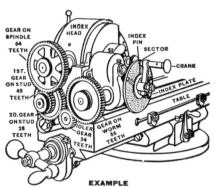

EXAMPLE
ARRANGED FOR PLAIN INDEXING 245 DIVISIONS.
49 HOLE CIRCLE, 8 HOLES
INCREASED BY COMPOUND GEARING TO 257 DIVISIONS.
(NO OF SECTOR GRADUATION 30)

FIG. 35. — DIFFERENTIAL INDEXING WITH COMPOUND GEARING.

Solution. — 240 − 233 = 7.   7 × 8 = 56 gear on spindle.

240 ÷ 40 = 6.   6 × 8 = 48 gear on worm.

One idler about 44.   See To Calculate Simple Gearing, p. **3**27.

Example 2. — *Compound gearing*, Fig. 35.   Desired number of divisions 257, selected number 245.

Solution. — 257 − 245 = 12. $\dfrac{12 \times 8}{6\frac{1}{8} \times 8} = \dfrac{96}{49}$.

245 ÷ 40 = 6⅛.

Factoring. — $\dfrac{96}{49} = \dfrac{16 \times 6}{7 \times 7} = \dfrac{64 \times 48}{28 \times 56}$.

Gear on spindle, 64.   First gear on stud, 48.

Second gear on stud, 28.   Gear on worm, 56.

One idler of about 24.   See To Calculate Compound Gearing, p. **3**33.

CALCULATING DIAMETER OF BLANK TO MILL SQUARE OR HEXAGONAL

60. Diameter to turn work to mill or file square is the product of diameter across the flats multiplied by 1.414.

Example. — Fig. 36.   What diameter must a piece be to mill square 1¼″ across the flats?

Solution. — 1.250″ × 1.414 = 1.767″ or 1$\frac{25}{32}$″ diameter of blank.

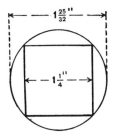

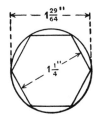

FIG. 36. — DIAGRAM TO TURN CYLINDER TO MILL SQUARE.

FIG. 37. — DIAGRAM TO TURN CYLINDER TO MILL HEXAGONAL.

61. Diameter to turn work to mill or file hexagonal is the product of the diameter across the flats multiplied by 1.155.

Example. — Fig. 37. What diameter must a piece be to mill hexagonal $1\frac{1}{4}''$ across the flats?

Solution. — $1.250'' \times 1.155 = 1.444''$ or $1\frac{29}{64}''$ diameter of blank.

62. To mill bolt head hexagonal with heading mills, universal or plain milling machine. Fig. 38.

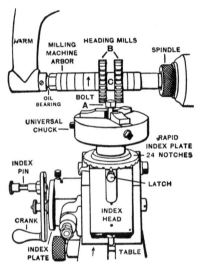

Fig. 38. — Milling Hexagonal Bolt Head or Nut with Heading Mills.

Specifications: $\frac{1}{2}''$ bolt turned and threaded, with $\frac{7}{8}''$ hexagonal head and nut to be milled to $1\frac{3}{16}''$.

Machines and tools: Universal or plain milling machine, arbor, index head and universal chuck, two high-speed steel heading mills, and $1\frac{3}{16}''$ spacing-collar, milling-machine nut mandrel.

Speed, 60 F.P.M., or 1st or 2d speed.

Feed, medium power feed, or $.010''$ per cutter revolution.

Lubricant, lard oil.

Time: for nut 4 min., bolt 5 min. with machine "set up."

SCHEDULE OF OPERATIONS, MACHINES AND TOOLS

I. Clean and oil machine including index head.

II. Preparatory adjustments. Place clean arbor in clean hole in spindle and drive lightly with lead hammer. Place mills B on arbor without key, to cut in direction of rotation as shown, with $1\frac{3}{6}''$ spacing collar C between them, and clamp hard. Oil end of arbor, adjust and clamp arm. Fasten index head on clean table. Set index spindle and chuck vertical, as shown. Raise table to bring chuck jaws about $\frac{1}{4}''$ below mills.

III. Arrange index mechanism. For rapid indexing, place latch in notch in index plate with index pin out. For plain indexing, place index pin in hole in 39-hole circle in index plate with latch out.

IV. Aline mills with chuck by nut mandrel, as on p. **102**0.

V. Mill nut hexagonal. Screw nut hard on milling-machine nut mandrel held between copper jaws in vise, place mandrel in chuck with nut $\frac{3}{8}''$ above jaws and two sides of nut parallel to mills, and clamp hard. Start machine, feed by hand and mill only enough to take measurement. Move table back until nut clears mill. Measure across flats with $1''$ micrometer. Reading should be from $.8125''$ to $.8155''$. If reading is not within limit, make corrections. See p. **102**1, IV. Index a sixth of a revolution each time and mill remaining sides. Use power feed and adjust table dog to trip feed automatically; or trip feed by hand at end of cut.

VI. Mill bolt head hexagonal. Clamp bolt by body in chuck very hard, in same position as nut and mill same as nut.

Attention. — To use rapid indexing, see p. **101**9, revolve index head spindle four notches in 24-notch plate after each cut.

To use plain indexing, see pp. **102**1–**102**3, adjust index crank pin to fit a 39-hole circle; adjust sector to span 26 spaces or 27 holes in 39-hole circle. After each cut, revolve crank six turns and 26 spaces.

Information. — To center chuck accurately with heading mills, move cross feed "out" then "in" to remove back lash. Mill one side of trial piece held in chuck with inside of outer mill, index twenty turns and mill opposite side. Measure with micrometer, then set dial at zero and move cross feed inward one-half the difference between size of trial piece and space between cutters. Example. Size of trial piece $\frac{5}{8}''$, space between cutters $1\frac{3}{8}''$; $1\frac{3}{8}'' - \frac{5}{8}'' = \frac{3}{16}''$; $\frac{3}{16}'' \div 2 = \frac{3}{32}''$ or $.094''$ distance to move cross feed inward.

63. To mill square head with end mill. Fig. 39.

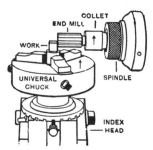

FIG. 39. — MILLING SQUARE HEAD WITH END MILL.

Specifications: 1″ reamer blank, mill head on shank ⅝″ square and 1 1/16″ long.

Machines and tools: Universal or plain milling machine, bolt-heading machine, collet, index head and universal chuck, 1⅛″ high-speed steel end mill, 1″ micrometer, lead hammer.

Speed, 60 F.P.M., or 3d speed.

Feed, medium power, or .010″ per cutter revolution.

Lubricant, lard oil.

Time: 15 min. with machine "set up."

SCHEDULE OF OPERATIONS, MACHINES AND TOOLS

I. Clean and oil machine including index head.

II. Preparatory adjustments. Clean shank of end mill and place in clean hole in collet, and place clean collet in clean hole in spindle and drive lightly with lead hammer.

Fasten index head on clean table. Set index spindle and chuck vertical as shown. Raise table to bring chuck jaws about ¼″ below end mill.

III. Arrange for indexing, rapid or plain.

IV. To mill square head. Place 1″ reamer or tap blank in chuck with end of shank ⅞″ above jaws and clamp hard.

Start machine, move work inward to take cut about ⅛″ deep and down to line for length of square head. Feed by hand against up-cutting side of end mill as shown by arrows, and mill one side.

Move table back until work clears mill, revolve work one-half revolution and mill opposite side.

Stop machine, move table back and measure work with micrometer, and from reading of micrometer subtract the finish size and divide the remainder by two, which will give the amount to move work inward to obtain correct measurement. Example: Trial cut may measure .750″, finish measurement is .625″ (.750″ — .625″ = .125,″ .125″ ÷ 2 = .0625″, the amount to move work inward). Set cross-feed dial at zero and move work inward .0625.″ Mill side, index and mill opposite side, measure and make correction, if necessary. Mill remaining sides with one cut each.

Attention. — File slight bevels on edges of squared end to remove burr.

MILLING ON CENTERS

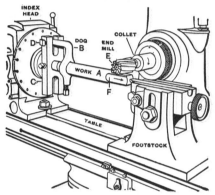

Fig. 40. — Milling End of Shaft Square Leaving Curved Shoulder.

64. To mill end of shaft square with curved shoulder, Fig. 40. — Set index head at zero, clamp and set footstock spindle in alinement. Mount work *A* on centers, clamp dog *B* to face plate *C* with screw *D*. Arrange for indexing, mount end mill *E* in collet and set dog to trip feed at *F*. Take trial cuts on opposite sides and measure with micrometer. Set vertical feed dial at zero, move work away from mill and elevate work by vertical feed shaft one-half its oversize. Mill sides in order with one cut each.

65. Steady rest to support slender work, *A*, Fig. 41. — To mill the grooves in rose reamer, use convex cutter *B* and a special angular cutter for teeth at end.

66. To flute or groove small reamers, taps, drills, etc., hold work in chuck and support outer end by steady rest with a V-grooved head elevating screw.

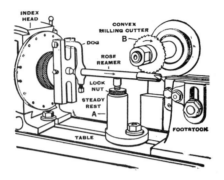

FIG. 41.—SUPPORTING SLENDER WORK, FLUTING ROSE REAMER, WITH STEADY REST.

67. To mill a number of nuts at once, string on a gang mandrel and clamp between shoulder and nut. Mount mandrel on centers and mill nuts similarly to milling square end of a shaft as in Fig. 40. Use end, plain, or heading mills.

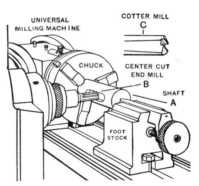

FIG. 42. — MILLING KEYWAY IN SHAFT WITH CENTER-CUT END MILL OR COTTER MILL.

68. To mill keyway or slot in shaft with center-cut end mill, Fig. 42. — Aline index centers and cone spindle by zero lines on knee and column. Mount one end of shaft *A* in chuck and the other end on footstock center. Mill keyway with center-cut end mill *B*. If cotter mill *C* is used, drill a hole in shaft in which to start mill.

GROOVING TAPS

69. To groove or flute taps, reamers, cutters, etc. — Taps, reamers, cutters, etc., are grooved or fluted by milling, to provide cutting edges and places to receive and discharge chips. There are no standards for depth of groove, width of land, or number of grooves.

70. Taps, number of grooves, rake of teeth. Figs. 43, 44. — Four grooves are given taps up to 3″ in diameter, and five or more for larger diameters. Hobs for tapping threading dies are usually given from six to eight grooves.

Fig. 43. — Tap Grooved with Special Double-Angle Tap Cutter, Radial Teeth.

Fig. 44. — Commercial Tap Grooved with Convex Cutter.

For tapping all kinds of metal, the two forms of grooves in Figs. 43, 44 are largely used, and each cutter is suitable for several diameters of taps. The teeth of special taps for brass are often given a negative rake (to prevent chattering) by the shape of the grooves, or by grinding the face of the teeth.

To give parallel lands on plug or bottoming taps, set index centers in alinement; but on taper taps, lower index head slightly to give parallel or slightly taper lands.

71. To groove a tap. Direct method. Double-angle formed tap cutter. — Figs. 45, 46.

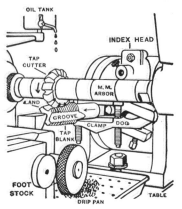

Fig. 45. — Grooving Tap.

Specifications: $\frac{3}{4}''$ × 10 U.S.S. taper tap blank threaded, and with end of shank milled square. Four grooves.

Machines and tools: Universal or plain milling machine, arbor, index head and centers, No. 5 high-speed steel double-angle tap cutter, clamp dog or special driver.

Speed, 60 F.P.M., or 1st or 2d speed.

Feed, medium power feed, or .007″ per cutter revolution.

Lubricant, lard oil.

Time: 32 min. with machine "set up."

SCHEDULE OF OPERATIONS, MACHINES AND TOOLS

I. **Clean and oil machine and index head.**

II. **Preparatory adjustments.** Insert clean arbor in clean hole in spindle. Drive arbor lightly with lead hammer. Place cutter on arbor without key and clamp hard. Oil end of arbor and adjust arm to bearing. Fasten index head and footstock on clean table. Set centers in horizontal aline- ment, then lower index head one-third degree to give slightly taper lands.

III. **Arrange for indexing, rapid or plain.**

IV. **Mount tap blank on centers.** Fasten clamp dog or special driver on tap, oil dead center and mount on centers. Clamp tail of dog in face plate to avoid backlash. With index latch in notch or index pin in hole, lightly drive face plate or driver with lead hammer until side of square end of tap is perpendicular to table, testing with square resting on table. Start machine and power feed and adjust table dog to stop feed automatically when center of cutter has passed last thread on tap, see p. 10̇14.

V. To obtain depth of setting. Raise knee until highest part of tap blank touches largest diameter of revolving cutter.

A piece of white paper held between revolving cutter and tap blank will indicate touch. Then set vertical feed dial at zero.

VI. To obtain side setting. Move cross feed until outside of tap and outside of cutter are in same vertical plane, testing by square resting on table. Then set cross-feed dial at zero.

VII. Move both feeds required amount. With long-feed handle, move tap away from cutter and with cross-feed handle move tap

outward .150," see *A*, Fig. 46. Then with vertical feed raise tap .175," see *B*, Fig. 46. Use rapid method of indexing and mill four grooves, one cut for each groove.

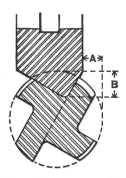

Fig. 46. — Diagram for Setting Double-Angle Tap Cutter.

Attention. — Data may be obtained for grooving taps of all diameters by the direct method by taking experimental cuts on trial blanks.

---

72. To groove tap, trial method. — Set cutter approximately central to footstock center. Use trial blank and take short light cut. Then index quarter turn and take similar cut and measure land. Raise knee a small amount and repeat until land is proper width. Remove trial blank. Move cross feed and with rule set small angle of cutter in alinement with center, as in Fig. 47, that is radial or slightly ahead of radial, if desired.

Fig. 47. — Setting Double-Angle Cutters to Mill Radial Teeth.

FLUTING REAMERS

73. Fluting reamers. — For reaming all kinds of metal, use the form of flute in Fig. 48, which gives radial cutting edges. Each cutter may be used for several diameters.

The cutting edges of brass reamers are often made to recede from a radial line giving a negative rake from 5° to 10°, Fig. 49.

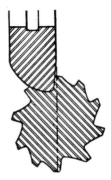

Fig. 48. — Form of Flute in Reamer for General Work. Teeth Radial. Irregularly Spaced.

Fig. 49. — Form of Flute in Reamer for Brass. Teeth with Negative Rake, Irregularly Spaced.

Space the reamer irregularly to avoid chattering, with an even number of teeth and with opposite cutting edges in the same diameter so that it can be measured with micrometer. The irregularity, 1° to 4° from equal spacing, is obtained by subtracting or adding a predetermined number of holes to or from the regular indexing for each pair of flutes diametrically opposite.

74. Principle of indexing and fluting a 1″ hand reamer to avoid chattering. — The teeth are spaced two holes in the 20-hole circle progressively wider, and to obtain lands of uniform width the flutes are milled .007″ progressively deeper from 1 to 5 and from 1′ to 5′.

75. To flute hand reamer, Fig. 50.

Specifications: 1″ hand reamer blank with end of shank milled square. Ten flutes. See Fig. 48.

Machines and tools: Universal or plain milling machine, arbor, index head and centers, high-speed steel special formed reamer cutter, clamp dog or special driver.

Speed, 60 F.P.M., 1st or 2d speed.

Feed, medium power feed, or .007″ per cutter revolution.

Lubricant, lard oil.

Time: 50 min. with machine "set up".

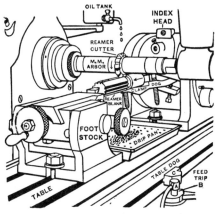

FIG. 50. — FLUTING HAND REAMER.

SCHEDULE OF OPERATIONS, MACHINES AND TOOLS

I. Clean and oil machine and index head.

II. Preparatory adjustments. Insert clean arbor in clean spindle. Drive arbor lightly with lead hammer. Place cutter on arbor without key and clamp hard.

Oil end of arbor and adjust arm to bearing. Fasten index head and footstock on clean table. Set centers in horizontal alinement.

III. Arrange for plain indexing 10 teeth with a 20-hole circle.

IV. Mount reamer blank or trial blank on centers. Fasten clamp dog or special driver on square end of reamer blank, oil dead center and mount reamer blank (or trial blank) on centers, see Fig. 50. To avoid backlash, clamp tail of dog in face plate with set screw.

Start machine and power feed and adjust table dog to stop the feed automatically when the center of the cutter has reached the corner of the blank. See p. **101**4.

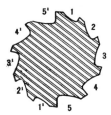

FIG. 51. — DIAGRAM OF REAMER SHOWING PRINCIPLE OF IRREGULAR SPACING.

V. To obtain width of land with trial blank. It is safer to obtain width of land and setting of cutter with a trial blank the same diameter as reamer blank. Set cutter approximately central with footstock center, and mill two flutes to give desired width of land as follows:

Raise table and take a light short cut in 1, Fig. 51. Move work clear of cutter, then index three turns and eighteen holes. Raise .007" and take a cut in 2. Repeat cuts in 1 and 2 until land is $\frac{1}{16}$" wide, keeping flute 2, .007" deeper than 1.

VI. Set cutter radial. Remove trial blank. Move cross feed and with rule set angle of cutter in alinement with center (that is radial), see Fig. 47, or set cutter ahead of radial, if it is desired to have face of teeth ahead of radial.

Or set cutter to give a negative tooth if it is desired to have teeth with negative rake for brass.

VII. Mill flutes. Mount reamer blank on centers, and mill flutes as follows: Begin at 1 with cutter set at depth of 1, with vertical dial set at zero and mill flute 1. Index twenty turns and mill flute 1'. Raise to .007", index three turns and eighteen holes and mill flute 2.'

Index twenty turns and mill flute 2. Raise to .014," index four turns and mill flute 3. Index twenty turns and mill flute 3.' Raise to .021," index four turns and two holes and mill flute 4.' Index twenty turns and mill flute 4. Raise to .028," index four turns and four holes and mill flute 5.

Index twenty turns and mill flute 5'.

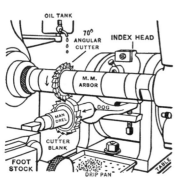

FIG. 52. — MILLING TEETH OF PLAIN MILLING CUTTER.

76. Electric alarm to call operator when feed stops and terminates cut, may be arranged, as at A, B, Fig. 50. Metal cap A, insulated with fiber or vulcanite, is placed on feed trip B. One wire is attached to cap and runs to battery in column. The other wire connects machine, bell and battery. When table dog C, strikes cap A, the bell rings.

MILLING TEETH OF MILLING CUTTER

77. To mill teeth of plain milling cutter, Fig. 52.

Specifications: $2\frac{1}{2}'' \times \frac{1}{2}''$ plain milling cutter blank, carbon or high-speed steel turned and squared, 18 teeth, width of land $\frac{1}{32}''$. See pp. **122**4, **122**5.

Machines and tools: Universal or plain milling machine, arbor, 70° angular high-speed steel cutter, index head and centers, mandrel, dog or special driver.

Speed, 60 F.P.M., or 1st or 2d speed.

Feed, medium power feed, or .007″ per cutter revolution.

Lubricant, lard oil.

Time: 30 min. with machine "set up."

SCHEDULE OF OPERATIONS, MACHINES AND TOOLS

I. Clean and oil machine and index head.

II. Preparatory adjustments. Insert clean arbor in clean hole in spindle. Drive arbor lightly with lead hammer. Place cutter on arbor without key and clamp hard. Oil end of arbor and adjust arm to bearing.

Fasten index head and footstock on clean table. Set centers in horizontal alinement.

III. Arrange plain indexing for 18 teeth.

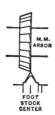

FIG. 53.—ALINING CENTER TO ANGULAR CUTTER TO OBTAIN RADIAL TEETH.

IV. Aline cutter and center. Set cutter radial, as shown in Fig. 53.

V. Mount work on centers. Force mandrel into cutter blank (or use gang mandrel), fasten dog on mandrel, oil dead center and mount on centers. To avoid backlash, clamp tail of dog in face plate with set screw. Start machine and power feed, and adjust table dog to stop feed automatically when center of cutter has passed the corner of blank.

VI. Obtain width of land by trial cuts in two grooves. Raise table and take a light trial cut less than required depth with power feed. Index cutter blank forward for the next tooth and take similar cut. Then raise table and take trial cuts in second groove until desired width of land $(\frac{1}{32}'')$ is obtained.

VII. Mill teeth. Index and mill all teeth with one cut each.

78. To mill teeth of angular cutter, Fig. 54. — Mount angular cutter A, on milling machine arbor B. Mount angular cutter blank C, on stub arbor D, which fits the index head spindle. Raise index head spindle to approximate angle and take trial cuts and adjust angle to obtain parallel lands. Then raise table to obtain desired width of land.

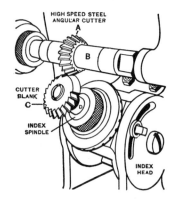

Fig. 54. — Milling Teeth in an Angular Cutter.

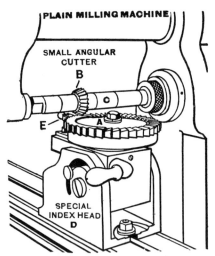

Fig. 55. — Milling Radial Teeth of Side Milling Cutter. Indexing from Teeth.

79. To mill radial teeth in side milling cutter, Fig. 55. — Mill face teeth of the side milling cutter *A*, in the same manner as a plain mill, with small angular cutter *B*, on arbor *C*.

Set special index head *D* vertically, then adjust to give teeth with parallel lands. Index the cutter by the face teeth with latch at *E*.

80. To level or set taper work with surface gage, as the taper tap or reamer blank *A*, Fig. 56, mount on centers and test with surface gage *B*. Cut deeper at one end, as shown by dotted line *CD*, so that lands will be parallel or nearly so. Adjust scriber to touch work at *E*, and insert metal piece equal in thick-

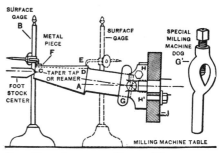

FIG. 56. — LEVELING TAPER WORK WITH SURFACE GAGE.

ness to difference in depth of cut at *F*. Adjust work by raising dead center or lowering index head.

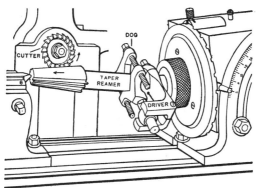

FIG. 57. — MILLING TAPER WORK. SPECIAL MILLING MACHINE DOG AND DRIVER.

81. Special milling machine dogs for taper work. — A regular dog will spring taper work unless the screw confining the tail is released before rotating work.

Figs. 56, 57, show two types of special milling machine dogs which give desired motion and avoid springing work.

To divide taper work exactly, the axis of index head, spindle and axis of dead center must be in alinement. The footstock center on some machines can be adjusted to obtain this alinement. Taper attachments are obtainable provided with index centers arranged so that axes of centers are always in alinement.

82. To graduate with a milling machine. — In the absence of a dividing engine and diamond cutter, or a sharp-edge wheel which will make a smooth line by pressure, use an index head and cutter as in Fig. 58 to graduate dials, disks and similar work. Block the spindle and feed work to cutter by hand. Rules and scales may be clamped to table and graduated in this manner.

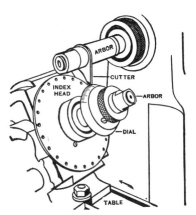

FIG. 58. — GRADUATING A DIAL.

CIRCULAR MILLING

83. Circular milling. — To mill end of rectangular piece with end mill, Fig. 59, secure work on nut mandrel and hold in chuck. Feed work to cut by rotating crank. To mill a portion of the circumference of a piece, as in Fig. 60, use raising block, set index head at right angles to table and feed work to cut by crank. In Fig. 61 is shown how a cam cutting attachment may be used to mill a bevel gear blank.

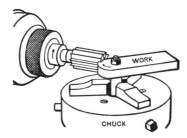

FIG. 59. — MILLING END OF WORK CIRCULAR.

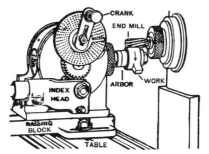

FIG. 60. — MILLING CIRCUMFERENCE OF WORK.

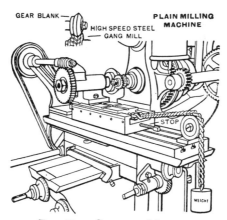

FIG. 61. — CIRCULAR MILLING.

84. Tee slots are milled more rapidly than planed. To mill Tee slots, in circular table, Fig. 62, first mill portion *A* with an end mill and then portion *BB'* with a Tee-slot cutter.

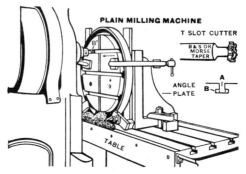

FIG. 62. — MILLING TEE SLOTS IN A CIRCULAR TABLE.

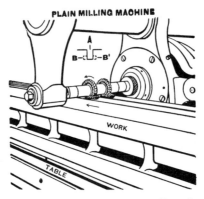

FIG. 63. — MULTIPLE MILLING. MILLING TWO SLOTS AT ONE OPERATION.

To mill Tee slots in heavy work such as milling-machine table, Fig. 63, first mill portion *A* with a side-milling cutter, then mill portion *BB'* with a Tee-slot cutter. Use a vertical attachment, or clamp work flat on table and mill.

VERTICAL MILLING

85. A vertical milling machine has advantages over the horizontal particularly on work that can be milled with end mills. Fig. 64, shows an inserted rake-tooth face mill A,

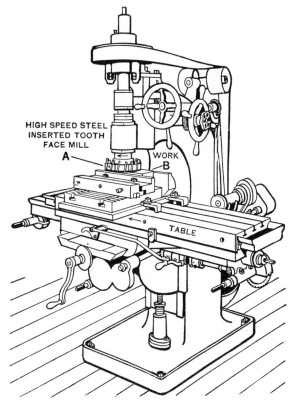

HIGH SPEED STEEL
INSERTED TOOTH
FACE MILL
A

WORK
B

TABLE

FIG. 64. — VERTICAL MILLING MACHINE. FACE MILLING.

milling a plain surface on work B in a vertical milling machine. The work is held in a vise.

86. Vertical and circular milling attachments. — Circles, circular slots and segments of circles may be milled by converting a horizontal milling machine into a vertical milling machine by vertical attachment as in Fig. 65. The work is held and operated by fixture and circular milling attachment. Vertical spindle A operates end mill B to mill a segment of a

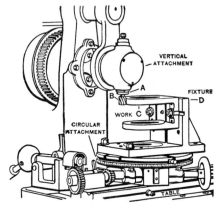

FIG. 65. — MILLING MACHINE WITH VERTICAL AND CIRCULAR ATTACHMENTS.

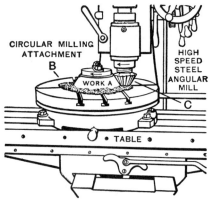

FIG. 66. — BEVEL CIRCULAR MILLING WITH VERTICAL MILLING MACHINE.

circle on work C, held by fixture D bolted to circular attachment. The feed is obtained by rotating circular table by hand or power.

See Inspection of a Vertical Milling Machine, p. **14**14.

87. Bevel circular milling. — To mill work A, Fig. 66, (rotary cone of a circular attachment) bolt to table of circular attachment B and mill bevel with angular mill C.

88. Milling an inside boss, Fig. 67. — Hold swinging arm A (casting) in vise. Mill edge B, set stop C and feed mill down

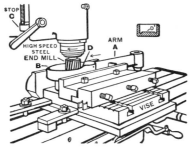

FIG. 67. — MILLING AN INSIDE BOSS. VERTICAL MILLING MACHINE.

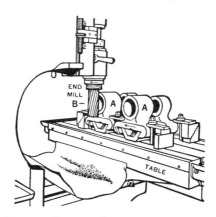

FIG. 68. — MILLING ENDS OF SLIDING HEADS WITH A VERTICAL
MILLING MACHINE.

until stop strikes, then mill inside boss DD', operating both cross and long. table feed by hand.

89. Milling ends of sliding heads, Fig. 68. — Two sliding head castings A, A are bolted to fixture clamped to table and the ends of boxes are milled with side of end mill B with power feed.

PROFILING

90. Profiling machines, Fig. 69. — Mill work to irregular shapes by means of profiling forms.

In Fig. 69 work A is clamped to fixture B and face C is milled with cutter D while form pin E is kept in contact with profiling form F by moving table G with long.- and cross-feed handles together by hand. See plan at A', C', D', E' and F'.

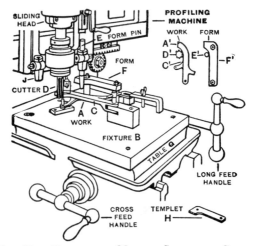

Fig. 69. — Profiling. Milling Irregular Shapes.

91. To make profiling form, Fig. 69. — Templet H is made and clamped in position of work A, and the form blank in its position F on fixture. Straight form pin E is fastened in block J and held in contact with templet, and a taper cutter in spindle generates form F. The taper form and taper pin allow for reduction in diameter of cutter by grinding, by raising pin to allow cutter to resume its previous relation to work.

Rake Tooth Milling Cutters

92. Coarse tooth milling cutters give wide space between the teeth for chips, and are adapted to the heavier classes of milling.

93. Rake tooth spiral milling cutter with coarse teeth.— Fig. 70 shows an improved spiral milling cutter A. The teeth are undercut giving about 10° rake, as at B, and also spiral giving an effective shearing angle. Each tooth has a true cutting action like a lathe tool.

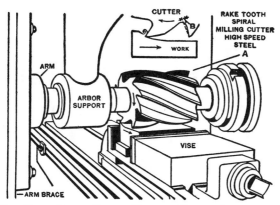

FIG. 70.—RAKE TOOTH SPIRAL MILLING CUTTER MILLING STEEL. HEAVY WORK.

94. Helical tooth milling cutter A, Fig. 71, is a shear type of cutter made with two or three teeth in the form of a screw thread, either singly or interlocking right and left cutters, as at B.

The teeth are ground on the front and on the top and have a rake of 8° to 15° and a clearance of 5° to 7°. When milling steel it removes a gimlet-shaped chip in the direction of the axis of the cutter, consumes less power than the ordinary cutter, does not spring the arbor and largely eliminates chattering.

It is adapted to mill broad steel surfaces accurately, smoothly, and rapidly, and is particularly useful for milling thin castings that are likely to chatter. See Shear Planer Tool, p. 908.

95. Thread miller (gear headstock, cone drive), Fig. 72, is used for milling right and left straight and tapered threads, English or Metric system, on taps, hobs, worms, precision screws and spiral gears. It is also used for milling oil grooves and splines in shafting.

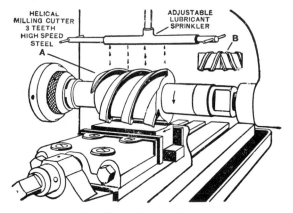

FIG. 71. — MILLING STEEL.

96. To mill a thread with a thread miller, Fig. 72. — Mount screw blank A on center or, preferably, hold one end in a collet and the other end on dead center. Fasten cutter on arbor of cutter head B and tilt arbor to the proper angle for lead of thread as shown at A' and B'. Arrange the change gears for the pitch of the screw. Move cutter forward by cross slide until a tooth of the cutter touches the blank, and set micrometer head on cross screw at zero. Move carriage until cutter clears end of blank, and move cutter to required depth of thread with micrometer cross-slide screw. Set knock-off feed at desired point. Arrange speed. Start machine. Lubricate cutter freely, and mill thread with one cut, automatically. An index and tables are furnished with each type of machine giving the change gears for single or multiple threads, the angles for setting cutter head, and the depths for setting the cutters.

97. Collets. — It is best to hold the screw blank with a collet instead of on centers. One end of the screw blank is held firmly in a collet in the headstock spindle and the other end is held on the footstock centers if the work is long and slender and needs support.

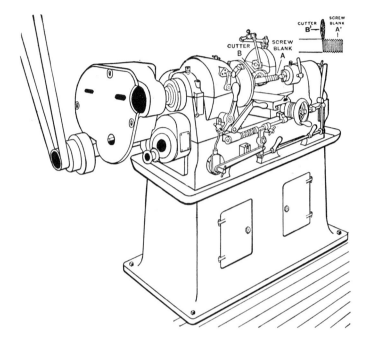

FIG. 72. — MILLING THREAD WITH AUTOMATIC THREAD MILLING
MACHINE.

98. To mill multiple threads, notched index rings are used
for indexing the different cuts.

99. To mill internal threads, special thread milling cutters
are obtainable.

100. To sharpen thread milling cutters, an automatic
grinder is used.

101. Follow-rests. — To cut screws of small diameter
which would be likely to spring away from the cut, follow-rests,
with interchangeable bushings for various diameters of screw
blanks, are furnished.

ADVANCED MACHINE WORK

SECTION 11

GEAR CUTTING

Spur Gearing: Formulas in Spur Gearing, Cutting Spur Gears, Generating Spur Gears, Cutting Internal Gears, Cutting Racks. Spiral Gearing: Formulas in Spiral Gearing, Cutting Spiral Gears. Worm Gearing: Formulas in Worm Gearing, Cutting Worm Gears, Threading Worms. Hobbing Gears. Bevel Gearing: Formulas in Bevel Gearing, Cutting Bevel Gears, Planing Bevel Gears. Testing Gears. Bevel Protractor. Compound Rest.

SPUR GEARING

1. Spur gears are used for transmitting positive and uniform rotary motion from one shaft to another shaft which is parallel to the first.

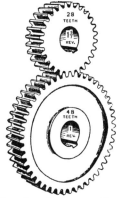
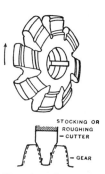

FIG. 1. — INVOLUTE GEAR CUTTER.

FIG. 2. — INVOLUTE GEAR AND PINION. RATIO $1\frac{5}{7}$ TO 1.

FIG. 3. — ROUGHING OR STOCKING STEPPED GEAR CUTTER.

2. Systems of gearing. — There are two systems of gearing: The involute or single curve form of tooth, as shown in Fig. 2, and the epicycloidal (cycloidal) or double curve form of tooth. The epicycloidal form of tooth requires an exact center distance, but with the involute form of tooth, the center distance can be varied slightly without affecting the running of the gears.

1101

The involute form has replaced the epicycloidal to such an extent that the latter is now rarely used.

3. Processes of cutting gears. — At *A*, *B*, *C*, *D*, and *E*, Fig. 4, are shown the different processes used in cutting gears.

Cutting a spur, bevel, or spiral gear with a formed milling cutter is shown at *A*; generating a spur, spiral, or worm gear

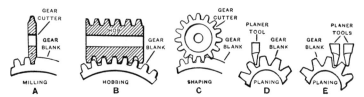

FIG. 4. — PRINCIPAL PROCESSES OF CUTTING GEARS.

with a hob in a hobbing machine, at *B*; generating a spur gear in a gear shaper, at *C*; generating a bevel or spur gear with a single tool in an automatic generating bevel-gear shaper, at *D*; and planing a bevel or spur gear with two tools in an automatic generating bevel-gear planer, at *E*.

Beside these principal processes, small internal gears can be broached. See p. 543. Cluster gears are often cut in a gear shaper or slotter; thin gears are stamped out of sheet metal on presses; and small gears, for light transmission, are often die-cast.

Attention. — Gears may be cut in milling machines; but, if required in quantities, they should be cut with automatic machines.

Information. — Spur gears may be cut over-size in standard gear-cutting machines, heat treated, and then finish ground to size in special machines. Cast gears are finish ground in the same manner.

4. Formed gear cutters in standard diametral and circular pitches are obtainable. One cutter will answer for several numbers of teeth of same pitch without producing any material error. Any gear will mesh into and run properly with any other gear or rack of the same pitch.

Warning. — Gear cutters should be kept sharp. A dull cutter will quickly destroy itself. See pp. 823, 824.

5. Fly cutters, as on p. 1007, may be "home-made" and used for a special job of hobbing and gear cutting.

6. Cutters for the two systems of gearing. — In the *involute system*, 8 cutters are required to cut all gears of any given pitch from a pinion gear of 12 teeth to a rack, and are numbered from 1 to 8. Cutters are obtainable in half-numbers for a more accurate form of tooth. Special cutters may be ordered for exact numbers of teeth. *The epicycloidal system* requires 24 cutters lettered from A to X.

7. Involute cutters.

No. ON Cutter.	Number OF Teeth.	No. ON Cutter.	Number OF Teeth.	No. ON Cutter.	Number OF Teeth.	No. ON Cutter.	Number OF Teeth.
1	135 to rack	3	35 to 54	5	21 to 25	7	14 to 16
$1\frac{1}{2}$	80 to 134	$3\frac{1}{2}$	30 to 34	$5\frac{1}{2}$	19 to 20	$7\frac{1}{2}$	13
2	55 to 134	4	26 to 34	6	17 to 20	8	12 to 13
$2\frac{1}{2}$	42 to 54	$4\frac{1}{2}$	23 to 25	$6\frac{1}{2}$	15 to 16	...	....

Attention. — Involute cutters are obtainable for circular pitch and for Metric gears.

Exception. — Involute cutters are made with shoulder for cutting small brass gears. They are known as "topping" cutters and have no clearance.

8. Epicycloidal cutters.

Letter ON Cutter.	Number OF Teeth.	Letter ON Cutter.	Number OF Teeth.	Letter ON Cutter.	Number OF Teeth.	Letter ON Cutter.	Number OF Teeth.
A	12	G	18	M	27 to 29	S	60 to 74
B	13	H	19	N	30 to 33	T	75 to 99
C	14	I	20	O	34 to 37	U	100 to 149
D	15	J	21 to 22	P	38 to 42	V	150 to 249
E	16	K	23 to 24	Q	43 to 49	W	250 or more
F	17	L	25 to 26	R	50 to 59	X	Rack

Note. — Duplex or gang gear cutters to cut two or more teeth at once are obtainable.

Information. — Epicycloidal gears must be cut accurately to depth so that pitch lines will coincide. Epicycloidal cutters are made with a shoulder which limits depth of tooth.

9. Limit in sizing gear blanks. — In commercial work, spur gear blanks from 4 to 6 pitch may vary from calculated outside diameter to .005″ small; 6 to 10 pitch, to .004″ small; and 12 to 20 pitch, to .003″ small.

10. Allow for error of blank in depth of teeth. — Measure blank with micrometer and if under size, deduct one-half the error from setting for depth of tooth.

Example. — If gear blank, Fig. 11, is .002″ undersize, deduct .001″ from setting of .216″.

Solution. — .216 − .001 = .215″, depth of space or tooth.

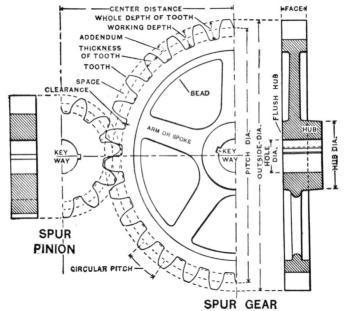

Fig. 5. — Diagram of Gear Teeth and Parts.

FORMULAS IN SPUR GEARING

11. Diametral pitch is the number of teeth to each inch of pitch diameter.

Formula. — Diametral pitch = $\dfrac{\text{number of teeth}}{\text{pitch diameter (inches)}}$.

Example. — To find diametral pitch, given number of teeth 28, pitch diameter 2.8″.

Solution. — $\dfrac{28}{2.8} = 10$, diametral pitch.

Attention. — The word "diameter" applied to gears means the *pitch* diameter. See Fig. 5.

Formula. — Diametral pitch $= \dfrac{\text{number of teeth} + 2}{\text{outside diameter}}$.

Example. — To find diametral pitch, given outside diameter 5", number of teeth 48.

Solution. — $\dfrac{48 + 2}{5} = 10$, diametral pitch.

Formula. — Diametral pitch $= \dfrac{3.1416}{\text{circular pitch}}$.

Example. — To find diametral pitch, given circular pitch and constant 3.1416.

Solution. — $\dfrac{3.1416}{.31416} = 10$ diametral pitch.

12. Circular pitch is the distance from the center of one tooth to the center of the next tooth measured along the pitch circle.

Formula. — Circular pitch $= \dfrac{3.1416}{\text{diametral pitch}}$.

Example. — To find circular pitch, given constant 3.1416 and diametral pitch 10.

Solution. — $\dfrac{3.1416}{10} = .31416''$, circular pitch.

13. Outside diameter or size of gear blank, Fig. 5. —

Formula. — Outside diameter $= \dfrac{\text{number of teeth} + 2}{\text{diametral pitch}}$.

Example. — To find outside diameter, given number of teeth 28, diametral pitch 10.

Solution. — $\dfrac{28 + 2}{10} = 3''$ outside diameter of gear blank.

14. The pitch diameter, Fig. 5, is the diameter of pitch circle.

Formula. — Pitch diameter $= \dfrac{\text{number of teeth}}{\text{diametral pitch}}$.

Example. — To find pitch diameter, given number of teeth 48, diametral pitch 10.

Solution. — $\dfrac{48}{10} = 4.8''$, pitch diameter.

15. Number of teeth on gear.

Formula. — Number of teeth = (outside diameter × diametral pitch) − 2.

Example. — To find number of teeth, given outside diameter 5″, diametral pitch 10.

Solution. — (5 × 10) − 2 = 48, number of teeth.

Formula. — Number of teeth = pitch diameter × diametral pitch.

Example. — To find number of teeth given pitch diameter 2.8″, diametral pitch 10.

Solution. — 2.8 × 10 = 28, number of teeth.

16. To find distance between centers of two gears.*

Formula. — Distance between centers =

$$\frac{\text{number of teeth one gear} + \text{number of teeth other gear}}{2 \times \text{diametral pitch}}.$$

Example. — To find the distance between centers of two gears, given the number of teeth in both gears, 48 and 28, and the diametral pitch 10.

Solution. — $\dfrac{48 + 28}{2 \times 10} = 3.8000''$ center distance.

Formula. — Distance between centers =

$$\frac{\text{diameter of pitch circle of gear} + \text{diameter of pitch circle of pinion}}{2}.$$

Example. — To find the distance between centers of two gears, given the diameter of the pitch circles of both gear and pinion.

Solution. — $\dfrac{4.8 + 2.8}{2} = 3.8000''$ center distance.

17. To find whole depth of milled gear tooth using constant 2.157. —

Formula. — Whole depth of milled gear tooth =

$$\frac{2.157}{\text{diametral pitch}}.$$

---

* Often called center to center (C to C).

Example. — Find whole depth of tooth in gear of 10 pitch.

Solution. — $\dfrac{2.157}{10}$ = .2157″ or .216″, depth of space or tooth.

18. To find thickness of tooth at pitch line, Fig. 5, given circular pitch .314″.

Formula. — Thickness of tooth at pitch line = $\dfrac{\text{circular pitch}}{2}$

or = $\dfrac{1.57}{\text{diametral pitch}}$.

Example. — To find thickness of tooth, given diametral pitch 10.

Solution. — $\dfrac{1.57}{10}$ = .157″, thickness of tooth.

19. Clearance, see Fig. 5, at bottom of tooth is made equal to $\frac{1}{10}$ thickness of tooth at pitch line.

The clearance of 10 pitch gear (gear used in Schedule, p. 11₁₆) = .0157″.

20. Working depth of tooth, see Fig. 5, is equal to the whole depth minus the clearance.

For 10 pitch .2157 − .0157 = .2000″.

21. Addendum, see Fig. 5, is the distance from the pitch circle to outside circle.

Formula. — Addendum = $\dfrac{1}{\text{diametral pitch}}$.

Example. — To find addendum, given diametral pitch 8.

Solution. — $\frac{1}{8}$ = .125″ addendum.

Dedendum is the depth below the pitch line to which the tooth of the mating gear extends.

22. To find whole depth of gear-shaper gear tooth. Constant 2.250″. —

Formula. — Whole depth of gear-shaper gear tooth = $\dfrac{2.250}{\text{diametral pitch}}$.

Example. — Find depth of tooth for 10 pitch gear.

Solution. — $\dfrac{2.250}{10}$ = .225″, depth of space or tooth.

23. Table of Tooth Parts.

DIAMETRAL PITCH IN FIRST COLUMN

DIAME-TRAL PITCH.	CIRCULAR PITCH.	THICKNESS OF TOOTH ON PITCH LINE.	ADDEN-DUM AND MODULE.	WORKING DEPTH OF TOOTH.	DEPTH OF SPACE BE-LOW PITCH LINE.	WHOLE DEPTH OF TOOTH.
P	P′	t′	a	F	a + c	F + c
$\frac{1}{2}$	6.2832	3.1416	2.0000	4.0000	2.3142	4.3142
$\frac{3}{4}$	4.1888	2.0944	1.3333	2.6666	1.5428	2.8761
1	3.1416	1.5708	1.0000	2.0000	1.1571	2.1571
$1\frac{1}{4}$	2.5133	1.2566	.8000	1.6000	.9257	1.7257
$1\frac{1}{2}$	2.0944	1.0472	.6666	1.3333	.7714	1.4381
$1\frac{3}{4}$	1.7952	.8976	.5714	1.1429	.6612	1.2326
2	1.5708	.7854	.5000	1.0000	.5785	1.0785
$2\frac{1}{4}$	1.3963	.6981	.4444	.8888	.5143	.9587
$2\frac{1}{2}$	1.2566	.6283	.4000	.8000	.4628	.8628
$2\frac{3}{4}$	1.1424	.5712	.3636	.7273	.4208	.7844
3	1.0472	.5236	.3333	.6666	.3857	.7190
$3\frac{1}{2}$	.8976	.4488	.2857	.5714	.3306	.6163
4	.7854	.3927	.2500	.5000	.2893	.5393
5	.6283	.3142	.2000	.4000	.2314	.4314
6	.5236	.2618	.1666	.3333	.1928	.3595
7	.4488	.2244	.1429	.2857	.1653	.3081
8	.3927	.1963	.1250	.2500	.1446	.2696
9	.3491	.1745	.1111	.2222	.1286	.2397
10	.3142	.1571	.1000	.2000	.1157	.2157
11	.2856	.1428	.0909	.1818	.1052	.1961
12	.2618	.1309	.0833	.1666	.0964	.1798
13	.2417	.1208	.0769	.1538	.0890	.1659
14	.2244	.1122	.0714	.1429	.0826	.1541

Module is used in the metric system of gear teeth and is equal to the pitch diameter in millimeters divided by the number of teeth in the gear.

Table of Tooth Parts — *Concluded.*

DIAMETRAL PITCH IN FIRST COLUMN

DIAME-TRAL PITCH.	CIRCULAR PITCH.	THICK-NESS OF TOOTH ON PITCH LINE.	$\frac{1}{P}$ OR THE AD-DENDUM OR MODULE.	WORKING DEPTH OF TOOTH.	DEPTH OF SPACE BELOW PITCH LINE.	WHOLE DEPTH OF TOOTH
P	P′	t′	a	F	a + c	F + c
15	.2094	.1047	.0666	.1333	.0771	.1438
16	.1963	.0982	.0625	.1250	.0723	.1348
17	.1848	.0924	.0588	.1176	.0681	.1269
18	.1745	.0873	.0555	.1111	.0643	.1198
19	.1653	.0827	.0526	.1053	.0609	.1135
20	.1571	.0785	.0500	.1000	.0579	.1079
22	.1428	.0714	.0455	.0909	.0526	.0980
24	.1309	.0654	.0417	.0833	.0482	.0898
26	.1208	.0604	.0385	.0769	.0445	.0829
28	.1122	.0561	.0357	.0714	.0413	.0770
30	.1047	.0524	.0333	.0666	.0386	.0719
32	.0982	.0491	.0312	.0625	.0362	.0674
34	.0924	.0462	.0294	.0588	.0340	.0634
36	.0873	.0436	.0278	.0555	.0321	.0599
38	.0827	.0413	.0263	.0526	.0304	.0568
40	.0785	.0393	.0250	.0500	.0289	.0539
42	.0748	.0374	.0238	.0476	.0275	.0514
44	.0714	.0357	.0227	.0455	.0263	.0490
46	.0683	.0341	.0217	.0435	.0252	.0469
48	.0654	.0327	.0208	.0417	.0241	.0449
50	.0628	.0314	.0200	.0400	.0231	.0431
56	.0561	.0280	.0178	.0357	.0207	.0385
60	.0524	.0262	.0166	.0333	.0193	.0360

24. Table of Tooth Parts.

CIRCULAR PITCH IN FIRST COLUMN

CIRCULAR PITCH.	THREADS OR TEETH PER INCH LINEAR.	DIAMETRAL PITCH.	THICKNESS OF TOOTH ON PITCH LINE.	ADDENDUM AND MODULE.	WORKING DEPTH OF TOOTH.	DEPTH OF SPACE BELOW PITCH LINE.	WHOLE DEPTH OF TOOTH	WIDTH OF THREAD-TOOL AT END.	WIDTH OF THREAD AT TOP.
P'	$\frac{1''}{P'}$	P	t'	a	F	a+c	F+c	P'×.3095	P'×.3354
2	1/2	1.5708	1.0000	.6366	1.2732	.7366	1.3732	.6190	.6707
1 7/8	8/15	1.6755	.9375	.5968	1.1937	.6906	1.2874	.5803	.6288
1 3/4	4/7	1.7952	.8750	.5570	1.1141	.6445	1.2016	.5416	.5869
1 5/8	8/13	1.9333	.8125	.5173	1.0345	.5985	1.1158	.5029	.5450
1 1/2	2/3	2.0944	.7500	.4775	.9549	.5525	1.0299	.4642	.5030
1 7/16	16/23	2.1855	.7187	.4576	.9151	.5294	.9870	.4449	.4821
1 3/8	8/11	2.2848	.6875	.4377	.8754	.5064	.9441	.4256	.4611
1 1/3	3/4	2.3562	.6666	.4244	.8488	.4910	.9154	.4127	.4471
1 5/16	16/21	2.3936	.6562	.4178	.8356	.4834	.9012	.4062	.4402
1 1/4	4/5	2.5133	.6250	.3979	.7958	.4604	.8583	.3869	.4192
1 3/16	16/19	2.6456	.5937	.3780	.7560	.4374	.8154	.3675	.3982
1 1/8	8/9	2.7925	.5625	.3581	.7162	.4143	.7724	.3482	.3773
1 1/16	16/17	2.9568	.5312	.3382	.6764	.3913	.7295	.3288	.3563
1	1	3.1416	.5000	.3183	.6366	.3683	.6866	.3095	.3354
15/16	1 1/15	3.3510	.4687	.2984	.5968	.3453	.6437	.2902	.3144
7/8	1 1/7	3.5904	.4375	.2785	.5570	.3223	.6007	.2708	.2934
13/16	1 3/13	3.8666	.4062	.2586	.5173	.2993	.5579	.2515	.2725
4/5	1 1/4	3.9270	.4000	.2546	.5092	.2946	.5492	.2476	.2683
3/4	1 1/3	4.1888	.3750	.2387	.4775	.2762	.5150	.2321	.2515
11/16	1 5/11	4.5696	.3437	.2189	.4377	.2532	.4720	.2128	.2306
2/3	1 1/2	4.7124	.3333	.2122	.4244	.2455	.4577	.2063	.2236
5/8	1 3/5	5.0265	.3125	.1989	.3979	.2301	.4291	.1934	2096
3/5	1 2/3	5.2360	.3000	.1910	.3820	.2210	.4120	.1857	.2012
4/7	1 3/4	5.4978	.2857	.1819	.3638	.2105	.3923	.1769	.1916
9/16	1 7/9	5.5851	.2812	.1790	.3581	.2071	.3862	.1741	.1886

Table of Tooth Parts — *Concluded.*

CIRCULAR PITCH IN FIRST COLUMN

CIRCULAR PITCH.	THREADS OR TEETH PER INCH LINEAR.	DIAMETRAL PITCH.	THICKNESS OF TOOTH ON PITCH LINE.	ADDENDUM AND MODULE.	WORKING DEPTH OF TOOTH.	DEPTH OF SPACE BELOW PITCH LINE.	WHOLE DEPTH OF TOOTH.	WIDTH OF THREAD TOOL AT END.	WIDTH OF THREAD AT TOP.
P'	$\frac{1''}{P'}$	P	t'	a	F	$a+c$	$F+c$	$P' \times .3095$	$P' \times .3354$
1/2	2	6.2832	.2500	.1592	.3183	.1842	.3433	.1547	.1677
4/9	2 1/4	7.0685	.2222	.1415	.2830	.1637	.3052	.1376	.1490
7/16	2 2/7	7.1808	.2187	.1393	.2785	.1611	.3003	.1354	.1467
3/7	2 1/3	7.3304	.2143	.1364	.2728	.1578	.2942	.1326	.1437
2/5	2 1/2	7.8540	.2000	.1273	.2546	.1473	.2746	.1238	.1341
3/8	2 2/3	8.3776	.1875	.1194	.2387	.1381	.2575	.1161	.1258
4/11	2 3/4	8.6394	.1818	.1158	.2316	.1340	.2498	.1125	.1219
1/3	3	9.4248	.1666	.1061	.2122	.1228	.2289	.1032	.1118
5/16	3 1/5	10.0531	.1562	.0995	.1989	.1151	.2146	.0967	.1048
3/10	3 1/3	10.4719	.1500	.0955	.1910	.1105	.2060	.0928	.1006
2/7	3 1/2	10.9956	.1429	.0909	.1819	.1052	.1962	.0884	.0958
1/4	4	12.5664	.1250	.0796	.1591	.0921	.1716	.0774	.0838
2/9	4 1/2	14.1372	.1111	.0707	.1415	.0818	.1526	.0688	.0745
1/5	5	15.7080	.1000	.0637	.1273	.0737	.1373	.0619	.0671
3/16	5 1/3	16.7552	.0937	.0597	.1194	.0690	.1287	.0580	.0629
2/11	5 1/2	17.2788	.0909	.0579	.1158	.0670	.1249	.0563	.0610
1/6	6	18.8496	.0833	.0531	.1061	.0614	.1144	.0516	.0559
2/13	6 1/2	20.4203	.0769	.0489	.0978	.0566	.1055	.0476	.0516
1/7	7	21.9911	.0714	.0455	.0910	.0526	.0981	.0442	.0479
2/15	7 1/2	23.5619	.0666	.0425	.0850	.0492	.0917	.0413	.0447
1/8	8	25.1327	.0625	.0398	.0796	.0460	.0858	.0387	.0419
1/9	9	28.2743	.0555	.0354	.0707	.0409	.0763	.0344	.0373
1/10	10	31.4159	.0500	.0318	.0637	.0368	.0687	.0309	.0335
1/16	16	50.2655	.0312	.0199	.0398	.0230	.0429	.0193	.0210
1/20	20	62.8318	.0250	.0159	.0318	.0184	.0343	.0155	.0168

CHORDAL THICKNESS AND CORRECTED ADDENDUM

25. To measure gear teeth accurately, the chordal thickness and corrected addendum * must be used. — In gears of precision allowance must be made for the curve of the circle of the pitch line. The curve of the top of the tooth also affects the distance at which the thickness of the tooth is measured.

Example. — When measuring a tooth, instead of using the distance a, Fig. 6, which is the addendum of a gear tooth as given in tables on pp. **110**8–**111**1, use distance a' which is the corrected addendum ($a + H = a'$). For the thickness of the tooth at the pitch line, measure the chordal thickness or straight line measurement, instead of using the thickness of the tooth as given in tables on pp. **110**8–**111**1, which is measured on an arc or part of a circle.

Notations: — t' = chordal thickness
a' = corrected addendum
a = addendum
H = height of arc
D' = pitch diameter
β = $\frac{1}{4}$ of angle subtended by circular pitch.

Formulas: —

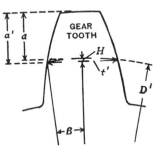

$$t' = D' \sin \beta$$

$$H = \frac{D'(1 - \cos \beta)}{2}$$

$$a' = a + H = a + \frac{D'(1 - \cos \beta)}{2}.$$

To eliminate the necessity of making extended calculations, the table on p. **111**4 is arranged. To obtain t' and a' for any diametral pitch, it is necessary to divide the values found opposite the number of teeth by the diametral pitch.

Example. — 8 diametral pitch, 32 teeth.

Fig. 6. — Diagram of Gear Tooth Showing Chordal Thickness.

* Corrected addendum is often called *chordal addendum.*

From the table we find that the values opposite 32 teeth are $t' = 1.5702$ and $a' = 1.0193$.

For 8-pitch gear, $t' = 1.5702 \div 8 = .1963''$ chordal thickness, and $a' = 1.0193 \div 8 = .1274''$, corrected addendum.

For any circular pitch the values of a' and t' may be obtained by multiplying the figures in table on p. **111**4, opposite the required number of teeth, by the value of a taken from the table of tooth parts on pp. **111**08–**111**1.

Example. — $\frac{5}{8}''$ circular pitch — 20 teeth. $a = .1989''$.

$t' = 1.5692 \times .1989 = .3121''$, chordal thickness.

$a' = 1.0308 \times .1989 = .2050''$, corrected addendum.

26. Gear tooth vernier. — These dimensions are measured with the gear tooth vernier, as in Fig. 7. To test a tooth, two

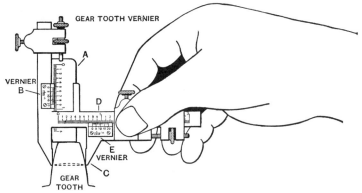

FIG. 7. — MEASURING CHORDAL THICKNESS WITH GEAR TOOTH VERNIER.

spaces are cut in the gear blank, then the vertical scale A, which is graduated to read by means of vernier B to thousandths of an inch, is set by the adjusting screws to the dimension of the corrected addendum. The caliper is then placed over the tooth and the sliding jaw C is moved against the tooth and the chordal thickness measured by reading the graduated bar D and vernier E. See Vernier Principle, pp. **2**11, **2**12.

Reading gear-tooth vernier for 2 pitch 20 teeth, Fig. 7.

Corrected addendum $(a') = 1.0308 \div 2 = .5154''$.

Chordal thickness $(t') = 1.5692 \div 2 = .7846''$.

27. Table of chordal thickness of gear teeth.

Number of Teeth.	t'	a'	Number of Teeth.	t'	a'	Number of Teeth.	t'	a'
6	1.5529	1.1022	50	1.5705	1.0123	94	1.5707	1.0066
7	1.5568	1.0873	51	1.5706	1.0121	95	1.5707	1.0065
8	1.5607	1.0769	52	1.5706	1.0119	96	1.5707	1.0064
9	1.5628	1.0684	53	1.5706	1.0117	97	1.5707	1.0064
10	1.5643	1.0616	54	1.5706	1.0114	98	1.5707	1.0063
11	1.5654	1.0559	55	1.5706	1.0112	99	1.5707	1.0062
12	1.5663	1.0514	56	1.5706	1.0110	100	1.5707	1.0061
13	1.5670	1.0474	57	1.5706	1.0108	101	1.5707	1.0061
14	1.5675	1.0440	58	1.5706	1.0106	102	1.5707	1.0060
15	1.5679	1.0411	59	1.5706	1.0105	103	1.5707	1.0060
16	1.5683	1.0385	60	1.5706	1.0102	104	1.5707	1.0059
17	1.5686	1.0362	61	1.5706	1.0101	105	1.5707	1.0059
18	1.5688	1.0342	62	1.5706	1.0100	106	1.5707	1.0058
19	1.5690	1.0324	63	1.5706	1.0098	107	1.5707	1.0058
20	1.5692	1.0308	64	1.5706	1.0097	108	1.5707	1.0057
21	1.5694	1.0294	65	1.5706	1.0095	109	1.5707	1.0057
22	1.5695	1.0281	66	1.5706	1.0094	110	1.5707	1.0056
23	1.5696	1.0268	67	1.5706	1.0092	111	1.5707	1.0056
24	1.5697	1.0257	68	1.5706	1.0091	112	1.5707	1.0055
25	1.5698	1.0247	69	1.5707	1.0090	113	1.5707	1.0055
26	1.5698	1.0237	70	1.5707	1.0088	114	1.5707	1.0054
27	1.5699	1.0228	71	1.5707	1.0087	115	1.5707	1.0054
28	1.5700	1.0220	72	1.5707	1.0086	116	1.5707	1.0053
29	1.5700	1.0213	73	1.5707	1.0085	117	1.5707	1.0053
30	1.5701	1.0208	74	1.5707	1.0084	118	1.5707	1.0053
31	1.5701	1.0199	75	1.5707	1.0083	119	1.5707	1.0052
32	1.5702	1.0193	76	1.5707	1.0081	120	1.5707	1.0052
33	1.5702	1.0187	77	1.5707	1.0080	121	1.5707	1.0051
34	1.5702	1.0181	78	1.5707	1.0079	122	1.5707	1.0051
35	1.5702	1.0176	79	1.5707	1.0078	123	1.5707	1.0050
36	1.5703	1.0171	80	1.5707	1.0077	124	1.5707	1.0050
37	1.5703	1.0167	81	1.5707	1.0076	125	1.5707	1.0049
38	1.5703	1.0162	82	1.5707	1.0075	126	1.5707	1.0049
39	1.5704	1.0158	83	1.5707	1.0074	127	1.5707	1.0049
40	1.5704	1.0154	84	1.5707	1.0074	128	1.5707	1.0048
41	1.5704	1.0150	85	1.5707	1.0073	129	1.5707	1.0048
42	1.5704	1.0147	86	1.5707	1.0072	130	1.5707	1.0047
43	1.5705	1.0143	87	1.5707	1.0071	131	1.5708	1.0047
44	1.5705	1.0140	88	1.5707	1.0070	132	1.5708	1.0047
45	1.5705	1.0137	89	1.5707	1.0069	133	1.5708	1.0047
46	1.5705	1.0134	90	1.5707	1.0068	134	1.5708	1.0046
47	1.5705	1.0131	91	1.5707	1.0068	135	1.5708	1.0046
48	1.5705	1.0129	92	1.5707	1.0067	......	.....	.....
49	1.5705	1.0126	93	1.5707	1.0067	......	.....	.....

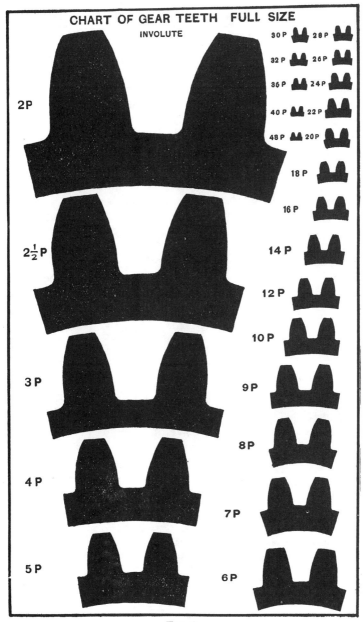

FIG. 8.

28. Gear rule. — Fig. 9 shows a gear rule with graduations for sizing gear blanks of a large variety of pitches.

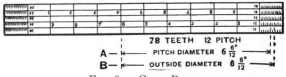

A — |◄──── PITCH DIAMETER 6$\frac{6''}{12}$ ──────►|

B — |◄──── OUTSIDE DIAMETER 6$\frac{8''}{12}$ ──────►|

Fig. 9. — Gear Rule.

Example. — To set the calipers for a gear blank, 78 teeth, 12 pitch. Divide the number of teeth by the pitch, as $78 \div 12 = 6\frac{6}{12}''$ which is the pitch diameter; then take 6 of the blank inches and 6 of the 12th graduations which gives $6\frac{6}{12}''$, as at *A*. To this add 2 of the 12ths which gives $6\frac{8}{12}''$, as at *B*, the setting of calipers for outside diameter of blank.

PREPARING A SPUR-GEAR BLANK

29. To prepare spur-pinion blank, 10 pitch, 28 teeth, involute. Fig. 10. Also see Specifications, p. 111s.

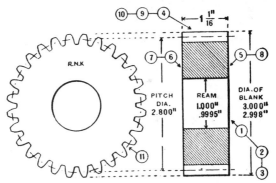

Fig. 10. — Schedule Drawing of Spur Gear.

Specifications: Material, iron casting $\frac{1}{8}''$ large; weight 2 lbs. 8 oz. Hardness, 29 to 31 (scleroscope).

High-speed steel or stellite cutting tools.

Time to prepare 3″ gear blank: Study drawing and schedule in advance, 5 min. — Oil lathe, 6 min. — Chuck, 20 min. — Rough turn, 19 min. — Finish turn, 11 min. — File and scrape, 7 min. — Polish, 11 min. — Clean lathe, 5 min. — Total, 1 h. 24 min.

SCHEDULE OF OPERATIONS, MACHINES AND TOOLS

OPERATIONS.	MACHINES, SPEEDS, FEEDS.	TOOLS.
Snag casting.		
Set dead center in approximate alinement.	Engine lathe, 12″ to 16″.	
Mount casting in chuck and true up.	3d or 4th speed.	Independent chuck, chalk.
Drill and power ream hole (**1**), (**2**). See pp. **4**09, **4**17.	1st or 2d speed, or 115 R.P.M. Hand feed.	Centering tool, $\frac{43}{64}''$ twist drill and holder, fluted chucking reamer, .005″ small.
Hand ream hole. Start reamer in lathe, pull belt by hand, finish at reaming stand (**3**). See p. **4**16, Figs. 22, 23.	Reaming stand.	1″ hand reamer, reamer wrench.
Oil mandrel and press into hole.	Mandrel press.	1″ mandrel.
True live center. Set dead center in accurate alinement.	3d speed, or 60 F.P.M.	Center truing tool, center gage.
Rough turn $3'' + \frac{1}{64}''$, (**4**). See p. **3**07.	Engine lathe, 12″ to 16″, 3d or 4th speed, back gears in, or 40 F.P.M. Medium power feed — 80 to 1″.	Holder and cutter 15° rake, calipers.
Rough faces to $1\frac{1}{16}'' + \frac{1}{64}''$, feed inward, (**5**), (**6**). See p. **3**07.	4th speed, back gears in, or 40 F.P.M. Medium power feed — 80 to 1″.	Holder and cutter 15° rake, calipers, rule.
Finish to size, feed outward, (**7**), (**8**). See p. **4**25.	3d speed or 60 F.P.M. Hand or fine power feed — 140 to 1″.	Holder and cutter 15° rake, or facing tool, calipers, rule.
Finish turn to 3″ + .004″, (**9**).	2d or 3d speed or 60 F.P.M. Fine power feed — 140 to 1″.	Holder and cutter 15° rake, 3″ micrometer or 3″ caliper gage.
File to 3″ + .0015″, (**10**).	Engine lathe, 2d or 3d speed, or 175 F.P.M.	8″ or 10″ mill-bastard file, file card.
Polish (**7**), (**8**). Polish (**10**) to limit.	Speed lathe, highest speed.	Polishing stick, 60, 90 emery cloth, lard oil.
Stamp name.	Anvil, piece of sheet copper.	$\frac{1}{8}''$ steel letters, hammer.
Cut teeth (**11**). See p. **111**8.		

Exception. — Filing, scraping and polishing are usually omitted on large gears and on gears that are not to be exposed.

To prepare 5″ gear blank (see Specifications, p. **111**8 and Fig. 2) use same schedule as 3″ gear blank, pp. **111**6, **111**7, but use lowest speeds given.

Time to prepare a 5″ gear blank: Study schedule in advance, 5 min. — Oil lathe, 6 min. — Chuck, 20 min. — Rough turn, 21 min. — Finish turn, 21 min. — File and scrape, 7 min. — Polish, 12 min. — Paint, 3 min. — Clean lathe, 5 min. — Total, 1 h. 40 min.

CUTTING SPUR GEARS

30. To cut spur gear and pinion. Plain or universal milling machine, Fig. 11.

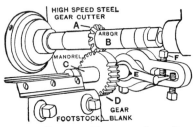

Fig. 11. — Cutting Teeth in Spur Pinion.

SPECIFICATIONS FOR CUTTING SPUR GEAR AND PINION, FIG. 10.

NAME.	GEAR.	PINION.	NAME.	GEAR.	PINION.
Material ...	C. iron	C. iron	Depth of space	.216″	.216″
W — Web..	Web	Plain	Thickness of tooth at P. L.	.157″	.157″
Style.......	Spur	Spur	Depth of tooth at P. L. (Addendum)....	.100″	.100″
Hole.......	1″	1″	Cutter marked .	/3–10 P. 35 to 54 involute.	/4–10 P. 26 to 34 T, involute.
O. dia......	5″	3″	Center distance	3.8″	3.8″
Face.......	1 1/16″	1 1/16″	Keyway.......	1/4″ × 1/8″	1/4″ × 1/8″
Hub dia....	2″	......	....	....	....
Hub proj'n.	Flush	Flush	Speed........	1st or 2d speed, or 60 F.P.M.	
Teeth......	48	28	Feed.........	Medium power feed — .014″ per cutter revolution	
Pitch......	10	10	Lubricant.....	Dry	Dry
P. dia......	4.8″	2.8″			

Time to cut gear and pinion: Study drawing and schedule in advance, 5 min. — Oil machine, 5 min. — "Set up" machine, 25 min. — Cut 5″ gear with machine "set up," 48 min. — Cut 3″ pinion with machine "set up," 28 min. — Clean machine, 5 min.

SCHEDULE OF OPERATIONS FOR PINION

I. Clean and oil machine and index head.

II. Preparatory adjustments. Insert clean arbor in clean spindle. Drive arbor lightly with lead hammer.

Place cutter *A* on arbor *B* to cut in direction of rotation and clamp hard. Oil end of arbor and adjust arm to bearing. Fasten index head and footstock on clean table.

Set centers in horizontal alinement. On universal milling machine, set swivel table at zero.

III. Arrange for plain indexing, 28 teeth.

IV. Set gear cutter central (Approximate Method). Raise knee

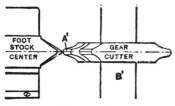

Fig. 12. — SETTING GEAR CUTTER CENTRAL.

and move cross slide until footstock center coincides with center line on cutter *A'* and clamp cross slide, then lower knee to leave room enough to mount work.

V. Mount gear blank on centers. Press mandrel *C* firmly into blank *D*, oil footstock center and mount mandrel on centers securing dog *E* in driver by screw *F*.

VI. Set machine for depth of teeth. Start machine. Move table by hand long. feed until blank *D* is under revolving cutter *A*. Hold .001″ tissue paper on top of blank and raise knee until revolving cutter cuts paper. Then set vertical dial at zero and move blank back until clear of cutter. Raise knee .216″, required depth of space or tooth. See *Attention*.

VII. Mill trial tooth. Move table by hand long. feed until cutter nearly touches blank, then throw in power feed and mill first space, one cut. Set table dog to trip feed when center of cutter has passed far corner of gear blank. At end of cut, move table back until blank is clear of cutter; index and cut second space to form first tooth. Measure tooth and make corrections, if necessary. See Figs. 7, 52.

VIII. Mill 28 teeth. With setting correct, mill teeth all-around blank, one cut each.

Warning. — When raising knee to obtain touch, cutter should be revolving, otherwise it may cut too deeply, as cutters do not always run true. It is best to stop the machine before moving the table back, as the revolving cutter may scrape the teeth more or less.

Attention. — For greater accuracy, two spaces are first cut *less* than the required depth, and the tooth measured with a vernier caliper.   See Chordal Thickness, pp. **111**2–**111**4, and Figs. 6, 7.   Then the spaces are cut deeper until the measurement shows that the tooth is of correct thickness.

Information. — To cut two or more gears, see Gang Mandrels, p. **4**22.

31. To set gear cutter central (Accurate Method). — Fast-running gears are noisy unless teeth are exactly central.   Set cutter central, as at *A'*, Fig. 12 (this method will do for slow-running gears), and cut space in trial blank.   Take dog off mandrel, remount mandrel reversed and free on centers.

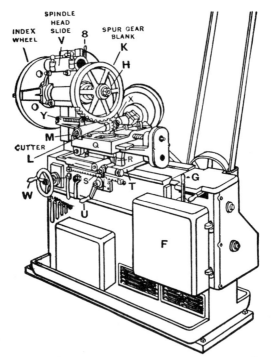

Fig. 13. — Cutting Spur Gear in Automatic Gear–Cutting Machine.

Move blank toward back of teeth until cutter is in the space. Pull belt to revolve cutter and if it cuts at top on one side of space and at bottom on the other, it is not central and blank should be moved slightly away with cross feed from side of cutter that cuts side of space at top. Repeat process at other places on trial blank until cutter passes through a reversed space without cutting and is central.

Note. — The information given in Schedule for cutting spur gear in a milling machine is needed when cutting a similar spur gear in an automatic gear-cutting machine and is, practically, used in the same manner.

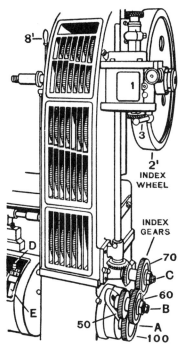

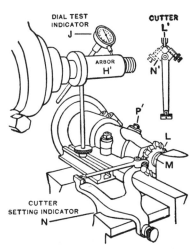

Fig. 14. — Indexing Mechanism. Automatic Spur– and Bevel-Gear Cutting Machine.

Fig. 15. — Testing Arbor and Setting Cutter Central. Automatic Spur– and Bevel-Gear Cutting Machine.

CUTTING SPUR GEAR IN AUTOMATIC GEAR–CUTTING MACHINE

32. Automatic spur- and bevel-gear cutting machines are used for cutting spur gears (see Fig. 13), bevel gears (see Fig. 54), and internal gears (see Fig. 19).

33. To cut spur gear in an automatic machine, the gear is mounted upon an arbor in the headstock spindle and the cutter upon a cutter spindle, as in Fig. 13.

The indexing is automatic and the machine stops when the last tooth is cut. This is obtained by an arrangement of change gears and indexing mechanism, as in Fig. 14. Data for indexing any number of teeth is obtained from tables supplied with gear-cutting machines. Adjustments for setting the gear and cutter are practically the same as those given in schedule for cutting a spur gear in a milling machine, pp. **111**8, **111**9.

34. To cut spur gear and pinion. Automatic spur- and bevel-gear cutting machine, Figs. 13, 14, 15.

SPECIFICATIONS FOR CUTTING SPUR GEAR AND PINION, FIGS. 13, 11, 15

NAME.	GEAR.	PINION.	NAME.	GEAR.	PINION.
Material....	C. iron	Steel	Whole depth of tooth........	.1798″	.1798″
Design......	Web	Plain	Thickness of tooth at P. L.	.1309″	.1309″
Style.......	Spur	Spur	Depth of tooth at P. L. (Addendum).....	.0833″	.0833″
Hole........	$\frac{3}{4}''$	1″	Chordal thickness.........	.13088″	.13085″
			Corrected addendum......	.0842″	.0848″
O. diam.....	5.167″	3.167″	Cutter marked	#2–12 P. 55 to 134 T., involute.	#3–12 P. 35–54 T., involute.
Face........	$\frac{3}{4}''$	$\frac{3}{4}''$			
Hub. diam..	$1\frac{1}{2}$	2	Center distance	4″	4″
Hub. proj'n..	$\frac{3}{4}''$	$1\frac{1}{4}''$	Keyway........	$\frac{1}{8}'' \times \frac{1}{16}''$	$\frac{3}{16}'' \times \frac{3}{32}''$
No. of teeth	60	36	Revs. of cutter..	60	60
Pitch.......	12	12	Feed per min...	2″	2″
P. diam.....	5″	3″	Lubricant......	Dry	Lard oil

SCHEDULE OF OPERATIONS FOR GEAR

Arrange index change gears for 70 teeth.

Place gear 100, Fig. 14, on locking disk shaft A; gear 50, 1st gear, on arm stud B; gear 60, 2d gear, on arm stud B; gear 70, 3d gear, on bevel gear stud C; then set knob D in No. 1 slot to give one turn of locking disk E for indexing.

Arrange speed and feed change gears inside box F, Fig. 13.

Set lever G, Fig. 13, central (out of mesh).

Mount arbor H', Fig. 15, in spindle. Test with dial test indicator J and correct error by tissue paper between spindle and arbor.

Mount gear blank K, Fig. 13, on arbor H.

Mount cutter LL', Fig. 15, on cutter spindle M and set central, testing with indicator N, clamped to bed. Indicate opposite sides of cutter about on pitch line by reversing indicator arm, as at N'. Adjust cutter by screw P', Fig. 15, until readings are equidistant from zero line on plate then clamp spindle.

Set cutter slide Q, Fig. 13, at zero on graduated arc R by pinion S, and clamp.

Move carriage T under blank by crank handle U, and clamp.

Loosen bolts on spindle-head slide V, and lower blank slowly to touch revolving cutter (use .002″ white paper to indicate touch) by hand wheel W.

Set dial X at zero.

Run cutter clear of blank.

Caliper diameter of blank, make allowance for error (see p. **11**04), lower blank a little more than depth of cut, to take up back lash of screw, then raise to exact depth = .431″ + .002″ for white paper, and clamp slide. Adjust rim rest Y against blank to take thrust when cutting teeth, and test, to prevent excessive friction, by disengaging worm 1, Fig. 14, from index wheel 2′ by lever 3, and rotate wheel by hand.

When correct, engage index wheel. Set reversing dogs to give length of cut.

Start machine, throw lever G, Fig. 13, forward which starts automatic feed cutter slide and index mechanism.

Cut two spaces and run cutter slide back by hand. Index gear part way around by lever 8′, Fig. 14, then measure thickness of tooth at pitch line with gear tooth vernier. Make corrections by changing depth of cut.

When correct tooth is obtained, continue cutting teeth all around gear by automatic index, and feed.

Note. — After first cut, readjust reversing dogs if necessary.

GENERATING SPUR GEARS

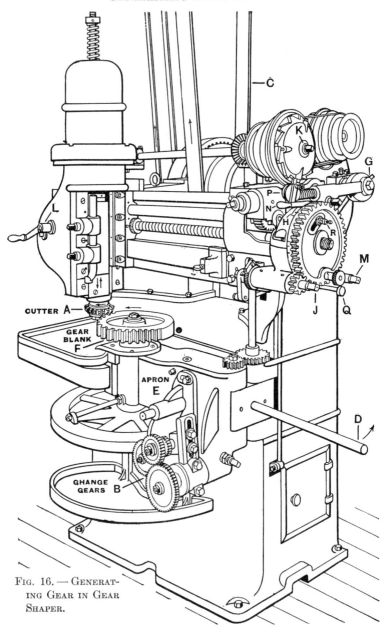

CUTTER A

GEAR
BLANK
F

APRON
E

CHANGE
GEARS B

FIG. 16. — GENERAT-
ING GEAR IN GEAR
SHAPER.

35. Generating spur gears in a gear shaper, Fig. 16. — This gear shaper cuts spur gears, involute teeth, by a generating process.

The high-speed or carbon-steel cutter is a gear generated by a grinding process controlled by two metal tapes and a rolling cylinder which produces an involute curve. One cutter is all that is required for each pitch up to the capacity of the machine. The action of this gear-like cutter is similar to that of a shaper tool and, for ordinary spur gears, cuts on the up or "pull" stroke. As the cutter-slide moves up and down, both cutter and gear blank rotate together in the proper ratio as determined by the number of teeth in the cutter and the number of teeth to be cut, which is controlled by a train of change gears and an indexing mechanism.

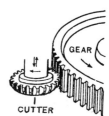

FIG. 17. — CUTTER ACTION OF A GEAR SHAPER.

In one revolution of the gear, the cutter forms tooth spaces and teeth of the proper shape to mesh with any gear of the same pitch. For cutting teeth in cluster gears having two or more diameters on the same piece, or close to a shoulder which is larger than the root diameter of the gear, the cutter is reversed and the cut made on the "push" stroke. Whenever the *direction of the cut is changed,* the mechanism for relieving the work from the cutter must be *reversed;* this is done by changing the stud, in the internal driving gear operating the plungers at the rear of the machine, to the hole marked "push stroke" or "pull stroke."

This relieving mechanism on some styles of machines draws the cutter from contact with the gear on the return stroke; and on others, the gear is withdrawn from the cutter.

To set the machine for cutting spur gears, place cutter *A,* Fig. 16, on cutter spindle and fasten it securely. Arrange change gears *B* for the number of teeth to be cut. (The chart furnished with the machine gives the change gears required for any number of gear teeth.) Pull belt *C* by hand until the plungers on rear of machine are raised clear of the lever on the back of the apron; then, with detachable lever D, swing apron E open to receive the work. Place gear blank F on arbor and fasten it securely. Swing apron into place and again pull belt by hand until the plungers run back into the rolls on apron lever. To set cutter stroke, turn crank-adjusting screw G until pointer is on graduation for width of gear blank. The mechanism allows for $\frac{3}{32}$ of an inch over travel at beginning of cut and $\frac{3}{32}$ of an inch at the end. To set machine for pitch of gear to be cut, turn stud J backward then forward to take up backlash until the pitch line on pitch dial H coincides with the zero line on the machine. Now set cutter slide on the "up" stroke and with crank handle on squared end of upper index shaft K, rotate cutter until one of its teeth is central with work spindle. This is determined by the centering gage provided with the machine. Place crank handle on stud L and move cutter toward the work until the thin setting gage (.019″ thick) will pinch between cutter and gear blank. This completes the setting. Now start the machine and set the timing mechanism by turning stud M. Start the feed by throwing in starting lever N and the cutter will feed into the blank until it reaches the proper depth, when the depth-feed automatically stops and the rotary feed starts; and after the gear is completed the rotary feed stops and bell P rings. The machine can be set by means of stop pins Q and K to finish the gear in two cuts.

Attention. — This machine is also used to cut sprockets, ratchets, and cams, and for graduating.

CUTTING INTERNAL GEARS

36. Internal gear, Fig. 18, may be cut with formed milling cutters by making the gear in the form of a ring which is secured to a backing plate after the teeth have been cut.

37. To cut internal gear with an automatic gear-cutting machine. — Hold internal gear blank A, Fig. 19, in special chuck B. Mount cutter C upon the projecting arm of attachment D. Cutter is driven by the main cutter spindle through a train of gears so that it enters the internal gear and works upon its inner periphery.

The automatic indexing of the blank and the return of the cutter is the same as for spur gears.

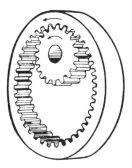

FIG. 18. — INTERNAL GEAR AND PINION. RATIO 41 TO **19.**

The shape of the space in an internal gear is the same as the tooth of a spur gear and requires a special cutter. For ordinary work use the same cutter as for spur gears.

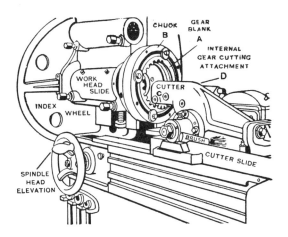

FIG. 19. — CUTTING INTERNAL GEAR WITH AN ATTACHMENT IN AUTOMATIC GEAR-CUTTING MACHINE.

38. Generating internal gear, Fig. 20. — In this case the teeth are cut on the inner surface of a ring.

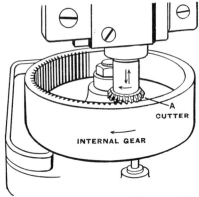

FIG. 20. — GENERATING INTERNAL GEAR IN GEAR SHAPER.

The cutter A is used in the inverted position so as to cut on the push stroke. An intermediate gear is introduced into the train of change gears, to make the gear and cutter rotate in the proper direction.

39. Spur helical gears, Fig. 21, are spur gears with the teeth cut at an angle to the axis.

These gears are mounted on parallel shafts and operate like ordinary spur gears except that the engagement of the teeth is continuous which tends toward noiseless action. They may be cut in a milling machine or in a helical gear shaper. The principle of the helical gear shaper is the same as that of the gear shaper for generating spur gears. The helical cutter is generated like the spur gear cutter and is finished by a similar grinding process. The only difference in the mechanism of this machine is in the design of the cutter slide and the helical guides for giving the helical motion to the cutter spindle as it moves up and down for cutting and return strokes.

One helical guide is mounted on the cutter spindle and its mating guide fastened in the hub of the upper index-wheel.

These guides, like the cutters, can be changed when changing from one helix to another, or for cutting a right-hand or a left-hand gear. This shaper, like the spur gear shaper, will cut teeth into a narrow clearance recess making it easy to cut teeth close to a shoulder.

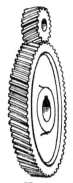

FIG. 21. — HELICAL GEARS.   FIG. 22. — HERRINGBONE GEARS.

Note. — Internal helical gears can be cut on this machine.

40. Herringbone gears may be cut, as in Fig. 23. A nar-

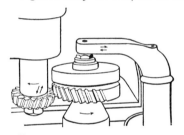

FIG. 23. — CUTTING TEETH OF HERRINGBONE GEAR.

row groove in which to run the cutter, from $\frac{3}{16}''$ to $\frac{5}{16}''$ wide depending on the pitch and helix angle, is cut in the center of the blank.

CUTTING RACKS

41. Racks, Fig. 24, are straight bars with teeth cut on the sides or edges, to mesh and operate with gears and pinions to produce reciprocating motions. Common examples are the racks and pinions of the feed mechanism of engine lathes, and the racks and gear drives of planers.

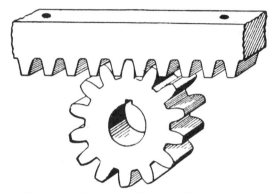

FIG. 24. — RACK AND PINION. RATIO 14 TO 1.

Racks are usually cut in machines that automatically index from one tooth to the next. Short racks may be cut in a milling machine, Fig. 25, or planed. They may be held in a vise set parallel to a regular milling machine arbor. Fasten rack-cutting attachment A, Fig. 25, to a plain milling machine. Clamp rack blank B in shoe C parallel to cutter arbor D. Set long.

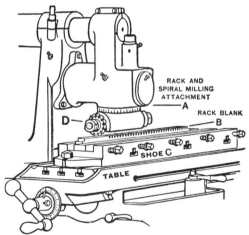

FIG. 25. — CUTTING TEETH IN RACK. PLAIN MILLING MACHINE.

feed dial at zero and move table one pitch for each cut.

Attention. — Gang cutters are often used for milling racks.

SPIRAL GEARING

Fig. 26. — Spiral Gear and Pinion. Ratio 1.6 to 1.

42. Spiral gears are used to transmit rotary motion from one shaft to another not parallel, and have teeth like a screw, Fig. 26. They are desirable for high speeds where the load is light, and smooth and noiseless action is a factor; for this reason, spiral gears are often used instead of bevel gears. The absence of noise is due to the sliding action between the teeth. The tooth curves are the same as those used in spur gearing, but the teeth are so placed that their center lines are sections of helices. The lead or pitch of these helices is dependent on the angle between the shafts. They should be designed so that they may be cut with standard spur-gear cutters. Spiral gears may be cut with a milling machine or gear-hobbing machine, see pp. 1148–11**52**.

Examples of spiral gears may be found in automobile and stationary gas engine timing gears, and in centrifugal machinery including extractors, separators, and dryers. See Spiral-Type Bevel Gears, pp. 11**76**, 11**77**.

Spiral gears are often used in the place of spur gears, the shafts are then parallel and the angles of the teeth of both gears are alike. They are usually made not over 20°.

Important. — In order to understand spiral gearing one should have a good knowledge of spur gearing.

43. Normal, Fig. 27. — As the teeth of spiral gears are at an angle with the shaft, as at *A*, the true shape of the teeth is at the *normal*, as at *B*, which is a line at right angles to the teeth, as at *C, D*. The shape of the teeth at the sides of the gear is distorted, as at *E*.

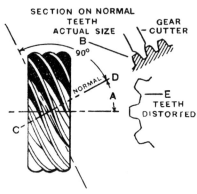

FIG. 27. — DIAGRAM OF A SPIRAL GEAR SHOWING THE NORMAL AND TRUE SIZE OF TEETH.

FORMULAS IN SPIRAL GEARING

44. Spiral gear data. — Data for spiral gears vary. In designing a machine in which spiral gears are to be used space should be allowed for a center distance for which a pair of spiral gears can be readily designed; otherwise, gears will have to be calculated to fill the space available, which will be much more difficult and expensive both to design and to cut.

Example. — A pair of spiral gears is required to fulfill the following conditions:

Ratio. — 1.6 to 1. Driver is 800 revolutions per minute; driven 500 revolutions per minute.

Center distance. — From available space, center distance not over $4\frac{5}{8}''$ and may be as small as $4\frac{3}{8}''$.

Pitch (approximate). — About 12 normal diametral pitch selected to satisfy the conditions of power and load.

Angle. — The driver to be greater than the driven.

Pitch diameter. — Restricted by available space.

Outside diameter of driving gear must not be over 4″.

45. Trial calculations. —

To find the trial pitch diameter.

Solution. — 4 − 2 (12-pitch addendums) = $4 - \frac{2}{12} = 4 - \frac{1}{6} = 3\frac{5}{6} = 3.833''$.

$3.833''$ = trial pitch diameter of driver.

$(4.625 \times 2) = 9.25 - 3.833 = 5.417''$ maximum pitch diameter of driven.

$(4.375 \times 2) = 8.75 - 3.833 = 4.917''$ minimum pitch diameter of driven.

To find the trial angle of the driver.

Formula. —

$$\text{Tangent of angle of driver} = \frac{\text{pitch diameter of driver}}{\text{pitch diameter of driven}} \times \frac{\text{R.P.M. driver}}{\text{R.P.M. driven}}.$$

Solution. — $\dfrac{3.833}{5.417} \times \dfrac{800}{500} = 1.132$ tangent * of angle of driver.

$1.1321 = 48° 33'$ angle of driver.

$90° - 48° 33' = 41° 27'$ angle of driven.

To find the number of teeth in driver and driven.

Formula. —

Pitch diameter × normal diametral pitch × cosine of angle = number of teeth.

Solution. —

Driver = $3.833 \times 12 \times .66197 = 30.45$

Driven = $5.417 \times 12 \times .74953 = 48.72$.

Consider a 12-pitch driving gear of 30 teeth and a driven gear of 48 teeth which are in the ratio of 1.6 to 1.

---

* For Tables of Natural Trigonometrical Functions, see any Engineers' Handbook.

To find the pitch diameters of driver and driven.

Formula. — Pitch diameter of a spiral gear =
$$\frac{\text{number of teeth}}{\text{cosine angle} \times \text{normal diametral pitch}}.$$

Solutions. — $\dfrac{30}{.66197 \times 12} = 3.7766''$ pitch diameter of driver.

$\dfrac{48}{.74953 \times 12} = 5.3367''$ pitch diameter of driven.

To find center distance.

Formula. — Center distance =
$$\frac{\text{pitch diameter of driver} + \text{pitch diameter of driven}}{2}.$$

Solution. — $\dfrac{3.7766 + 5.3367}{2} = 4.5566''$ which is within the requirements $(4\frac{5}{8}, 4\frac{3}{8})$.

By adding two 12-pitch addendums to the pitch diameter of the driver, we find that we have a gear that is less than 4″ as required.
$$3.7766 + (2 \times .0833) = 3.9433''.$$

Note. — If these trial calculations do not meet requirements, compute the work again and continue this process until sizes are obtained that do meet requirements.

46. To calculate the true size of a pair of spiral gears for the ratio of 1.6 to 1. —

To find the angle of the driver.

By substituting the *actual* pitch diameters in place of the *trial* diameters, we get the same angles for drivers and driven that we did in the trial calculations.

From Art. 45 we know that tangent of angle of driver =
$$\frac{3.7766}{5.3367} \times \frac{800}{500} = 1.132.$$

$1.132 = 48° 33'$, angle of driver.

$90° - 48° 33' = 41° 27'$ angle of driven.

To find the outside diameters of driver and driven.

Formula. — Outside diameter = pitch diameter + (2 × nor. addendum).

Solution. — $3.7766 + (2 \times .0833) = 3.9433''$ outside diameter of driver.

$5.3367 + (2 \times .0833) = 5.5033''$ outside diameter of driven.

To find the normal circular pitch of driver and driven.

Formula. — Normal circular pitch =

$$\frac{\text{pitch diameter} \times \text{cosine angle} \times 3.1416}{\text{number of teeth}}.$$

Solution. — $\dfrac{3.7766 \times .66197 \times 3.1416}{30} = .2619''$, normal circular pitch of driver.

$\dfrac{5.3367 \times .74953 \times 3.1416}{48} = .2618''$, normal circular pitch of driven.

To find the normal diametral pitch.

Formula. — Normal diametral pitch = $\dfrac{3.1416}{\text{normal circular pitch}}$.

Solution. — $\dfrac{3.1416}{.2618} = 12$, normal diametral pitch.

To find the normal thickness of tooth.

Formula. — Normal thickness of tooth = $\dfrac{\text{normal circular pitch}}{2}$.

Solution. — $\dfrac{2618}{2} = .1309''$, normal thickness of tooth.

To find normal addendum.

Formula. — Normal addendum = $\dfrac{1}{\text{normal diametral pitch}}$.

Solution. — $\dfrac{1}{12} = .0833''$ normal addendum.

To find normal depth of tooth.

Formula. — Normal whole depth of tooth (depth of space) =

$$\frac{2.15708}{\text{normal diametral pitch}}.$$

Solution. — $\dfrac{2.15708}{12} = .1798''$ normal depth of tooth.

*To find the number of cutter, see p. **11**59.*

Formula. — Range of cutter =

$$\frac{\text{actual number of teeth in spiral gear}}{\text{cosine}^3 \text{ of angle}}.$$

Solution. — $\dfrac{30}{.66197^3} = 103 +$ teeth = No. 2, 12-pitch cutter.

$\dfrac{48}{.7495^3} = 114 +$ teeth = No. 2, 12-pitch cutter.

To find the lead of driver and driven.

Formula. — Lead $= \dfrac{\text{pitch diameter} \times 3.1416}{\text{tangent of angle}}$.

Solution. — $\dfrac{3.7766 \times 3.1416}{1.1323} = 10.4786''$ lead of driver.

$\dfrac{5.3367 \times 3.1416}{.8832} = 18.9833''$ lead of driven.

To find gears for lead of driver.

Formula. — $\dfrac{\text{lead of driver}}{\text{lead of machine}} = \dfrac{\text{gear on worm} \times \text{2d gear on stud}}{\text{1st gear on stud} \times \text{gear on screw}}$.

Exact lead $= 10.4786''$
Approximate lead $= 10.477''$

Solution. — $\dfrac{10.477}{10} = \dfrac{48 \times 44}{28 \times 72}$ gears for driver.

To find gears for lead of driven.

Exact lead $= 18.9833''$
Approximate lead $= 19.000''$

Solution. — $\dfrac{19.000}{10} = \dfrac{38 \times 64}{40 \times 32}$ gears for driven.

CUTTING SPIRAL GEAR

47. To cut spiral gear and pinion, universal milling machine, Fig. 28.

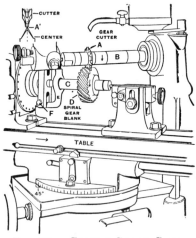

FIG. 28. — CUTTING SPIRAL GEAR.

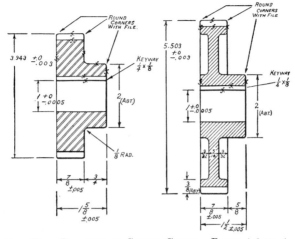

FIG. 29. — DRAWING OF SPIRAL GEARS. RATIO 1.6 TO 1.

SPECIFICATIONS FOR PAIR OF SPIRAL GEARS
OTHER THAN 45°, Fig. 29

NAME.	DRIVER.	DRIVEN.	NAME.	DRIVER.	DRIVEN.
Material.........	Machine steel	Cast iron semi-steel or phos-phor bronze	Cutter number	2	2
Speed ratio: driver 800 revolutions, driven 500 revolutions					
Hole...........	1″	1″	Exact lead of		
O. dia..........	3.9433″	5.5033″	spiral......	10.4786″	18.9833″
Face...........	$\frac{7}{8}$″	$\frac{7}{8}$″	Approx. lead		
No. of teeth.....	30	48	of spiral....	10.477″	19.000″
Hand..........	Right	Right	Gears for lead		
Pitch dia.......	3.7766″	5.3367″	of spiral:		
Angle..........	48° 33′	41° 27′	Gear on worm	48	38
Normal diametral			1st gear on stud	28	40
pitch........	12	12	2d gear on stud	44	64
Normal cir. pitch	.2618″	.2618″	Gear on screw	72	32
Normal thickness.	.1309″	.1309″	Speed........	1st or 2d speed, or 60 F.P.M.	
" addendum.	.0833″	.0833″	Feed.........	Medium power feed	
" depth.....	.1797″	.1797″		— .014″ cutter revolution	

SPECIFICATIONS FOR PAIR OF SPIRAL GEARS OTHER THAN 45°, Fig. 29. *Concluded.*

NAME.	DRIVER.	DRIVEN.	NAME.	DRIVER.	DRIVEN.
Center distance..	4.5566″	4.5566″	Keyway......	$\frac{1}{4}″ \times \frac{1}{8}″$	$\frac{1}{4}″ \times \frac{1}{8}″$
Cutter pitch.....	12	12	Lubricant....	Lard oil	Lard oil Dry for C. I.

SCHEDULE OF OPERATIONS FOR GEAR

I. Preparatory adjustments. Place cutter A on arbor B in direction of rotation, and clamp hard. Fasten index head and footstock in horizontal alinement on table.

II. Set cutter central. Set swivel table at zero and cutter central, as at A', Fig. 28, and clamp cross slide.

III. Arrange gears and index. Arrange gears for spiral, set index for 21 teeth, withdraw stop so that index plate can revolve.

IV. Mount gear blank on centers. Press mandrel C hard into blank D and mount on centers securing dog E in driver by screw F.

V. Set machine for depth of teeth. Start machine; move table by hand long. feed until blank D is under revolving cutter A. Hold .001″ tissue paper on top of blank and raise knee until revolving cutter cuts paper. Then set vertical dial at zero and move blank back until clear of cutter. Stop machine. Set swivel table or milling machine attachment, see p. 1130, for tooth angle and clamp. Raise knee for depth of

space, and again set vertical dial at zero.

VI. Mill trial tooth. Start machine; move table by hand long. feed until cutter nearly touches blank, then throw in power feed, and mill tooth, one cut. Set table dog to trip feed when center of cutter has passed far corner of gear blank. At end of cut, lower knee and move table back until blank is clear of cutter; or, sufficient clearance may be obtained by pulling belt until no tooth of cutter is on center of the gear, then table may be moved back without lowering the knee.

VII. Mill all teeth by same process.

Information. — Before first cut, raise knee until revolving cutter touches blank, then move table by hand to trace spiral line on work which will show if lead and angle are correct.

Warning. — Press m a n d r e l harder into blank for spiral milling as the tendency to slip by pressure of cut is greater.

Attention. — Cut pinion by same process as gear.

48. If the pitch diameters of spiral gears are the same ratio as the number of teeth, the spiral angle of each is **45°.**

SPECIFICATIONS FOR PAIR OF SPIRAL GEARS, 45° ANGLE

Name.	Gear.	Pinion.	Name.	Gear.	Pinion.
Material........	Phosphor bronze	Machine steel	Cutter number	2	2
Speed ratio: Driver 800 revolutions, driven 500 revolutions			Exact lead of spiral......	11.6616″	8.330″
Hole...........	1″	1″	Approx. lead of spiral....	11.667″	8.333″
O. dia.........	3.962″	2.902″			
Face..........	1¼″	1¼″	Gears for lead		
No. of teeth.....	21	15	of spiral:		
Hand..........	Right	Right	Gear on worm.	56	48
Pitch dia........	3.712″	2.652″	1st gear on stud	32	32
Angle of teeth with axis......	45°	45°	2d gear on stud	48	40
Normal diametral			Gear on screw	72	72
pitch..........	8	8			
Normal circular			Speed........	1st or 2d speed, or 60 F.P.M.	
pitch.........	.3927″	.3927″	Feed.........	Medium power feed — .014″ per cutter revolution	
Normal thickness.	.1963″	.1963″			
Normal addendum.........	.1250″	.1250″	Keyway......	¼″ × ⅛″ Lard oil	¼″ × ⅛″ Lard oil
Normal depth....	.2696″	.2696″	Lubricant....		
Center distance..	3.182″	3.182″			
Cutter pitch.....	8	8			

WORM GEARING

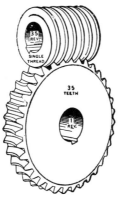

Fig. 30. — Worm and Worm Gear. Ratio 35 to 1.

49. Worm gearing, Fig. 30, is used to transmit rotary motion from one shaft to another not parallel, and usually at right angles. While worm gearing is generally used where the load is heavy and for relatively large reductions, there are numerous exceptions where multiple-threaded worms are used, to reduce the pressure on the teeth and also to increase the efficiency. The thread on the worm, Fig. 30, is similar to the 29° thread. The teeth of the gear are cut at an angle with the axis of the shaft, usually from 55° to 65° depending on the lead of the worm.

Cutters of 50° included angle, having straight sides, are commonly used, producing a worm tooth which is curved both on the axis and on the normal, and the hob for cutting the wheel is made of the same form. Worm gears are sometimes roughed out with a gashing cutter, but they are usually finished with a hob which gives a bearing on the worm thread the full length of the teeth.

Worms and worm gears are usually designed so that available hobs may be used. This may be according to diametral pitch or circular pitch.

Worm gears may be found in the power feed driving mechanisms of many machine tools including engine lathes, drill presses, and boring mills. They are also used for worm wheel drives for automobiles and trucks, steering mechanisms, and for other purposes where a large reduction in speed is desired. By increasing the pitch and number of threads, this mechanism becomes one of spiral gears.

Important. — In order to understand worm gearing one should have a good knowledge of spur gearing.

FORMULAS IN WORM GEARING

50. Circular pitch (circumferential) is the distance from the center of one tooth to the center of the next tooth measured along the pitch circle, as in spur gears. Either the circular pitch or its equivalent in diametral pitch is usually given.

51. The pitch diameter of a worm gear, Fig. 31, is the diameter of the pitch circle.

Formula. — Pitch diameter = $\dfrac{\text{circular pitch} \times \text{number of teeth}}{3.1416}$.

Example. — To find the pitch diameter, given circular pitch .3333″ and number of teeth 35.

Solution. — $\dfrac{.3333 \times 35}{3.1416} = 3.7136″$ pitch diameter.

52. The lead of the worm. —

Formula. — The lead $= \dfrac{1}{\text{number of turns per 1″ in worm}}$.

Example. — To find the lead of worm, given number of teeth 1 and number of turns per inch in worm 3.

Solution. — $\frac{1}{3} = .3333″$, lead of worm.

53. Outside diameter of gear may be calculated, but as the points of these gears are usually made with a slight land to give strength, a measurement taken from a drawing is sufficient.

54. Diametral pitch (size of tooth), see Spur Gears, p. **11**04.

55. Throat diameter is equivalent to the outside diameter of a spur gear of equal pitch, Fig. 31.

Formula. — Throat diameter = pitch diameter + 2 × addendum.

Example. — To find throat diameter, given pitch diameter 3.7136″ and addendum .1061″.

Solution. — 3.7136 + 2 (.1061) = 3.9258″ throat diameter.

56. Throat radius, see Fig. 31, is radius of curve at top of tooth.

Formula. — Throat radius $= \dfrac{\text{outside diameter of worm}}{2} -$ 2 × addendum.

Example. — To find throat radius, given outside diameter of worm 1.75″ and addendum .1061″.

Solution. — $\dfrac{1.75}{2} - 2\,(.1061) = .6628″$, throat radius.

57. Addendum is the distance from the pitch line to the throat on the gear.

Formula. — Addendum = circular pitch ÷ 3.1416.

Example. — To find the addendum, given circular pitch .3333″.

Solution. — $\dfrac{.3333}{3.1416}$ = .1061″, addendum.

58. The chordal addendum (corrected pitch depth) is also necessary for measurement of worm gear tooth thickness (.1079″ in this case). See pp. **11**12–**11**14.

59. Thickness of tooth at pitch line. —

Formula. — Thickness of tooth at pitch line = $\dfrac{\text{circular pitch}}{2}$.

Example. — To find thickness of tooth at pitch line, given circular pitch .3333″.

Solution. — $\dfrac{.3333}{2}$ = .1667″, thickness of tooth at pitch line.

60. Working depth of tooth. —

Formula. — Working depth of tooth = 2 × addendum.

Example. — To find working depth of tooth, given addendum .1061″.

Solution. — .1061 × 2 = .2122″, working depth of tooth.

61. Whole depth of tooth. —

Formula. — Whole depth of tooth = 2 × addendum + clearance.

Example. — To find whole depth of tooth, given addendum .1061″ and clearance .0067″.

Solution. — 2 × .1061 + .0067 = .2189″, whole depth of tooth.

62. Clearance between the worm and the gear. —

Formula. — Clearance at bottom of tooth = whole depth of tooth − working depth.

Example. — To find clearance between worm and gear, given whole depth of tooth .2289″ and working depth .2212″.

Solution. — .2289 − .2212 = .0067″, clearance at bottom of tooth.

63. The worm thread tool has an included angle of 29° like the 29° thread when the standard pressure angle of $14\frac{1}{2}$° is used. The width of the tool at the end is obtained by the following formula:

Formula. — Constant .3095 × circular pitch = width of end of worm thread tool.

Example. — To find width of worm thread tool at end, given constant .3095 and circular pitch .3333″.

Solution. — .3095 × .3333 = .10316″, width of worm thread tool at end.

64. Angle of worm thread tooth (gashing angle) is the angle of thread with line at right angles to worm axis.

Formula. — Tangent of angle $= \dfrac{\text{lead of thread}}{\text{pitch circumference of worm}}$.

Example. — To find angle of worm thread tooth (gashing angle), given lead of .3333″ and pitch diameter 1.5378″.

Solution. — Tangent $= \dfrac{.3333}{1.5378 \times 3.1416} = .06891″$.

Tangent .06891 = 3° 56′ 50″.

65. Width of top of worm thread. —

Formula. — Width of top of worm thread = constant .3354 × circular pitch.

Example. — To find width of top of worm thread, given constant .3354 and circular pitch .3333″.

Solution. — .3354 × .3333 = .1118″, width of top of worm thread and width of end of hob thread.

66. Center distance is the distance between axis of worm and axis of worm gear.

Formula. — Center distance =

$$\frac{\text{pitch dia. worm gear + pitch dia. worm}}{2}.$$

Example. — To find center distance, given pitch diameter of worm gear 3.7136″ and pitch diameter of worm 1.5378″.

Solution. — $\dfrac{3.7136 + 1.5378}{2} = 2.6257″$, center distance.

67. Using diametral pitch system. — In threading a worm in an engine lathe, the thread tool is made according to diametral pitch sizes, and sizes of the gear parts may be taken directly from the table of spur gear parts for diametral pitch. See Table of Tooth Parts, pp. 1108–1111.

SPECIFICATIONS FOR WORM AND WORM GEAR, Fig. 31

Name.	Worm.	Gear.	Name.	Worm.	Gear.
Material.........	Machine steel	Phosphor bronze	Thickness of tooth at P.L.	.1667″	.1667″
Hole...........	¾″	1″	Normal thickness of tooth at P.L.......	.1663″	.1663″
Face...........	2″	⅞″	Corrected addendum for worm gear...	....	.1079″
No. of teeth (gear ratio).........	1 (single thread)	35	Clearance at bottom of tooth......	.0067″	.0067″
Pitch dia........	1.5378″	3.7136″	Working depth	.2122″	.2122″
Lead of the worm.	.3333″ R.H. single	....	Whole depth of tooth......	.2189″	.2189″
Threads per 1″ (see p. 320)....	3	....	Width of end of worm thread tool, also width at end of hob tooth..	.10316″	....
Circ. pitch of gear and linear pitch or axial pitch of worm.........	.3333″	.3333″	Angle of worm thread*tooth (gashing angle).......	3° 56′ 50″	3° 56′ 50″
O. dia. (taken as nearest 32d calculated value)..	1.75″	4.1035″ or 4³⁄₃₂″	Width of top of worm thread	.1118″	....
Diametral pitch..	9.4248″	9.4248″	Center distance	2.6257″	2.6257″
Throat dia.......	....	3.9258″	Hob marked...	....	.3333″ P. singleR.H.
Throat radius....	....	.6628″			
Clearance radius (½ dia. of hob-clearance).....	....	.8917″	Hob (outside) dia..........	....	1.7634″
Addendum..:.....	.1061″	.1061″	Lubricant.....	Lard oil	Dry, or lard oil
			Keyway.......	₁⁄₁₆″×³⁄₃₂″	¼″×⅛″

* If the angle is more than 18°, it is advisable to use the formulas for spiral gears, although gear manufacturers do not always follow this rule.

CUTTING WORM GEARS

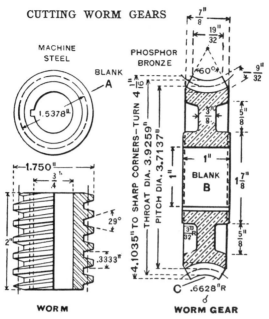

FIG. 31. — DRAWING OF WORM AND WORM GEAR. RATIO 35 TO 1.

68. To prepare worm-gear blank $1'' \times 4.1035''$ **diameter, 35 teeth; material, phosphor bronze.** Fig. 31.—True casting in chuck, drill, ream, press mandrel into hole, mount on centers, square and turn to size. Turn concave surface, or throat, with forming tool, or rough out and finish with hand tools to templet. When a worm gear is to be cut with a hob in a hobbing machine, see pp. **11**48, **11**52, the turning of the concave surface is usually omitted as the hob will cut the throat of the wheel to the proper size and shape. Bevel corners and cut keyway.

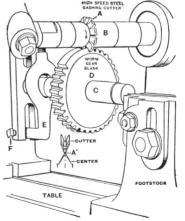

FIG. 32. — GASHING WORM GEAR.

69. To gash worm gear, universal milling machine. Fig. 32.

Specifications: Material phosphor-bronze worm-gear blank, finished
$1'' \times 4.1035''$ diameter for 35 teeth.

Machines and tools: Universal milling machine, index head and centers, gashing cutter.

Speed: 2d or 3d speed, or 60 F.P.M. Hand vertical feed.

Lubricant: Dry or lard oil.

SCHEDULE OF OPERATIONS

I. Preparatory adjustments.
Place gashing cutter A on arbor
B to cut in direction of rotation
and clamp hard. Fasten index
head and footstock in horizontal
alinement on table.

II. Set cutter central. Set swivel
table at zero and cutter central,
as A', Fig. 32.

III. Arrange index. Arrange index for 35 teeth.

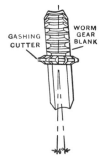

GASHING CUTTER

WORM GEAR BLANK

Fig. 33. — Setting Swivel Table
to Gashing Angle for Worm
Gear.

IV. Mount gear blank on centers.
Press mandrel C into blank D
and mount on centers, securing
dog E in driver by screw F. Raise
knee and move table until cutter
fits concave part of blank. Lower
knee and set swivel table at
$4°$, the angle of teeth. Clamp
table.

V. Gash blank. Start machine
and raise knee until gear blank
touches revolving cutter. Set
vertical dial at zero, raise knee
slowly, sinking cutter to depth
of roughing cut .220''. Set vertical stop. Lower knee, index
and raise knee to stop and cut
next gash. Continue process until all teeth are gashed.

Attention. — If the machine has
a vertical stop, set it for depth of
gash after the first cut and feed
up against it for each gash; otherwise, use the vertical dial to obtain
the depth.

HOBBING WORM GEAR

70. To hob worm gear, universal or plain milling machine.
Fig. 34.

Specifications: Material, phosphor-bronze worm-gear blank, finished
$1'' \times 4.1035''$, with 35 teeth gashed.

Machines and tools: Universal or plain milling machine, index head
and centers. $1\frac{3}{4}'' \times 2''$ right-hand hob, 3 threads per $1''$.

Speed: 1st or 2d speed, or 40 F.P.M.

Lubricant: Dry or lard oil.

SCHEDULE OF OPERATIONS

I. Preparatory adjustments. Place hob A on arbor B to cut in direction of rotation, and clamp hard. Fasten index head and footstock in horizontal alinement on table. Set swivel table at zero. Set top of knee to center line on vertical slide which alines the index centers with machine spindle, then set vertical dial at zero.

II. Mount gear blank on two dead centers.

Press mandrel C into blank D, oil both center holes, lower knee and mount on two dead centers without dog so that hob will drive gear.

III. Set hob central. Raise knee and adjust table by long. and cross feeds until hob is central by measurement with gashed gear, then clamp long. and cross feeds.

IV. Obtain depth of teeth. Start machine and raise knee slowly until hob begins to cut and after each revolution of gear raise knee a little. Repeat until distance between top of knee and center on vertical slide equals center distance of worm and gear. This distance may be measured with a rule from top of knee to center line on vertical slide or, more accurately, with rule for the $2''$ and vertical dial for the decimal $.6258''$.

Information. — Hobs with relieved teeth same as formed milling cutters are obtainable to order only. Depth of teeth equals working depth of worm plus one clearance, plus allowance for grinding.

Attention. — In Fig. 34, gear is driven by hob making it necessary to gash gear, but on gear-hobbing machines or attachments where gear is driven by hobbing-machine spindle, gashing may be omitted. See Fig. 35.

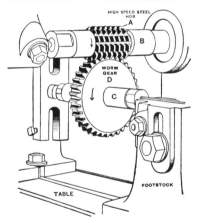

Fig. 34. — Hobbing Worm Gear in Milling Machine.

THREADING WORMS

71. To make worm $1\frac{3}{4}'' \times 2''$, 3 threads to $1''$; material, machine steel (to be case-hardened), Fig. 31. — True blank in chuck, drill, ream, press mandrel into hole, mount on centers, square and turn to size. Arrange gears for 3 threads to 1″. As worm is same angle as 29° thread, rough thread with 29° rough-threading tool similar to square thread tool. To finish thread, use 29° finish-threading tool, with point .1033″ wide. Cut keyway.

Attention. — Worm threads are milled in milling machines or thread millers with formed cutters, see Thread Miller, p. 1048; or cut in a hobbing machine.

HOBBING GEARS

72. Automatic gear hobbing machines are used to cut and generate worm, spur, and spiral gears. — The spiral hob, as

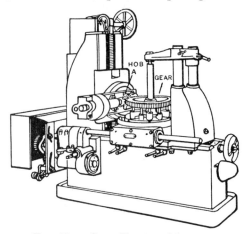

shown at *A*, Fig. 35, by which the gear is cut and generated, may be considered a rack. Each row of teeth when set at the helix angle, forms the shape of a rack tooth with straight flanks. By rotating the hob in unison with the gear blank, so as to rotate the gear blank the exact distance of the pitch to be

Fig. 35. — Gear-Hobbing Machine.

cut for each revolution of the hob, the curve of the tooth is formed or generated. The pitch or lead of the thread on the hob is the same as the pitch of the required gear. Only one hob is necessary for all gears and pinions of the same pitch. The hob spindle and table are connected by change gears to give the proper ratio between pitch or lead of hob and pitch of required gear.

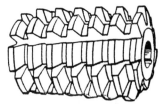

Fig. 36. — Hob for Generating Worm, Spur and Spiral Gears.

73. A standard hob is shown in Fig. 36. The high-speed or carbon-steel hob blank may be threaded in a lathe or thread miller. See p. **10**48.

For standard involute gear teeth, the hob is cut with a 29° threading tool or cutter; and for standard stub teeth, with 40° threading tool or cutter.

The hob is grooved in a milling machine and the teeth "backed off" or relieved in a backing-off or relieving machine, or lathe with a relieving attachment. See pp. **12**82–**12**86. The first two or three teeth of a hob are naturally weak and they are rounded off to prevent breaking. The hob is hardened and tempered, and may be sharpened. See p. **8**24.

74. Grinding hobs. — For very accurate work the teeth are ground, after hardening, with a special grinding machine.

75. Worm gears (often called worm wheels) are hobbed or generated, as in Fig. 37.

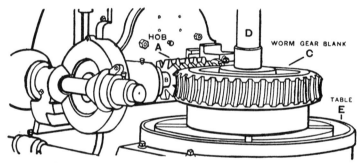

FIG. 37. — HOBBING WORM GEAR IN AUTOMATIC GEAR-HOBBING MACHINE.

Hob A is keyed to horizontal hob spindle B at right angles to the axis of the gear blank. One tooth in the hob is set in line with the center axis of the gear blank, by a centering gage.

Worm gear blank C is fastened securely on vertical arbor D and rests on table E.

The table carrying gear blank is driven by a large worm gear below the table. An automatic gear feed moves the table and gear blank toward the hob. The hob spindle and work table are connected by a train of change gears which determines the exact ratio of speeds between gear blank and hob. During the cutting process all moving parts revolve steadily in one direction.

Tables and formulas are used to obtain the proper change gears for any combination of hob and gear and feed.

Attention. — The operation is practically the same as hobbing a worm gear in a milling machine, see Fig. 34, except that the positive gear drive of the gear blank makes gashing the gear blank unnecessary. It also makes it unnecessary to turn the concaved surface or throat, as the hob cuts it to the proper diameter and shape.

Information. — "Slewing the hob." If the correct hob for a gear is not obtainable, a worm wheel may be cut by using the available hob at a slight angle.

76. Tapered hobs, Fig. 38. — Set worm-gear blank *A* to the exact depth of the teeth or center distance, and feed hob *B* across the blank thereby generating the gear. The roughing

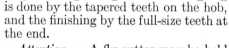

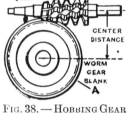

FIG. 38. — HOBBING GEAR WITH TAPERED HOB.

is done by the tapered teeth on the hob, and the finishing by the full-size teeth at the end.

Attention. — A fly cutter may be held in an arbor and fed across in the same manner as the tapered hob. See Fly Cutters, pp. **11**03, **10**07.

77. Lubricant for gear hobbing. — Cast-iron gears are hobbed dry. Steel or bronze gears are lubricated with lard oil or a lubricating compound.

78. Spur gears can be hobbed or generated, as in Fig. 39. — Hob spindle *A* to which hob *B* is keyed, is tilted to make the thread vertical on the front side of the hob, by saddle *C*. This setting is obtained by means of a special gage. Gear blank *D* is clamped tightly to vertical arbor *E* to prevent shifting. The

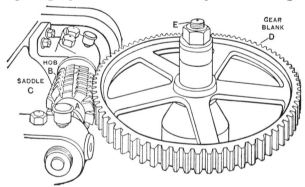

FIG. 39. — HOBBING SPUR GEAR IN AUTOMATIC GEAR-HOBBING MACHINE.

hob spindle and work table are connected by a train of change gears which determines the exact ratio of speeds between hob and gear.

The hob is set above the top of gear blank at the proper tooth depth. The gear blank is revolved and the hob fed downward upon the vertical ways of the housing. The hob is fed downward for each rotation of the gear blank from .002″ to .015″ depending upon the pitch of the gear. When the center line of the hob has passed the bottom edge of the gear, the gear is finished. A number of spur gears, up to the capacity of the machine, may be hobbed at one time.

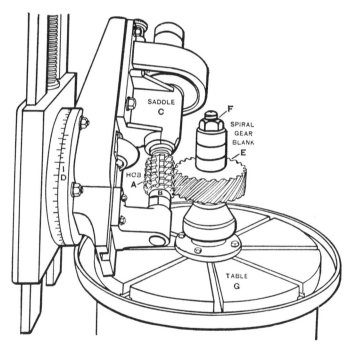

FIG. 40. — HOBBING SPIRAL GEAR IN AUTOMATIC GEAR-HOBBING MACHINE.

79. Spiral gears can be hobbed or generated, as in Fig. 40.— Hob A is keyed to horizontal spindle B on saddle C which is tilted to the right or left to correspond with the tooth angle. The saddle is graduated to degrees and with the aid of a vernier

D can be read to 5 minutes which facilitates accurate settings The spiral gear blank E is fastened to vertical arbor F on table G.

The principle of "setting up" and "gearing up" is practically the same as that used on the universal milling machine to cut spirals of any kind. See pp. **12**39–**12**47.

To generate the required tooth angle, the table is geared to gain or lose one complete revolution while the hob is fed a distance equal to the *lead* of the spiral gear.   After the machine is set up for hobbing a spiral gear, the feed and general processes are the same as for hobbing a spur gear.   Tables and formulas are used to obtain the proper change gears for indexing, for lead, and for feed.   Any number of spiral gears up to the capacity of the machine may be hobbed at one time.

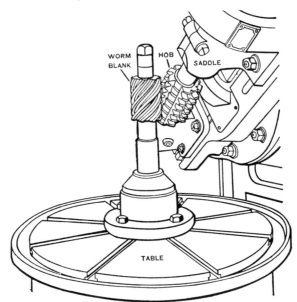

WORM BLANK   HOB   SADDLE

TABLE

Fig. 41. — Hobbing Worm in Automatic Gear-Hobbing Machine.

80. Worms are practically spiral gears, and they may be hobbed, as in Fig. 41.   Worms may also be milled by a single cutter with an automatic gear-hobbing machine.

BEVEL GEARING

81. Bevel gears are used to transmit rotary motion from one shaft to another not parallel and usually at right angles, as in Fig. 42. All parts of a bevel-gear tooth are assumed to converge at the apex where center lines of the shafts intersect except when the gears are cut with rotary cutters, then the bottom of the tooth is made parallel to the face of the mating gear. See Drawing of Bevel Gears, Fig. 43. The tooth curves and other specifications at large end of bevel-gear teeth are the same as

FIG. 42. — BEVEL GEAR AND PINION. RATIO 2 TO 1.

those of a spur gear of the same pitch whose radius is equal to the back-cone radius of a bevel gear, as at *A*, Fig. 43. See

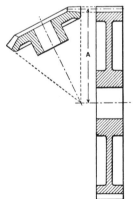

FIG. 43. — RELATIVE SIZES OF BEVEL AND SPUR GEAR LEAVING TEETH OF SAME DIMENSIONS.

also Fig. 46, p. **11**58. As the dimensions of the teeth are proportionally smaller toward the inner end, the cutting of an accurate tooth is difficult.

82. Bevel gears that are to run at uniform speeds or to transmit a large amount of power are first rough cut with a formed milling cutter, then finish planed in bevel-gear generating machines. See pp. 11̲72–11̲74.

Bevel gears may be cut with formed milling cutters in a milling machine or automatic machine, after which the teeth are corrected by filing.

Information. — When bevel or miter gears are to be used principally to transmit motion and not a large amount of power, filing the teeth may be avoided by making the faces narrower or by cutting the spaces deeper at the small end, or by doing both.

Important. — In order to understand bevel gearing, one should have a good knowledge of spur gearing.

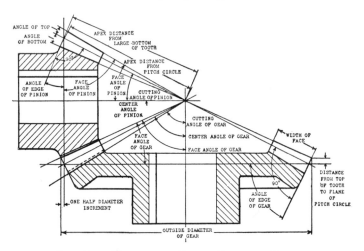

Fig. 44. — Diagram of Bevel Gear and Pinion.

FORMULAS IN BEVEL GEARING

83. Angle of shafts equals 90°. Number of teeth in gear 32, in pinion 16; pitch (size of teeth) 8.

84. Pitch diameter. —

Formula. — Pitch diameter $= \dfrac{\text{number of teeth}}{\text{diametral pitch}}$.

Example. — To find the pitch diameter, given number of teeth 32 and diametral pitch 8.

Solution. — $\frac{32}{8} = 4''$, pitch diameter.

85. Apex distance from the pitch circle, Fig. 44. —

Formula. — Apex distance $= \dfrac{1}{2 \times \text{diametral pitch}}$

$\sqrt{\text{number of teeth in gear}^2 + \text{number of teeth in the pinion}^2}$.

Example. — To find apex distance, given diametral pitch 8, number of teeth in gear 32, and number of teeth in pinion 16.
Solution. — $\frac{1}{16}$ × 35.7771 = 2.23606″, apex distance.

86. Face is usually made from $\frac{1}{4}$ to $\frac{3}{10}$ of the apex distance.

87. The angle of the edge of the gear or pinion is made at right angles with the pitch cone. This angle is used in turning the gear blank and may be measured from the pitch line, and when measured from the pitch line it is equal to the center angle.

88. Diametral pitch at large end (size of tooth) is equal to that of a spur gear of the same diametral pitch whose radius is equal to the back-cone radius of the bevel gear. See pp. 1108–1114.
For an 8-pitch gear the diametral pitch = 8.
For an 8-pitch gear the addendum (gear or pinion) = .1250″.
For an 8-pitch gear the corrected addendum (gear) = .1261″.
For an 8-pitch gear the corrected addendum (pinion) = .1295″.
For an 8-pitch gear the corrected dedendum (gear) = .1457″.
For an 8-pitch gear the corrected dedendum (pinion) = .1489″.
For an 8-pitch gear the clearance = .0196″.
For an 8-pitch gear the working depth of tooth = .2500″.
For an 8-pitch gear the whole working depth of tooth = .2696″.
For an 8-pitch gear the thickness of tooth at pitch line (gear or pinion) = .196″.
For an 8-pitch gear of 32 teeth, the chordal thickness (gear) = .1961″.
For an 8-pitch gear of 32 teeth, the chordal thickness (pinion) = .1961″.

Information. — The corrected addendum is found by dividing the number of teeth in the bevel gear by the cosine of the center angle and taking the resulting value as the number of teeth of the spur gear that is to determine the corrected addendum; i.e., the number of teeth in a spur gear having a pitch diameter equivalent to the back-cone circle of the bevel gear.

Note. — The chordal thickness of a bevel gear is generally taken as that of a spur gear of the same number of teeth and pitch.

89. Diametral pitch at small end (size of tooth) equals $\frac{1}{4}$ to $\frac{3}{10}$ of the dimensions at the large end. Therefore, given the size at the large end, the dimensions at the small end are readily found.

90. Angle of top or addendum angle is added to the pitch-cone angle and will give the angle of the face of the gear with axis.

Formula. — Tangent of addendum angle $= \dfrac{\text{addendum.}}{\text{apex distance}}$.

Example. — To find tangent of addendum angle, given addendum .1250″, and apex distance 2.2361″.

Solution. — $\dfrac{.1250}{2.2361} = .0559″ =$ tangent of angle of top. Tangent .0559 $= 3° 12'$, angle of top or addendum angle.

Note. — The face angle is often given on gear blank drawings from the pitch line to facilitate turning the angle of the blank with a compound rest, see p. **11**89.

91. Angle of bottom (dedendum angle). —

Formula. — Tangent of angle of bottom $= \dfrac{\text{dedendum.}}{\text{apex distance}}$.

Example. — To find tangent of angle of bottom, given dedendum .1446″, and apex distance 2.23606″.

Solution. — $\dfrac{.1446}{2.23606} = .06467,″$ tangent of angle of bottom. Tangent .06467 $=$ the angle of $3° 42'$, angle of bottom.

92. The cutting angle. —

Formula. — Cutting angle $=$ center angle $-$ angle of bottom.

Example. — To find cutting angle, given center angle 26° 34′ and angle of bottom 3° 42′.

Solution. — $26° 34' - 3° 42' = 22° 52'$, cutting angle.

93. Outside diameter. —

Formula. — Outside diameter = pitch diameter $+ 2 \times$
$$\frac{\text{cosine pitch-cone angle}}{\text{diametral pitch}}.$$

Example. — To find outside diameter, given pitch diameter 4″, cosine of pitch-cone angle .44724, and diametral pitch 8.

Solution. — $4 + 2 \times \dfrac{.44724}{8} = 4.112″$, outside diameter.

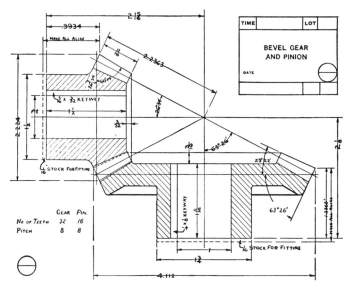

FIG. 45. — DRAWING OF BEVEL GEARS. RATIO 2 TO 1.

94. To find the center angle or angle of pitch cone, Fig. 44.—

Formula. — Tangent of center angle of pinion =

$$\frac{\text{pitch diameter of pinion}}{\text{pitch diameter of gear}}, \quad \text{also} = \frac{\text{number of teeth of pinion}}{\text{number of teeth of gear}}.$$

Example. — To find tangent of center angle, given pitch diameter of pinion 2″ and pitch diameter of gear 4″.

Solution. — $\frac{2}{4}$ = .5000 tangent of center angle.   Referring to tables of natural trigonometrical functions, we find opposite .5000 the angle of 26° 34′, center angle of pinion.

Formula. — Center angle of the gear =

$$\frac{180° - 2 \times \text{center angle of pinion}}{2}, \text{ also } 90° - \text{center angle of pinion.}$$

Example. — To find center angle of gear, given angle of 180° and center angle of pinion 26° 34′.

Solution. $\dfrac{180° - 2 \times 26° 34′}{2} = 63° 26′$, center angle of gear.

Note. — In compiling these formulas, Brown & Sharpe's *Formulas In Gearing* was used as standard authority.

Miter gears. — A pair of bevel gears whose diameters, angles, and number of teeth are the same, are called *miter* gears and both may be cut with the same cutter.

CUTTING BEVEL GEARS

95. To select cutter for cutting bevel gears. — When ordering cutters for bevel gears, give exact specifications: Pitch, number of teeth in bevel gear and pinion, length of face, and angle of shafts, Fig. 46.

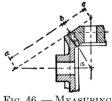

Measure back-cone radius *AB* for the gear, or *BC* for the pinion.   This is equal to the pitch radius of a spur gear the number of teeth in which would determine the cutter to be used.   Hence, twice *AB* times the diametral pitch equals the number of teeth for which the cutter should be selected for the gear.

FIG. 46. — MEASURING BACK CONE.

Example. — To find the number of cutter for a bevel gear of 32 teeth, 8 pitch with a back-cone radius of 4.4375″.

Solution. — 2 × 4.4375″ × 8 = 71   teeth.   From table of range of cutters it will be seen that 71 comes between 55 and 134 teeth, the range covered by a No. 2 cutter.

The cutter for pinion can be selected by using the same formula.

Example. — To find the number of cutter for a bevel pinion of 16 teeth, 8 pitch with a back-cone radius of 1.125''.

Solution. — 2 × 1.125'' × 8 = 18 teeth. From table of range of cutters, it will be seen that 18 comes between 17 and 20 teeth, the range covered by a No. 6 cutter.

INVOLUTE CUTTERS FOR BEVEL GEARS

No. 1 will cut wheels from 135 teeth to a rack.

" 2 " " 55 " " 134 Teeth.

" 3 " " 35 " " 54 "

" 4 " " 26 " " 34 "

" 5 " " 21 " " 25 "

" 6 " " 17 " " 20 "

" 7 " " 14 " " 16 "

" 8 " " 12 " " 13 "

These involute bevel-gear cutters (no other system is used for cutting bevel gears with rotary cutters) are .005'' thinner than the space at the small end of the tooth, or thin enough to cut bevel gears whose faces are not over one-third the apex distance. The tooth at the inner end is not less than two-thirds the standard thickness and height.

The number of teeth for which the cutter should be selected can also be found by the following formula: $\tan \alpha = \dfrac{Na}{Nb}$.

Number of teeth to select cutter for gear $= \dfrac{Na}{\cos \alpha}$.

Number of teeth to select cutter for pinion $= \dfrac{Nb}{\sin \alpha}$.

96. Table for obtaining set-over or trial distance for cutting bevel gear.

No. of Cutter.	3/1	3¼/1	3½/1	3¾/1	4/1	4¼/1	4½/1	4¾/1	5/1	5½/1	6/1	7/1	8/1
1	.254	.254	.255	.256	.257	.257	.257	.258	.258	.259	.260	.262	.264
2	.266	.268	.271	.272	.273	.274	.274	.275	.277	.279	.280	.283	.284
3	.266	.268	.271	.273	.275	.278	.280	.282	.283	.286	.287	.290	.292
4	.275	.280	.285	.287	.291	.293	.296	.298	.298	.302	.305	.308	.311
5	.280	.285	.290	.293	.295	.296	.298	.300	.302	.307	.309	.313	.315
6	.311	.318	.323	.328	.330	.334	.337	.340	.343	.348	.352	.356	.362
7	.289	.298	.308	.316	.324	.329	.334	.338	.343	.350	.360	.370	.376
8	.275	.286	.296	.309	.319	.331	.338	.344	.352	.361	.368	.380	.386

To obtain set-over by above table use this formula:

$$\text{Set-over} = \frac{Tc}{2} - \frac{\text{factor from table}}{P}.$$

P = diametral pitch of gear to be cut.

Tc = thickness of cutter used, measured at pitch line.

Rule. — Find the factor in the table corresponding to the number of the cutter used and to the ratio of apex distance to width of face; divide this factor by the diametral pitch and subtract the quotient from half of the thickness of the cutter at the pitch line.

This amount can also be found by the following formula:

$$\text{Ratio of apex distance to width of face} = \frac{\text{apex}}{\text{face}}.$$

97. Set-over for a gear, 32 teeth, 8 pitch. —

Example. — A bevel gear of 32 teeth, 8 pitch, 63° 26′ pitch-cone angle, $1\frac{1}{8}$ face. These dimensions call for a No. 2 cutter and an apex distance of 2.2363″.

In order to get the factor from the table, the ratio of apex distance with length of face must be known. This ratio for the above mentioned bevel gear is $\dfrac{2.2363}{.6875} = \dfrac{3.25}{1}$ or $\dfrac{3\frac{1}{4}}{1}$. The factor in the table for this ratio with a No. 2 cutter is .268.

Next measure the cutter at the proper depth $a + c$ for 8 pitch which is found in the column marked "depth of space below pitch line" in the regular table of tooth parts, pp. **11**08–**11**11, or dividing 1.157″ by the diametral pitch. Thus $a + c = .1446″$. Suppose it is found by measurement that the thickness of the cutter at this depth is .1259″. This dimension will vary with different cutters and will vary in the same cutter as it is ground away, since formed bevel-gear cutters are commonly provided with side relief. By substituting these values in the formula $\dfrac{Tc}{2} - \dfrac{\text{factor from table}}{P}$ the following result is obtained, $\dfrac{.1259}{2} -$

$\dfrac{.268}{8} = .02945″$ set-over which is the required dimension.

98. The set-over for a pinion may be obtained by using the same formula.

Example. — A bevel pinion of 16 teeth, 8 pitch 23° 22′ pitch-cone angle, $\frac{11}{16}″$ face. These dimensions call for a No. 6 cutter and an apex distance of 2.2363″.

The ratio for this pinion is $\dfrac{2.2363}{.6875} = \dfrac{3.25}{1}$ or $\dfrac{3\frac{1}{4}}{1}$.

The factor in the table for this ratio with a No. 6 cutter is .318.

Assuming the cutter to measure .1259″ at the proper depth, by substituting these values in the formula the following result is obtained:

$$\frac{.1259}{2} - \frac{.318}{8} = .0232″ \text{ set-over.}$$

99. Rule for obtaining rotation of bevel-gear blank. — Some consider it safer to take several trimming cuts on the teeth of a trial blank to obtain correct settings which may be noted and duplicated on the gear; others obtain approximate rotation as follows:

Finish one side of tooth, move table an amount equal to twice the set-over and rotate blank to follow cutter an amount sufficient to thin the tooth to the required thickness. Applying this to the gear we have $.02945 \times 2 + .1963 = .2552″$, or the amount which the pitch line must be rotated.

A rotation of 13 holes in a 16-hole circle gives a movement at the pitch line of the gear of .2553″.

For the pinion we have $(.0232 \times 2) + .1963 = .2427″$ or the amount which the pinion must be rotated, and a movement of .2427″ is obtained by one turn and 36 holes in a 66-hole circle.

100. To cut bevel gears with formed gear cutters it is necessary to know the following data: — Pitch and number of teeth; the whole depth of tooth spaces at both ends of teeth; the thickness of teeth at the pitch line at both ends; the height of teeth above the pitch line (or addendum) at both ends; the cutting angle, that is the angle at which to set index head on milling machine, or cutter slide carriage on automatic gear cutting machine; number of cutter; approximate set-over of blank and approximate rotation of blank; and whether milling machine or gear cutter is to be used.

101. Cutting angle. — It may be noted, Fig. 44, that the cutting angle of the gear, 60° 14′, measured from the horizontal or axis of gear, is the same as face angle of pinion; and, inversely, the cutting angle of pinion, 23° 22′, is the same as face angle of gear. See p. 1153. This gives a parallel clearance, that is, the line of the bottom of the tooth is parallel to the face of the mating gear and it does not pass through the cone apex or common point of the axes. The process, the number of cutter, and the approximate set-over must also be known.

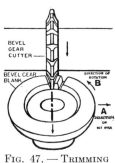

FIG. 47. — TRIMMING SIDE OF TOOTH.

The process of cutting a bevel gear with a formed gear cutter: First measure outside diameter of blank, angles and length of face, and if correct, securely fasten blank on mandrel or arbor, then scratch whole depth of space at large end with depth of gear tooth gage, see Fig. 50. Fasten cutter on arbor central, and cut one space; then index and cut second space which produces a tooth which is a little too thick. Now move the blank the trial distance off center *A*, Fig. 47, then rotate the blank in the same direction as the set-over, as at *B*, until the cutter just touches the side of the tooth at the small end, and allow the cutter to pass through again. See pp. 1160, 1161, 1165, 1167.

Index blank for next space and make the same settings on the other side of center and trim the second side of tooth. Measure thickness of tooth at both ends and make corrections. If by measurement both ends are a little too thick, but proportional, rotate blank and make trial cuts until tooth is of correct thickness at both ends. If too thick at large end and right at small end *increase* set-over, but if too thick at small end and right at large end, the set-over must be *decreased* and the process repeated on another tooth.

After determining the proper amount to set blank out of center, the teeth can be finished, without making a central cut, by cutting all the teeth with the blank set out of center, first on one side and then on the other.

To prevent the teeth being too thin at either end, it is important, after cutting once around with blank out of center, to give careful attention to the rotative adjustment of the gear blank when setting the cutter for trimming the opposite sides of the teeth.

The finished spaces of teeth are not always of the form that the cutter might be expected to make them, for the reason that the blank must be set off center and rotated to make the teeth and spaces right at each end.

Warning. — Movement made on one side of tooth must be made on the other side, otherwise the tooth will not be central.

Information. — On cast-iron gears, 5-pitch and coarser, and on all steel gears it is best to make one central cut all around the blank, then two trimming cuts one on each side of teeth.

Attention. — As cutting the first tooth is largely experimental, it is best for a student to use a trial blank and make a record of data obtained.

102. To file bevel-gear teeth, it is necessary to file the faces of the teeth slightly above the pitch line at the small ends, as the cutter is selected for the outline of tooth at large end thus leaving the tooth at the small end too thick at the point. Use 5″ or 6″, number 0 or 1, *barrette* (smooth-back), or half-round 2d cut or bastard file. Assume that the tooth curve AB, Fig. 48, at the large end is correct and the small end CD is correct up to pitch line. File both sides of each tooth circular and tapering as at AE and $A'E'$ from top at large end down to, or nearly to, pitch line at small end.

Mount gears in testing machine or fit two shafts to an angle plate. Place gears on shafts and run by hand. Use a little

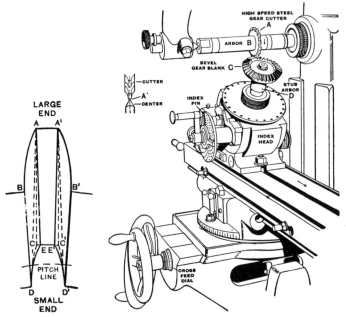

FIG. 48. — DOTTED LINES   FIG. 49. — CUTTING BEVEL GEAR.
SHOWING WHERE TO FILE
BEVEL-GEAR TOOTH.

Prussian blue or a mixture of red lead and oil on teeth and make further corrections by filing high places as indicated by bearing. Repeat process until bearing is along pitch line, then fit the gears to place in machine by squaring off hubs. See Testing Gears, p. **116**82.

103. To cut bevel gear and pinion with milling machine, Fig. 49.

SPECIFICATIONS FOR CUTTING BEVEL GEAR AND PINION, Fig. 49

Name.	Gear.	Pinion.	Name.	Gear.	Pinion.
Material ...	C. iron	C. iron	Thickness of tooth at P. L. large end....	.196″	.196″
W — Web..	Web	Plain	Thickness of tooth at P. L. small end....	.137″	.137″
Style.......	Bevel	Bevel	Depth of tooth at P. L. large end. (Addendum).........	.125″	.125″
Hole.......	1″	$\frac{3}{4}$″	Depth of tooth at P. L. small end. (Addendum)........	.0866″	.0866″
O. dia......	4.111″	2.224″	Chordal thickness at large end........	.1961″	.1961″
Face.......	$\frac{11}{16}$″	$\frac{11}{16}$″	Chordal thickness at small end........	.1358″	.1358″
Hub dia....	$1\frac{3}{4}$″	$1\frac{1}{2}$″	Corrected addendum at large end........	.1261″	.1293″
No. of teeth.	32	16	Corrected addendum at small end........	.0873″	.0895″
Diametral pitch.....	8	8	Center angle...	63° 26′	26° 34′
P. dia......	4″	2″	Face angle.....	23° 22′	60° 14′
Depth of tooth at large end.	.2696″	.2696″	Cutting angle..	60° 14′	23° 22′
Depth of tooth at small end	.1867″	.1867″	Apex distance at P. L........	2.2363″	2.2363″
			Cutter marked	#2–8 P.	#6–8 P.
			Speed........	1st or 2d speed or 60 F.P.M.	
Approx. setover.....	.0294″	.0232″	Feed..........	Medium power feed — .007″ per cutter revolution	
Approx. index setting	7 holes in 17-hole circle	32 holes in 43-hole circle	Keyway.......	$\frac{1}{4}$″ × $\frac{1}{8}$″	$\frac{3}{16}$″ × $\frac{3}{32}$″
			Lubricant.....	Dry	Dry

Attention. — It is best to leave bevel-gear hubs long to allow for squaring to place.

SCHEDULE OF OPERATIONS FOR GEAR

I. Preparatory adjustments. Place cutter A on arbor B to cut in direction of arrow *so as to cut from small end to large end of tooth,* then clamp hard. Fasten index head in table. Set swivel table at zero.

II. Set cutter central. Place pointed center in index head and set cutter central, as at A', Fig. 49, and cross-feed dial at zero.

III. Mount trial blank on arbor. Remove center and mount trial blank C on stub arbor D and drive arbor lightly into index-head spindle.

IV. Arrange index. Arrange index for 32 teeth and set index head at angle 60° 14'.

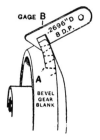

Fig. 50. — Marking Depth on Bevel-Gear Blank.

V. Obtain depth of teeth. Mark depth line A with gage B, Fig. 50, for large end of teeth. Start machine, elevate knee and with power feed cut first space to depth, two or three cuts. Adjust table dog to trip feed at end of cut. Move blank back until clear of cutter, index and cut second

space. Cutting two spaces central produces a tooth that is a little too thick.

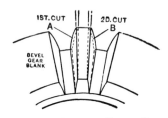

Fig. 51. — Dotted Lines Showing Where to Trim Bevel-Gear Tooth.

VI. Trimming bevel-gear tooth. Move table back until blank is clear of cutter. Move cross slide inward .0294″, revolve blank outward toward cutter, 7 holes in 43-hole circle and trim first side of tooth. See A, Fig. 51. Move blank back clear of cutter and cross slide back to zero; index to first space, then move outward .0294″ and revolve blank inward toward cutter, 4 holes second side of tooth. See B, Fig. 51.

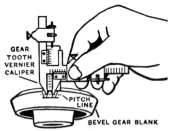

Fig. 52. — Measuring Thickness of Bevel-Gear Tooth with Vernier Caliper.

VII. Test thickness of tooth. Caliper tooth at both ends with gear-tooth vernier, Fig. 52, or gage, Fig. 53. To measure thickness of gear tooth with vernier, Fig. 52, adjust until vertical scale is the addendum distance from ends of jaws, then adjust sliding jaw to tooth and the horizontal scale will show thickness of tooth at pitch line. See Vernier Principle, pp. 211, 212, 1113.

Make corrections if necessary by changing set-over or index setting, or both, and take trial cuts until tooth is correct at both ends.

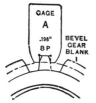

FIG. 53. — TESTING THICKNESS OF BEVEL-GEAR TOOTH.

VIII. Cut 32 teeth. Remove trial blank from arbor and mount gear blank. Move cross slide inward .0294″ off center, or the amount found to be correct, and cut all around. Move table back until blank is clear of cutter. Move cross slide outward the same amount on other side of center, revolve blank inward toward cutter 14 holes and cut all around second time, which completes gear.

Warning. — To obtain accurate settings, move table to correct position by always keeping the backlash on the same side, move table out a half-turn or so and then advance it to the correct dial reading when necessary.

Information. — For coarse pitches or very accurate gears, allowance should be made for the curve of the pitch circle when measuring thickness of tooth by obtaining the "corrected addendum" and "chordal thickness."

104. Trial method of obtaining data for set-over and rotation for cutting bevel gears. — In the absence of data, cut two spaces; then, take as a trial distance for set-over from one-sixth to one-eighth thickness of tooth on pitch line at large end. Rotate blank and move table until stationary cutter enters space and touches side of tooth at small end and cut through, then return and index.

Move blank same amount out of center on other side and rotate blank until cutter touches other side of tooth at small end. Cut through and measure tooth at both ends then proceed as on p. 1163.

CUTTING BEVEL GEAR IN AUTOMATIC GEAR–CUTTING
MACHINE

**105. To cut bevel gear and pinion with automatic spur- and
bevel-gear cutting machine,** Fig. 54. — The information given
in the schedule for cutting a bevel gear in a milling machine is
needed when cutting a similar bevel gear in an automatic gear-
cutting machine and is, practically, used in the same manner

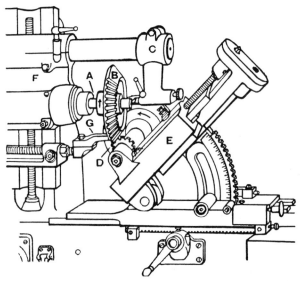

Fig. 54. — Cutting Bevel Gear. Automatic Spur- and Bevel-
Gear Cutting Machine.

except that the cutter-slide carriage on the automatic gear-
cutting machine is set to the cutting angle instead of the index head on the milling machine, and the cutter is set off center
instead of moving the gear blank. See pp. 1159–1168. This
type of machine is also used to "rough out" bevel gears that
are to be finished by planing or by generating processes. See
pp. 1171–1177.

SPECIFICATIONS FOR CUTTING BEVEL GEAR AND PINION,
Fig. 54

NAME.	GEAR.	PINION.	NAME.	GEAR.	PINION.
Material......	Machine steel	Machine steel	Depth of tooth at P.L. large end. (Addendum)...	.1000″	.1000″
W — Web.....	Web	Plain	Depth of tooth at P.L. small end. (Addendum)...	.0675″	.0675″
Style.........	Bevel	Bevel	Circular pitch, large end...	.314″	.314″
Hole..........	1″	$\frac{5}{8}$″	Chordal thickness at large end........	.157″	.157″
O. diam.......	5.074″	2.186″	Chordal thickness at small end........	.106″	.106″
Face..........	$\frac{7}{8}$″	$\frac{7}{8}$″	Corrected addendum at large end...	.100″	.103″
Hub diam.....	2″ (rough)	$1\frac{3}{8}$″	Corrected addendum at small end...	.0675″	.0695″
Backing-tip of tooth to back of hub......	$1\frac{3}{16}$″	$1\frac{1}{16}$″	Center angle..	68° 12″	21° 48′
Teeth.........	50	20	Face angle....	19° 41′	66° 5′
Diametral pitch at large end .	10	10	Cutting angle.	66° 5′	19° 41′
Approx. diametral pitch at small end.	14.8	14.8	Apex. distance at P.L......	2.6926″	2.6926″
P. diam.......	5″	2″	Cutter marked	#2–10 P. Bevel	#5–10 P. Bevel
Depth of tooth at large end..	.216″	.216″	Rev. for cutter	60	60
Depth of tooth at small end..	.146″	.146″	Feed per min..	2″	2″
Thickness of tooth at P.L., large end....	.1571″	.1571″	Keyway......	$\frac{1}{4}$″ × $\frac{1}{8}$″	$\frac{3}{16}$″ × $\frac{3}{32}$″
Thickness of tooth at P.L., small end....	.1060″	.1060″	Lubricant....	Cutting Oil or compound	

SCHEDULE OF OPERATIONS FOR BEVEL GEAR

Arrange index change gears.

Test work arbor A, Fig. 54.

Mount gear blank B.

Mount and adjust the supporting arm C.

Mount cutter D and set central with center indicator. See p. **11**21.

Set cutter slide E to cutting angle.

Lower head F approximately.

Mark depth of tooth with gage.

Arrange rim rest G.

Arrange speed and feed gears.

Arrange reversing dogs to give length of cut.

Start machine. Throw in feed and move carriage until cutter cuts to depth line (two or three cuts). Clamp carriage before each trial cut.

Adjust stop that limits carriage travel.

Throw in automatic index and feed. Cut two spaces to produce a tooth. Throw out index and feed.

Move index gear around by hand.

Measure dimensions of tooth at both ends with gear tooth vernier caliper. Make corrections by setting cutter off center first on one side then on the other, revolving gear (by moving index wheel) and taking trial cuts until tooth is of correct thickness. See pp. **11**21, **11**62.

After the amount of set-over of cutter and index wheel rotation are obtained, cut gear with two cuts. First cut all around blank with cutter set over to right of central line, then take second cut all around with cutter set to left of central line with proper index wheel rotation. See pp. **11**21, **11**62.

Attention. — Pinion is cut in the same manner as gear.

PLANING BEVEL GEARS

106. The planing process is largely used to produce bevel gears of precision. — In a correctly formed bevel-gear tooth, the curvature of the sections from end to end is not uniform; therefore, a *formed* milling cutter will not produce an accurate tooth. The planing process shapes each tooth correctly to the cone lines, thus insuring an accurate tooth-contour. The tool always travels at the correct angle of the gear from the top of the tooth to the root. The small end of the tooth is, therefore, in proportion to the large end.

107. To rough out teeth of bevel-gear blanks. — While bevel-gear generators are capable of planing finished gears from smooth blanks, they are, primarily, precision machines that take the finishing cuts to produce accurate tooth-contour; therefore, it is best to first rough out the teeth with a milling cutter in an automatic gear-cutting machine, or in a milling machine.

FIG. 55. — GENERATING BEVEL GEAR IN AUTOMATIC FORMER BEVEL-GEAR PLANER.

108. Single-tool automatic former bevel-gear planer, Fig. 55, shows the method of planing the teeth of bevel gears by the former or templet principle. The formers or templets are made in a former generating machine. The tool is reciprocated by a quick return crank motion.

To finish plane the curved sides of the teeth of bevel-gear blank A, templet B guides tool C to plane the surplus stock from the side of one tooth. The machine then automatically indexes to the next tooth, and so on until one side of each tooth is finished. The reverse templet D is then brought into position, and the opposite side of each tooth is finish planed to correct thickness.

Attention. — Straight templet E may be used to rough out the teeth.

109. Action of tool and gear blank. — The gear blank remains stationary. The templet holder F, Fig. 55, is set to the cutting angle. The templet curve is larger than the tooth curve to allow for the distance it is from the cone center. Tool C', Fig. 56, guided by its slide travels toward the apex of the cone. The tool cuts with its point only.

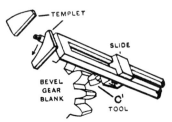

FIG. 56. — GENERATING ACTION OF SINGLE-TOOL AUTOMATIC BEVEL-GEAR FORMER PLANER.

110. The forming process is obtained by means of a cam and a slotted link motion; the head is swung vertically upon its axis, and simultaneously, the roll runs on the templet swinging the tool about its horizontal axis which reproduces the desired form on the tooth.

BEVEL-GEAR GENERATING MACHINES

111. Two-tool automatic generating bevel-gear planer, Fig. 57. — The teeth of blank A are finished with a pair of cutting tools which plane simultaneously opposite sides of the same tooth. After one tooth is finished, the machine automatically indexes to the next.

112. The action of gear blank and tools. — The tools, A, A, Fig. 58, represent a cross section of two adjacent teeth of a circular rack or crown gear.

The tools are guided by the slides so that they travel toward the apex of the gear cone. The motions of gear blank and tools, as indicated by arrows B and C, cause the tools to move "out" and "in" as the rolling motion passes the center, thereby generating the curves on both sides of a gear tooth. While one tool cuts on the forward stroke, the other cuts on the backward stroke, as at DE.

113. The generating process of the two-tool automatic generating bevel-gear planers. — Bevel-gear blank A, Fig. 57, is mounted upon a spindle which is permanently attached to a semicircular arm. To this arm a removable segment of a master gear is fastened which meshes with a permanent sector

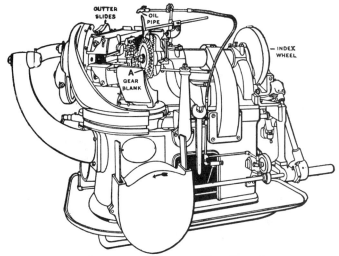

FIG. 57. — GENERATING GEAR IN TWO-TOOL AUTOMATIC GENERATING BEVEL-GEAR PLANER.

of a circular rack or crown gear; this crown gear is connected with the cutter slides and in line with them. These cutter slides can be set to any angle and size of tooth.

The rolling-generating motion of the arm and gear blank is obtained from a cam mechanism and a slotted link. The cutters have a crank-shaper motion. Automatic indexing is obtained by a worm and worm wheel and change-gear mechanism.

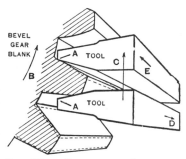

FIG. 58. — GENERATING ACTION OF TWO-TOOL AUTOMATIC GENERATING BEVEL-GEAR PLANER.

114. Automatic generating bevel-gear shaper, Fig. 59. — This machine consists of two principal parts, — shaper A which operates cutting tool C, and generating head B which controls the movements of gear blank D.   Cutting tool C represents one tooth of a circular rack or crown gear and is given a reciprocating motion similar to the cutting tool of a shaper.

115. The generating process of the single-tool automatic generating bevel-gear shaper, Fig. 59. — The rolling-generating motion of the bevel-gear blank past the cutting tool is obtained by a conical swinging motion of the arbor simultaneously with the rotation of the blank produced by two steel tapes stretched in opposite directions and fastened to a segment of a master-cone corresponding to the pitch cone of the gear blank and attached to the arbor.   The other ends of the tapes are attached to the frame work of the generating head.

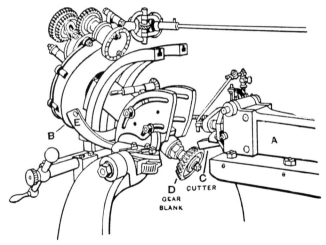

FIG. 59. — AUTOMATIC BEVEL-GEAR GENERATOR SHAPER.
SINGLE TOOL.

The combination of these movements generates the involute curve.

The feed mechanism gives a slow intermittent movement to the semicircular plate which supports the inclined arbor and produces a progressive rolling of the blank while the tool planes the side of a tooth. Automatic indexing is obtained by worm and wheel and change gears.

GENERATING BEVEL GEARS WITH MILLING CUTTERS

116. Automatic bevel-gear generating machine. Radial teeth milling cutters, Fig. 60. — Bevel gears can be cut or generated with milling cutters providing the cutters are made to follow the path of a circular rack or crown gear tooth.

In this machine two milling cutters, with staggered interlocking radial teeth, cutting simultaneously, are used.

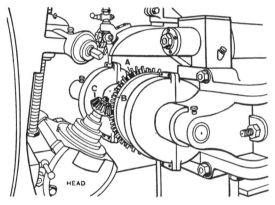

Fig. 60. — Generating Bevel Gear with Rotary Cutters.

117. The generating process of the automatic bevel-gear generating machine which cuts bevel-gear teeth with milling cutters. — The interlocking cutters A and B, Fig. 60, are mounted upon separate spindles at the desired pressure angle and also at the required angle, to form the sides of the two adjacent teeth of the bevel gear. The bevel-gear blank C is fastened to a mechanism similar to a milling machine dividing head which is mounted upon a sector and set to the cutting angle. The blank is fed to the revolving cutters to the proper depth, then rolled to the right and left past these revolving cutters to generate the tooth curves. The rolling-generating motion is obtained by gears and a crank mechanism. The cutters remain stationary.

The work spindle is provided with an indexing mechanism which operates automatically.

118. Spiral bevel gears, Fig. 61, are used for automobile drives, and machines where high speeds and extremely quiet running are desired. The engagement of the teeth is gradual. Instead of striking a full line contact at once, they roll on to each other: One pair of teeth is engaging while the preceding pair is still in contact, similar to other forms of spiral gearing.

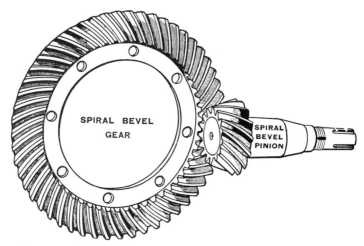

Fig. 61. — Spiral Bevel Gear and Pinion. Ratio $3\frac{8}{15}$ to 1.

119. Automatic spiral-type bevel-gear generating machine.— Fig. 62 shows the bevel-gear blank and cutter in position. The cutter is in the form of a face mill, and is set at the same radius as the gear teeth, and passes through the blank in a curved path. Spiral bevel pinion A, with teeth roughed out, is mounted upon its arbor for finishing. The rotary cutter head has 20 staggered inserted cutters B, B, B, which in cross section are similar to the teeth of a circular rack or crown gear. They are set at a radius to give the required curve to the teeth — ten for the curve of the outside of a tooth and ten for the curve of the inside of a tooth. To generate the tooth curve, both cutter head and gear blank roll together by means of a geared drive. The machine is set to cut one side of one tooth after which it automatically indexes to the next tooth and continues this process until all are cut. Then the machine is set to cut the other side of a tooth to the required thickness; and so on, until all are automatically finished in the same manner.

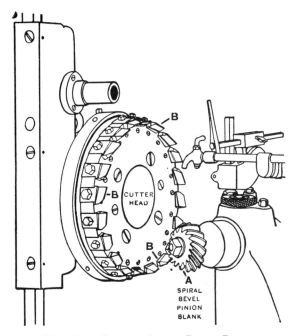

FIG. 62. — CUTTING SPIRAL BEVEL PINION.

120. To " rough out " or " stock out " spiral bevel gears, special cutters are needed. The generating motion is used in rough-cutting the pinion. In the gear, the sides of the teeth are so flat and so nearly straight that the generating motion is omitted.

121. Comparison between standard involute teeth and stub teeth is shown in Figs. 63, 64. — Involute teeth have a pressure angle of $14\frac{1}{2}°$, as in Fig. 63, in which two standard involute spur gears are shown in mesh, as at A, and a gear engaging a rack, as at B.

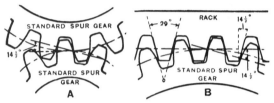

FIG. 63. — STANDARD INVOLUTE SPUR GEARS AND RACK.

Stub-tooth spur gears generally have a pressure angle of 20°, as in Fig. 64, in which two stub-tooth spur gears are shown in mesh as at A, and gear engaging rack, as at B.

These gears are largely used in automobile transmissions where strength is the main consideration. A table of stub-tooth proportions is shown on p. **117**9.

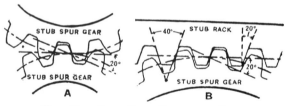

FIG. 64. — STUB SPUR GEARS AND RACK.

122. Table of Tooth Dimensions, 20° stub-tooth system. — The figures denoting the diametral pitch of the stub tooth are stated in the form of a fraction, the first figure indicating the pitch or thickness of the tooth, and the second the depth. For example, a $\frac{6}{8}$ stub tooth has the thickness of a 6 pitch tooth of $14\frac{1}{2}°$ and the depth of an 8 pitch tooth, the pressure angle generally being 20°.

DIAMETRAL PITCH.	THICKNESS OF TOOTH.	ADDENDUM.	CLEARANCE.	WHOLE DEPTH OF TOOTH.
$\frac{4}{5}$	.3927	.2000	.0500	.4500
$\frac{5}{7}$	.3142	.1429	.0357	.3214
$\frac{6}{8}$	.2618	.1250	.0312	.2812
$\frac{7}{9}$	.2244	.1111	.0278	.2500
$\frac{8}{10}$	.1963	.1000	.0250	.2250
$\frac{9}{11}$	.1745	.0909	.0227	.2045
$\frac{10}{12}$	.1571	.0833	.0208	.1875
$\frac{12}{14}$	.1309	.0714	.0179	.1607

123. A pair of stub-tooth sliding spur gears is shown in Fig. 65. —

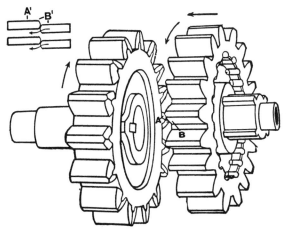

FIG. 65. — STUB-TOOTH SLIDING SPUR GEARS WITH TEETH ROUNDED AT ENDS. USED FOR AUTOMOBILE TRANSMISSION.

These gears are largely used in automobile transmissions and in some machine tools. They are made of steel and are heat treated. See *Principles of Machine Work.* The ends of the teeth, as at A and B, are rounded with a cutter in a milling machine or with a special cutter in a tooth-rounding machine. The rounded teeth readily permit an end or sliding engagement while in motion, as shown at A' and B'. The center distance is not changed as is the case with pointed gears.

124. Pointed-tooth gears, Fig. 66, are designed to engage readily with each other while in motion. They are usually made of steel and heat treated.

Gear *A* is moved toward gear *B*. The points of gear *A* readily enter the spaces, or mesh with gear *B*. Under a light load, these

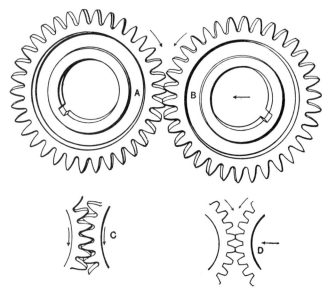

FIG. 66. — POINTED-TOOTH GEARS. EASY ENGAGEMENT. ADJUSTABLE CENTER.

gears may be run at variable center distances. They are often called adjustable spur gears. At *C* the gears are shown at full mesh, running with their pitch circles tangent to each other.

Warning. — Do not attempt to·make standard gears mesh while in motion as the back gears of an engine lathe, for the wide tops of the teeth may come together, as at *D,* which usually breaks the teeth and ruins the gears.

NOISELESS GEARS

125. Noiseless Gears. — While gearing is undoubtedly an ideal method of power transmission, the noise produced by the metal-to-metal contact of high-speed gear transmissions is a source of annoyance. This noise and vibration can be greatly

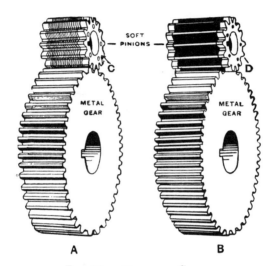

FIG. 67. — NOISELESS GEARS.

reduced by the substitution of a "noiseless pinion" made of rawhide, compressed cotton, or fiber, as at *A* and *B*, Fig. 67. The contact of metal with a soft material is nearly noiseless, or at the most produces only a low hum. These gears are made with steel or bronze flanges at the sides, as at *C* and *D*, which may or may not form part of the working surface, and often with metal centers. The flanges and soft material are drilled and riveted or bolted together. The hole is bored and reamed, the blank turned in the lathe, and the teeth cut by the same process as metal gears.

TESTING GEARS

126. To test a spur gear in the spur-gear testing machine,
Fig. 68. —

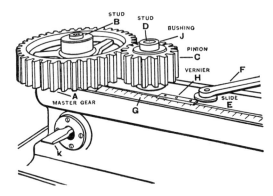

FIG. 68. — TESTING CENTER DISTANCE OF SPUR GEAR AND
PINION IN SPUR GEAR TESTING MACHINE.

This machine is used to test spur gears to determine whether
they have correct bearing and run quietly at the correct center
distance for which they were designed.

The gear to be tested is adjusted to mesh with a master-gear,
that is, a gear that is known to be accurately cut or ground.

Mount master-gear A on stud B and spur pinion C on stud D.
Rub a little Prussian blue or red lead on the teeth of the master-
gear and bring the gears together by moving slide E and clamping with handle F. Rotate master-gear and feel for eccentricity, noting the amount of bearing on the teeth. Read slide
G and vernier H, which reads to thousandths of an inch, to see
if center distance is within the required limit. If the center
distance is too great, the spaces of the pinion may be cut deeper.
A very small amount of eccentricity in the pinion may be corrected by filing the teeth as indicated by the bearing on the teeth,

see pp. **11**63, **11**64. A large amount of eccentricity shows that the arbor on which the pinion was cut was not true. Hardened and ground steel bushings J and studs are used for gears with holes of different sizes. The studs are released by handle K.

Attention. — Bevel and miter gears may be tested as in Fig. 69, and spiral and worm gears as in Fig. 70.

127. Power driven bevel gear testing machines for bearing and quietness. —

Heat treated and pack hardened bevel gears may be tested for bearing and quietness with the bevel-gear testing machine, as shown in Fig. 69. These two conditions go together in straight-tooth gears.

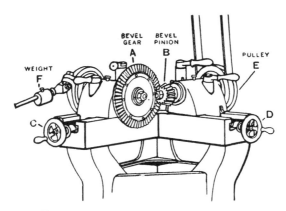

Fig. 69. — Testing Hardened Bevel Gears for Bearing and Quietness.

Mount gear A and pinion B on suitable arbors on the machine and adjust to proper positions so that the theoretical cone apices coincide. If the blanks have been turned within close limits, set the back angles flush with each other and, by eye, adjust for depth engagement using the hand wheels C and D, then lock the heads. Rub a marking of Prussian blue, or red lead, on the tooth surfaces. Apply power to pulley E on pinion spindle and adjust weight F on lever of brake on gear spindle to obtain desired load. If the pinion tooth has been topped off too much in cutting, it will have to be moved nearer to run quietly. If the pinion is flanked out too much, it will have to be moved outward. The amount of variation permissible for high-speed gears is from .002″ to .003″.

128. Lapping gears. — Gears made of suitable steel and cut accurately in suitable machines and properly heat treated, seldom need correction; but gears made of unsuitable steel, or improperly heat treated, may warp excessively in pack hardening.

Slight irregularities in a pair of hardened gears are sometimes corrected by running them together with a fine abrasive and oil.

Information. — Different types of test indicators are often fastened to the movable slide to determine eccentricity or tooth irregularities.

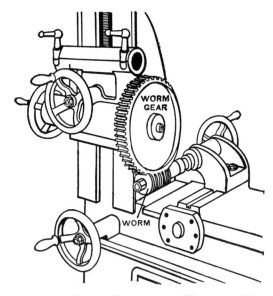

Fig. 70. — Testing Center Distance of Worm and Worm Gear.

BEVEL PROTRACTOR

129. Comparative Table of Tooth Dimensions, gears cut in a gear shaper, and milled gears, $14\frac{1}{2}°$ system.

DIAMETRAL PITCH.	THICKNESS OF TOOTH.	ADDENDUM.	CLEARANCE.		WHOLE DEPTH.	
			GEAR SHAPER GEAR.	MILLED GEAR.	GEAR SHAPER GEAR.	MILLED GEAR.
1	1.5708	1.0000	.2500	.1571	2.2500	2.1571
$1\frac{1}{2}$	1.0472	.6667	....	.1047	....	1.4381
2	.7854	.5000	....	.0785	....	1.0785
$2\frac{1}{2}$	.6283	.4000	....	.0628	....	.8628
3	.5236	.3333	....	.0524	....	.7190
4	.3927	.2500	.0625	.0393	.5625	.5393
5	.3142	.2000	.0500	.0314	.4500	.4314
6	.2618	.1667	.0417	.0262	.3750	.3595
7	.2244	.1429	.0357	.0224	.3214	.3081
8	.1963	.1250	.0312	.0196	.2812	.2696
9	.1745	.1111	.0278	.0175	.2500	.2397
10	.1571	.1000	.0250	.0157	.2250	.2157
12	.1309	.0833	.0208	.0131	.1875	.1798
14	.1122	.0714	.0179	.0112	.1607	.1541
16	.0982	.0625	.0156	.0098	.1406	.1348
18	.0873	.0555	.0139	.0087	.1250	.1198
20	.0785	.0500	.0125	.0078	.1125	.1079
22	.0714	.0455	.0114	.0071	.1023	.0980
24	.0654	.0417	.0104	.0065	.0938	.0899
26	.0604	.0385	.0096	.0060	.0865	.0829
28	.0561	.0357	.0089	.0056	.0804	.0770
30	.0524	.0333	.0083	.0052	.0750	.0719
32	.0491	.0312	.0078	.0049	.0703	.0674

Thickness of tooth......... = 1.5708 ÷ diametral pitch.
Addendum................ = 1.0000 ÷ diametral pitch.
Clearance, gear shaper gear .. = 0.2500 ÷ diametral pitch.
Clearance, milled gear....... = 0.1571 ÷ diametral pitch.
Whole depth, gear shaper gear = 2.2500 ÷ diametral pitch.
Whole depth, milled gear .... = 2.1571 ÷ diametral pitch.

130. Bevel protractor. — A, Fig. 71, is used to establish and test angles. It is graduated into 360 divisions called degrees. The degree is the unit of angular measurement, $\frac{1}{360}$ part of a circle. For calulation a degree is divided into 60 parts called minutes and a minute is sub-divided into 60 parts called seconds.

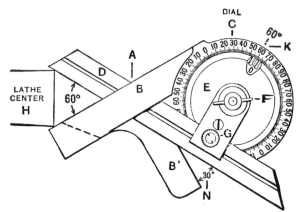

FIG. 71. — MEASURING ANGLE OF LATHE CENTER. BEVEL PRO-
TRACTOR.

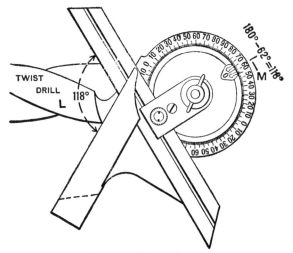

FIG. 72. — MEASURING ANGLE OF TWIST DRILL. BEVEL PROTRACTOR

The protractor consists of beam B, graduated dial C and
blade D which is connected to rotating disk E by thumb-nut
F and clamp G. Disk E carries zero point (0). Blade D is
adjustable and reversible and may be clamped at G. Dial
C is divided into four quadrants of 90°. When zero lines

on dial C and disk E coincide, the beam and blade are parallel and the angle is 180°.

131. To measure angle with bevel protractor.— For included angles less than 180° and greater than 90°, loosen thumb-nut F slightly and bring beam B and blade D in contact with work, as twist drill L, Fig. 72, and subtract reading from 180°, as at M. For angles less than 90°, as lathe center H, Fig. 71, read directly as at K.

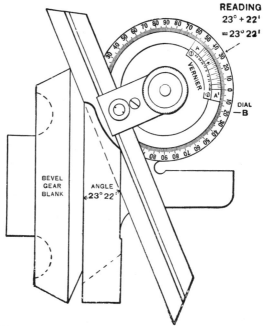

Supplementary beam B', Fig. 71, is convenient for measuring small angles, the angle being obtained by subtracting reading from 90°, as 90° − 60° = 30°, as at N, Fig. 76.

132. Vernier bevel protractor, Fig. 73, is used to measure parts of a

Fig. 73. — Measuring Angle of Bevel Gear Blank with Vernier Bevel Protractor. Reads to 5 Minutes or $\frac{1}{12}$ of Degree.

degree. A, A' is a double vernier divided into 12 equal parts each side of zero. When reading angles less than 180° and greater than 90°, vernier A is used; for angles less than 90°, vernier A' is used.

The 12 divisions on vernier occupy same space as 23° on dial B. This divides a degree into 12 parts, each part equals 60 ÷ 12 = 5 minutes. See Vernier Principle, pp. 2₁₁, 2₁₂.

133. Table of tapers per foot with corresponding angles and tapers per inch.

Taper per Foot.	Included Angle.		Angle with Center Line.		Taper per Inch.	Taper per Inch from Center Line.
	Deg.	Min.	Deg.	Min.		
1/8	0	36	0	18	.0104	.0052
3/16	0	54	0	27	.0156	.0078
1/4	1	12	0	36	.0208	.0104
5/16	1	30	0	45	.0260	.0130
3/8	1	48	0	54	.0312	.0156
7/16	2	6	1	3	.0365	.0182
1/2	2	24	1	12	.0417	.0208
9/16	2	42	1	21	.0469	.0234
5/8	3	0	1	30	.0521	.0260
11/16	3	16	1	38	.0573	.0286
3/4	3	34	1	47	.0625	.0312
13/16	3	52	1	56	.0677	.0338
7/8	4	10	2	5	.0729	.0365
15/16	4	28	2	14	.0781	.0391
1	4	46	2	23	.0830	.0417
1¼	5	58	2	59	.1042	.0521
1½	7	10	3	35	.1250	.0625
1¾	8	20	4	10	.1458	.0729
2	9	32	4	46	.1666	.0833
2½	11	52	5	56	.2083	.1042
3	14	16	7	8	.2500	.1250
3½	16	36	8	18	.2916	.1458
4	18	56	9	28	.3333	.1667
4½	21	14	10	37	.3750	.1875
5	23	32	11	46	.4166	.2083

134. To find angles for given taper per foot. —

Rule: Divide one-half the taper in inches per foot by 12 for the tangent of the angle with the center line. The included angle is twice the angle with the center line.

Example. — Given 6″ taper per foot. To find angle.

Solution. — $\dfrac{6}{2 \times 12}$ = .25000 tangent of angle with center line.

The angle whose tangent is .25000 (see Table of Tangents in any Engineers' Handbook) equals 14° 2′. Included angle = 14° 2′ × 2 = 28° 4′.

135. To find taper per inch. — Divide taper per foot by 12 for included angle and by 24 for angle with center line.

COMPOUND REST

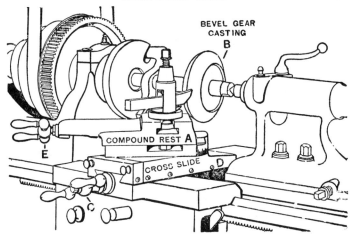

FIG. 74. — TURNING AN ANGLE WITH COMPOUND SLIDE REST.

136. Compound rest, Fig. 74, is used to turn or bore steep tapers or large angles. It may be a permanent part of cross slide, as in Fig. 74, or an attachment to interchange with tool block. When the zero lines coincide, as at F, Fig. 75, it

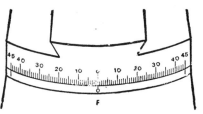

FIG. 75. — COMPOUND SLIDE REST GRADUATION.

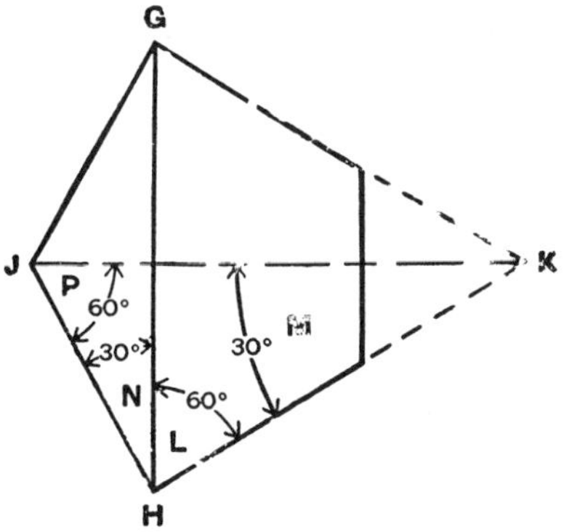

is in the same plane as regular cross slide. To set at other angles, find whether angle is measured from perpendicular to axis, as line GH, Fig. 76, or parallel to axis, as line JK. If from perpendicular, as angle L, set rest at 60°. If parallel, as angle M, set at 90° − 30° = 60°. If 30°, as at N, set direct. If 60°, as at P, set at complemen-

FIG. 76. — DIAGRAM FOR READING ANGLES.

tary angle 30°. If angle or bevel on drawing is not given in degrees, measure with protractor or transfer angle with universal bevel. Hold head of bevel against face plate of lathe and adjust slide of compound rest to coincide with blade of bevel.

137. To set compound rest to turn taper or bore taper hole when taper is given in inches per foot. —

First Method. — Consult table of tapers and angles and find angle corresponding with desired taper.

Example. — To bore short taper hole $\frac{5}{8}''$ taper per foot.

Solution. — Angle with center line for taper $\frac{5}{8}''$ per foot = 1° 30'. Complement of angle is 90° − 1° 30' = 88° 30' which is setting for this taper.

Second Method. — In absence of table set rest approximately, and take trial cuts until exact taper is obtained.

138. To make bevel gear blank 8 pitch, 32 teeth. — For specifications, see Figs. 45 and 74, and p. 1165. True casting in chuck, drill, ream, press mandrel into hole, mount on centers, square and turn to size. Set compound rest *A*, Fig. 74, for angle of face *B* which is 23° 22' from line perpendicular to axis, therefore, set compound rest *A* at 23° 22' directly. Move carriage under blank and with handle *C* move cross slide *D* to approximate depth of cut and clamp carriage. Take one or two cuts operating compound rest by handle *E*, then test bevel with bevel protractor, as in Fig. 73, and if necessary, adjust compound rest and repeat cuts. Set compound rest to 63° 26' and turn outside bevel on back of gear and with same setting turn inside bevel to make face *B* $\frac{11}{16}''$ wide. Set compound rest at zero. Filing and polishing may be omitted.

139. Fine tool adjustments with compound rest. — While a finer tool adjustment is obtained with the compound rest than with the cross slide due to its finer actuating screw, a still finer adjustment may be obtained by setting the compound rest at an angle. For example, we assume that the compound rest is set at 30°. If the tool is moved .001'', according to the micrometer dial, the actual distance the tool moves into the work is in proportion to the cosine of the angle or .001 × cos 30° = .001 × 87 = .00087''. At 45° the depth is .00071'', and at 60° the depth is .0005''.

ADVANCED MACHINE WORK

SECTION 12

TOOL MAKING

Introduction. Making Mandrel. Testing Lathe Work With Indicators.
Thread Micrometer Calipers. Making Taps. Making Plain Milling
Cutter. Making Hand Reamer. Lapping. Making Plug Gage.
Measuring Machine. Spiral Milling. Cutting Teeth In Spiral
Mill. Grooving Twist Drill. Making Twist˙ Drill.
Precision Methods of Locating Holes For Jigs and
Accurate Machine Parts — Plug Method, Button
Method. Sine Bar.

INTRODUCTION

1. Tool making is the fine art of machine building and
manufacturing. It consists of designing, making and repairing small tools, such as taps, dies, reamers, twist drills, mandrels, arbors, counterbores, milling cutters and gages.

Standard small tools are obtainable, but many others, special
in diameter, length or shape including punches and dies, jigs
and fixtures, needed in machine building, manufacturing and
experimental work, are not obtainable and must be *made* to
meet the requirements.

The processes of making tools and making machine parts are
similar, but greater knowledge and accuracy are generally required in making tools than in making machine parts.

While schedules of making a number of different types of tools
are given in this section, a large number of tool-making processes may be found in other sections of both books (*Principles
of Machine Work* and *Advanced Machine Work*), such as Hardening and Tempering, Accurate Filing and Scraping, Drilling,
Internal, Surface, and Cutter Grinding, Planing, Grooving and
Fluting Taps and Reamers, and Cutting Teeth In Milling
Cutters, Heat Treatment of Steel and Autogenous Cutting.

2. Selecting sizes of steel for tools.—As the surface of carbon-steel bar stock is decarbonized to some extent, and will not
harden properly, select annealed carbon steel $\frac{1}{16}''$ to $\frac{1}{8}''$ large for
tools that are to be hardened, and high-speed steel $\frac{1}{16}''$ to $\frac{1}{8}''$ large.

MAKING MANDREL

3. To prepare $\frac{9}{16}''$ standard mandrel blank for hardening, tempering, and grinding, Fig. 1.

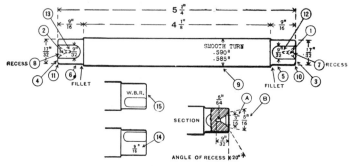

Fig. 1. — Schedule Drawing.

Specifications: Material, annealed carbon steel $\frac{1}{8}''$ large; weight, 8 oz. Hardness, 22 to 23 (scleroscope).

For heat treatment see *Principles of Machine Work.*

True live center. Set dead center in approximate alinement.

Machine dry or use lard oil.

High-speed steel or stellite cutting tools.

Time: Study drawing and schedule in advance, 5 min. — Oil lathes, 6 min. — Prepare blank, 55 min. — Harden, 10 min. — Polish and color temper, 25 min. (or oil temper, 5 min.) — Lap center holes, 8 min. — Clean lathes, 5 min. — Total, 1 h. 54 min.

SCHEDULE OF OPERATIONS, MACHINES AND TOOLS

Operations.	Machines, Speeds, Feeds.	Tools.
Center to $\frac{3}{16}''$, (1), (2). See (A), Fig. 1.	Centering machine. Drill 1000 R.P.M., countersink 400 R.P.M.	$\frac{5}{64}''$ drill, or combination drill and countersink, 60°. Lard oil.
Rough square to $5\frac{1}{4}'' + \frac{1}{64}''$, (3), (4).	Engine lathe, 12″ to 16″, 2d or 3d speed, or 35 F.P.M. Hand feed.	Dog, side tool, or holder and cutter, 30° rake, calipers, rule.
Recenter to $\frac{3}{16}''$, (1), (2).	Speed lathe, 2d or 3d speed.	Drill, countersink, lard oil.
Finish square to length, (3), (4).	Engine lathe, 3d or 4th speed, or 50 F.P.M.	Dog, side tool or holder and cutter, 30° rake, calipers, rule.

Rough turn reduced portions to $\frac{17}{32}'' + \frac{1}{64}''$ (5), (6), one or two cuts. Do not square shoulders but leave fillets as shown to avoid cracking in hardening.	2d or 3d speed, or 30 F.P.M. Medium power feed — 80 to 1″.	Copper under set screw of dog, diamond-point tool, or holder and cutter, 30° rake, calipers, rule.
Recess countersinks to $\frac{5}{16}''$. Angle 20°, (7), (8). See (B), Fig. 1.	3d or 4th speed, or 50 F.P.M. Speed lathe, 2d or 3d speed.	Side tool or drill and special countersink.
Recenter to $\frac{3}{16}''$, (1), (2).		Drill, countersink.
Finish turn reduced portions, (5), (6), one cut.	Engine lathe, 3d or 4th speed, or 50 F.P.M. Fine power feed — 140 to 1″.	Copper under set screw of dog, diamond-point tool, or holder and cutter, 30° rake, calipers, rule.
Smooth turn body Turn half way, reverse, and turn other half, (9), one or two cuts.	2d or 3d speed, or 30 F.P.M. Fine power feed — 140 to 1″.	Micrometer, copper under set screw of dog, diamond-point tool, or holder and cutter, 30° rake, calipers, rule.
Round corners, (10), (11).	Speed lathe, 3d or 4th speed, or 200 F.P.M.	Graver or file.
File reduced portions.	3d or 4th speed, or 175 F.P.M.	6″ or 8″ mill bastard file.
Polish reduced portion, also ends and recesses except countersinks, (5), (6). See pp. 432–434.	Speed lathe, highest speed.	Lard oil, 90 and 120 emery cloth, polishing stick.
Mill or file and polish flats, (12), (13).	Milling machine, 3d speed, medium power feed.	*To mill*, use milling machine vise, parallel piece, 1″ end mill, rule, lead hammer, oil. *To file*, use 8″ or 10″ hand smooth and 5″ half-round, 2d cut files. *To polish*, use oil, 90 and 120 emery cloth.
Stamp size $\frac{9}{16}''$ (14).	Vise.	$\frac{1}{16}''$ steel figures, $\frac{1}{4}''$ chisel, hammer, copper jaws.
Stamp name or initials (15). (File off burr.)	. . . .	Steel name stamp.

Harden in water (test with file). See p. **12**17 and *Principles of Machine Work*.	Gas furnace or forge, 1325° F. to 1350° F.	Tongs, water, dead-smooth file.
Temper in oil, or	Oil, tempering furnace, 425° F. to 435° F.	
Polish reduced portions and ends and temper to dark straw.	Gas furnace or forge.	Red-hot iron rings, tongs.
To lap center holes, see p. **12**28.		
To grind, see p. **12**06.		

Important. — The hardness of both hardened and tempered work should be tested with a fine file; or, preferably, given a scleroscope or Brinell hardness test. See *Principles of Machine Work.*

Attention. — The hand tools may be carbon steel.

Note. — If not equipped with cylindrical grinding machine, harden and temper ends only, turn body slightly taper and file to fit standard hole.

4. Table of standard mandrel dimensions.

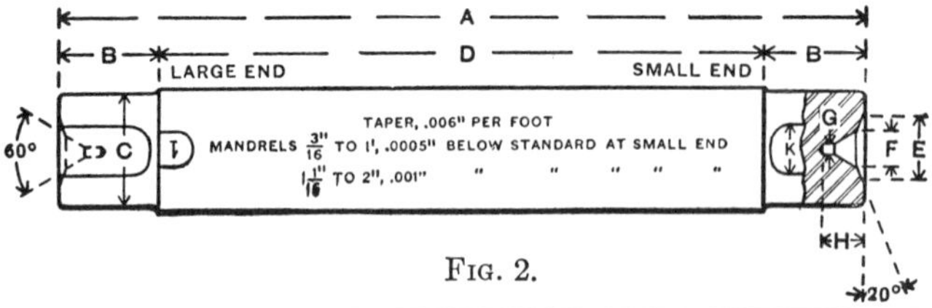
A
B
D
B
LARGE END
SMALL END
TAPER, .006" PER FOOT
MANDRELS 3/16" TO 1", .0005" BELOW STANDARD AT SMALL END
1 1/16" TO 2", .001" " " " " "
60°
C
K
G
F
E
H
20°

Fig. 2.

Nominal Diam.	Total Length.	Length of Ends.	Diam. of Ends.	Length of Taper.	Diam. of Recess.	Diam. of Countersinks.	Drill Size Fraction and Number.		Depth of Drilled Hole.	Width of Flat.
	A	B	C	D	E	F	G		H	K
$\frac{3}{16}$	$3\frac{1}{4}$	$\frac{5}{16}$	$\frac{5}{32}$	$2\frac{5}{8}$	......	$\frac{5}{64}$	......	56	$\frac{5}{32}$	$\frac{1}{8}$
$\frac{1}{4}$	$3\frac{3}{4}$	$\frac{7}{16}$	$\frac{7}{32}$	$2\frac{7}{8}$	$\frac{11}{64}$	$\frac{1}{8}$	$\frac{1}{16}$	52	$\frac{7}{32}$	$\frac{5}{32}$
$\frac{5}{16}$	4	$\frac{7}{16}$	$\frac{9}{32}$	$3\frac{1}{8}$	$\frac{7}{32}$	$\frac{1}{8}$	$\frac{1}{16}$	52	$\frac{7}{32}$	$\frac{3}{16}$
$\frac{3}{8}$	$4\frac{1}{4}$	$\frac{1}{2}$	$\frac{11}{32}$	$3\frac{1}{4}$	$\frac{1}{4}$	$\frac{5}{32}$	$\frac{1}{16}$	52	$\frac{1}{4}$	$\frac{7}{32}$
$\frac{7}{16}$	$4\frac{1}{2}$	$\frac{1}{2}$	$\frac{13}{32}$	$3\frac{1}{2}$	$\frac{1}{4}$	$\frac{5}{32}$	$\frac{1}{16}$	52	$\frac{1}{4}$	$\frac{1}{4}$
$\frac{1}{2}$	5	$\frac{9}{16}$	$\frac{15}{32}$	$3\frac{7}{8}$	$\frac{5}{16}$	$\frac{3}{16}$	$\frac{5}{64}$	47	$\frac{9}{32}$	$\frac{1}{4}$
$\frac{9}{16}$	$5\frac{1}{4}$	$\frac{9}{16}$	$\frac{17}{32}$	$4\frac{1}{8}$	$\frac{5}{16}$	$\frac{3}{16}$	$\frac{5}{64}$	47	$\frac{9}{32}$	$\frac{9}{32}$
$\frac{5}{8}$	$5\frac{1}{2}$	$\frac{5}{8}$	$\frac{19}{32}$	$4\frac{1}{8}$	$\frac{13}{32}$	$\frac{7}{32}$	$\frac{5}{64}$	47	$\frac{5}{16}$	$\frac{9}{32}$
$\frac{11}{16}$	$5\frac{3}{4}$	$\frac{5}{8}$	$\frac{21}{32}$	$4\frac{1}{2}$	$\frac{3}{8}$	$\frac{7}{32}$	$\frac{5}{64}$	47	$\frac{5}{16}$	$\frac{5}{16}$
$\frac{3}{4}$	6	$\frac{3}{4}$	$\frac{23}{32}$	$4\frac{1}{2}$	$\frac{7}{16}$	$\frac{1}{4}$	$\frac{3}{32}$	42	$\frac{3}{8}$	$\frac{5}{16}$
$\frac{13}{16}$	$6\frac{1}{4}$	$\frac{3}{4}$	$\frac{25}{32}$	$4\frac{3}{4}$	$\frac{15}{32}$	$\frac{1}{4}$	$\frac{3}{32}$	42	$\frac{3}{8}$	$\frac{5}{16}$
$\frac{7}{8}$	$6\frac{1}{2}$	$\frac{13}{16}$	$\frac{29}{32}$	$4\frac{7}{8}$	$\frac{17}{32}$	$\frac{9}{32}$	$\frac{3}{32}$	42	$\frac{13}{32}$	$\frac{3}{8}$

Nominal Diam.	Total Length.	Length of Ends.	Diam. of Ends.	Length of Taper.	Diam. of Recess.	Diam. of Countersinks	Drill Size Fraction and Number.		Depth of Drilled Hole.	Width of Flat.
	A	B	C	D	E	F	G		H	K
$\frac{15}{16}$	$6\frac{3}{4}$	$\frac{7}{32}$	$\frac{29}{32}$	5	$\frac{17}{32}$	$\frac{9}{32}$	$\frac{3}{32}$	42	$\frac{13}{32}$	$\frac{3}{8}$
1	7	$\frac{7}{8}$	$\frac{31}{32}$	$5\frac{1}{4}$	$\frac{9}{16}$	$\frac{5}{16}$	$\frac{3}{32}$	42	$\frac{7}{16}$	$\frac{3}{8}$
$1\frac{1}{16}$	$7\frac{1}{4}$	$\frac{7}{8}$	1	$5\frac{1}{2}$	$\frac{9}{16}$	$\frac{5}{16}$	$\frac{3}{32}$	42	$\frac{7}{16}$	$\frac{7}{16}$
$1\frac{1}{8}$	$7\frac{1}{2}$	1	$1\frac{1}{16}$	$5\frac{1}{2}$	$\frac{9}{16}$	$\frac{5}{16}$	$\frac{7}{64}$	38	$\frac{7}{16}$	$\frac{7}{16}$
$1\frac{3}{16}$	$7\frac{3}{4}$	1	$1\frac{1}{8}$	$5\frac{3}{4}$	$\frac{5}{8}$	$\frac{11}{32}$	$\frac{7}{64}$	38	$\frac{15}{32}$	$\frac{7}{16}$
$1\frac{1}{4}$	8	1	$1\frac{3}{16}$	6	$\frac{5}{8}$	$\frac{11}{32}$	$\frac{7}{64}$	38	$\frac{15}{32}$	$\frac{1}{2}$
$1\frac{5}{16}$	$8\frac{1}{4}$	1	$1\frac{1}{4}$	$6\frac{1}{4}$	$\frac{11}{16}$	$\frac{3}{8}$	$\frac{7}{64}$	38	$\frac{1}{2}$	$\frac{1}{2}$
$1\frac{3}{8}$	$8\frac{1}{2}$	$1\frac{1}{8}$	$1\frac{5}{16}$	$6\frac{1}{4}$	$\frac{11}{16}$	$\frac{3}{8}$	$\frac{7}{64}$	38	$\frac{1}{2}$	$\frac{1}{2}$
$1\frac{7}{16}$	$8\frac{3}{4}$	$1\frac{1}{8}$	$1\frac{3}{8}$	$6\frac{1}{2}$	$\frac{3}{4}$	$\frac{3}{8}$	$\frac{7}{64}$	38	$\frac{1}{2}$	$\frac{9}{16}$
$1\frac{1}{2}$	9	$1\frac{1}{8}$	$1\frac{7}{16}$	$6\frac{3}{4}$	$\frac{3}{4}$	$\frac{3}{8}$	$\frac{1}{8}$	28	$\frac{1}{2}$	$\frac{9}{16}$
$1\frac{9}{16}$	$9\frac{1}{4}$	$1\frac{1}{8}$	$1\frac{1}{2}$	7	$\frac{13}{16}$	$\frac{13}{32}$	$\frac{1}{8}$	28	$\frac{17}{32}$	$\frac{9}{16}$
$1\frac{5}{8}$	$9\frac{1}{2}$	$1\frac{3}{16}$	$1\frac{9}{16}$	$6\frac{3}{8}$	$\frac{13}{16}$	$\frac{13}{32}$	$\frac{1}{8}$	28	$\frac{17}{32}$	$\frac{5}{8}$
$1\frac{11}{16}$	$9\frac{3}{4}$	$1\frac{3}{16}$	$1\frac{5}{8}$	$6\frac{5}{8}$	$\frac{7}{8}$	$\frac{13}{32}$	$\frac{1}{8}$	28	$\frac{17}{32}$	$\frac{5}{8}$
$1\frac{3}{4}$	10	$1\frac{3}{16}$	$1\frac{11}{16}$	$6\frac{7}{8}$	$\frac{7}{8}$	$\frac{13}{32}$	$\frac{1}{8}$	28	$\frac{9}{16}$	$\frac{5}{8}$
$1\frac{13}{16}$	$10\frac{1}{4}$	$1\frac{1}{4}$	$1\frac{3}{4}$	$7\frac{1}{4}$	$\frac{7}{8}$	$\frac{7}{16}$	$\frac{1}{8}$	28	$\frac{19}{32}$	$\frac{11}{16}$
$1\frac{7}{8}$	$10\frac{1}{2}$	$1\frac{1}{4}$	$1\frac{13}{16}$	8	$\frac{15}{16}$	$\frac{7}{16}$	$\frac{1}{8}$	28	$\frac{19}{32}$	$\frac{11}{16}$
$1\frac{15}{16}$	$10\frac{3}{4}$	$1\frac{1}{4}$	$1\frac{7}{8}$	$8\frac{1}{4}$	$\frac{15}{16}$	$\frac{7}{16}$	$\frac{1}{8}$	28	$\frac{19}{32}$	$\frac{11}{16}$
2	11	$1\frac{1}{4}$	$1\frac{15}{16}$	$8\frac{1}{2}$	$\frac{15}{16}$	$\frac{7}{16}$	$\frac{1}{8}$	28	$\frac{19}{32}$	$\frac{3}{4}$

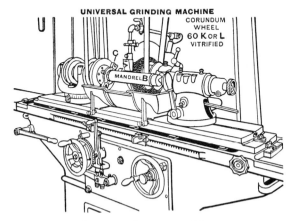

UNIVERSAL GRINDING MACHINE

CORUNDUM WHEEL
60 K or L
VITRIFIED

MANDREL B

FIG. 3. — GRINDING A MANDREL.

5. To grind $\frac{9}{16}''$ **standard mandrel, taper .006" to 1', on two dead centers, Figs. 3, 4.**

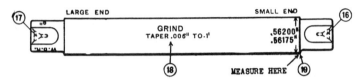

FIG. 4. — SCHEDULE DRAWING.

Specifications: Mandrel blank, carbon steel smooth turned $\frac{.590''}{.585''}$ diameter. Stamped, hardened and tempered. Center holes (**16**) (**17**), lapped.

Machines and tools: Universal grinding machine.

Grinding wheel 12" × ½", No. 60, Grade K or L, vitrified, grinding dog. Speed of wheel, 5000 F.P.M. Speed of work, 50 F.P.M.

Feed ⅓ width of wheel per revolution.

Oil bearings and two dead centers with machine oil.

Lubricant, solution of sal soda and water.

Time: Study drawing and schedule in advance, 5 min. — Oil machine and true wheel, 7 min. — Grind mandrel with machine "set up," 20 min. — Clean grinder, 3 min. — Total, 35 min. ("Set up" machine to grind correct taper, 20 min. extra.)

SCHEDULE OF OPERATIONS

I. Preparatory Adjustments. True wheel (see p. 709) and centers (see p. 710). Set zero lines on headstock and swivel table to grind desired taper approximately. See p. 712. Arrange water guards. See Fig. 10, p. 715. Move grinding wheel back with hand cross-feed wheel 8, to allow space to mount work. Mount mandrel on centers with grinding dog on stamped end and without copper under screw. Start grinding wheel, work and feed, 1, 2, 3, and adjust table dogs 6, 7, to obtain length of stroke and avoid wheel striking dog or footstock. Set automatic cross feed 11, 12, to feed .0005" to .001" (2 to 4 teeth on ratchet) at each end of stroke.

II. Grind Trial Diameter (19), Fig. 4. (Grind out tool marks only.) Move revolving wheel to lightly cut revolving work with cross-feed hand wheel 8. Turn on water and throw in feed pawl 9 to mesh with ratchet wheel 10, and allow automatic cross feed to take three or four trial cuts whole length of work to grind out tool marks. Then throw out pawl 9 and, without moving cross feed allow wheel to pass over work several times until cutting nearly dies out.

Stop work, feed and wheel, 2, 3, 1, **with wheel at footstock end.**

III. Correct Taper (18), Fig. 4. Measure work at two places 4″ apart with 1″ micrometer, and if taper is not .002″ in 4″, move swivel table (see p. **712**). Take one or two cuts with automatic cross feed. Measure, and repeat until taper is correct. If there is danger thereby of grinding mandrel too small, use trial piece same length.

IV. Find Amount Oversize. Measure small end, subtract .562″ + .002″ for finishing from reading of micrometer.

For example, reading of micrometer may be .574″, then .574″ − .564″ = .010″, or 40 teeth for rough grinding.

V. Set Automatic Cross Feed. Raise perpendicular latch 13, in head 14, throw in pawl 9 and without moving cross feed, move shield 15 to right or left until end of shield and pawl are 40 teeth apart, then drop latch 13 in ratchet wheel 10.

VI. Rough Grind (19), .002″ Large, as follows: Start wheel, work and feed, 1, 2, 3, and rough grind automatically until shield 15 lifts pawl 9. Then lock latch head 14 with horizontal latch 16. Throw out pawl 9 and allow wheel to pass over work several times until cutting nearly dies out. Stop work with wheel at footstock end and measure. From reading subtract .562″. (.564″ − .562″ = .002″.)

VII. Finish Grind (19), Fig. 4, with pinch feed ($\frac{1}{4}$/1000″ in diameter of work), as follows:

Start work. Take two cuts $\frac{1}{2}$/1000″ each (two pinches at each end of work). Stop work and measure (.5635″). Take one cut $\frac{1}{4}$/1000″ (one pinch at footstock end of work) and measure, and repeat until work measures .562″.

Limit .56200″ to .56175″.

Clean machine with waste.

Attention. — Machine steel is often used for making mandrels. To prevent the center holes wearing excessively, the ends are case-hardened after which the mandrel may be ground or turned and filed to size. Large mandrels are frequently made without hardening. To avoid excessive wear at the center holes, the ends of the mandrel may be drilled to receive hardened plugs which contain the center holes.

6. To make $\frac{1}{2}'' \times 13$ U. S. S. nut mandrel. See Fig. 5.

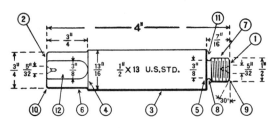

FIG. 5. — SCHEDULE DRAWING.

Specifications: Material, machine steel $\frac{1}{16}''$ to $\frac{1}{8}''$ large; weight, 12 oz.

Hardness, 15 to 18 (scleroscope).

High-speed steel, or stellite cutting tools.

Time: Study drawing and schedule in advance, 5 min. — Oil lathe, 4 min. — Make mandrel blank, 1 h. 13 min. — Case-harden and polish, 10 min. — Clean machine, 3 min. — Total, 1 h. 35 min.

SCHEDULE OF OPERATIONS, MACHINES AND TOOLS

OPERATIONS.	MACHINES, SPEEDS, FEEDS.	TOOLS.
Center. See that live center is nearly true and dead center in approximate alinement.	Engine lathe, 12'' to 16''.	
Rough square $\frac{1}{64}''$ long, (**1**), (**2**).	3d or 4th speed, or 50 F.P.M. Hand feed.	Dog, holder and cutter 35° rake, calipers, rule.
Recenter. **Finish square** (**2**), (**1**).		
Rough turn $\frac{1}{64}''$ large, (**3**), one cut. Turn half way, reverse and turn other half.	2d or 3d speed, or 50 F.P.M. Medium power feed — 80 to 1''.	Holder and cutter 35° rake, copper, calipers, rule.

Finish turn (3), one cut. Turn half way, reverse and turn other half.	3d or 4th speed, or 70 F.P.M. Fine power feed — 140 to 1″.	
Draw lines (4), (5), for length of reduced parts (6), (7).	Vise, copper jaws.	Copper sulphate, rule, scriber.
Rough and finish turn (6), one or two cuts. Do not square shoulder, leave fillet.	Engine lathe, 3d or 4th speed, or 70 F.P.M. Fine power feed — 140 to 1″.	Copper, holder and cutter 35° rake, calipers, rule.
Cut groove (8), to $\frac{3}{8}$″ diameter and within $\frac{1}{64}$″ of line (5).	1st or 2d speed, or 30 F.P.M. Hand feed.	$\frac{3}{32}$″ round end grooving-tool, calipers, rule, oil.
Rough and finish turn (7), terminating in groove (8), three cuts.	3d or 4th speed, or 70 F.P.M. Fine power feed — 140 to 1″.	
Chamfer end to 30°, (9).	3d speed, or 70 F.P.M.	Center gage.
File (3), (6), and round corner (10).	4th speed, or 175 F.P.M.	8″ or 10″ mill-bastard file, file card.
Thread (7), to fit $\frac{1}{2}$″ × 13 U. S. S. nut.	Arrange for 13 threads. 1st or 2d speed, or 20 F.P.M.	13 pitch U. S. S. thread tool, center gage, thread calipers, rule, oil.
Rechamfer end (9).		
Finish square shoulder (11).	3d speed, or 70 F.P.M. Hand feed.	Holder and cutter 35° rake.
Mill flat (12), or	Milling machine. 3d speed. Medium power feed.	1″ end mill, milling machine centers or vise, oil.
File flat, (12).	Vise, copper jaws.	8″ or 10″ hand-smooth and 5″ half-round 2d-cut files, file card.
Stamp size $\frac{1}{2}$″ × 13 U. S. S. and on opposite side stamp name.	Vise, copper jaws.	$\frac{1}{16}$″ and $\frac{1}{8}$″ steel figures and letters, hammer.
File (3), to remove burr from stamping.	Hold mandrel in hand or vise.	8″ or 10″ hand-smooth file, file card.
Case-harden (6), (7) and clean. (Polish (3) optional.)	Gas furnace or forge, 1325° F. to 1350° F.	Tongs, cyanide of potassium or prussiate of potash, water, hand-smooth file, waste.

Exception. — Finish turn and file (3) may be omitted and ground after case-hardening.

TESTING LATHE WORK WITH INDICATORS

7. Test Indicators, as in Figs. 6, 7, are used to determine the degree of accuracy of machine parts by enlarging the error by a multiplying mechanism so that $\frac{1}{1000}''$ will register about $\frac{1}{16}''$ on scale or dial which is easily read and fractions thereof readily estimated.

They are used to test the truth and alinement of machine parts on either interior or exterior work; as, the truth of live centers, the alinement of dead centers to turn straight or taper, the truth of mandrels and arbors, to set finished work in chuck by hole or side, to set jig work on face plate of lathe, to test eccentricity of shaft in straightening, to set cathead on shaft to turn spot for steady rest, to test taper hole in lathe or other spindles, to aline lathe heads in machine building, to aline cross rail of planer, to aline work on planer table, to test the accuracy of table with drill spindle, to test the truth of milling machine arbor, to aline angle plates and vises on milling and planing machines, to set center punch mark on work true to axis of rotation on face plate of lathe.

Attention. — In alining and setting machine tool fixtures for very accurate work, such as fine tool and jig making, the vernier height gage and the universal sine bar are used. See pp. **12**49, **12**63–**12**65.

8. Lathe test indicators. — Fig. 6 shows an indicator testing the truth of a mandrel mounted on lathe centers. Holder A is held in tool post B, the cross slide is moved inward until feeler or contact point C touches mandrel D as it is being revolved by hand and the pointer indicates on the scale E the amount, if any, in thousandths of an inch or fraction thereof, that the mandrel is out of true. Feeler C is for testing mandrels, centers, etc., and is removable.

Feeler F is for testing interior or exterior work and registers either a horizontal or perpendicular movement. Feeler G is broad faced and is used for testing the accuracy of a shaft when filing. Feeler H is for small work, either internal or external, and for narrow spaces.

See Indicating Gages, p. **13**29.

Important. — When setting the indicator, it is best to move feeler *C* against the work until the pointer on the scale is at zero (0).

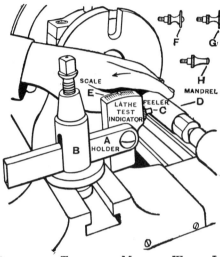

Fig. 6. — Testing the Truth of a Mandrel With a Lathe Test Indicator.

Fig. 7 shows the method of setting a center punch mark on work true to the axis of rotation. The work is clamped lightly to the face plate in an approximate position. Spring plunger *A* is inserted in center punch mark *B* and mounted on dead center *C*. The work *D* is revolved by hand and the truth of the plunger is tested with the indicator. The work is adjusted by rapping until plunger *A* is motionless when the work is revolved.

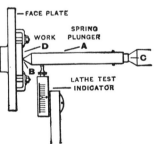

Fig. 7. — Setting a Center Punch Mark True to Axis of Rotation with Lathe Indicator.

9. Dial test indicator. — Fig. 8 shows a dial test indicator. To enlarge the hole in gear *A* which is mounted in chuck *B*, the indicator is used to test the truth of hole preparatory to boring and reaming. Holder C is held in tool post D. The long feed is moved and the rise and fall rest adjusted until feeler of hole attachment E touches wall of hole in gear A. The gear is revolved by hand and the pointer on dial F indicates the axial truth of gear. The gear is adjusted in chuck and wall of hole tested until pointer on dial F is motionless or within a reasonable limit, as $\frac{1}{1000}$ of an inch or a fraction thereof. The face of the gear is also tested by using indicator without attachment E. The dial is divided into 100 spaces of one-thousandth of an inch each.

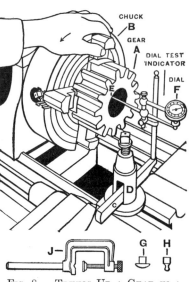

FIG. 8. — TRUING UP A GEAR IN A CHUCK WITH A DIAL TEST INDICATOR.

Different feelers, as at G and H, are for various classes of work. Clamp J is to fasten indicator to lathe and planer tools and milling machine arbors.

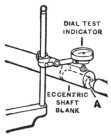

FIG. 9. — COMPARING THE THROW OF BOTH ENDS OF AN ECCENTRIC SHAFT BLANK WITH A DIAL TEST INDICATOR.

Fig. 9 shows a dial test indicator comparing the throw of an eccentric shaft. The diameter of shaft must be the same at both ends. One end is tested, then shaft is reversed and second end tested. Slight corrections are made by scraping eccentric center holes, one of which is shown at A, with center scraper. See p. 631.

Attention. — To test side of gear, remove hole attachment E, Fig. 8, set dial vertical and use indicator direct.

10. The lathe axis indicator, shown at *A* in Fig. 10, is used to test the axial truth of a center punch mark on work held in a chuck or bolted to a face plate such as the engine crank shaft center fixture *B*, which is laid out and marked at *C* and *D* to be drilled, bored, and reamed exactly at these points.

The fixture is bolted to the face plate with mark *C* approximately central and counterbalance *E* opposite. The point of indicating needle *F*, which is pivoted in the universal joint *G*, is placed in center mark *C*.

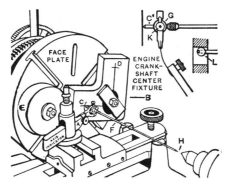

FIG. 10.—TESTING AXIAL TRUTH OF CENTER PUNCH MARK WITH INDICATOR.

As the lathe is rotated by hand, the needle point at *H* will revolve in a circle, exaggerating the error at *C*. The fixture is moved by tapping until needle point *H* remains stationary, when the lathe is rotated.

This indicator may be used to locate holes as well as center marks by first tightening nut *K*, which converts the universal into a single joint with a vertical movement, and then placing spherical head *L* upon point *C*, and using it against the upper wall of the hole. In this way the fixture may be used also to test the truth of a live center or a shaft turning on centers.

Attention. — Paper is sometimes placed between the face plate and smooth work to prevent the work slipping.

THREAD MICROMETER CALIPERS

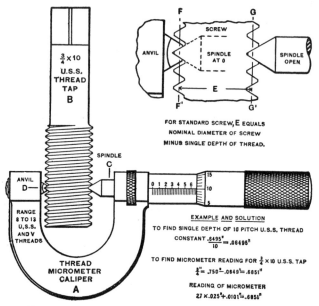

FIG. 11. — MEASURING WITH THREAD MICROMETER.

11. Thread micrometer caliper, Fig. 11, is used to measure U. S. S. and V thread screws, taps, thread gages, etc., as tap *B*. Spindle *C* and anvil *D* are relieved at points so that contact is only on sides of thread and measures diameter of screw thread independently of outside diameter of thread.

12. Principle of thread micrometer. — When spindle is in contact with anvil, reading is zero; when opened and applied to screw, it measures pitch diameter of screw thread *E* as shown by lines *FF'* and *GG'*. For a standard screw, reading for *E* is nominal diameter of screw minus single depth of thread.

13. To find depth of thread and reading for micrometer. — Divide constant by number of threads to 1″ and subtract quotient from nominal diameter of tap. Remainder is reading for micrometer.

See Thread Measuring Wires, p. **14**31.

Constant for single depth of U. S. S. thread, .6495; V thread .866. See *Example* and *Solution*, Fig. 11. Spindle *C* is suited to a wide range of pitches but anvil *D* is limited, different anvils being substituted for each range of pitches.

14. Tables of thread micrometer readings for U. S. S. and V threads.

U. S. STANDARD THREADS.				V THREADS.			
DIAM.	PITCH.	CALIPER READING OR PITCH DIAM.	THREAD, DEPTH.	DIAM.	PITCH.	CALIPER READING OR PITCH DIAM.	THREAD, DEPTH.
D	P	$D - \frac{.6495}{P}$	$\frac{.6495}{P}$	D	P	$D - \frac{.866}{P}$	$\frac{.866}{P}$
$\frac{3}{16}$	24	.1604	.0270	$\frac{3}{16}$	24	.1514	.0360
$\frac{1}{4}$	20	.2176	.0324	$\frac{1}{4}$	24	.2139	.0361
$\frac{5}{16}$	18	.2765	.0360	$\frac{1}{4}$	20	.2067	.0433
$\frac{3}{8}$	16	.3344	.0406	$\frac{5}{16}$	20	.2692	.0433
$\frac{7}{16}$	14	.3911	.0464	$\frac{5}{16}$	18	.2644	.0481
$\frac{1}{2}$	13	.4501	.0499	$\frac{3}{8}$	18	.3269	.0481
$\frac{9}{16}$	12	.5084	.0541	$\frac{3}{8}$	16	.3209	.0541
$\frac{5}{8}$	11	.566	.0590	$\frac{7}{16}$	16	.3834	.0541
$\frac{3}{4}$	10	.6851	.0649	$\frac{7}{16}$	14	.3756	.0619
$\frac{7}{8}$	9	.8029	.0721	$\frac{1}{2}$	14	.4381	.0619
1	8	.9188	.0812	$\frac{1}{2}$	13	.4334	.0666
$1\frac{1}{8}$	7	1.0322	.0928	$\frac{1}{2}$	12	.4278	.0722
$1\frac{1}{4}$	7	1.1572	.0928	$\frac{9}{16}$	14	.5006	.0619
$1\frac{3}{8}$	6	1.2668	.1082	$\frac{9}{16}$	12	.4903	.0722
$1\frac{1}{2}$	6	1.3918	.1082	$\frac{5}{8}$	11	.5463	.0787
$1\frac{5}{8}$	$5\frac{1}{2}$	1.507	.1180	$\frac{5}{8}$	10	.5384	.0866
$1\frac{3}{4}$	5	1.6201	.1299	$\frac{11}{16}$	10	.6009	.0866
$1\frac{7}{8}$	5	1.7451	.1299	$\frac{3}{4}$	10	.6634	.0866
2	$4\frac{1}{2}$	1.8557	.1443	$\frac{7}{8}$	9	.7788	.0962
$2\frac{1}{2}$	4	2.3376	.1624	1	8	.8918	.1082
3	$3\frac{1}{2}$	2.8145	.1855	$1\frac{1}{8}$	8	1.0168	.1082
$3\frac{1}{2}$	$3\frac{1}{4}$	3.3002	.1998	$1\frac{1}{4}$	7	1.1263	.1237
4	3	3.7835	.2165	$1\frac{1}{2}$	6	1.3557	.1443

MAKING TAPS

15. The thread on taps must be smooth. — A spring threading tool holder and cutter, or a forged tool, may be used as at *A*, Fig. 12.

If a solid spring holder is used, as at *A*, Fig. 12, the top face *D* of the cutter *E* must not be set above the top of holder *C*.

The cutter may be removed by releasing nut F, then ground and replaced without disturbing holder and it will resume its cut.   All taps to be milled with a thread miller, see pp. **10**48, and fine thread taps of 20-pitch or finer, to be threaded in a lathe, as in B, Fig. 12, may be tapered and grooved before threading.   This eliminates the burr that grooving after threading forces between the threads which is difficult to remove, especially from fine threads.

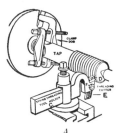   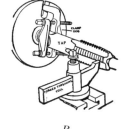

A Fig. 12. B

THREADING TAP **Before**   THREADING FINE THREAD TAP
TAPERING AND GROOVING.   **After** TAPERING AND GROOVING.

Taps often contract in length in hardening.   For very accurate work they may be threaded oversize and lapped to size with a disk charged with emery and oil driven from a drum countershaft.

For relieving or backing off taps, see pp. **12**76–**12**80.

Warning. — Avoid deep center holes, large countersinks and square corners, which might cause a tap or other tool to crack in hardening.

16. Outside and inside U. S. S. thread gages, Fig. 13,

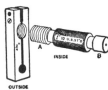

for standardizing taps and dies. — They are hardened, then thread is ground by a method which leaves pitch, angles and size correct.   Plug A is the standard to which the outside gage is adjusted.   The top of the thread is truncated, according

FIG. 13. — STANDARD   to the U. S. standard; the bottom is left
THREAD GAGES.   sharp to give clearance.   Part B is the standard root diameter of thread.

To test lead of thread, see Thread Lead Indicators **and** Testing Machines, p. **14**27.

17. To straighten hardened and tempered tools, Fig. 14.

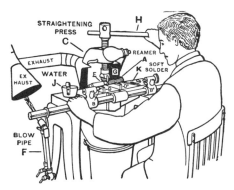

FIG. 14. — STRAIGHTENING HARDENED AND TEMPERED WORK.

SCHEDULE OF OPERATIONS

Taps, reamers, drills, mandrels, gages, and work of that class, often spring in hardening and tempering and have to be straightened.

Example. — Mount hardened and tempered reamer A on centers B, B' of straightening press C. Rotate with fingers, test with chalk, and note eccentricity.

Place reamer on supports D, D' with eccentric side up, and heat at most eccentric part E with blowpipe F.

Apply pressure with screw G operated by handle H to force reamer straight or slightly beyond straight as it may spring back some, then cool under tension with water from cup J or wet waste. Again test on centers, and repeat process if needed until reamer is true within grinding limit.

Caution. — To avoid drawing the temper while heating, test temperature occasionally by touching reamer with soft solder K. (Soft solder melts at 370° F.) If the solder melts readily, cool with water, for the temperature must not exceed 430° F.

Attention. — Large lots of tools of the above classes may be straightened rapidly by heating in an oil-tempering gas furnace to a temperature from 350° to 400° F. This temperature is not high enough to draw the temper, but is high enough to permit the work to be easily and safely straightened in a press.

18. To make $\frac{3}{4}''$ × 10 United States Standard tap, Fig. 15.

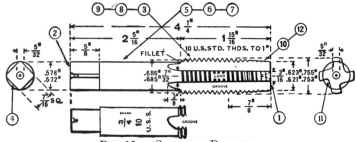

FIG. 15. — SCHEDULE DRAWING.

Specifications: Material, annealed carbon steel $\frac{1}{16}''$ to $\frac{1}{8}''$ large; weight, 12 oz. Hardness, 20 to 25 (scleroscope).

For heat treatment see *Principles of Machine Work.*

High-speed steel, or stellite cutting tools.

Time: Study drawing and schedule in advance, 5 min. — Oil machines, 10 min. — Prepare blank, **1 h.** 12 min. — Thread and turn taper, 38 min. — Mill grooves (with machine " set up "), file clearance, stamp and polish, 48 min. — Harden, 10 min. — Polish and color temper, 27 min. (or oil temper and polish, 15 min.) — Clean machines, 10 min. — Total, 3 h. 40 min.

SCHEDULE OF OPERATIONS, MACHINES AND TOOLS

OPERATIONS.	MACHINES, SPEEDS, FEEDS.	TOOLS.
Center. See that live center is nearly true and dead center in approximate alinement. **Rough square (1), (2).**	Engine lathe, 12'' to 16''. 3d or 4th speed, or 40 F.P.M. Hand feed.	 Dog, holder and cutter 30° rake, calipers, rule.
Recenter. **Finish square, (2), (1).** **Rough turn** to .755'' + .015'', (3), one or two cuts. Turn half way, reverse and turn other half.	 2d or 3d speed, or 40 F.P.M. Medium power feed — 80 to 1''.	 Holder and cutter 30° rake, 1'' micrometer.
Mill square (4), see p. **102**8.	*Note.* — If a milling machine is not available, rough and finish turn shank (5), file (6), and polish (7), index (4), see p. 352.	
Rough turn shank **(5)**, to .576'' + .015'', two cuts $2\frac{5}{16}''$ long.	File square (4). Use vise, copper jaws, 8'' or 10'' hand-smooth file.	
True live center. Set dead center in accurate alinement. **Finish turn** to .576'' + .003'', (6), one cut.	 3d speed, or 60 F.P.M. Fine power feed — 140 to 1''.	

File and polish (7) to limit.	Speed lathe, highest speed.	8″ or 10″ mill-bastard file, 60 and 90 emery cloth, polishing clamps, 1″ micrometer, oil.
Finish turn to .755″ + .003″, file to limit (8).	Engine lathe, 3d or 4th speed, or 60 F.P.M. Fine power feed — 140 to 1″.	Clamp dog, holder and cutter 30° rake, 8″ or 10″ mill-bastard file, 1″ micrometer.
Thread (9) to thread micrometer size .686″ to .685″. Take eight cuts .005″ each, four cuts .004″ each, three cuts, .002″ each, two cuts .001″ each, one cut .0005″, clean, oil, test and repeat cuts of .0005″ until thread measures within limit, or fits gage. Single depth of thread .067″. **Turn taper on tap $1\frac{27}{32}$″ to 1′ (10).**	Arrange for 10 threads. 1st speed, or 25 F.P.M.	10 pitch U. S. S. spring-thread tool and gage, center gage, thread micrometer (see pp. 1214–1216), oil.
Set over footstock $\frac{5}{16}$″, or use taper attachment.	....	Rule, dividers.
Turn taper to bottom of thread at end (10), four cuts.	3d speed, or 60 F.P.M. Fine power feed — 140 to 1″.	Holder and cutter 30° rake.
Mill grooves (11). See pp. 1031–1033. Mill grooves beyond thread $\frac{3}{8}$″.	....	No. 5 double angle tap-grooving cutter, oil.
File off burr produced by milling.	Vise, copper jaws.	2d cut three-square file.
Heat tap to blue, or coat with copper sulphate.	Blow pipe or gas furnace.	Tongs.
File clearance of 6° to 7° on top of taper threads only, (12). See Relieving or Backing Off Taps, pp. 1276–1280.	Vise. copper jaws.	6″ to 8″ pillar file No. 3, or smooth cut.
Stamp size and name on shank.	....	Hammer, $\frac{1}{16}$″ and $\frac{3}{32}$″ steel figures, and letters.
Harden in water. See *Principles of Machine Work*. **Temper** in oil (or polish and temper to straw color).	Gas furnace or forge 1325° F. to 1350° F. Oil-tempering furnace, 425° F. to 435° F.	Tongs, water. (Red-hot ring or collar).
Test hardness with scleroscope, or dead-smooth file. **Polish shank** and grooves after oil tempering, or before color tempering.	Speed lathe, highest speed. Vise, copper jaws.	90 emery cloth, polishing clamps, oil, 5″ half-round file.

Information. — Threads of taps are often given "body clearance" by setting the lathe to thread about .001″ to 1″ smaller at the shank.

Attention. — Grooves of taps may be ground after hardening and tempering, to remove scale and sharpen teeth. See p. 825.

19. To make $1\frac{1}{4}'' \times 5$, Square thread tap, Fig. 16.

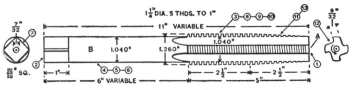

FIG. 16. — SCHEDULE DRAWING.

Specifications: Material, annealed carbon steel $\frac{1}{16}''$ to $\frac{1}{8}''$ large; weight, 4 lbs. 8 oz. Hardness, 20 to 25 (scleroscope). High-speed steel, or stellite cutting tools.
Time: 10 h.

SCHEDULE OF OPERATIONS, MACHINES AND TOOLS

OPERATIONS.	MACHINES, SPEEDS, FEEDS.	TOOLS.
Center. See that live center is nearly true and dead center in approximate alinement. **Rough square (1), (2).**	Engine lathe, 12″ to 16″. 2d or 3d speed, or 40 F.P.M. Hand feed.	Dog, holder and cutter 30° rake, calipers, rule.
Recenter. Finish square to length (2), (1). **Rough turn** to $1.260'' + .015''$, (3).	1st or 2d speed, or 40 F.P.M. Medium power feed — 80 to 1″.	Holder and cutter 30° rake, 2″ micrometer.
Rough turn to $1.040'' + .015''$, (4), one or two cuts. **True live center. Set dead center in accurate alinement.** Finish turn to $1.040'' + .003''$, (5), one cut.	3d speed, or 60 F.P.M. Fine power feed — 140 to 1″.	
File and polish (6), $1.040''$ to $1.038''$.	Speed lathe, highest speed.	8″ or 10″ mill bastard file, 60 and 90 emery cloth, polishing clamps, oil.
Mill square (7). See p. **10**28. **Finish turn** to $1.260'' + .004''$, (8).	Engine lathe, 3d speed or 60 F.P.M. Fine power feed — 140 to 1″.	Clamp dog, holder and cutter 30° rake, 2″ micrometer.
File (8), $1.260''$ to $1.262''$.	4th speed.	8″ or 10″ mill bastard file.

Rough thread to 1.040″ + .010″, (9), twenty-one cuts .005″ each. Depth of thread .105″.	Arrange for 5 threads. 1st speed, or 25 F.P.M.	Square thread roughing tool, width .090″, or holder and cutter, thin calipers, rule, oil.
Set finishing tool to cut on both sides of groove, (9).	. . . .	Square thread finishing tool, width .099″, or holder and cutter.
Finish thread to 1.040″, (10), twenty cuts .005″ each, two cuts .002″ each, six cuts .0005″ each. Depth of thread .110″. Turn taper on tap $1\frac{3}{32}$″ to 1′, (11).	1st speed, or 25 F.P.M.	Thin calipers, rule, oil.
Set over footstock $\frac{1}{2}$″, or use taper attachment.	. . . .	Rule, dividers.
Turn taper to bottom of thread at end (11), three or four cuts.	2d or 3d speed, or 60 F.P.M. Fine power feed — 140 to 1″.	Holder and cutter 30° rake.
Mill grooves (12). Lower index head spindle one-third degree below horizontal to produce taper lands. See pp. 1031-1033.	Universal or plain milling machine. 1st speed or 50 F.P.M. Medium power feed — .007″ per cutter revolution.	No. 6 tap grooving cutter, index centers, rule, oil.
File off burr.	Vise, copper jaws.	8″ or 10″ hand-smooth file, 6″ warding file.
Heat tap to a blue.	Blow pipe or gas furnace.	Tongs.
File clearance on top of taper threads only, (13). File clearance on sides of threads.	Vise, copper jaws.	8″ or 10″ hand smooth file. 6″ warding file with teeth ground off one side and one edge.
Stamp size and name on shank.	Vise, copper jaws.	Hammer, $\frac{1}{16}$″ and $\frac{1}{8}$″ steel figures and letters.
Harden in water.	Gas furnace or forge, 1325° F. to 1350° F.	Tongs, water.
Temper in oil (or polish and temper to straw color). Test hardness with scleroscope, or dead-smooth file.	Oil-tempering furnace, 425° F. to 435° F.	(Red-hot ring or collar.)
Polish shank and grooves after oil tempering, or before color tempering.	Speed lathe, highest speed. Vise, copper jaws.	90 emery cloth, polishing clamps, oil, 5″ half-round file.

Attention. — The set over for taper is correct for this length of **tap** only.

20. To make $1\frac{1}{4}'' \times 5$, 29° thread tap, Fig. 17.

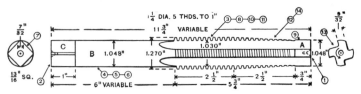

FIG. 17. — SCHEDULE DRAWING.

Specifications: Material, annealed carbon steel $\frac{1}{16}''$ to $\frac{1}{8}''$ large; weight, 5 lbs. Hardness, 20 to 25 (scleroscope).
High-speed steel, or stellite cutting tools.
Time: 10 h.

SCHEDULE OF OPERATIONS, MACHINES AND TOOLS

OPERATIONS.	MACHINES, SPEEDS, FEEDS.	TOOLS.
Center. See that live center is nearly true and dead center in approximate alinement.	Engine lathe, 12″ to 16″.	
Rough square, (1), (2).	2d or 3d speed, or 40 F.P.M. Hand feed.	Dog, holder and cutter 30° rake, calipers, rule.
Recenter. **Finish square** to length, (2), (1).		
Rough turn to 1.270″ + .015″, (3).	1st or 2d speed, or 40 F.P.M. Medium power feed — 80 to 1″.	Holder and cutter 30° rake, 2″ micrometer.
Rough turn to 1.048″ + .015″, (4), one or two cuts. **True live center. Set dead center in accurate alinement.**		
Finish turn to 1.048″ + .003″, (5), one cut.	3d speed, or 60 F.P.M. Fine power feed — 140 to 1″.	
File and polish, (6), 1.048″ to 1.046″.	Speed lathe, highest speed.	8″ or 10″ mill-bastard file, 60 and 90 emery cloth, polishing clamps, oil.
Mill square, (7). See p. 1028. **Finish turn** to 1.270″ + .004″, (8).	Engine lathe, 3d speed, or 60 F.P.M. Fine power feed — 140 to 1″.	Clamp dog, holder and cutter 30° rake, 2″ micrometer.
File, (8), 1.270″ to 1.272″.	4th speed.	8″ or 10″ mill-bastard file.

Rough and finish turn and file leader 1.048″ to 1.046″, **(9)**.	2d, 3d and 4th speeds. Medium and fine power feeds.	Holder and cutter 30° rake, 8″ or 10″ mill-bastard file, 2″ micrometer.
Rough thread to 1.030″ + .010″, **(10)**, twenty-three cuts .005″ each. Depth of thread .115″.	Arrange for 5 threads. 1st speed, or 25 F.P.M.	Square thread roughing tool width .060″, or holder and cutter, thin calipers, rule, oil.
Set finishing tool to cut on both sides of groove, **(10)**.	. . . .	29° thread tool, 29° thread tool gage.
Finish thread to size, (11), twenty-one cuts .005″ each, two cuts .002″ each, two cuts .001″ each, one cut .0005″, clean, oil, and measure and repeat cuts of .0005″ until thread fits gage. Depth of thread .120″.	1st speed, or 20 F.P.M.	29° thread micrometer or gage, oil.
Turn taper $1\frac{3}{32}$″ to 1′, **(12)**. **Set over footstock** $\frac{17}{32}$″, or use taper attachment.	. . . .	Rule, dividers.
Turn taper to diameter of leader **(12)**, three or four cuts.	2d or 3d speed, or 60 F.P.M. Fine power feed — 140 to 1″.	Holder and cutter 30° rake.
Mill grooves, (13). Lower index head spindle one-third of a degree below horizontal to produce taper lands. See pp. **103**1–**103**3.	Universal or plain milling machine. 1st speed, or 50 F.P.M. Medium power feed — .007″ per cutter revolution.	No. 6 tap grooving cutter, oil, index centers, rule.
File off burr.	Vise, copper jaws.	8″ or 10″ hand-smooth file, 6″ warding file.
Heat tap to a blue.	Blow pipe or gas forge.	Tongs.
File clearance on top of taper threads only, (14).	Vise, copper jaws.	8″ or 10″ hand smooth file.
File clearance on sides of threads.	. . . .	6″ warding file with teeth ground off one side and one edge.
Stamp size and name on shank.	Vise, copper jaws.	Hammer, $\frac{1}{16}$″ and $\frac{1}{8}$″ steel figures and letters.
Harden in water.	Gas furnace or forge, 1325° F. to 1350° F.	Tongs, water.
Temper in oil (or polish and temper to straw color).	Oil-tempering furnace, 425° F. to 435° F.	(Red-hot ring or collar.)
Test hardness with scleroscope or dead-smooth file.		
Polish shank and grooves after oil tempering or before color tempering.	Speed lathe, highest speed. Vise, copper jaws.	90 emery cloth, polishing clamps, oil.

Attention. — The set over for the taper is correct for this length of tap only.

MAKING PLAIN MILLING CUTTER

21. To make $2\frac{1}{2}'' \times \frac{1}{2}''$ plain milling cutter, 18 teeth, Fig. 18.

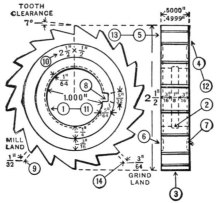

FIG. 18. — SCHEDULE DRAWING.

Specifications: Material, annealed carbon or high-speed steel disk $1\frac{1}{16}''$ large; weight, 14 oz. Hardness, 20 to 25 (scleroscope).
High-speed steel, or stellite cutting tools.

Time: Study drawing and schedule in advance, 5 min. — Oil machines, 12 min. — Chuck and jig ream, 26 min. — Turn, face, and cut grooves, 51 min. — Mill teeth with machine " set up " and stamp, 52 min. — Harden, 10 min. — Lap, 25 min. — Polish and color temper, 15 min. (or oil temper and polish, 7 min.) — Grind teeth twice around with machine " set up," 6 min. — Clean machines, 11 min. — Total, 3 h. 33 min.

SCHEDULE OF OPERATIONS, MACHINES AND TOOLS

OPERATIONS.	MACHINES, SPEEDS, FEEDS.	TOOLS.
Chip or file stem off side of disk. Set dead center in approximate alinement.	Engine lathe, 12″ to 16″.	
Mount disk in chuck, smooth side out; true up and drill hole (1) $\frac{15}{16}''$. See Chucking, pp. 408–412.	1st or 2d speed or 130 R.P.M. Hand feed.	Independent or universal chuck centering tool, $\frac{15}{16}''$ twist drill and holder, oil.
Bore hole (1) $\frac{63}{64}''$. See Boring, pp. 504–506. **Power ream.** See *Exception,** 1225.	1st or 2d speed or 40 F.P.M. Medium Power feed — 80 to 1″.	Boring tool 30° rake calipers, rule, oil. Fluted chucking reamer .005″ small, oil.

Jig ream, see pp. 415, 416.	Hand reaming jig, vise.	1″ spiral hand reamer, reamer wrench, oil.
Cut groove (2). See *Attention.*†	1st or 2d speed.	⅛″ grooving tool, rule, oil.
Oil mandrel and press into hole. **Rough turn** diameter to 2½″ + .015″ (3), two or more cuts.	Mandrel press. 1st speed, or 50 F.P.M. Medium power feed — 80 to 1″.	1″ mandrel, oil. Holder and cutter 30° rake, oil, 3″ micrometer.
Rough square sides to ½″ + .010″, (4), (5). See p. **12**26, Fig. 19. Square an equal amount from each side.	1st or 2d speed or 50 F.P.M. Medium power feed.	Holder and cutter 30° rake, 1″ micrometer, oil.
Finish square sides to ½″ + .00025″, (4), (5), see p. **12**26, Fig. 20. See *Exception.**	2d or 3d speed, or 150 F.P.M. Fine power or Hand feed.	Holder, high-speed steel or stellite cutter 30° rake, 1″ micrometer.
Cut grooves .020″ deep, (6), (7). **Cut keyway** (8). See pp. 926, 927, 936, 937. **Mill teeth** (9). See pp. 1036, 1037. Remove burr from sides.	1st or 2d speed, or 40 F.P.M. Hand feed. Keyway cutter, planer or shaper, vise.	3/16″ grooving tool, oil, rule, depth gage. Hub keyway tool, rule, oil.
Stamp size at (10), and name in same groove diametrically opposite. **Harden** in water.	Engine lathe, 2d speed. Lead block, sheet copper or thick paper on anvil. Gas furnace or forge 1325° F. to 1350° F.	8″ or 10″ mill bastard file. 1/16″ and ⅛″ steel figures and letters, hammer. Tongs, water.
Temper in oil (or polish and temper to straw color). **Test hardness** with scleroscope, or dead-smooth file. **Lap hole** to size. See pp. **12**30, **12**31. **Polish sides** (4), (5) slightly.	Oil-tempering furnace 425° F. to 435° F.	(Red-hot shaft.)
Grind teeth for clearance 7°, (14), see pp. 814, 815, diameter of cutter not important. See *Note.*‡	Speed lathe, 3d or 4th speed.	1″ mandrel, flour emery cloth, polishing stick.

* *Exception.* — If convenient to use a grinding machine to grind hole and sides, the cutter blank may be prepared as follows:
Drill, bore, and power ream .005″ small. Press on special mandrel, rough square ½″ + .010″ 4, 5. Cut grooves 6, 7 .025″ deep. Cut groove 2. Mill teeth 9, stamp size 10. Harden and temper. Grind hole 1 to size + .001″ to .0015″. See pp. 728, 729. Grind sides 4, 5 to dimensions. See method on p. 827.

† *Attention.* — Some prefer to defer grooving (2) until the sides are squared when the blank may be placed in a universal chuck and the groove cut central.

‡ *Note.* — Diameter of cutter is important only when used in gangs.

22. To rough face large work in steel, as milling cutter disk B, Fig. 19. — Use holder and high-speed steel cutter A. Grind point of cutter as at C for clearance and rake. Use oil and feed by power or hand in direction of arrow D.

To finish face, use high-speed steel cutter A or side tool, Fig. 20. Grind point of tool about $\frac{1}{32}''$ flat as at C with clearance as at C', and set parallel to cut or to drag slightly. Take light cuts (.0005' to .0020''), without oil, in direction of arrow D. Use very fine power or hand feed, and same speed as for filing, — 3d or 4th speed for $2\frac{1}{2}''$ disk.

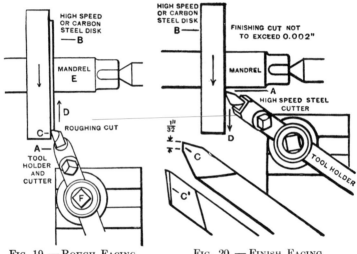

FIG. 19. — ROUGH FACING MILLING–CUTTER DISK.

FIG. 20. — FINISH FACING MILLING-CUTTER DISK.

MAKING HAND REAMER

23. To make hand reamer, Fig. 21.

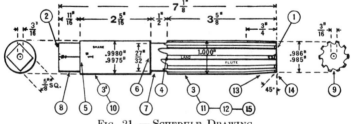

FIG. 21. — SCHEDULE DRAWING.

Specifications: Material, annealed carbon steel $\frac{1}{16}''$ to $\frac{1}{8}''$ large; weight, 1 lb. 14 oz. Hardness, 20 to 25 (scleroscope). High-speed steel, or stellite cutting tools. **Time:** 3 h. 40 min. which includes 15 min. for milling head square and 50 min. for milling flutes.

SCHEDULE OF OPERATIONS, MACHINES AND TOOLS

Operations.	Machines, Speeds, Feeds.	Tools.
Center. True live center. Set dead center in approximate aline- ment.	Engine lathe, 12'' to 16''.	
Rough square (1), (2).	2d or 3d speed, or 40 F.P.M. Hand feed.	Holder and cutter 30° rake, calipers, rule.
Recenter. **Finish square (2), (1).** **Rough turn** to 1.010'' (3), (3'), one or two cuts. Turn half way (3), reverse and turn other half (3').	1st or 2d speed, or 40 F.P.M. Medium power feed — 80 to 1''.	Holder and cutter 30° rake, 1'' mi- crometer.
Draw lines for length of flutes (4), square end (5), and shank (6).	Vise, copper jaws.	Copper sulphate, rule, scriber.
Rough and finish turn reduced part (7). Leave fillets at be- ginning and end of cut.	Engine lathe, 1st or 2d speed, or 40 F.P.M. Fine power feed — 140 to 1''.	Calipers, rule.
File reduced part (7).	4th speed, or 175 F.P.M.	5'' half-round 2d-cut file.
Mill square to $\frac{5}{8}''$, (8). See p. 1028. File slight bevel on edges of squared end to re- move burr.		
Stamp size and name on shank (3').	Vise, copper jaws.	$\frac{1}{8}''$ steel figures and letters, hammer.
Mill flutes (9). See pp. 1034– 1036.		
Harden in water.	Gas furnace or forge, 1325° F. to 1350° F.	Tongs, water.
Temper in oil (or polish and temper to straw color).	Oil-tempering fur- nace, 425° F. to 435° F.	(Red-hot ring or col- lar.)
Test hardness with scleroscope, or dead-smooth file.		
Lap center holes (1), (2). See p. 1228.	Speed lathe, highest speed.	Copper lap, flour emery, oil.
Grind shank to size (10). Grind fluted part, cylindri- cally, to 1.0004'', (11).	Universal grinding machine.	Dog, grinding wheel No. 60 K or L, vitrified. 1'' mi- crometer.

Grind clearance 4° on teeth to a cutting edge (12). Elevate center of wheel spindle $\frac{9}{32}''$ above centers. See pp. 811, 812. Grind 1° taper and 4° clearance on end (13). Measurement at end .986″ to .985.″ Grind bevel on end of teeth to 45° with 7° clearance and $\frac{1}{32}''$ wide.	Cutter grinder, tooth rest, centering gage.	8″ grinding wheel No. 60 *I*, vitrified, bevel protractor.
Oil stone reamer (15). See pp. 821, 822.	Vise, copper jaws.	Fine oil stone, oil.

Information. — If the reamer springs in hardening and tempering, an amount in excess of that allowed for grinding, it should be carefully straightened before grinding. See p. 12 17.

Attention. — As grooving the first reamer is largely experimental, the student should prepare a machine-steel trial blank and use it to obtain setting of cutter for milling flutes, and also to become familiar with the process of irregular spacing.

LAPPING

24. Lapping may be called refined grinding as it is used for the final finishing of hardened and unhardened work where smooth and accurate work is required.

A lap of lead, cast iron, copper or brass is made the desired shape and charged with emery and oil, the lathe revolved at a high speed and the work passed back and forth over lap, or inside of lap.

25. To make lap for center holes. — Place short piece of round copper rod in drill chuck of speed or hand lathe, and turn with graver to 60°. Test with center gage, and file smooth.

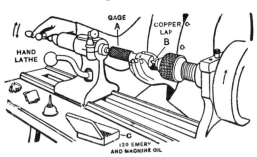

FIG. 22. — LAPPING CENTER HOLES.

26. To lap center holes in hardened gage A,

Fig. 22. — Shift belt to highest speed. Wipe center holes clean.

Start lathe. With the finger, spread on lap B a little of mixture of 120 grain emery and machine oil from box C. Mount work on dead center, with left hand slowly revolve work in opposite direction to revolution of lathe, and with right hand operating footstock handle, advance and recede work slightly to avoid cutting grooves in work. Wipe center hole and examine. Continue until center hole is clean and smooth. Repeat in opposite center hole.

27. To lap standard plug gage, Fig. 23.

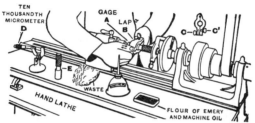

FIG. 23. — LAPPING STANDARD PLUG GAGE.

Specifications: Standard plug gage ground to 1.0004″.

Machines and tools: Speed or hand lathe, 1″ cast-iron outside lap, flour of emery, measuring machine or standard gage and ten-thousandth micrometer. Lubricant, machine oil.

Time: Study Fig. 23 and schedule, 5 min. — Oil lathe, 2 min. — Lap and measure gage, 50 min. — Clean lathe, 3 min. — Total, 1 h.

SCHEDULE OF OPERATIONS

Oil hand lathe and shift belt to highest speed.

Place dog on gage A with copper under set screw.

With the fingers spread a little machine oil and flour of emery on gage A.

Adjust cast-iron lap B to running fit on gage with screws CC'.

Mount freely on centers. Start lathe, and move lap back and forth as shown by arrows.

First lap gage straight, then lap to size. Wipe clean with waste, cool in water and measure.

To measure gage, use measuring machine, Fig. 29, or ten-thousandth micrometer D and 1″ standard reference gage E.

When within 1/10,000″ of diameter wipe lap and gage clean, oil gage and lap with machine oil; adjust lap closely and finish lapping without adding more emery, and a good finish will be obtained.

Limit 1.0000″ + or − 1/50,000″.

Clean lathe with waste.

Attention. — Gages are standardized at 62° F., therefore, heat produced by lapping, and heat of hands, will interfere with true size of gage. This is overcome by cooling in water or by leaving near measuring machine over night before taking final measurement.

Note. — For fine finish and for detecting soft spots on gage (defective hardening), use lead-lined lap.

28. To make lead lap for ring gages, Fig. 24. — Make lead lap A by turning a taper of $\frac{1}{2}''$ to 1' on piece of steel or cast iron, as mandrel B. Then mill groove C entire length. Set mandrel central in iron or wood mold and pour in molten lead. Turn to size. When lap wears, force mandrel into lead and turn to size.

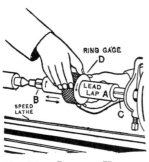

FIG. 24. — LAPPING HOLE IN 1″ STANDARD RING GAGE.

29. To lap 1″ standard ring gage (hole .9996″). — Fasten dog. Place ring gage D on lap and mount freely on centers. Spread a little of mixture of flour emery and machine oil on lap and run lathe at high speed.

Move gage slowly back and forth on lap and rotate slowly backward as indicated by arrows. To test, wipe hole clean and test with two plug gages — one small and the other standard.

When within 1/10000″ of diameter, wipe lap and gage clean, apply machine oil to both and continue lapping to size and to obtain fine finish. Limit 1.0000″+ or − 1/50000″.

Information. — As lapping has a tendency to make a hole larger at ends, a projection or lip is sometimes left on each end of gage which is ground off after lapping.

30. To lap hole in milling cutter A, Fig. 25. — Make lap B by turning and filing piece of cast iron to fit hole in cutter. Fasten dog and place cutter on lap, then mount loosely on centers. Spread mixture of 120 grain emery and machine oil from box C on lap.

Rotate cutter slowly in opposite direction of revolution of lathe and at same time press hard with thumbs and move cutter slowly back and forth as indicated by arrows. Wipe hole in cutter, clean and test with plug gage *D* or with inside micrometer *E*. Repeat until gage fits hole freely.

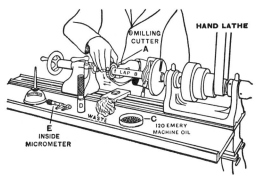

Fig. 25. — Lapping Hole in Milling Cutter.

Attention.— Holes in milling cutters shrink in hardening. In manufacturing cutters in large lots, the holes are usually ground.

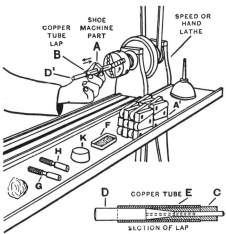

Fig. 26. — Lapping Holes in Machine Parts. Commercial Lapping.

Warning.—By placing cutter on lap with teeth toward back of lathe, injury to hands may be avoided.

31. Lapping holes in machine parts. Commercial lapping, *A, A,'* Fig. 26.— Holes in machine parts are often lapped to correct errors of under size, roughness, roundness, taper, etc. Make lap *B* as in detail. Turn and bore *C* and striker *D*. Saw three slots in *C* also copper tube *E*. Grip *C* in chuck and run lathe at high speed. Charge lap with 120 grain emery and machine oil from box *F* and lap hole to size by passing work back and forth on lap. Test work with limit gages *G, H.* When lap wears, expand by striking *D'* light blow with babbitt block *K.*

MAKING PLUG GAGE

32. To prepare standard plug-gage blank for grinding, Fig. 27.

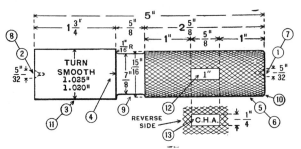

FIG. 27. — SCHEDULE DRAWING.

Specifications: Material, annealed carbon steel $\frac{1}{8}''$ large; weight, 1 lb. 8 oz. Hardness, 20 to 25 (scleroscope).

For **heat treatment** see *Principles of Machine Work.*

High-speed steel, or stellite cutting tools.

Time: Study drawing and schedule in advance, 5 min. — Oil lathe, 4 min. — Prepare gage blank, 1 h. 10 min. — Harden, 20 min. — Lap center holes and polish, 10 min. — Clean machines, 6 min. — Total, 1 h. 55 min.

SCHEDULE OF OPERATIONS, MACHINES AND TOOLS

OPERATIONS.	MACHINES, SPEEDS, FEEDS.	TOOLS.
Center to $\frac{9}{64}''$ **(1), (2).** See that live center is nearly true and dead center in approximate alinement.	Engine lathe, 12″ to 16″.	
Rough square $\frac{1}{64}''$ **long (1), (2).**	2d or 3d speed, or 40 F.P.M.	Dog, holder and cutter 30° rake, calipers, rule.
Recenter to $\frac{9}{64}''$. **Rough turn** whole length to $1\frac{1}{32}''$, **(3),** one cut. Turn half way, reverse and turn other half.	1st or 2d speed, or 40 F.P.M. Medium power feed — 80 to 1″.	Holder and cutter 30° rake, copper, calipers, rule.
Draw line (4), $1\frac{3}{4}''$ from gage end.	Vise, copper jaws.	Copper sulphate, rule, scriber.

Rough and finish turn handle (5), two cuts.	2d or 3d speed, or 40 F.P.M. Medium power feed — 80 to 1″.	
Nurl (6). See p. 636.	2d speed, or 40 F.P.M. Medium power feed — 80 to 1″.	Medium nurling tool, oil.
True live center. Recenter $\frac{5}{32}$″, (7), (8).	Speed lathe, 3d speed, or 400 R.P.M.	Drill chuck, combination drill and countersink.
Finish square to length 5″, (1), (2).	Engine lathe, 3d or 4th speed, or 60 F.P.M. Hand feed.	
Rough and finish turn reduced part (9), two or three cuts.	3d speed, or 40 F.P.M. Fine power feed — 140 to 1″.	
File (9).	4th speed, or 175 F.P.M.	7″ or 8″ pillar-smooth file.
Round (10).	. . . .	8″ or 10″ mill-bastard file.
Turn gage part (11), one or two cuts.	3d speed, or 60 F.P.M. Medium power feed — 80 to 1″.	2″ micrometer.
Polish ends (1), (2), and reduced part (9).	Speed lathe, highest speed.	90 and 120 emery cloth, thin polishing stick, oil.
File flats on opposite sides (12), (13), or Mill flats (12), (13).	Vise, copper jaws. Milling machine, 3d speed. Medium power feed.	7″ or 8″ pillar-smooth file. $\frac{5}{8}$″ milling cutter, milling machine vise or $\frac{5}{8}$″ end mill, index head with centers, rule.
Stamp size and name (12), (13).	Vise, copper jaws.	$\frac{1}{8}$″ steel figures and letters, hammer.
Harden gage. See *Information.*	Gas furnace, or forge, 1325° F. to 1350° F.	Tongs, water, sperm oil, hand-smooth file.
To lap center holes, see p. **12**28. Clean and polish ends (1), (2), reduced part (9), and flats (12), (13). To grind, see p. **12**34.		

Information. — To make a plug gage glass hard and the handle jet black, heat gage to a bright red and dip endwise (gage part only) in water and move about for eight seconds. Time may be estimated by counting as follows: one-thou-zan-*one*, one-thou-zan-*two*, one-thou-zan-*three*, one-thou-zan-*four*, one-thou-zan-*five*, one-thou-zan-*six*, one-thou-zan-*seven*, one-thou-zan-*eight*. Then plunge *whole* gage into oil and move about until cold.

33. To grind standard plug gage on two dead centers, Fig. 28.

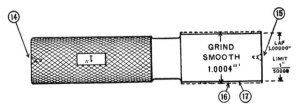

FIG. 28. — SCHEDULE DRAWING.

Specifications: Plug gage blank, carbon steel turned $\frac{1.025''}{1.020''}$ diameter, stamped, hardened and handle black.

Center holes, (**14**), (**15**), lapped.

Machines and tools: Universal grinding machine.

Measuring machine or $2''$ ten-thousandth micrometer.

Grinding wheel $12'' \times \frac{1}{2}''$, No. 80 K, vitrified, grinding dog.

Speed of wheel, 5000 F.P.M. Speed of work, 50 F.P.M.

Feed $\frac{1}{3}$ width of wheel per revolution.

Oil bearings and two dead centers with machine oil.

Lubricant, solution of sal soda and water.

Time: Study drawing and schedule in advance, 5 min. — Oil machine and true wheel, 7 min. — Grind gage, 25 min. with machine " set up." — Clean machine, 3 min. — Total, 40 min. (" Set up " machine to grind straight, 25 min. extra.)

SCHEDULE OF OPERATIONS

I. Preparatory Adjustments. True wheel (see p. 709) and centers (see p. 710). Set zero lines on headstock and swivel table to grind approximately straight. See p. 712. Arrange water guards. See p. 715. Move grinding wheel back with hand cross-feed wheel 8 to allow space to mount work. Use grinding dog with copper under set screw. Mount work on centers. Start grinding wheel, work and feed, 1, 2, 3, and adjust dogs 6, 7, to length of stroke. Set automatic feed 11, 12, to feed .0005″ to .001″ (2 to 4 teeth) at each end of stroke.

II. Grind Trial Diameter (16),

Fig. 28. (Grind out tool marks only.) Move revolving wheel to lightly cut revolving work with cross-feed hand wheel 8. Turn on water and throw in feed pawl 9 to mesh with ratchet wheel 10 and allow automatic cross feed to take four or five trial cuts whole length of work to grind out tool marks. Then throw out pawl 9 and without moving cross feed, allow wheel to pass over work several times until cutting nearly dies out.

Stop work, feed and wheel, 2, 3, 1, with wheel at footstock end.

III. Correct Straightness. Measure work at both ends with $2''$

micrometer or measuring machine, see pp. **12**36–**12**38, and if not straight, move swivel table (see p. 7**12**). Take one or two cuts with automatic cross feed. Measure and repeat until machine grinds straight. If there is danger thereby of grinding gage too small, use trial piece same length.

IV. Find Amount Oversize. Measure work with 2″ micrometer. Subtract rough diameter of work 1.0040″ from reading of micrometer. For example, reading of micrometer may be 1.0140″, then 1.0140″ − 1.0040″ = .0100″ or 40 teeth for rough grinding.

V. Set Automatic Cross Feed. Raise perpendicular latch 13 in head 14, throw in pawl 9 and without moving cross feed, move shield 15 to right or left until end of shield and pawl are 40 teeth apart, then drop latch 13 in ratchet wheel 10. See p. 7**15**.

VI. Rough Grind (16), .0040″ Large, Fig. 28, as follows: Start wheel, work and feed, 1, 2, 3, and rough grind automatically until shield 15 lifts pawl 9. Then lock latch head 14 with horizontal latch 16, throw out pawl 9 and allow wheel to pass over work several times until cutting nearly dies out. Stop work with wheel at footstock end and measure. For example, work may measure 1.004″.

VII. Finish Grind (17), Fig. 28, with pinch feed (¼/1000″ in diameter of work), as follows: Start work. Take two cuts ½/1000″ each (two pinches at each end of work). Stop wheel, work and feed, 1, 2, 3, and measure gage. It may measure 1.0030″. Also measure gage at both ends for straightness with micrometer, or take gage out of machine and measure with measuring machine, pp. **12**36–**12**38. Start wheel work and feed and pinch twice at footstock end. Measure = 1.0025″. Pinch twice and let wheel pass over work several times. Measure = 1.0020″. Pinch twice and let wheel pass over work several times and measure (1.0015″). Pinch twice and let wheel pass over work several times and measure (1.0010″). As there is some vibration in machine, it is best to grind to size as follows: Take one cut and measure 1.0007″. Then let wheel pass over work without moving cross feed and it may measure 1.0004″. If still large, again let wheel pass over work without moving cross feed. Limit 1.0004″ to 1.0003″. Clean machine with waste.

Attention. — A gage hardened, ground and lapped without seasoning may shrink about one-twenty-thousandth of an inch in six months. To season properly a gage should be hardened, then laid away from six months to a year to relieve internal stresses, and allowed to shrink; but if gage must be finished at once, heat to about 400° F. after hardening and cool in water. Do this three times.

Note. — Gages are often made as follows: Rough ground to 1.004″, seasoned, then finished ground to 1.0004″ and lapped to size.

MEASURING MACHINE

34. Twenty-four-inch standard measuring machine, Fig. 29.
This instrument is for originating standard gages and for other
measurements of precision.

The bed is supported on three points, with *V* ways for the
sliding and fixed heads; the spindles are provided with
measuring faces.

35. The sliding head carries the spindle Precision Screw,
linear scale and graduated index wheel. The Precision Screw
on some machines has 50 threads per inch, on others 25 threads.

36. For Precision Screw with 50 threads the turns of index
wheel are indicated on linear scale graduated into 50 divisions
to the inch; each $20/1000'' = .020''$. The fractional part
of a turn is indicated by graduations on index wheel which
has 400 divisions, each $1/20,000''$.

37. For Precision Screw with 25 threads the turns of index
wheel are indicated on linear scale graduated into 25 divisions
to the inch; each $40/1000'' = .040''$. The fractional part of a
turn is indicated by graduations on index wheel which has 400
divisions, each $1/10,000''$.

By estimation, a division may be subdivided to indicate
one-half, one-quarter or one-fifth of a division. Variation
within a limit of $1/100,000''$ may be determined.

Fixed head. — In the fixed head is a sliding spindle kept in
place by a helical spring. This spindle carries one of the
auxiliary jaws while the other jaw is fastened to the head.

Contact indicators. — To indicate when contact is made between measuring faces and work, a small cylindrical gage is
placed between the auxiliary jaws in a horizontal position.
When measurement is obtained, gage drops to a vertical
position.

In Fig. 29 the auxiliary jaws are connected with an electrical buzzer which announces contact of work and measuring
faces.

To clean machine. — With a soft woolen cloth wipe oil and
dirt from finished parts, and other parts with waste. Benzine may be used to clean plugs in reference bar, followed by

kerosene to prevent rusting. To remove dust from polished surfaces of graduated plugs use a camel's-hair brush. The lenses are cleaned with alcohol and chamois skin. When setting and measuring clean measuring faces with paper.

38. To measure standard plug gage with 24″ standard measuring machine. Precision Screw, 50 threads to 1″, Fig. 29.

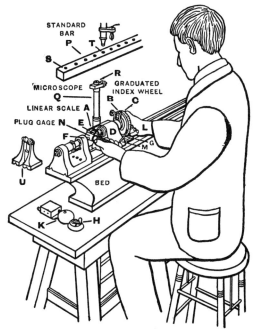

FIG. 29. — MEASURING ¾″ STANDARD PLUG GAGE.

SCHEDULE OF OPERATIONS

Oil slides and bearings with machine oil and *Precision Screw* with porpoise oil.

To set at zero.

Clean measuring faces with paper.

Run screw to zero on linear scale *A*, Fig. 29.

Adjust pointer *B*, graduated index wheel *C*, nearly to zero and slide head *D* until measuring faces *E*, *F*, are nearly in contact. Clamp fine adjustment *G*, throw in switch *H* which operates buzzer *K*, and adjust sliding head with screw *L* (not shown), until buzzer shows tendency to stop. Clamp sliding head with lever M and adjust index wheel until buzzer stops, then set pointer at zero on graduated index wheel.

Clean gage N and place it between cleaned measuring faces of machine.

Throw in switch and adjust index wheel until buzzer stops. Count whole number of divisions on linear scale and multiply by .020″ then add divisions on index wheel multiplied by 1/20000″ or .00005″ and the result will be the size.

For example: There are 37 divisions on the scale which gives $37 \times .020'' = .740''$. Then we read 201 divisions on index wheel which give $201 \times 1/20000''$. $201 \times .00005'' = .01005''$; adding we have $.740'' + .01005'' = .75005''$, or $.750'' + 1/20000''$.

Attention. — A variation of temperature of 1° in a 1″ gage will change its size about 1/100000″.

For example: Holding a 1″ gage a few minutes in the hand may increase its temperature 10° and its size 1/10000″.

39. To test $5\frac{1}{2}''$ end-measuring rod with 24″ standard measuring machine. Precision Screw, 25 threads to 1″, Fig. 29.

SCHEDULE OF OPERATIONS

Adjust machine as for one inch. Raise cover from standard bar P attached to back of machine and clean plug T with benzine.

To focus, lower microscope Q to stop collar, and clamp.

Adjust hair line in eyepiece R to coincide with hair line on hardened steel plug S.

Unclamp sliding head and move to fifth plug T without disturbing eyepiece. Clamp fine adjustment G and adjust sliding head with screw L until hair line in eyepiece coincides with hair line on bar. Clamp head with lever M.

Place end-measuring rod between measuring faces and read as follows: To the reading on standard bar (5″) add the following: Count number of whole divisions on linear scale (12 divisions) and multiply by .040″. Count divisions on index wheel (200 divisions) and multiply by 1/10000″ or .0001″ and result will be the length.

Example. — $5'' + (12 \times .040'')$ $= .480'' + (200 \times .0001'') = .0200''$. Adding we have $5'' + .480'' + .0200'' = 5.500''$.

Attention. — Two supports are provided for long work, as at U.

Information. — English machines are standard at 62° F. and metric machines at 0° C. For practical work, it is not necessary to use these machines at these temperatures as variations caused by heat or cold affect machine and work alike.

SPIRAL MILLING

40. Spirals are used on twist drills, milling cutters and counterbores, and in machine construction for spiral gears, cams, etc. To cut a spiral, the work requires rotative and longitudinal motion at the same time and also to be set at an angle to cutter.

41. To find angle to set saddle for cutting spirals either graphically, by simple calculation, or approximately from spiral line drawn upon work:

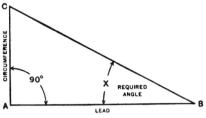

Graphically. — Draw a right triangle, *ABC*, to scale, Fig. 30. Draw *AB* equal to the lead, *AC* at right angle to *AB*, equal to circumference; connect

FIG. 30. — ANGLE DIAGRAM FOR TWIST DRILL.

CB. The angle *ABC* or *X* is the required angle. Measure angle with protractor.

Example. — Twist drill 1″ in diameter, 6″ lead, to find angle.

Solution. — *AB* = lead = 6″, *AC* = circumference = 3.1416; connect *CB* and measure; required angle $X = 27\frac{1}{2}°$.

By calculation. — Divide the circumference by the lead for the tangent. In table of tangents find the angle that corresponds to tangent.

Example. — Twist drill 1″ in diameter, 6″ lead, to find angle.

Solution. — Tan $X = 1 \times \dfrac{3.1416}{6} = .5236$. In table of tangents .5236 corresponds to 27° 38′ or $27\frac{1}{2}°$, the required angle.

Approximately. — Raise the knee until cutter touches work and with hand feed trace a light line upon the work, with the table set straight, and the machine geared for the spiral. Then swing table until spiral line is parallel to cutter.

Important. — As cutting the first tooth in a spiral mill or grooving the first drill is largely experimental, the student should prepare a machine-steel trial blank and use it to obtain correct setting of cutter.

Attention. — For right spirals, swivel the table to the *right;* and for left spirals, swivel to the *left,* and use an extra intermediate gear.

42. To calculate compound gears used in cutting spiral.

Example. — 1″ twist drill; 6″ lead, worm gear in dividing head 40 teeth, lead screw 4 *P.*, lead of machine equals $\frac{40}{4} = 10''$.

Solution. —
$$\frac{6}{10} = \frac{3 \times 2}{2 \times 5},$$

$$\frac{3}{2} \times \frac{24}{24} = \frac{72}{48} = \frac{\text{gear on worm } G}{\text{1st gear on stud } H},$$

$$\frac{2}{5} \times \frac{20}{20} = \frac{40}{100} = \frac{\text{2d gear on stud } J}{\text{gear on screw } K}.$$

When there is a fraction in the lead as in the case of spiral gears, as 8.333″:

Solution. — $8.333'' = \frac{8\frac{1}{3}}{10} = \frac{25}{30} = \frac{5}{6}$ and proceed as above.

Information. — If the fraction will not simplify easily, approximate the lead as follows: Exact lead 2.517″; approximate, 2.514.″   This is allowable for leads of cutters, drills and counterbores.   For mechanical motions, as spiral gears, change the diameter rather than lead.

43. To obtain lead for spiral milling cutter, given diameter and angle. —

Formula. — $\dfrac{\text{diameter} \times 3.1416}{\text{natural tangent of angle}} = \text{lead.}$

From lead calculate compound gears.

Attention. — Use above formula when changing pitch diameter of spiral gears to bring approximate lead within .005″ of exact lead.

44. Cutters for spiral mills are obtainable with either 40°, 48°, or 53° angle on one side and 12° on the other.

45. Cutters for two-groove twist drills are obtainable for drills $\frac{1}{16}''$ to $2\frac{1}{2}''$ in diameter.

CUTTING TEETH IN SPIRAL MILL

46. To cut teeth in spiral mill, universal milling machine. Fig. 31.

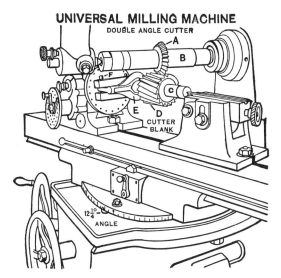

FIG. 31. — CUTTING TEETH IN SPIRAL MILL.

SPECIFICATIONS

Material..............	Carbon-steel cutter blank	Gears for spiral	
Hole.................	1″	Gear on worm..	72
Diameter of mill........	$2\frac{1}{2}″$	1st gear on stud	32
Length of mill.........	$2\frac{1}{2}″$	2d gear on stud	64
No. of teeth..........	18 R. H.	Gear on screw..	40
Width of land.........	$\frac{1}{32}″$	Speed	1st or 2d speed or 60 F.P.M.
Lead.................	36″		
Angle of spiral........	$12\frac{1}{4}°$	Feed..........	Medium power feed
Double-angle high-speed steel cutter..........	53°×12° R. H.		—.007″ per cutter revolution
Amount of set over.....	.208″	Lubricant.....	Lard oil

SCHEDULE OF OPERATIONS

I. Preparatory adjustments.
Place cutter A on arbor B to cut in direction of rotation, and clamp hard. Fasten index head and footstock in horizontal alinement on table.

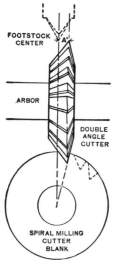

FOOTSTOCK CENTER

ARBOR

DOUBLE ANGLE CUTTER

SPIRAL MILLING CUTTER BLANK

Fig. 32. — Setting Double-angle Cutter Radial.

II. Set cutter radial. Set swivel table at zero, cutter central, and cross-feed dial at zero. Move cross slide inward $\frac{1}{12}$ diameter of blank (.208″) to give radial teeth, as at A', Fig. 32, then clamp cross slide.

III. Arrange gears, index and table. Arrange gears for spiral, set index for 18 teeth, withdraw stop so that index plate can revolve, set swivel table at $12\frac{1}{4}°$, Fig. 31, and clamp.

IV. Mount the blank on centers. Press mandrel C firmly into blank D and mount on centers, securing dog E in driver by screw F.

V. Obtain width of land by trial cuts. Start machine, raise knee and with power feed, take short trial cut less than required depth, then throw out power feed and stop machine. Set vertical dial at zero, lower knee, move table back until blank is clear of cutter and index for next tooth. Raise knee to zero and take another short cut.

Lower knee, move table back until blank is clear of cutter and measure width of land with rule.

If land is too wide, raise knee a little higher than before; repeat process until land is correct ($\frac{1}{32}″$ wide). Reset vertical dial at zero after each correction.

VI. Mill 18 teeth. With setting correct and table dog set to trip feed at end of cut, mill teeth all around blank, one cut each.

Warning. — To avoid cutter scraping work by back-lash in screw and gears, after obtaining depth of cut in spiral milling, set vertical dial at zero and at the end of each cut, lower knee and move table back to starting point, then raise knee to zero for next cut.

GROOVING TWIST DRILL

47. To groove twist drill, spiral milling, universal milling machine. Fig. 33.

UNIVERSAL MILLING MACHINE

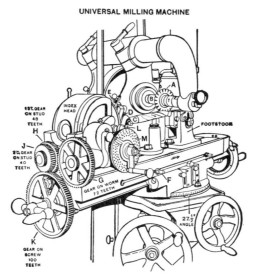

FIG. 33. — GROOVING TWIST DRILL.

SPECIFICATIONS

Material.........	Carbon-steel drill blank	Angle to set swivel bed for clearance of lands........	1°
Diameter of drill.	1.012″	Depth of clearance	$\frac{1}{64}$″
Length of drill...	11½″		
Length of grooves	6⅜″	**Gears for spiral**	
No. of grooves...	2	Gear on worm G ..	72
Thickness of point	⅛″	1st gear on stud H.	48
Lead	6″	2d gear on stud J .	40
Angle of spiral...	27½°	Gear on screw K..	100
Cutter..........	Special	Speed...........	3d speed, back gears in or 25 F.P.M.
Amount of set over	$\frac{1}{10}$ of dia. (.100″)	Feed............	.007″ per cutter revolution
Angle to lower index head......	2°	Lubricant........	Lard oil
		Time $\begin{cases} \text{grooving} .. \\ \text{clearance} .. \end{cases}$	1 h. 30 min.

SCHEDULE OF OPERATIONS

I. Preparatory adjustments.
Place cutter A on arbor B to cut
in direction of rotation, and clamp
hard. Fasten index head and
footstock in horizontal alinement
on table. Lower index head center 2° and clamp; this is to
thicken web and strengthen drill
which has a tendency to twist
when in use.

II. Set cutter radial. Set swivel

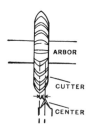

FIG. 34. — SETTING CUTTER OFF
CENTER FOR GROOVING
TWIST DRILL.

table at zero, cutter central, and
cross-feed dial at zero. Move
cross slide inward $\frac{1}{10}$ diameter of
blank (.100″), as at A, Fig. 34, and
clamp.

**III. Arrange gears, index and
table.** Arrange gears for spiral,
place index pin in any hole, then
withdraw stop L so that index
plate M can revolve; set swivel
table at $27\frac{1}{2}°$, and clamp. See
Fig. 35.

IV. Mount the blank on centers.
Mount blank C on centers and
secure dog D in driver by screw
E. Adjust V rest under middle
of blank to prevent springing.
Set table dog to trip feed at desired length of groove.

V. Obtain thickness of point.
Start machine, raise knee and take
short trial cut *less* than required
depth. Set vertical dial at zero,
lower knee, move table back until
blank is clear of cutter, index one-
half revolution, raise knee to zero
and take another short cut. Repeat until point is $\frac{1}{8}″$ thick.

VI. Mill grooves. Reset vertical dial at zero and mill grooves,
one cut each.

VII. Mill land clearance. Mount
drill A, Fig. 36, on centers to mill
lands B, with end mill C. Set
swivel table 1° (angle shown by
line D, E) to cut deepest at
F, F', and mill clearance $\frac{1}{64}″$ deep
leaving $\frac{3}{32}″$ lands, one or more
cuts.

Information. — Take trial cuts
and if lands are wider at shank
end of drill, raise index-head center 1° or 2° to make lands even
width. Oxidize small drills to
enable cut and land to be readily
seen.

Note. — Manufacturers of twist
drills use a special cutter for milling land clearances.

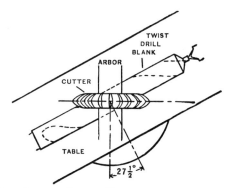

FIG. 35. — SETTING SWIVEL TABLE TO GROOVING ANGLE OF
TWIST DRILL.

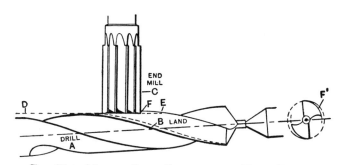

FIG. 36. — MILLING LAND CLEARANCE ON TWIST DRILL.

MAKING TWIST DRILL

48. To make taper shank twist drill, Fig. 37.

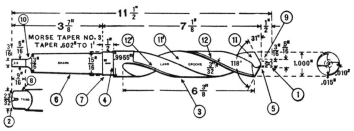

FIG. 37. — SCHEDULE DRAWING.

Specifications: Material, annealed high-speed or carbon steel $\frac{1}{16}''$ to $\frac{7}{8}''$ large; weight, 2 lb. 3 oz. Hardness, 20 to 25 (scleroscope). High-speed steel, or stellite cutting tools.

Time: 4 h. 45 min. which includes 15 min. for milling tang, 1 h. for milling spiral grooves and 30 min. for milling clearances.

SCHEDULE OF OPERATIONS, MACHINES AND TOOLS

OPERATIONS.	MACHINES, SPEEDS, FEEDS.	TOOLS.
Center.		
True live center. Set dead center in approximate alinement.	Engine lathe, 12″ to 16″.	
Rough square to $11\frac{1}{2}''$, (1), (2).	2d or 3d speed, or 40 F.P.M. Hand feed.	Dog, holder and cutter 30° rake, calipers, rule.
Recenter.		
Omit finish square.		
Rough turn to 1.012″, (3), one or two cuts. Turn half way, reverse and turn other half.	1st or 2d speed, or 40 F.P.M. Medium power feed — 80 to 1″.	Holder and cutter 30° rake, 1″ micrometer.
Draw lines for length of shank and body (5), (7).	Vise, copper jaws.	Copper sulphate, rule, scriber.
Turn taper shank .602″ to 1′, (6).		
Set over footstock $\frac{9}{32}''$, or use taper attachment.	Engine lathe.	Rule, dividers.
Rough turn taper (6) to $1\frac{3}{16}''$ at small end, one or two cuts.	1st or 2d speed or 40 F.P.M. Medium power feed — 80 to 1″.	Holder and cutter 30° rake, calipers, rule.
Take a light trial cut about .004″ to .005″. To correct this taper, see Schedule of Operations, p. 229.	3d speed or 60 F.P.M. Fine power feed — 140 to 1″.	Holder and cutter 30° rake, Morse taper-ring gage No. 3, chalk or Prussian blue.
Rough turn to .753″ + .020″ for grinding (6), one cut.	Medium power feed — 80 to 1″.	1″ micrometer.
Rough and finish turn reduced part (7), one or two cuts.	Fine power feed — 140 to 1″.	Calipers, rule.
Finish turn tang (8), one cut.		
Round end and corner (2).	Hand feed.	Forming tool.

SCHEDULE OF OPERATIONS, MACHINES AND TOOLS, *Concluded*

OPERATIONS.	MACHINES, SPEED, FEEDS.	TOOLS.
Rough turn stem $\frac{1}{2}''$ long (**9**), four or five cuts. Set tool at 59° and form point.	1st or 2d speed, or 40 F.P.M. Medium power feed — 80 to 1″.	Calipers, rule, and bevel protractor.
Stamp size and name on reduced part (**7**).	Vise copper, jaws.	$\frac{1}{8}''$ steel figures and letters, hammer.
Mill tang (**10**).	Universal milling machine, 2d or 3d speed, or 50 F.P.M. Medium power feed — .007″ per cutter revolution.	Index centers, 1″ end mill, calipers, rule, oil.
Mill grooves (**11**), (**11′**). See pp. **12**43–**12**45. **Mill land clearance** $\frac{1}{64}''$, (**12**), (**12′**). See pp. **12**44, **12**45. **Harden** in water.	Universal milling machine. Gas furnace or forge 1325° F. to 1350° F.	1″ end mill or special cutter, rule, oil. Tongs, water.
Temper in oil (or polish and temper to straw color). **Test hardness** with scleroscope, or dead-smooth file.	Oil-tempering furnace 425° F. to 435° F.	(Red-hot ring or collar).
Grind taper (**6**) to .753″ at end, or to fit Morse taper-ring gage. See pp. **7**12, **7**26, **8**30.	Universal grinding machine.	Grinding wheel No. 60, *K*, vitrified, Morse taper-ring gage, No. 3, chalk or Prussian blue, 1″ micrometer.
Set grinding machine to grind the taper for body clearance. Grind body to 1.000″ at point of drill. See pp. **12**06, **7**12. **Nick stem** (**9**), with grinding wheel; break off and grind point to 118°. See *Principles of Machine Work.*	Twist drill grinder.	Cup grinding wheel No. 46 *K*, vitrified, bevel protractor.

Attention. — If the drill springs in hardening and tempering an amount in excess of that allowed for grinding, it should be carefully straightened before grinding. See p. **12**17.

Information. — If not convenient to use a cylindrical grinding machine, the taper shank may be turned and filed to fit a Morse taper-ring gage, No. 3. The body may also be turned and filed to size.

PRECISION METHODS OF LOCATING HOLES FOR JIGS
AND ACCURATE MACHINE PARTS

49. The precise location of holes in jigs, fixtures, master plates and accurate machine parts is obtained by the plug method for large work and by the button method for small work. For ordinary machine work where an error of $\frac{5}{1000}''$ to $\frac{10}{1000}''$ is permissible, the usual method of locating holes with surface gage, dividers, rule, and center punch, and then drawing the drill, is sufficiently accurate. See Accurate Drilling, *Principles of Machine Work*. To locate holes of precision in jigs and fixtures where errors greater than $\frac{1}{1000}''$ are not permissible and an accuracy of $\frac{1}{20000}''$ is often required, the plug or button method is used.

50. Jigs and fixtures are usually designed by the draftsman and made by the toolmaker. Many toolmakers, however, are well qualified to both design and make jigs and fixtures. The shape and other requirements of the work determine whether a plate, solid, box, or rotary type of jig is required. See Drilling Jigs, pp. 601–610. The jig is designed by making a drawing having two or three views and the details. As an element of safety, it is best to make the design of such shape that it would be impossible to place the jig on the work or the work in the jig the wrong way.

After the position of the work in the jig is arranged, the locations of the holes and the clamping devices are determined. Wooden patterns are then made for the jig body and other parts that are to be made of cast iron and iron castings obtained.

51. Plane, mill or grind jig castings accurately. — The working surfaces of jigs and fixtures that are to be bored by precision methods must be planed, milled, or ground accurately and smoothly, and all sharp edges slightly beveled with a file. Box jigs and similar fixtures must have two or more surfaces or spots, exactly at right angles, from which all measurements are taken.

52. The vernier height gage, as shown in Figs. 40, 42, 43, 50 and 54, is a combination tool. As a height gage it is used for accurately measuring heights and spaces. As a caliper, it may be used in the same manner as the vernier caliper, see pp. 211, 212. It is an indispensable instrument for locating holes in jigs and fixtures by the plug or button method.

While the vernier height gage may be set to a ten-thousandth of an inch, accurate measurements and the comparison of distances and sizes also depend on the sense of touch or "feel" which is obtained by pressing the gage or caliper over or between the points being measured with a uniform pressure.

PLUG METHOD OF LOCATING HOLES

53. The plug method is a process of precisely locating the axis of the hole to be made with the axis of a standard plug held in the spindle of a universal or precision boring mill, or milling machine. With either machine the process is the same except for the variation in movements peculiar to the different makes of machines. The principle of the plug method of alining work in a boring mill is shown in Fig. 38. Measurements are taken from the hardened and accurately ground plug A, held in the spindle to the accurately scraped angle plate B which is clamped to the table parallel to the spindle, and from plug A to the table C.

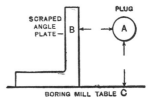

FIG. 38. — THE PRINCIPLE OF THE PLUG METHOD OF LOCATING HOLES.

54. To locate angle plate accurately on boring mill table. — To set angle plate A, Fig. 39, parallel to axis of spindle (assuming that the table is true) first place it in approximate position on table, test it with square, and clamp lightly. Then place drill chuck in spindle and fasten test indicator B in chuck. Test angle plate vertically with indicator by moving spindle up and down, and longitudinally by moving spindle in and out and adjusting angle plate by rapping it lightly with lead hammer. When angle plate is true, clamp hard and test again, as hard clamping has a tendency to change the location of angle plate.

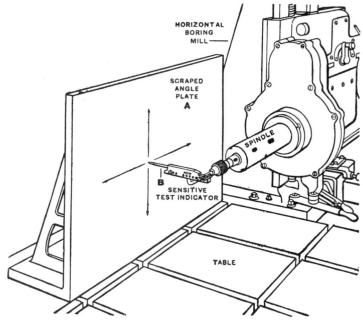

FIG. 39. — LOCATING ANGLE PLATE ACCURATELY ON BORING MILL TABLE.

Attention. — Vertical adjustments are made on some types of boring mills by raising or lowering spindle, see **Fig.** 39; and on others, by raising or lowering table, see p. **6**16.

55. To aline jig to angle plate. — Place drilling jig *A*, Fig. 40, against the tongue blocks *B* and *C*, clamp lightly and set parallel to the angle plate *D* by taking measurements at *E* and *F* with vernier height gage *G*. Make corrections by rapping jig lightly with soft hammer, then clamping it hard.

56. To true plug in spindle. — Taper shank plug *A*, Fig. 41, is placed in spindle *B* and tested with indicator *C* to see if it runs true. If the plug is hardened and ground accurately,

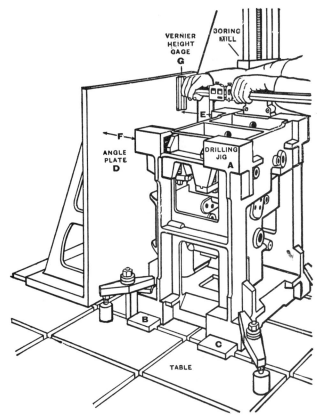

FIG. 40. — ALINING JIG TO ANGLE PLATE.

and care is used in placing it in the same position in the socket each time, it should run true; but, if it runs out a slight amount, the error is often corrected by removing and replacing the plug. When the plug runs true, the jig is lined up to it.

57. To set axis of the hole to be bored true to axis of the plug, two measurements are necessary. First set vernier height gage A, Fig. 42, the distance called for on the drawing from center of hole in boss B to bottom of jig C or to table D, minus one-half the diameter of plug E, then adjust spindle F,

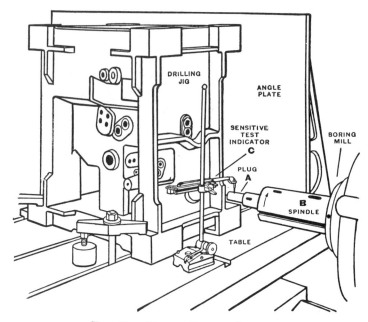

FIG. 41. —TRUING PLUG IN SPINDLE.

up or down, until distance is correct as determined by the vernier height gage A, Fig. 42.

Second, set vernier height gage A, Fig. 43, the distance called for on drawing from center of hole in boss B to side of jig C plus distance from jig C to angle plate D minus one-half diameter of plug E, then adjust table F until this distance is correct as determined by vernier height gage A, Fig. 43.

58. To bore hole in jig. — Place short drill A, Fig. 44, in the spindle and spot the work to make a true cavity, as at A', see also pp. **4**08, **4**09.   Then use two or three drills, as at B and C, the largest being about $\frac{1}{32}''$ smaller than the finished hole. Each drill removes but a small amount of stock, thus eliminating the danger of drilling the hole out of alinement.

After drilling, special boring tool D is used with a fine power feed.   Several boring tools may be used with the cutter set in each to remove but a small amount of stock.   The cutter in the last boring tool is set to bore the hole to the required size. See also Adjustable Boring Tool, Fig. 47.

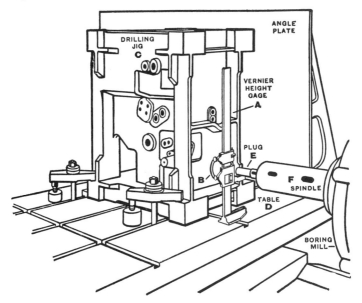

FIG. 42. — SETTING PLUG CORRECT DISTANCE FROM TABLE.

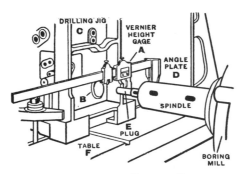

FIG. 43. — SETTING ANGLE PLATE CORRECT DISTANCE FROM PLUG.

Test hole with plug gage or measure with inside micrometer. Reamers are rarely used on this class of work as they have a tendency to throw the hole out of alinement. If the plug

gage fits rather tight after entering about halfway, a reamer
is used to relieve the back end of hole. In this operation

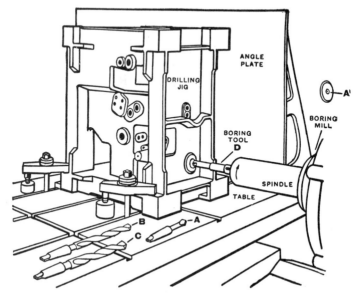

Fig. 44. — Boring Hole in Jig with Special Boring Tool.

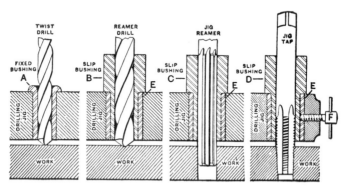

Fig. 45. — Bushings for Drilling Jigs.

the hole is not changed as the reamer removes so small an
amount of metal that it burnishes but does not cut.

Warning. — After boring the hole and before removing the jig, place a plug in the hole and test the measurements from gage to angle plate and table.

Information. — Additional holes in jig may be located by the same method.

Attention. — In locating and boring holes which pass through two opposite aces of a box jig, the boring is carried through the first face. The work is then reversed, relocated, and the second face bored. This avoids the error that may be caused by the spring of a long boring tool.

59. Bushings for drilling jigs. — As holes in cast iron or machine steel would soon wear out of true by the action of the tools which pass through them, holes in jigs are lined with hardened and ground steel bushings. The fixed drilling bushing, as shown at A, Fig. 45, is made of carbon steel hardened and tempered. The hole is ground and lapped to the required size, and the outside diameter ground to about .002″ larger than the bored hole and forced in with a press. Where more than one size of bushing must be used, they are called slip or removable bushings. A slip bushing for the drill is shown at B, for the reamer at C, and for the tap at D. The hole is also lined with a hardened and ground steel bushing as at E, forced in to receive the slip bushing. The tops of slip bushings are usually nurled so that they may be readily removed with the fingers, and they are often held in place with a binding screw, as at F.

60. To locate one hole accurately from another hole by the use of two plugs. — This method is used on jigs, fixtures, and accurate machine parts in the boring mill or milling machine. To bore holes 1 and 2 in lugs A and B, in a machine part, Fig. 46, as shown in detail at A' and B', clamp work to milling machine table, drill and bore hole 1 in lug A. Move table and place plug C in spindle, and plug D in hole 1, then adjust table and measure over plugs C and D with vernier caliper E or micrometer. This measurement should be the distance between centers of holes 1 and 2 plus half the diameter of each of the plugs C and D.

As holes 1 and 2 are not in horizontal alinement, measure distance F from plug C to the table with vernier height gage, then raise table the difference in alinement of holes 1 and 2, as shown in detail at *G*, and again measure with vernier

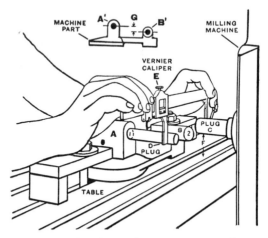

FIG. 46. — LOCATING ONE HOLE ACCURATELY FROM ANOTHER HOLE BY USING TWO PLUGS IN MILLING MACHINE.

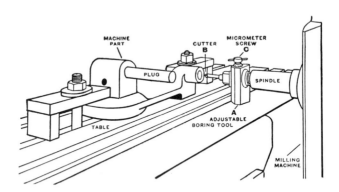

FIG. 47. — BORING WITH ADJUSTABLE TOOL.

height gage from plug *C* to table. This measurement should be distance *F* minus the difference in alinement. Hole 2 is then drilled and bored.

61. Adjustable boring tool shown at A, Fig. 47, is used for boring in milling machines, boring mills, drill presses and engine lathes.

The cutter B is adjusted by micrometer screw C, making it very useful for precision boring.

Attention. — If a milling machine is used, care must be taken to see that it is in good condition. If there is an appreciable sag of the knee, an allowance should be made in setting the work. In any machine used for accurate boring, the slides should be cleaned and oiled and the gib screws adjusted so that all movable parts fit closely but not too firmly.

BUTTON METHOD OF LOCATING HOLES

62. The button method is a process of precisely locating holes with hardened and ground buttons or collars often called toolmakers' buttons. It is a slower process than the plug method but it is used for small accurate work. It consists of three operations: First, locate the button precisely and clamp it to the face plate

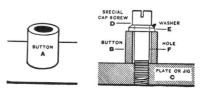

FIG. 48. — THE BUTTON, AND METHOD OF FASTENING THE BUTTON TO PLATE OR JIG.

or jig. Second, clamp the plate or jig to the face plate of the lathe, test the button with a sensitive test indicator and adjust plate or jig until the button is perfectly true to the axis of rotation. Third, drill and bore the hole.

The button used is shown at A, Fig. 48; it is about $\frac{1}{2}''$ long, and ground to some exact diameter, as .300″, .400″ or .500″, with perfectly square ends. Where only one hole is to be located, a button of an exact diameter is not important; but if two or more holes are to be located, they should all be of the *same* diameter. A sectional view of the button is shown at B, and the method of clamping to plate or jig C by special cap screw D' and washer E.

The screw may be about $\frac{1}{8}'' \times 40$, and, to permit adjustment, hole F in the button is about $\frac{1}{16}''$ larger than the screw.

63. To locate the button on a plate. — For hole A, Fig. 49, locate the button B on plate C the distances D and E from side and end of the plate by finding its approximate center with surface gage, rule, and square, and making a center punch mark. The center punch mark is usually within .010″ of being correct. Next drill and tap for the screw, file off burr and fasten button lightly to the plate with the screw.

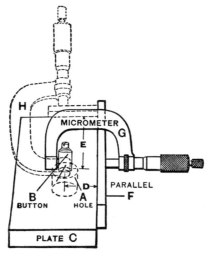

Fig. 49. — Locating the Button Accurately on a Plate.

Then hold parallel F against side of plate, measure with micrometer caliper G, and adjust the button by rapping it lightly with a lead hammer until this measurement equals the distance D plus thickness of parallel, plus half the diameter of the button. Repeat the process for distance E with micrometer as shown dotted at H. Measure the distances D and E alternately, until button is precisely located, then fasten button down hard. Clamp the plate lightly to an accurate face plate of an engine lathe, and test the button with a sensitive test indicator held in the tool post, and adjust the plate until the button is perfectly true to the axis of rotation, see Art. 65, pp. **12**59, **12**60. To bore the hole, see Art. 66, pp. **12**60, **12**61.

64. To locate the button on a drilling jig. — Hole A, Fig. 50, is to be located halfway between the legs of drilling jig B and at a given distance from the face. Locate center of hole with surface gage, rule, square, and center punch. Drill and tap hole, file off burr, wipe clean and fasten button C to the jig. Clamp jig to the accurately scraped angle plate D to give larger working distances. Set vernier height gage E

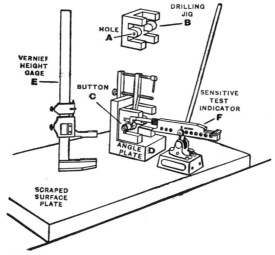

FIG. 50. — LOCATING THE BUTTON ON A DRILLING JIG.

to equal the distance from the edge of jig to the center of hole plus one-half the diameter of button plus thickness of the angle plate. Transfer this setting to sensitive test indicator F, with needle at center of scale. Adjust button until there is no change of needle-reading. Turn the jig down and repeat this process to locate button the correct distance from face of jig. Then fasten button on hard and test again, as hard clamping may change the position of button.

65. To set button true to axis of rotation. — Clamp drilling jig A, Fig. 51, lightly to a perfectly true face plate with button B approximately true to axis of rotation, then balance with counterweight C. See also pp. 407, 408. Rotate spindle by hand and test the location of button B by indicator D. Adjust jig by rapping it lightly with a lead hammer until the needle of indicator ceases to move, then tighten the clamping bolts. Test button again, as final clamping, may change location of button.

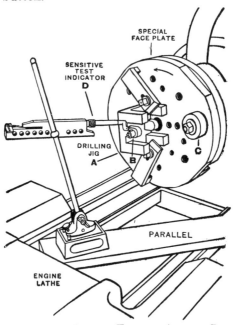

FIG. 51. — SETTING BUTTON TRUE TO AXIS OF ROTATION.

66. To bore the hole in jig. — Remove button and with flat drill A, Fig. 52 (or centering tool), held in a drill chuck fitted into footstock spindle, cut true cavity B to start drill. See also p. **4**09. Then use increasing sizes of drills until hole is within $\frac{1}{32}''$ of size. Bore hole C exactly to size with boring tool D taking two or more cuts and using a fine power feed. Test hole with plug gage or measure with inside micrometer. See Bushings for Drilling Jigs, p. **12**55.

Warning. — Care should be taken to see that the lathe spindle bearings are tight for any looseness will affect the truth of the bored hole. Also, see that the jig is accurately counterbalanced as the bored hole will be out of true if the jig has any springing or vibrating action.

If the shape of the work is such that the rotative speeds for drilling and boring will cause the spindle to spring and the work to vibrate, use the plug method and a boring mill, or milling machine, where the work will be stationary and the drills and boring tool will revolve.

Attention. — The *true* cavity starts the drill true and eliminates the eccentricity of the tapped hole.

Information. — The button method may be used in boring mill or milling machine by alining the button to a slip bushing on plug in the spindle.

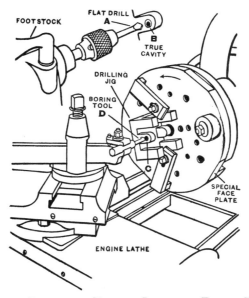

FIG. 52. — BORING THE HOLE IN JIG IN THE ENGINE LATHE.

67. To locate two holes by the button method usually requires that measurements be referred to the buttons themselves as well as to the edges of the work. In Fig. 53 button A for hole 1 is to be located with reference to edges CD and DE on plate F, and button B for hole 2, with reference to button A and edge DE. First locate both buttons approximately with reference to edges, by the method given on p. **12**58.

The precise locations of buttons A and B are determined with micrometers, as shown in Fig. 53.

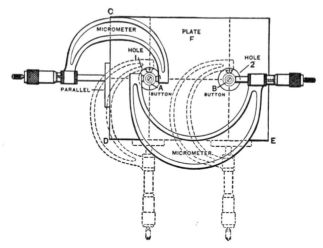

FIG. 53. — LOCATING TWO HOLES BY THE BUTTON METHOD.

68. To locate and bore holes an approximate distance apart with boring mill or milling machine. — On jigs and fixtures where the extreme accuracy of the plug or button method is not required, the holes may be located and bored approximately accurate a given distance apart by using the graduated dials, reading in thousandths of an inch, on the actuating screws of a boring mill or milling machine. See p. 616.

69. To locate and bore two or more holes an exact distance apart with boring mill or milling machine. — First, clamp the work to the table and adjust the table to move the work to the proper position, then drill and bore the first hole to the required size. Second, move the table by the actuating screw the required distance for the second hole as indicated on the micrometer dial. Third, drill and bore the hole $\frac{1}{16}''$ small and place a plug or proof bar in each hole. Fourth, test the distance between the bored holes by measuring over both plugs with a micrometer. If after subtracting half the diameter of each plug from this reading, the result is not the required distance, correct by moving the actuating screw the required number of thousandths of an inch or fraction of a thousandth, as indicated by the micrometer dial, then bore the hole to the required size.

SINE BAR

70. The sine bar is used for the precision measurement of angles, and for locating work at precise angles. It consists of a hardened and ground bar with hardened and ground plugs placed at a precise and known distance (usually 10″) apart, which by means of a vernier height gage and a table of sines and cosines may be set to any angle or used to measure any angle precisely. Some sine bars, as A, Fig. 54, are made with hardened bushings to which accurately ground removable plugs, B and C, are fitted; other sine bars, as B, Fig. 55, have fixed plugs, C and D.

Information. — The holes in the sine bar are located by either the plug method or the button method.

71. To set adjustable angle plate with the sine bar.— In Fig. 54 the sine bar A is used to set adjustable angle plate

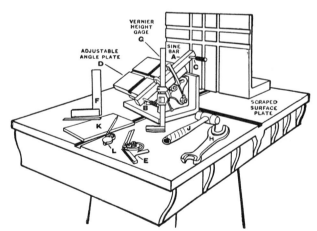

Fig. 54. — Setting Adjustable Angle Plate with the Sine Bar.

D at a precise angle: First, place angle plate on the scraped surface plate and set it approximately by bevel protractor E, then with square F set sine bar parallel to edge of angle plate, and clamp in position. Find difference in height of plugs B and C with vernier height gage G and obtain the precise angle. Adjust plate with wrench H and lead hammer J until difference in height of plugs is correct for the desired angle. For example, the angle plate is to be set at 30°. The sine of 30° is $\frac{1}{2}$. Therefore, with a 10″ sine bar the difference in height of the plugs should be $\frac{10}{2}$ or 5″.

Attention. — As each adjustment of angle plate changes the height of both plugs making the vernier readings difficult to remember, it is customary to make a record of the readings as suggested by the pencil and block of paper at K. The magnifying glass at L may be used to read the vernier.

72. To measure and originate tapers with the sine bar. — The sine bar may also be used for measuring and originating tapers.

In Fig. 55 plug gage A is to have a taper of .5161″ per foot. While this is a standard taper (Brown & Sharpe, No. 10) it may be assumed that there is no ring gage or templet available and the sine bar is used instead of a gage. The plug is placed in the fixture on the scraped surface plate and the sine bar B adjusted to the taper gage and clamped in position. The difference in height of the plugs C and D is found by vernier

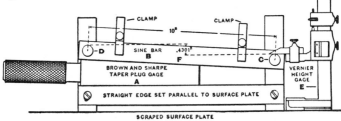

FIG. 55. — MEASURING A TAPER GAGE WITH THE SINE BAR.

height gage E. If the difference is not correct for the required taper, the gage is corrected by grinding or lapping and the process repeated. For example, gage A is to have a taper of .5161″ per foot or .5161″ ÷ 12 = .04301″ per inch.

Assuming a 10″ sine bar is used the difference in height of the plugs C and D should be .04301 × 10 or .4301″.

73. Other uses for the sine bar. — Besides being useful for the above processes, the sine bar is used for setting milling machine heads and tables, testing angle settings of vises, chucks, and other machine tools where precision is desired. It is also very useful in toolmaking for inspection purposes.

RELIEVING FORMED MILLING CUTTERS, TAPS, DIES, COUNTERBORES AND HOBS

74. Relief or clearance on formed milling cutters. — The clearance on lathe and planer cutting tools is produced by hand or machine grinding. See pp. 135–142. The clearance on plain, radial and angular milling cutters (also some form milling cutters) and reamers, is produced by grinding a flat clearance on the teeth with cutter grinding machines. See pp. 808–832.

FIG. 56. — FORMED MILLING CUTTER.

Formed milling cutters, as in Fig. 56, for milling surfaces of irregular outline and gear cutters, see p. 1101, also taps, dies, counterbores, hobs, and similar rotary cutters, are given an eccentric relief or clearance. A hand method of producing this relief is to file an increasing amount of metal away from the cutting edge to the heel of the teeth. This is not satisfactory as formed cutters are sharpened by grinding the teeth radially, see pp. 823, 824, and the grinding will change the outline of the teeth.

75. Machine relief. — Formed milling cutters are given an eccentric relief or clearance (the land back of the cutting edge is curved, not flat) with a cutter-relieving or backing-off attachment in an engine lathe, see p. **12**71, or with a cutter-relieving machine, see pp. **12**86.

To relieve very large cutters, or to make cutters in quantities, special cutter-relieving machines are generally used.

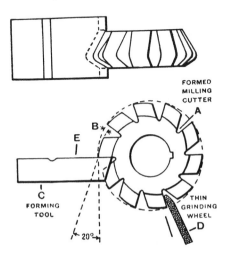

FIG. 57. — SHOWING THE PRINCIPLE OF RELIEVING FORMED MILLING CUTTERS.

76. The principle of relieving formed milling cutters is shown in Fig. 57. — The teeth of formed milling cutter A are relieved or backed off to give an eccentric relief, as at B. See Art. 89, p. **12**74. It corresponds to the clearance below the cutting edge of turning tools. See Cutting Tools, pp. **1**25–**1**35, **3**01–**3**09, and Cutter Grinding, pp. **8**08–**8**27. This relief is obtained by giving forming tool C a forward and backward motion from zero to $\frac{1}{4}''$ or more, according to the amount of relief desired. Each tooth is given the same amount of eccentric relief. After hardening and tempering, each tooth is ground radially with a thin grinding wheel, as at D. This grinding does not change the curvature or outline of the cutter which is of the greatest importance in producing duplicate work and work of precision.

77. The process of making forming tools for relieving formed milling cutters. — Simple forming tools for relieving

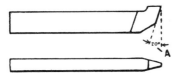

FIG. 58. — FORMING TOOL — 29° THREAD RELIEVING TOOL.

straight outlines or those which do not have to be geometrically correct, may be forged and ground to the desired shape like ordinary lathe tools, as shown in Fig. 58, but with a clearance angle, as at A, from 20° to 25°.

For cutters of irregular outline, the forming tool should be a flat cutter, as shown in Fig. 59, and secured in a suitable tool-

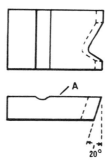

FIG. 59. — FLAT FORMING CUTTER.

holder. This tool is sharpened by grinding on the top surface A, which must be kept parallel with the shank. See p. **8**25.

78. Master-tools or scrapers, are used to produce these forming tools. — The master-tool is made the reverse of the shape of the forming tool: A flat piece of tool steel about ⅛ of an inch thick, as shown in Fig. 60, is accurately lined out, planed or milled, and carefully filed to the exact form desired with no clearance.   It is then hardened, tempered, ground, and lapped to correct errors.

CUTTING
EDGE

FIG. 60. — MASTER-TOOL OR SCRAPER.

Attention. — No straight cutting edge of a forming tool should be designed to cut perpendicularly to the axis of the cutter, but inclined about 5° for clearance.

79. Shaping the forming tool with a shaper. — The operation of shaping the forming tool is shown in Fig. 61.   The forming tool *A* is first machined and filed nearly to shape, then held in fixture *B*, which is clamped in the shaper vise at the desired clearance angle 20°, as in detail at *C*.

A clearance of about twenty degrees is given forming tools for eccentrically relieving formed milling cutters, taps, hobs, etc., having a moderate relief.   For a very large amount of relief, this clearance may have to be increased.

Forming tools for concentric forming need not have more than from 12° to 15° clearance.

The master-tool *D* on shaper is clamped in special tool-holder *E*, in alinement with the forming tool, or 180°, as at *F*, otherwise the master-tool will not cut the correct shape in the forming tool.   Supports, as at *G*, prevent the master-tool from springing.   The shaper must be run at a slow speed.   The forming tool should be so near to the desired shape that the master-tool will have to take but a few scraping cuts to produce the correct shape.   The forming tool is hardened and tempered, and ground.

Information. — Master-tools and forming tools for accurate work should be made of a high-grade carbon or tool steel that will shrink but a minimum amount in hardening and tempering.

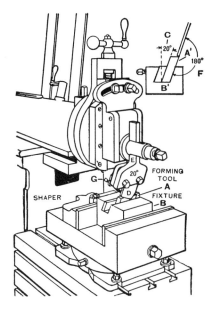

FIG. 61. — ACCURATE PROCESS OF SHAPING FORMING TOOL.

80. To set forming tools in the lathe or cutter relieving machine. — The bottom of forming tools and tool holders should be machined to facilitate accurate setting. To relieve work of straight outline, the tools are set at height of center, clamped in the tool post, then adjusted to suit the work.

Forming tools for work of precision of irregular outline, must be set at the height of centers and at right angles to the axis of the cutter. This may be done with reasonable accuracy by shaping the flat cutter to a known relation to an edge or a side, then setting this edge parallel or the side perpendicular to the axis of the work. For greater precision, the forming tool may be adjusted to the correct relation to the work axis with a test indicator. See Test Indicators, pp. **12**10–**12**13. Straight and curved parts of the forming tool should be set within .001″, and angular parts within five minutes.

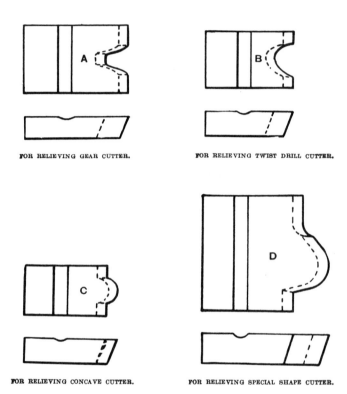

FOR RELIEVING GEAR CUTTER.

FOR RELIEVING TWIST DRILL CUTTER.

FOR RELIEVING CONCAVE CUTTER.

FOR RELIEVING SPECIAL SHAPE CUTTER.

FIG. 62. — EXAMPLES OF FLAT FORMING TOOLS FOR RELIEVING FORMED MILLING CUTTERS ARE SHOWN AT A, B, C, AND D

Information. — Some toolmakers match the form to the cutter by the use of a feeler gage or by placing a sheet of white paper under the tool and cutter and testing the accuracy of the setting by the light line between the edge of the tool and cutter when they are brought close together.

81. Cutter relieving attachment for engine lathes, Fig. 63. —
The attachment is bolted to lathe bed at A, and tool slide B is
fastened to cross slide in place of regular tool block or compound
rest.

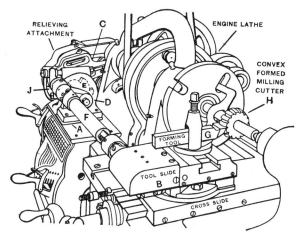

FIG. 63. — RELIEVING CONVEX FORMED MILLING CUTTER WITH
RELIEVING ATTACHMENT IN ENGINE LATHE.

Motion for the actuating mechanism is obtained from the
feed stud gears the same as for thread cutting. Sector C is
used to bring the gearing into mesh with gear on end of cam
shaft D.

To the sleeve of cam shaft D, a four-lobe cam E is keyed.
The gearing and cam impart an oscillating motion to the universal joint shaft F; and by means of an eccentric at tool-slide
end of shaft, a reciprocating motion is given forming tool G.

The amount of relief may be varied for different types of
cutters by a toothed coupling at J on ends of universal joint
oscillating shaft and cam lever. This permits a change of the
relative position of the eccentric in tool slide to cam lever, and
lengthens or shortens the reciprocating travel of tool. The
toothed coupling gives a relieving or backing-off adjustment
from zero to $\frac{1}{4}''$ on the different sizes of lathes.

A table of change gears for the number of teeth or flutes per revolution, is furnished with cutter relieving attachments and cutter relieving machines.

Attention. — On some cutters the relieving attachment is placed at the back of the lathe.

82. Speed for relieving should be slower than for ordinary turning or threading.

For narrow cuts, approximately 170 teeth may be relieved per minute; but, for very wide forming tools the speed may have to be reduced to 8 or 10 teeth per minute depending on the width of the cut. The tool slide should be adjusted to move freely but not loosely.

83. Feed for relieving may be considerably more than .001″ for each revolution of the cutter for the first cuts; but for the last few finishing cuts, the feed should be less than .001″ to insure smoothness.

84. Lubricant. — All work is relieved dry.

85. Coloring the cutter. — To plainly see the cutting action of the tool, and to know when to stop relieving, the cutter should be colored by dipping it into a strong solution of copper sulphate, or by heating it to a blue.

86. To prepare and flute formed milling cutter blank. — Chuck out an annealed high-speed or carbon-steel disk.

Press a mandrel firmly into the hole. Face both sides and rough turn curved parts to approximate shape.

Fasten forming tool, that is, the tool to be used in relieving, in tool post, at right angles to axis of work, and adjust it to height of centers and finish form cutter blank to shape. Next, mill flutes in blank with an angular cutter (30° to 45° with points of teeth rounded) to give the required number of teeth and ample chip room. The cutter is now ready to be relieved.

Attention. — Eccentrically relieved cutters are usually made with fewer and thicker teeth than ordinary cutters; this prolongs the life of the cutters as they may be repeatedly sharpened and used until the teeth are very thin and liable to break.

87. To relieve formed milling cutter, Fig. 63. — Press mandrel firmly into cutter as the cutter must not slip on mandrel. Fasten dog firmly on mandrel, and wedge or otherwise confine tail of dog in the slot in face plate to avoid back lash. Clamp forming tool *G*, or a flat forming cutter of the desired shape, firmly in a tool holder in tool post, at right angles to axis of cutter *H*, and at height of lathe centers. Adjust the tool by means of cross slide and carriage, to fit the cutter, then clamp carriage to lathe bed.

Select the change gears to drive the attachment from the table to index the required number of teeth or flutes in cutter. Arrange gears on stud and cam shaft. To adjust movement of tool slide so that forward movement of tool will meet cutting edge of tooth and promptly return after passing heel of tooth, drop a tooth in change gears. Then adjust tool to aline with the cutting edge of tooth, and rotate the lathe spindle forward a few revolutions by hand and observe action of tool in relation to the teeth of cutter. If tool does not meet the cutting edge of a tooth and promptly return after leaving the heel, repeat the adjusting operation. See p. 3³².

When the movement of the tool in relation to the teeth is timed correctly, the process of relieving the teeth may begin.

Move tool forward to touch a tooth near the heel. Start the lathe and, after each revolution of cutter, move tool forward one-thousandth of an inch or less by advancing the cross feed. Each cut begins nearer the front of the teeth. When the cutter reaches the cutting edge of the teeth, the relieving process is finished. After hardening and tempering, the cutter may be sharpened. See pp. 8²³, 8²⁴.

88. To measure relief of formed milling cutter place the cutter on a mandrel, mount on centers, revolve by hand and with a dial test indicator, see p. **121**² (or with tool in tool post and rule), measure the distance between the cutting edge and heel of a tooth. This difference is the relief or clearance. For general purposes the relief may be about .010″ for each $\frac{1}{8}$″ of thickness of tooth measured on the circumference.

Example. — A tooth $\frac{1}{4}''$ thick would have .040″ relief; a tooth $\frac{5}{16}''$ thick, .050″; a tooth $\frac{3}{8}''$ thick, .060″.

Attention. — The teeth of formed milling cutters are sometimes remilled after relieving which gives them a sharper edge and reduces the amount of grinding.

89. To relieve a large concave formed cutter, A, Fig. 64. — The process of relieving this cutter is the same as for the concave cutter, Fig. 63, except that the speed must be much slower and the cuts lighter.

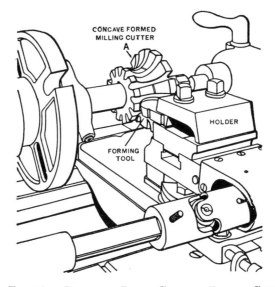

CONCAVE FORMED
MILLING CUTTER
A

HOLDER

FORMING
TOOL

Fig. 64. — Relieving Large Concave Formed Cutter.

90. To relieve side of angular formed cutter, A, Fig. 65. — The cutter relieving attachment can be used not only to relieve the teeth of cutters parallel to the axis, but also, to give side relief (90°), or relief at any angle to the axis, as in Fig. 65. The swivel tool slide is set perpendicularly to the surface to be relieved. The proper gears for indexing are selected and arranged. The relieving tool is fastened in the tool post, and the process of operating is the same as for the convex cutter, Fig. 63.

Information. — Formed milling cutters need not be made to exact diameters but the curvature should be accurate.

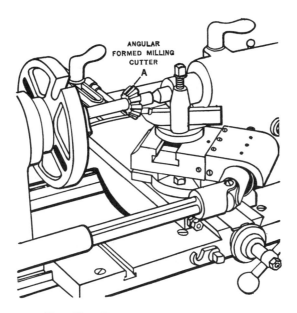

FIG. 65. — RELIEVING ANGULAR CUTTER.

91. To relieve a counterbore (side relief). — The method of giving side relief to the cutting teeth of a counterbore is shown at A, Fig. 66. The swivel tool slide is set perpendicularly to the ends of the teeth, and two universal joints B and C, are added to the oscillating shaft which allows the tool slide to be swiveled to a 90°-angle.

92. Hand taps, as shown in Fig. 67, are made in sets of three and are used for general purposes in machine building. The tapered tap A is turned taper from 5 to 8 threads from the bottom of the thread at the end to the full diameter. Some taper taps are made with leaders or guides; that is, 3 or 4 threads at the point are turned off, straight, down to the bottom of the thread so as to enter the drilled hole, then 5 or 6 threads are turned taper.

The plug tap B is tapered or chamfered from 2 to 4 threads at the point, and one thread of the bottoming tap C is chamfered.

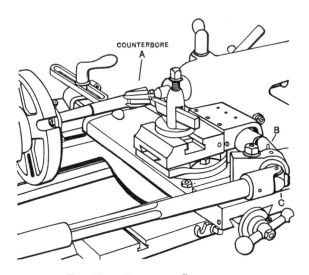

FIG. 66. — RELIEVING COUNTERBORE.

Taps for special purposes are obtainable or may be made, such as long pulley taps, long nut taps for tapping nuts in tapping machines, hob, or master-taps made slightly taper for tapping dies, taps for turret lathes and screw machines with special shanks to fit holders, staybolt taps for boiler work, pipe taps both straight and taper, and machine screw taps.

93. Hand relieved taps. — A hand method of relieving or backing off a tap is to color the tap, Art. 85, p. **12**72, in order to see more clearly where the file is cutting. Then, with a fine file of suitable shape, start at the heel of a tooth and file the metal away from about two-thirds of its top and sides and from the bottom of the space to give a circular relief or clearance from 5° to 10°, leaving one-third of the tooth back of the cutting edge full size. Repeat this process on each tooth. The top of the tapered threads at the end of a tapered plug or bottoming tap, as shown in Fig. 67, are filed flat to give a full relief or clearance from 5° to 10° up to the cutting edge of the teeth to facilitate starting the tap. The angle in the thread of the tapered part of a tap is often given a relief.

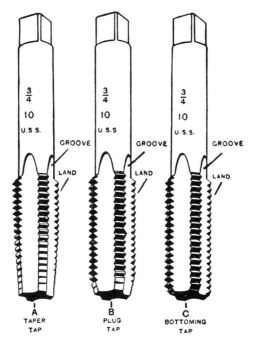

Fig. 67. — Set of Hand Taps.

The thread relief on small taps up to $\frac{1}{2}''$ in diameter with fine pitches is often omitted. By filing the tops of a few teeth tapering at the end, and with a sharp, flat relief, the tap will cut satisfactorily. Body relief, or back taper, is often given both large and small taps by setting the lathe to thread the tap at a slight taper toward the shank of about .001″ to 1″ and then filing a sharp, flat relief or clearance on the top of the tapered threads at the end.

94. Machine relieved taps. — The most modern method of relieving taps is to use a cutter relieving attachment, as on .p. **12**71, or a special relieving machine, and to relieve but $\frac{2}{3}$ of a tooth on top and sides and at bottom of space, as in Fig. 68,

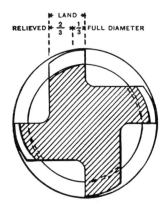

Fig. 68. — Section of Tap Showing Relief Given Teeth.

leaving $\frac{1}{3}$ of the tooth back of the cutting edge full size. This method prolongs the accurate life of a tap, as it may be repeatedly reground radially without reducing its maximum cutting diameter. See p. **8**25.

The tops of the tapered teeth at end may be relieved with the relieving attachment, or they may be milled or filed.

95. To relieve a right-hand tap. — The method of relieving a right-hand tap is shown at A, Fig. 69. The tap is made in the regular way by threading, grooving, and tapering the end. See pp. **12**15–**12**23.

Use the same gears as for threading; also select and arrange the gears to drive the attachment to index the number of grooves or "lands" of the tap. Set the threading tool the same as for threading. Drop a tooth in change gears and adjust tap until tool will accurately fit a thread space. See Art. 65, p. **3**32. Then arrange motion of tool slide so that the forward movement of the tool will meet front of teeth and promptly return after leaving heel of teeth.

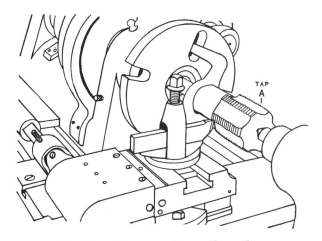

FIG. 69. — RELIEVING RIGHT HAND TAP.

The process of relieving the tap is similar to that of threading a tap: Arrange the thread stop. Set tool to cut near the heel of a tooth. Move tool forward one-thousandth of an inch or less, then start lathe and at the end of the cut withdraw tool and reverse lathe. See Catching the Thread, pp. **3**36, **3**37. Repeat this process until about two-thirds of a tooth is relieved which leaves one-third of a tooth back of the cutting edge the full size.

96. To relieve left-hand taps. — Left-hand taps may be relieved by either of two different methods.

The first method is to leave an extension at the end of tap. The tap is then placed on centers, reversed with the extension on the live center, and the process is then the same as for relieving a right-hand tap.

The second method is to mount tap on centers in the same way that a right-hand tap is mounted, but set travel of tool slide the same as for *inside relief*, see Fig. 70, and start tool at heel of tooth terminating cut at the front of the tooth, the tool moving *outward* during the cut. This is obtained, on some types of relieving attachments, by adjusting the toothed coupling of cam lever and oscillating shaft beyond the zero mark (clockwise) to secure the desired movement of tool slide; and on other types, by moving connecting-rod bolt to the opposite side of center in the radial slot in disk.

97. To relieve threading dies (inside relief) or hollow mills. — The method of inside relieving for threading dies, or hollow mills, is shown at *A*, Fig. 70.

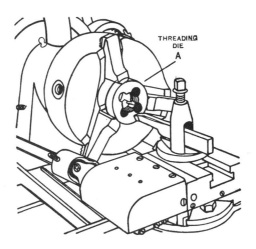

FIG. 70. — RELIEVING THREADING DIE.

The die is turned, threaded, and grooved in the regular way. Set die true in the chuck testing it with a test indicator. See pp. **12**10–**12**13. Gear lathe the same as for relieving a tap, and set inside threading tool the same as for threading and adjust it to fit a thread space. Set eccentric controlling travel of tool slide, to move *away from* instead of *toward* axis. On some types of relieving attachments this is obtained by rolling the toothed coupling beyond the zero mark (clockwise); on other types by setting connecting rod bolt beyond center in rocker-head slot.

The process of operating the lathe to relieve the die is the same as for a tap.

Warning. — The position of the opposing spring inside tool-slide hood, must be changed so that it will press against the end of slide to prevent tool jumping into work when in cut.

98. To relieve a straight fluted hob. — Hobs of fine pitches are generally fluted straight.

The set-up, arrangement of gears, and process for relieving straight fluted hobs, is similar to the method used in relieving straight grooved taps, Fig. 69.

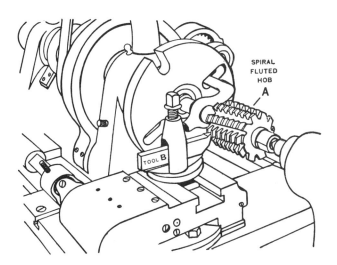

Fig. 71. — RELIEVING HOB WITH SPIRAL FLUTES.

99. To relieve spiral fluted hobs and taps. —The method of relieving a spiral fluted hob is shown in Fig. 71. The gears to index the number of flutes in hob may be obtained from the table, but the compensating gears for the angle of the spiral must be calculated and added as a compound. See pp. **3**33, **3**35.

Spiral fluted hob *A* may be single or multiple threaded and may have been threaded in a thread miller, see pp. **10**48, **10**49, or in an engine lathe.

Gear lathe for threading the required pitch of hob, then gear attachment for the number of flutes and angle of spiral. Set tool *B* to fit the thread spaces, and correctly time movement of tool in relation to teeth of hob.

After the lathe attachment and tools are properly set up, the process of relieving a spiral fluted hob is the same as for a straight fluted hob or for a tap.

Attention. — For large hobs, the top and sides of the teeth and the bottom of the spaces must be relieved by suitable tools and separate operations.

100. Spiral fluted hob, Fig. 72, is fluted at right angles (90°) to the thread, which makes the cutting action of the tooth more effective than a straight flute.

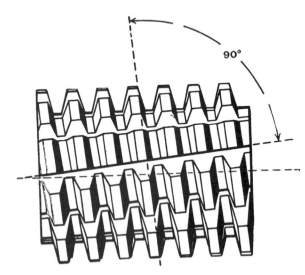

Fig. 72. — Hob Spirally Fluted at Right Angles to Thread.

The gears for a straight fluted hob or tap may be obtained from the index or table furnished with the attachment, but to relieve a spiral fluted hob or tap additional compensating or compound gears must be obtained by calculation and used in conjunction with the index gears to compensate for the angle of the spiral flute. In the calculation, however, it is usually necessary to slightly modify the angle of the flute in order to use available gears.

101. Method of computing change gears used on lathe attachment for relieving spiral fluted hobs and taps.

Deduction of formulas. — To obtain the lead of the spiral which is at right angles to the thread, the following symbols and formulas are used, Fig. 73.

$C = ab$ = circumference of hob at pitch line.
$L = bc$ = lead of thread.
$P = ad$ = lead of spiral or flute.

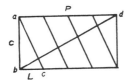

FIG. 73. — DIAGRAM OF HOB.

Note. — Fig. 73 represents a unit section of hob with ca equaling one complete turn of spiral, ad unit of straight flute, and ac representing the thread. By plane geometry (similar triangles) it can be proved that $\dfrac{\overline{ab}^2}{bc} = ad$ or $P = \dfrac{C^2}{L}$.

Gear ratio. — In calculating the change gears it will first be assumed for simplicity, that there is only one spiral flute.

For a straight flute, the number of teeth represented by ad would equal the length of the hob divided by lead of thread, or it may be expressed by the fraction $\dfrac{P}{L}$. When a spiral flute (bd) is considered, it can be seen that there is one more tooth on spiral flute than on straight flute.

Hence, number of teeth for straight flute $= \dfrac{P}{L} = M$, and number of teeth for spiral flute $= \dfrac{P}{L} + 1 = N$.

Therefore, the ratio of gears required for compensating for the spiral flute is $\dfrac{N}{M}$.

Note. — To flute a right-hand tap right-handed, the formula for the ratio $N = \dfrac{P}{L} + 1$ becomes $N = \dfrac{P}{L} - 1$.

For any number of flutes, the compensating gears are used in compound with those called for on the index for the required number of flutes.

Example. — A hob has a pitch circumference of 3.464″, a single thread of .750″ lead, and 6 flutes.

Solution. — By formula

$$\text{lead of spiral} = \frac{C^2}{L} = \frac{(3.464)^2}{.750} = \frac{12}{.750} = 16.000.$$

Then $\quad M = \dfrac{P}{L} = \dfrac{16}{.75} \quad$ and $\quad N = \dfrac{P}{L} + 1 = \dfrac{16}{.75} + 1$

$$= 21\tfrac{1}{3} \qquad\qquad\qquad = 22\tfrac{1}{3}$$

$$\text{Ratio} = \frac{22\tfrac{1}{3}}{21\tfrac{1}{3}} = \frac{67}{64}.$$

There the compensating gears can be 67 and 64 respectively, 67 being the driver.

For 6 flutes the index calls for a 60-gear on stud and 40 on cam shaft. Placing the compensating gears on the radius bar, the train will be as follows:

Stud, 60.

Intermediate, 64–67.

Cam shaft, 40.

Oftentimes the gear ratio will be so near unity that if only two gears were used they would be excessively large. To overcome this, trial calculations must be made to establish a train of gears to approximate the required ratio.

Example. — A hob has a pitch diameter of 2.00″, a single thread of .3145″ lead, and 10 flutes.

Solution. — By formula pitch circumference = pitch diameter × π. $C = 2.00 \times 3.1416 = 6.2832$

Hence,   lead of spiral $= \dfrac{C^2}{L} = \dfrac{(6.2832)^2}{.3145}.$

Then $M = \dfrac{125.50}{.3145};$ $N = \dfrac{125.50}{.3145} + 1$

$$\frac{N}{M} = \frac{\dfrac{125.50}{.3145} + 1}{\dfrac{125.50}{.3145}} = 1 + \frac{.3145}{125.50} = 1.0025064.$$

As this ratio is so near unity, the use of two gears is prohibitive. Therefore, a suitable train of gears to produce this ratio must be found. In determining these gears, first assume four gears of a size supplied with lathe and which will readily fill the space between stud and cam-shaft gears. In this instance 2–40's and 2–60's are assumed, making the ratio $\dfrac{40 \times 60}{40 \times 60}$. Then the gears forming the M portion of the ratio are increased and reduced in size until a ratio is obtained approximate to the one desired. In this instance it was found by trials that M gears will be 38 and 63 teeth. Therefore, the gear train will be $\dfrac{40 \times 60}{38 \times 63}$, making the ratio 1.0025062, an error of only 2 parts in the 8th place.

For 10 flutes the index plate calls for 70 gears on stud and 28 gears on cam shaft.

Hence, the gear train will be as follows:

Stud, 70.

Intermediate, 40–38–60–63.

Cam shaft, 28.

It is evident that as lead of the hob thread is reduced as for fine threads, the gear ratio must approach unity. To facilitate calculation in these instances, the gear ratio is equal to $1 + \tan^2 \alpha$ where α is the angle of hob thread or angle *bac*, see Fig. 73.

102. Cutter relieving machine, Fig. 74. — This machine is designed to give straight angular or end relief to formed cutters of all sizes up to $13\frac{1}{2}$ inches in diameter and with any number of flutes from 1 to 28, and any even number of flutes from 28 to 42.

A single-throw cam giving relief from zero to $\frac{9}{16}''$ is carried by a cam slide which is adjustable to any angle desired. By means of a double adjustable tool block, the forming tool may readily be placed in proper relation to the cutter. Adjustment is provided for bringing the cutter in proper relation to the forming tool without loosening the cutter on the arbor or readjusting the change gears. A table of change gears furnished with the machine, gives the gears for indexing different numbers of flutes in cutters.

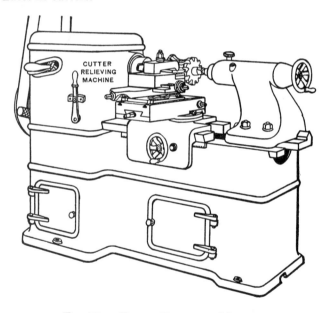

Fig. 74. — Cutter Relieving Machine.

GREEK LETTERS USED IN FORMULAS.

LETTER.		NAME.	LETTER.		NAME.
A	α	Alpha	N	ν	Nu
B	β	Beta	Ξ	ξ	Xi
Γ	γ	Gamma	O	o	Omicron
Δ	δ	Delta	II	π	Pi
E	ϵ	Epsilon	P	ρ	Rho
Z	ζ	Zeta	Σ	σ	Sigma
H	η	Eta	T	τ	Tau
Θ	θ	Theta	Υ	υ	Upsilon
I	ι	Iota	Φ	ϕ	Phi
K	κ	Kappa	X	χ	Chi
Λ	λ	Lambda	Ψ	ψ	Psi
M	μ	Mu	Ω	ω	Omega

ADVANCED MACHINE WORK

SECTION 13

INSPECTION, LIMIT SYSTEM OF MANUFACTURING, INSPECTION OF MACHINE PARTS

Introduction. Limit System of Manufacturing. Limit System on Drawings. Equipment for Inspection. Limit Gages. Combination Gage Blocks. Construction of Gages. Indicating Gages. Fluid Gage. Metric Micrometer. Inspection of Machine Parts. Inspection Organization. General Inspection Methods. Special Forms of Inspection. Devices and Machines for Inspecting Small Parts.

INTRODUCTION

1. Inspection may be called an official examination of articles or things to determine quality, accuracy, finish and interchangeability, and dates back to prehistoric times.

Primitive man inspected the natural materials which he used in obtaining his living, building his home and the rafts and boats which he utilized for transportation. In fact, there is nothing useful or ornamental in the life of man that has not been subjected to some form of inspection.

While machines and machine parts were inspected to some extent by manufacturers as far back as 1800, it is only in recent years that inspection has been given the attention and the importance it deserves.

The old way of building a machine was to fit one part to another and, if it worked well, it was considered good enough. Little attempt was made to establish strict standards of measurement, or interchangeability of parts.

2. Early forms of inspection. — Inspection of machines for accuracy was the work of the first inspectors. The various parts were fitted together and assembled to give a smooth-running unit. The completed machine was then inspected for accuracy with reference to performing the work for which it was designed.

Fig. 1. — Group of Bench Inspectors.

3. The installation of inspection systems has enabled manufacturers to increase production and promote interchangeability of machine parts. The use of machines which now enters into every phase of modern life, made it necessary to develop methods by which machine parts could be manufactured in large quantities and assembled, or repaired by substituting new parts for old ones, without the help of machines or skilled mechanics. Furthermore, machines fabricated in places other than where they are manufactured, require that the parts have a high degree of interchangeability.

Production inspection is now one of the most important

operations in modern manufacturing, and this importance is emphasized by the necessity and difficulty of maintaining and controlling *quality* while keeping up *quantity* production.

Information. — While the larger part of inspection is for interchangeability of machine parts, some classes of work, such as assembled machine tools, are inspected for the accuracy and quality of the work they will produce. Inspection for quality of finish is also made.

4. Combination of limit and inspection systems. — The adoption of a suitable limit system together with proper inspection methods will permit the employment of unskilled labor for many operations with satisfactory results. Generous or large limits whenever practicable give the workman sufficient leeway, make possible the acceptance of more of his work, and allow him to produce work at a rate of speed which would otherwise be unattainable. See Dimension Limit System, p. **2**13.

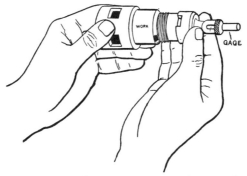

Fig. 2. — Testing Short Taper with Special Gage.

THE LIMIT SYSTEM OF MANUFACTURING

5. The limit system of manufacturing is a method of producing work just accurate enough to be suitable for the purpose for which it is to be used. This system recognizes that it is impossible to produce work exactly to the nominal size either by hand processes or machine processes for there is always some difference between the actual and the desired dimension; exact duplication of parts is likewise impossible. Therefore, some definite leeway is allowed, that is, maximum and minimum limits are given within which it is only necessary to keep the dimensions to make the work suitable for use.

The limit system is now recognized as the only method by which duplicate and interchangeable parts can be commercially produced.

Attention. — Unnecessary and unreasonable accuracy complicates tool making and manufacturing and increases the cost of production.

6. Limit defined. — Limit is the dimension beyond which the work must not extend. For example, if the nominal

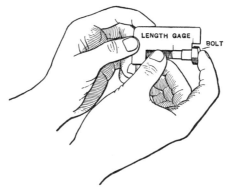

Fig. 3. — Inspecting Bolt With Length Gage.

diameter of a shaft is one inch and it is planned to fix the oversize at 0.001″, and the undersize at 0.001″, the maximum limit will be 1.001″ and the minimum limit 0.999″.

7. Tolerance defined. — Tolerance is the difference between the maximum and minimum limits (sizes). For example, if the maximum limit of a dimension is 1.001″ and the minimum limit 0.999″, the tolerance would be 0.002″.

The difference between the nominal size and maximum limit is the plus tolerance, and between the nominal size and minimum size, the minus tolerance. Plus and minus tolerances may or may not be equal.

8. Allowance defined. — Allowance is the difference in diameter between the hole and shaft for fits, such as forcing, driving, running, and sliding fits. For example, if the nominal size of a hole is one inch and the allowance of the shaft for a running fit is 0.001″, the shaft will be 0.999″ in diameter. Again, if the size of the hole is one inch and the allowance for a forcing fit is 0.002″, the shaft will be 1.002″ in diameter. See Fits in Machine Construction with Tables of Allowances, p. 214.

Information. — Allowances are subject to maximum and minimum limits, and should be distinguished from tolerances.

9. Limit of accuracy, or permissible error for machine parts in manufacturing, depends upon the kind and character of the work.

In manufacturing a machine or its component parts, definite requirements, as set forth by specifications, drawings, or models, are necessary. Since nothing can be made exactly to the nominal size, it is determined beforehand how large or

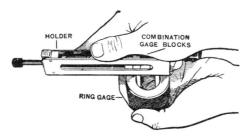

Fig. 4. — Testing Ring Gage with Gage Blocks.

how small parts may be and yet be suitable for the work for which they were designed. These limits and tolerances are not left to the skill and judgment of the workman as in former times, but are passed upon by the engineering department and management before the actual manufacturing begins.

10. To establish limits and tolerances. — First, the service which the machine is to perform, or the work it is to produce, should be considered; second, the method of building the machine together with the making of the component parts, especially in relation to the limits and tolerances permissible should be given attention; then, the limits and tolerances for the different machine parts may be established. Limits and tolerances are also determined by the quality of the work required to facilitate quantity production and still maintain the accuracy desired. For some parts it may be necessary to make the tolerance as small as 0.0001″; for the other parts the tolerances may be 0.0005″, 0.001″, 0.002″, 0.003″, 0.004″, 0.005″, 0.01″, $\frac{1}{16}$″, or any amount that is found reasonable. Close limits are expensive to produce and wide limits may prove destructive to the result desired.

Limits, once established, should be maintained, otherwise a repair part, made this year to replace one made five years ago, will not be a duplicate, and much " fitting-time " may be found necessary in order to utilize the new part.

Information. — Tolerances should be made as large as possible rather than as small as possible.

THE LIMIT SYSTEM ON DRAWINGS

11. The limit system of dimensioning drawings. — The ordinary method of dimensioning drawings by the use of a single dimension does not express limits; two dimensions are necessary.

Various methods are used for expressing limits but the plus and minus methods, the double-dimension method, and the combination method, are the ones most available.

12. The plus and minus methods express the nominal dimension and the tolerance in two forms.

The *first* form uses the tolerance as a decimal *after* the nominal dimension, and is expressed as 2.000 ± or $\pm$ (plus or minus) 0.002, see Fig. 5. When the plus and minus tolerances are unequal the dimension is expressed as $2.000 {}^{+\ 0.003}_{-\ 0.001}$.

The *second* form expresses the tolerance as a single digit *after* the nominal dimension, as in Fig. 6. This single figure is the number of units for the tolerance expressed in the last figure of the nominal dimension.   For example, 0.875 ±2 shows that the upper limit is 0.877″ and the lower limit 0.873″ and

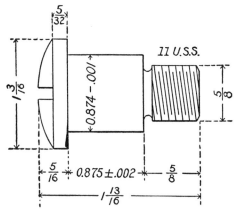

Fig. 5. — Expressing Limits by the Plus and Minus Decimal Method.

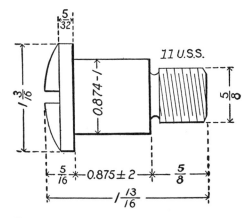

Fig. 6. — Expressing Limits by the Plus and Minus Single-Digit Method.

that the tolerance is 0.004″.   Again, if the dimension is expressed as 0.9827 − 2, the upper limit is 0.9827″, the lower limit is 0.9825″ and the tolerance is 0.0002″.

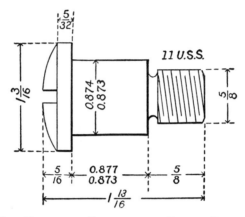

FIG. 7. — EXPRESSING LIMITS BY THE DOUBLE–DIMENSION
METHOD.

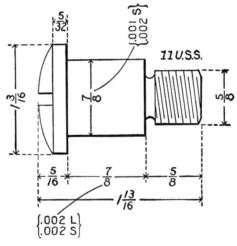

FIG. 8. — EXPRESSING LIMITS BY THE COMBINATION METHOD.

13. The double-dimension method, Fig. 7, expresses the dimension by giving the upper and lower limits on the drawing, one above the other. For example, if the dimension is expressed as $\frac{1.0025}{1.0020}$ the upper limit is 1.0025″, the lower limit 1.0020″ and the tolerance is 0.0005″

14. The combination method, Fig. 8, designates dimensions by binary fractions and expresses the tolerances by decimals, adding L for large and S for small. For example, in Fig. 8 the diameter of the body of screw is given as $\frac{7}{8}\begin{Bmatrix}0.001\\0.002\end{Bmatrix}S$. This means that the upper limit is 0.875″ minus 0.001″ or 0.874″, and the lower limit is 0.875″ minus 0.002″ or 0.873.″ See Dimension Limit System, p. 213.

15. To measure work produced by the limit system requires some form of measuring tool or gage which will show or indicate whether or not the part is within the designated limits and may be accepted, or above or below these limits and must be rejected. Since there are two dimensions for each nominal size, it is, therefore, necessary to have *two gages* for each nominal dimension. Instruments commonly used for this work are called " limit gages," and are obtainable in various forms. See Standard and Limit Gages, p. 223.

EQUIPMENT FOR INSPECTION

16. The general equipment of tools usually found in a modern machine-shop tool room, and which is shown in the charts, Figs. 9–14, may be used for inspecting machine parts and assembled machines. Most of the tools shown in the charts are of standard make, readily obtainable, and will be found sufficient for the inspection of many parts of all machines.

There are machine parts, however, which cannot be inspected or tested with standard tools. For these parts single-purpose or special tools and gages are designed and made when manufacturing is started.

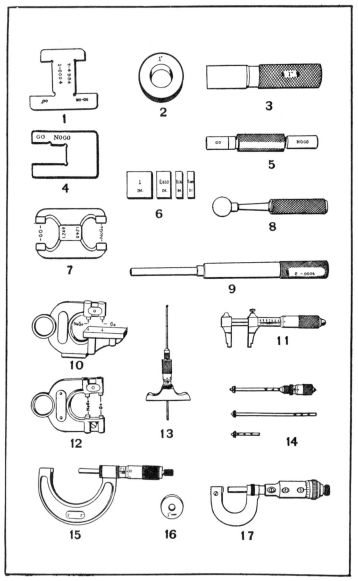

Fig. 9.

17. Names and uses of inspection tools, gages, and instruments in charts, pp. 1310 — **13**18.

No.	Name.	Use.
1	Double-ended external "go" and "no-go" limit gage.	To gage lengths in production work. See p. **13**19.
2	Standard ring gage.	To establish standards and test external work. See p. **2**31.
3	Standard plug gage.	To establish standards and test diameters of holes.
4	Single-ended external "go" and "no-go" limit gage.	To gage diameters or lengths in production work.
5	Double-ended "go" and "no-go" plug gage.	To test hole diameters in production work.
6	Gage blocks.	To set and test standard and limit gages. See p. **13**22.
7	External limit gage.	Check and reference gage for production work.
8	Ball plug gage.	To test diameters of holes. Spherical surface. Allows quick adjustment.
9	Concentricity or line gage.	To test alinement of holes of two diameters.
10	Rapid inspection limit gage (adjustable).	To gage work rapidly at machine or bench. Used in production work.
11	Inside micrometer caliper.	To measure diameters of small holes, from 0.200″ to 2.00″. See p. **5**02.
12	Adjustable limit snap gage (standard pattern).	To gage work at machine or bench. Used in production work. See p. **13**21.
13	Micrometer depth gage.	To measure depth of holes.
14	Inside micrometer caliper.	To measure diameters of holes 2″ and larger.
15	Two-inch standard micrometer caliper.	To measure thicknesses and diameters from 1″ to 2″. See Micrometer Calipers, p. **2**07.
16	One-inch standard gage.	To set or test micrometer.
17	One-inch direct-reading micrometer caliper.	Can be read in thousandths of an inch to exact figures without calculations from graduated lines.
18	Universal dial test indicator.	To test accuracy of all classes of work and to aline machines. May be held in tool post, attached to surface gage, or standard angle or surface plate. See p. **13**30.
19	Set of standard thickness or feeler gages.	To establish small distances, such as clearance between parts, etc.
20	Test indicator.	To test and set up work and machines, and to aline jigs and fixtures.
21	Standard test mandrel.	To test and aline machine parts. Used with No. 20.
22	Test indicator.	To test accuracy of revolving work usually in a lathe. See p. **12**10.
23	Set of standard fillet and radius gages.	To test small fillets and radius curves.

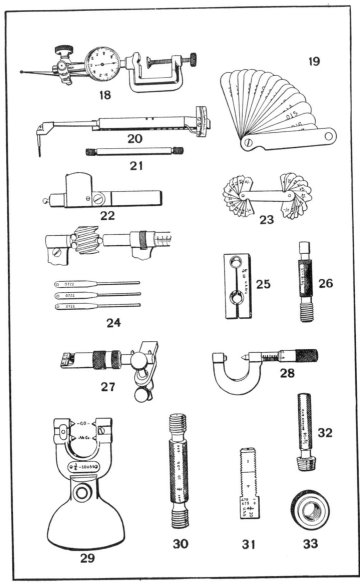

Fig. 10.

No.	Name.	Use.
24	Thread measuring wires.	To measure accurately the pitch diameter of a screw thread by three-wire method. See p. **14**31.
25	Standard adjustable external templet thread gage.	To establish standard thread size, on screws.
26	Standard internal plug thread gage.	To establish standard thread sizes for holes. Used to set gage No. 25. Plain end gives root diameter.
27	Thread lead indicator.	To test accuracy of lead of threads. See p. **14**27.
28	Thread micrometer caliper.	To measure pitch diameter of threads. See p. **12**14.
29	Adjustable thread limit snap gage.	To test pitch diameter of screw thread. See p. **2**25.
30	Double-ended thread limit plug gage.	To test internal threads.
31	Thread gage-setting plug.	To set gage No. 29.
32	Standard pipe thread plug-gage.	To establish standard internal pipe threads.
33	Standard pipe thread ring gage.	To establish standard external pipe threads.
34	Universal sine bar.	To measure angles and tapers using vernier height gage or gage blocks. See p. **12**63.
35	Standard tape plug gage.	To establish, measure, and test taper holes.
36	Plumb-bob.	To locate or establish perpendiculars.
37	Standard wire gage.	To measure diameter of wire and thickness of sheet metal. See Wire Gages, *Principles of Machine Work.*
38	Standard taper ring gage.	To establish, measure, and test external tapers. See p. **2**26.
39	Vernier caliper.	To measure accurately inside and outside dimensions. See p. **2**11.
40	Gear tooth depth micrometer.	To describe a line for full depth of teeth on a gear blank. Used also to draw other accurate lines.
41	Gear tooth vernier.	To measure accurately the thickness of gear teeth. See p. **11**13.
42	Dial cylinder gage.	To measure diameter and accuracy of engine cylinders.
43	Vernier height gage.	To measure accurately heights and distances. See p. **12**49.
44	Universal bevel protractor.	To establish and measure angles. See p. **11**85.
45	Prestometer or fluid gage.	To measure and indicate accurately, inspect, and check thickness and diameter. See p. **13**30.
46	Machinists' level.	To establish levels and horizontals. See p. **14**17.

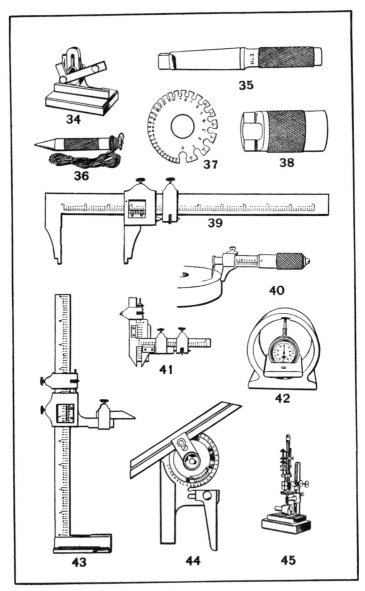

Fig. 11.

No.	Name.	Use.
47	Magnifying glass.	To examine surfaces and read fine graduations. See p. **13**42.
48	Outside spring calipers.	To measure outside distances.
49	Inside spring calipers.	To measure inside distances.
50	Spring dividers.	To describe circles and space off distances.
51	Scleroscope.	To test hardness of metals and other materials. See *Principles of Machine Work.*
52	Smooth or dead-smooth file.	To test steel to determine whether hardened or not, and to test the temper of steel.
53	Brinell Testing Machine.	To test hardness of metals.
54	Toolmakers' bridge. All faces scraped.	To inspect and lay out work.
55	Speed indicator.	To determine speed of shafts, etc.
56	Machinists' parallel clamp.	To clamp work to table, vise, etc.
57	Toolmakers' clamps.	To clamp light accurate work to fixtures.
58	Screw clamp.	To clamp rough work.
59	Try square.	To establish right angles (90°) and test squareness of surfaces. See p. **14**19.
60	Surface gage.	To test surfaces, also to lay out and set up work. See p. **9**21.
61	Standard steel rule.	To measure distances for ordinary work.
62	Dial test indicator.	To measure and test accurately the thickness and diameter of work, and the accuracy of surfaces. See p. **13**30.
63	Steel straight edge.	For fine work where a straight edge (180°) angle is required. See p. **14**08.
64	Scraped straight edge.	To test scraped ways of a machine.
65	Pair of steel or cast-iron parallels ground or planed to equal widths and thicknesses.	To set up and level work for inspecting or machining. See p. **14**10.
66	Box parallel with finished surfaces.	To mount and aline work for inspecting or machining. See p. **14**15.
67	Toolmakers' knee.	To hold small work by clamping for inspecting either on machine table or surface plate.
68	Standard cast-iron surface plate.	To establish and test plane surfaces. When not scraped, used as leveling plate for laying out work.
69	Pair of *V* blocks.	To hold cylindrical work in accurate alinement for inspecting or machining.
70	Bench centers.	To hold centered work so that it may be revolved and tested for trueness with chalk or test indicator. See p. **12**17.

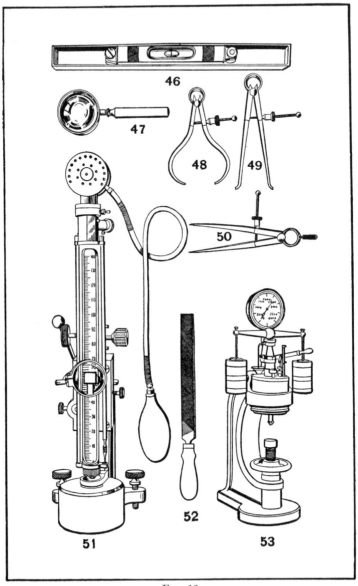

Fig. 12.

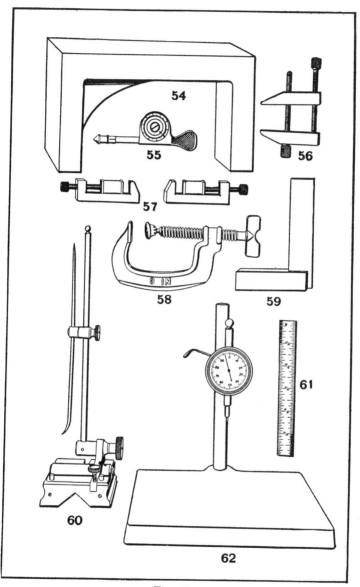

Fig. 13.

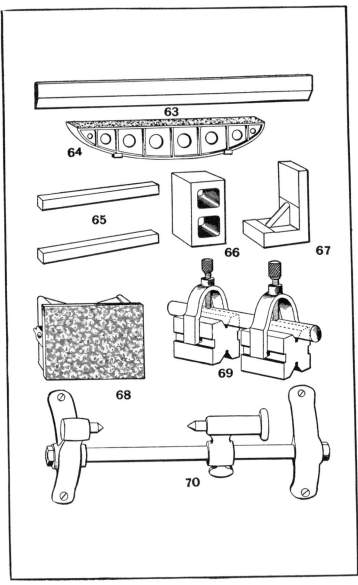

FIG. 14.

LIMIT GAGES

18. Limit or " go " and " no-go " gages are the most efficient tools for gaging or measuring work produced by the limit system. They have made inspection practical owing to the high rate of speed with which they may be operated, and the superior quality of the results produced.

Since these gages do not measure the exact size of the work, it is only necessary to test the work to see if the size of the part is within the predetermined limits, as shown in Figs. 15, 16. See also p. 224.

The limit gage is a combination of two gages, one made to the maximum limit and the other to the minimum limit.

External limit gages are known as limit snap gages; internal gages, as limit plug gages. Both types are commonly called " go " and " no-go " gages. See Inspection of Threads, p. 1421.

19. Limit snap gages are used for measuring or inspecting the external diameter or length of work. For example, the

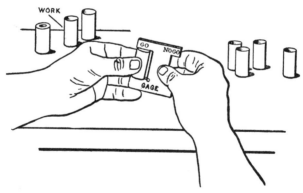

FIG. 15. — INSPECTING SMALL PARTS WITH "GO" AND "NO-GO" GAGES.

nominal diameter of a shaft is 1″ and the tolerance 0.002″ with the plus and minus tolerances equal.

The gage for measuring or inspecting this shaft may be the double-ended solid type, as shown in Fig. 15 and Nos. 1, 7,

Chart, Fig. 9; the single-ended type, as shown in No. 4, Chart, Fig. 9; or the adjustable type, as shown in Fig. 16 and Nos. 10, 12, Chart, Fig. 9. One pair of measuring surfaces or points would be made or set to 1.001″ and called the " go " gage. The other would be made or set to 0.999″ and called the " no-go " gage.

If the shaft is small enough *to allow* the " go " gage to pass over it and large enough *not to allow* the " no-go " gage to pass over it, the shaft is within the working limits and of the proper size.

20. Limit plug gages are used principally for measuring or inspecting the diameters of holes. For example, the nominal diameter of a hole is 2″ and the tolerance 0.001″ with the plus tolerance 0.00075″ and the minus tolerance 0.00025″. The gage for measuring or inspecting this hole may be the double-ended plug type, as shown in Fig. 38 and No. 5, Chart, Fig. 9; or the single-ended type which has a single plug or gage with two diameters made to the maximum and minimum limits. Sometimes two separate plug gages similar to No. 3, Chart, Fig. 9, are used. One plug would be made to 1.99975″ and called the "go" gage; the other, to 2.00075″ and called the " no-go " gage. If the hole is of the proper size to *admit* the " go " gage and *not to admit* the " no-go " gage, it is within the working limits.

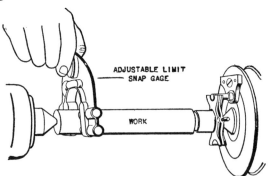

Fig. 16. — Calipering Work with Adjustable Limit Snap Gage.

21. Adjustable limit gages. — From inspecting a large number of parts, gages become worn and inaccurate and must be repaired or discarded. Repairing gages is a slow and expensive process for they must be annealed and made over again. To avoid this, adjustable limit snap gages, Figs. 16, 17, may be used. The measuring points or anvils are adjusted with a screw-driver the same as the anvil of a micrometer. See p. 225.

Information. — The accuracy of gages may be affected by the heat of the hand. To eliminate this, gages with insulated grips are obtainable, see Figs. 16, 17.

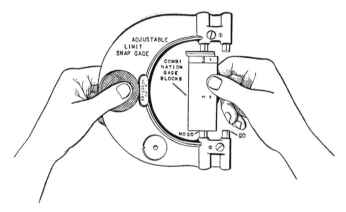

Fig. 17. — Testing Adjustable Limit Snap Gage with Combination Gage Blocks.

22. To set adjustable limit gages and establish other odd-sized dimensions, some standard is required. Ordinary standard gages or blocks which are accurately made to nominal sizes such as 1.00″, 1.50″, etc., have only a restricted use in the limit system.

Inspection or working gages are very seldom set at nominal sizes, but more often from one-half to three- or four-thousandths above or below a nominal dimension. For this work a special set of gages or gage blocks must be used.

COMBINATION GAGE BLOCKS

23. Johansson combination standard gage blocks, Fig. 18. — These combination gage blocks are used for accurate measurements, in making, checking, and inspecting limit or standard gages, and to establish standard sizes. They are so designed that by placing various combinations of blocks

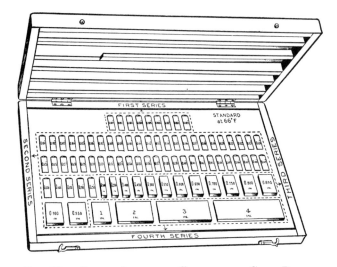

FIG. 18. — SET OF JOHANSSON COMBINATION GAGE BLOCKS.

together, it is possible to obtain any dimensions within the capacity of the set, advancing by increments of one ten-thousandth of an inch.

While these blocks are especially useful for establishing odd sizes, they may be used for establishing standard sizes and for checking precision measurements in all kinds of toolmaking and laboratory work. They are obtainable in the English and Metric systems.

These gages are hardened steel blocks ground and accurately lapped to a fine finish. The two opposite broad faces of each block are perfectly parallel, and the distance between them is equal to the size marked upon the block. Blocks up to 6″ in length, at 66° F., are said to have no greater error than 0.00001″ (one hundred-thousandth part of an inch).

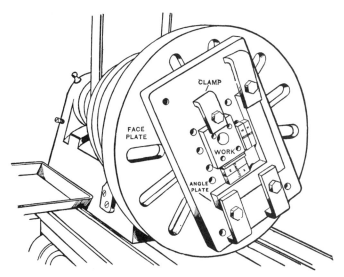

FIG. 19. — LOCATING WORK ACCURATELY ON FACE PLATE WITH GAGE BLOCKS.

24. To combine two or more gage blocks to obtain a size not given by any one block. — Hold a block in each hand. Wipe the face of the block held in the left hand on the fleshy part of thumb or on the wrist of the right hand, as in Fig. 20.

Similarly, wipe a face of the block held in the right hand, as in Fig. 21; then, with a little pressure, slide one block upon the other, as in Fig. 22. This expels the air from between the two surfaces so that the blocks will hold together. Several blocks may be assembled in this manner and used as a solid gage. See Figs. 17, 23.

Note. — Gage blocks should not be left in combination after using, but separated and the surfaces wiped off with a piece of chamois.

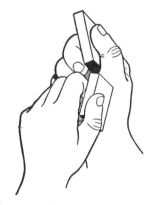

FIG. 20. — CLEANING SURFACE OF
BLOCK HELD IN LEFT HAND.

FIG. 21. — CLEANING SURFACE OF
BLOCK HELD IN RIGHT HAND.

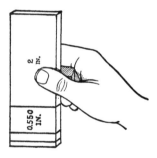

FIG. 22. — JOINING TWO BLOCK.

FIG. 23. — SEVERAL BLOCKS
IN COMBINATION.

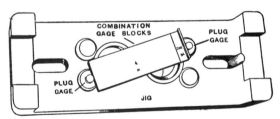

FIG. 24. — TESTING CENTER DISTANCE OF HOLES IN JIG WITH
GAGE BLOCKS AND STANDARD PLUGS.

25. Sets of combination standard gage blocks. — These blocks may be obtained singly, in special groups, or in standard combination sets. One of the largest and most practical sets contains 81 blocks, as shown in Fig. 18. The blocks are divided into four general series or groups from which the various combinations are built. With this set 300,000 different sizes are obtainable.

First Series. — 9 blocks with sizes as follows:

0.1001	0.1002	0.1003	0.1004	0.1005
	0.1006	0.1007	0.1008	0.1009

Second Series. — 49 blocks with sizes as follows:

0.101	0.102	0.103	0.104	0.105	0.106	0.107
0.108	0.109	0.110	0.111	0.112	0.113	0.114
0.115	0.116	0.117	0.118	0.119	0.120	0.121
0.122	0.123	0.124	0.125	0.126	0.127	0.128
0.129	0.130	0.131	0.132	0.133	0.134	0.135
0.136	0.137	0.138	0.139	0.140	0.141	0.142
0.143	0.144	0.145	0.146	0.147	0.148	0.149

Third Series. — 19 blocks with sizes as follows:

0.050	0.100	0.150	0.200	0.250	0.300	0.350
0.400	0.450	0.500	0.550	0.600	0.650	0.700
		0.750	0.800	0.850	0.900	0.950

Fourth Series. — 4 blocks with sizes as follows:

1, 2, 3, 4″.

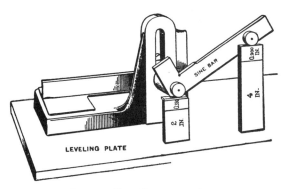

Fig. 25. — Setting Sine Bar with Gage Blocks.

26. To select gages for any desired size. —

Example. — To make up a set of blocks for 2.7834″. First eliminate the ten-thousandths by selecting block 0.1004″ from series 1;
then the thousandths by selecting block 0.133″ from series 2. This
block should be so chosen that it will leave the hundredths figure 0 or
5. Blocks 0.550″ and 2″ complete the combination. See Fig. 18.

Solution. —

Desired size	2.7834	
First block,	0.1004	0.1004
	2.6830	
Second block,	0.133	0.133
	2.550	
Third block,	0.550	0.550
	2.000	
Fourth block,	2.000	2.000
	0.000	2.7834″

Information. — It is not necessary to use a particular combination
to make up a gage of a desired size, but it is advisable to use the least
number of blocks possible.

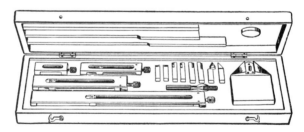

Fig. 26. — Set of Accessories for Gage Blocks.

27. Accessories for gage blocks. — To enable the blocks to
be used for different purposes, such as calipering and for height
gages, a set of accessories and attachments are obtainable.
The larger sets contain 4 holders, 3 pairs of jaws, 1 straight
edge, 1 base to convert holders into height gages, 1 scriber,
and 1 center.

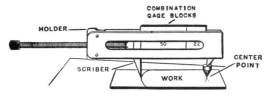

FIG. 27. — DESCRIBING ACCURATE CIRCLES AND ARCS WITH GAGE
BLOCKS.

CONSTRUCTION OF GAGES

28. Material for gages. — Solid gages are usually made
from tool or alloy steel, machined, hardened, tempered, seasoned, ground and lapped to size. Some manufacturers
make these gages, especially if large, from machine steel and
case-harden them before grinding and finishing to size. See
Hardening and Tempering, *Principles of Machine Work.*

Solid gages may be cut from plate or bar stock, or made
from drop forgings or castings depending upon the style
required. Adjustable gages have their frames made from
drop forgings, malleable castings or iron castings, and their
gaging points of tool or alloy steel which are finished by grinding and lapping.

Information. — The frames of gages often strike the work. This
hammering or peening effect may destroy the accuracy of the gage.
By making the gages of rigid design, and attaching soft bumpers, peening may be avoided to a large extent.

Warning. — If a gage falls to the floor, it should not be used again
until it is tested for accuracy.

Note. — Soft gages may be used for testing a few pieces; but, as they
wear rapidly, should only be used in emergencies.

29. Tolerance for wear of gages. — In the design of snap,
plug, and special " go " and " no-go " gages, attention must
be given to the wear of the measuring surfaces. Allowance
must be made for this by slightly varying the size of the gage.
This tolerance for wear also varies with manufacturing tolerances, and the greater the difference between maximum and
minimum limits, the greater the tolerance for wear that can

be allowed. To provide for this wear, the gages are made slightly smaller for external gages and slightly larger for internal gages, and the operator or inspector works closer to the nominal dimension. For example, a steel stud with a nominal diameter of 1″, has a tolerance of ±0.001″. The gage for the maximum dimension would be 1.0008″ while the gage for the minimum dimension would be 0.9992″, allowing 0.0002″ for wear.

Information. — For tolerance for wear of the gage, 10 per cent of the tolerance of the work is permissible.

30. The effect of rough and finished surfaces on the life of gages. — The condition of the surface of the metal gaged has an effect on the life of the gage. Rough-machined surfaces will wear a gage about three times as fast as smooth surfaces.

Information. — Ground surfaces usually retain some of the abrasive used in grinding which is also a factor in the life of gages.

31. The effect of different metals on the life of gages. — The kind of metal of which the machine parts are made determines, approximately, the number of pieces that can be inspected before the gage is worn beyond its tolerance for wear.

Complicated parts to which it is somewhat difficult to fit gages, and work that requires the use of thin gages, cause excessive wear and shorten the life of the gages.

Cast iron and aluminum have a large lapping effect which causes considerable wear on the measuring surfaces of gages.

Cast iron, due to the presence of crystals of cementite or carbide of iron in the metal, is liable to wear the surfaces of a gage at the rate of 0.00025″ for about 10,000 parts inspected.

Aluminum which contains alumina or aluminum oxide in crystalline form, may be expected to wear a gage 0.00025″ for about 11,000 parts tested.

Steel and brass have only a small lapping effect and produce only a slight wear on gages. Before wearing an amount exceeding 0.00025″, a gage will inspect about 30,000 steel parts, or 50,000 brass parts.

32. Tolerances for plated work. — If parts are to be plated, they should be machined to allow for the thickness of the deposited metal. For ordinary work, from 0.0005″ to 0.001″ may be allowed on each surface for copper, zinc, or nickel plating.

Information. — Worn gages are sometimes temporarily repaired by plating the measuring surfaces.

INDICATING GAGES

33. Indicating gages are often used in inspecting and testing work as they tell at a glance whether or not the work is within the required limits and often show the actual error. There are many types of these gages, but dial reading indicators are generally used for ordinary work.

34. The comparative type of indicators is used in inspection for testing alinements, accuracy of machine parts, surfaces, etc., more than for the determination of actual measurements or differences of measurements. See Testing Lathe Work with Indicators, p. **12**10.

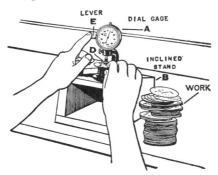

Fig. 28. — Measuring Thin Work with Dial Gage.

35. Direct reading dial indicators, Figs. 28, 29, have the movement of the measuring points multiplied by a train of gears to a needle or pointer which indicates on a dial. These gages may be used to measure the exact size of the work, or may be used as comparators.

Dial gage A, Fig. 28, is attached to a special, inclined stand B for quick reading. The needle of indicator C is set at zero when measuring point D is in contact with face of the stand.

To gage work, raise point D by lever E. Place work F under the spindle and lower point to work by releasing lever E. The reading on the dial gives the exact thickness of the disk. If necessary several readings may be taken on the same work by pressing and releasing lever E.

Universal dial test indicator A, Fig. 29, clamped to rod B, is set for measuring the error of work C. With the contact point D touching a standard gage block or a standard piece of work, the needle is set at zero. When the work is moved beneath the gage the needle shows the amount of oversize or undersize.

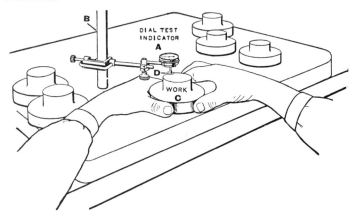

Fig. 29. — Testing Thickness of Flanges with Universal Dial Indicator.

By attaching these gages to different stands or fixtures, they have almost unlimited uses. See Testing Lathe Work with Indicators, p. **12**10, and Inspection of Machine Tools, p. **14**01.

36. Prestometer or fluid gage, Fig. 31, is an instrument used as a comparator, or for the determination of fine variations in sizes of duplicate parts with reference to a master block or standard part. The operation of the instrument is based upon the deflection of a diaphragm which creates a varying pressure on a colored liquid and causes it to rise or fall in a glass tube. The principle of operation of the fluid gage is shown in Fig. 30.

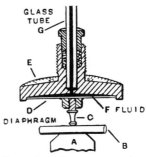

Stationary anvil *A* is used to hold work *B*. Moving anvil *C* is attached to diaphragm *D* held in chamber *E*. A piece of work passed between the two anvils raises moving anvil *C* and causes diaphragm *D* to deflect upward, forcing the fluid *F* up into glass tube *G*. A variation in the size of work raises or lowers anvil *C* and causes a variation in height of liquid in tube.

FIG. 30. — SECTION OF DIAPHRAGM AND FLUID CHAMBER OF PRESTOMETER.

Calibrated tube *G* is used with a scale graduated to agree with the tube calibration. Any variation in the height of fluid column due to variation in the size of work is read on scale.

Information. — Calibrated tubes with different-sized bores and with properly graduated scales are obtainable for changing the magnifying power of the instrument. The fluid used in the instrument is pure alcohol colored with aniline dye.

37. To set and use fluid gage. — Place standard gage, or a piece of work of the required size, *A*, Fig. 31, on solid anvil *B*. Loosen clamp nut *C* and lower clamping head *D* with rack *E* and pinion *F* until movable anvil *G* almost touches work. Clamp in this position by tightening clamp nut *C*. Screw diaphragm chamber *H* downward so that moving anvil *G* presses on work sufficiently hard to deflect diaphragm and raise top of liquid in tube *J* about one-half inch. Bring scale *K* which may be read to 0.0001″, to a position where the top of fluid column will register at the zero line in center. Remove work from instrument. Adjust pointer *L* to the maximum limit of work size, and pointer *M* to the minimum limit. Set pointer N to top of fluid column with work removed. Check setting by passing the standard between anvils a few times to make sure fluid column registers at zero, on the scale. The gage is now ready for use.

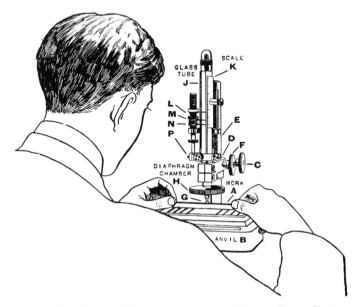

FIG. 31. — GAGING WORK WITH PRESTOMETER OR FLUID GAGE.

Attention. — If more fluid is added to the tube, or the temperature change affects the length of the fluid column, the accuracy of the instrument is not affected as long as pointer N is set at the normal fluid level. All the pointers may be moved simultaneously by means of pointer carriage P.

Information. — The normal fluid level should be about half-way up in the tube. More liquid may be added with a dropper. Excess liquid may be removed by pressing on diaphragm to force liquid up into a reservoir at top of tube and removing the excess amount with a piece of blotting paper.

38. "Feel" or touch indicators are operated by the use of pins, or collars in jigs, or box gages, and determine accuracy by noting whether or not the "telltale" is flush with a surface.

This type of indicator has only a limited use and is seldom employed except for special work.

39. Combination scale indicators based upon the micrometer and vernier principles are often attached to machines, tools and fixtures for obtaining fine adjustments and measurements, or for testing and inspecting.

METRIC MICROMETER CALIPERS

40. Metric micrometer caliper, Fig. 33, measures to 25 millimeters. It is similar to a 1″ micrometer caliper, and is used for measuring work by the Metric system. See Micrometer Calipers, p. 207. This micrometer consists of frame *A* and barrel *B*. The threaded portion of precision screw *C*, which is concealed within barrel *B*, fits a nut in the frame.

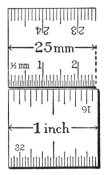

41. Principle of metric micrometer. — The pitch of the screw is 0.5 mm. and barrel *B* is graduated into 0.5 mm. divisions reading from zero to 25 mm., as at *D*. The 0.5 mm. divisions are shown *above* line *E*; and the 1.0 mm. divisions, below line *E* with the designating figures. *One* revolution of the screw *C* opens or closes the caliper from a division above or below line *E* to the next division above or below line *E*, or 0.5 mm. *Two* revolutions will move the caliper from one division below line *E* to the next division below line *E*, or 1.0 mm. Thimble *F* is graduated into 50 divisions from zero to 50. Rotating the thimble from one graduation to the next moves the caliper $\frac{1}{50}$ of 0.5 mm., or 0.01 mm. Every fifth mm. on the barrel is numbered 5, 10, 15, etc., and may be read as 5 mm., 10 mm., etc.

Fig. 32.—Metric and English Systems Compared. 25 mm. and 1 Inch. Full Size.

42. To read metric micrometer. — In Fig. 33 the figure 10 on the barrel may be read as 10 mm., and the five additional divisions, counting both above and below line *E*, as 2.5 mm.

Then adding 23 divisions, or 0.23 mm., on thimble makes the complete reading $10 + 2.5 + 0.23 = 12.73$ mm. which is the diameter of work G. This is approximately one-half inch. See Tables of Inches with Equivalents in Millimeters, *Principles of Machine Work.*

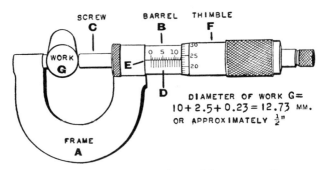

Fig. 33. — Measuring with Metric Micrometer Caliper.

Information. — Metric micrometers are graduated to read to one-hundredth of a millimeter, but half, quarters and tenths of this division may be easily estimated.

INSPECTION OF MACHINE PARTS

43. Requirements for inspecting machine parts. — The best results are obtained by organizing a suitable system for the work at hand. The gages should be accurately designed and constructed, the work of the inspectors properly laid out, and the routing or movement of the parts from inspector to inspector, or from department to department, carefully planned.

INSPECTION ORGANIZATION

44. Inspection staff. — The organization of the inspection department should be separate and distinct from the production department, and should be responsible to the management only.

In large plants there is a chief inspector who has under him foremen inspectors for the various departments each in charge of his own inspection staff which includes traveling, field or operation inspectors, and bench or crib inspectors.

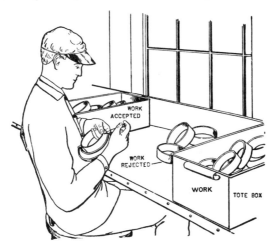

Fig. 34. — Inspecting with Micrometer Calipers.

45. The chief inspector is the executive head of the inspection department. He is responsible for the quality of the entire production and reports directly to the management. He should have the ability to organize the various departments and to supervise and direct the subordinate branches. This work requires a large knowledge of machine-shop practice and tool and gage making. It is also desirable that the chief inspector be acquainted with materials of machine construction, testing of materials, metallurgy, hardening and tempering, and heat treatment of steel.

46. The foremen inspectors have charge of departments and supervise the actual inspection operations performed within these departments which include both the work of traveling, field, or operation inspectors and that of the bench or crib inspectors. Their duties also include checking all gages used by the various inspectors in the department.

A foreman inspector should be acquainted with general machine-shop practice and possess the ability to organize as well as to supervise, the department.

47. Traveling, field or operation inspectors watch the new set-up of a machine and check the first few pieces to see that production starts correctly. They also circulate among the different machines and inspect one or two pieces from each machine at stated intervals — ten minutes or longer depending upon the class of work — to see that the tools are properly adjusted and maintain their settings, and to detect errors before a large number of parts are machined. A traveling inspector should have a comprehensive knowledge of machine-shop practice and an intimate acquaintance with different types of gages and methods of inspection.

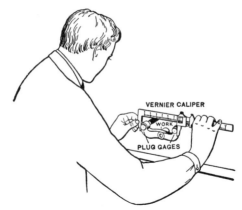

Fig. 35. — Testing Center Distance with Vernier and Plug Gages.

48. Bench or crib inspectors inspect the parts after each, or a series of operations. The work is usually carried on in an inspection crib or room, under the supervision of the foreman inspector. Bench inspectors should, after proper instruction, be able to use the various types of gages, to read and use micrometer and vernier calipers, and to read blue prints. It is also important that bench inspectors have keen powers of observation in order that they may readily detect errors or defects which may not be directly connected with the inspection operation.

A bench inspector usually inspects only one dimension, or one operation, at one time, passing the part on to the next inspector for further examination should the work require it.

Information. — A bench inspector may inspect certain classes of small parts at the rate of from one thousand to fifteen hundred pieces per hour.

49. Inspection cribs or rooms are spaces usually screened or partitioned off from the rest of the departments. They should be conveniently located and large enough to contain

Fig. 36. — Testing Taper with Special Gage.

ample storage space for work in progress of construction as well as the necessary room for the inspectors and their equipment. Work may be stored and transported in tote boxes or racks, Fig. 36, each containing a definite number of parts. This will allow a quick survey of the number of parts awaiting inspection, parts ready for further operations, parts completed and parts rejected. The crib should act as a clearing house, no work remaining outside unless in process of machining. This gives a constant check on production and trouble is easily located.

GENERAL INSPECTION METHODS

50. Systems of inspection. — There are three general systems of inspection: Random, Complete and Combination.

51. Random or percentage inspection is based upon the principle of inspecting a certain definite number of parts and then judging the quality of the entire output on the results obtained. If the number of parts selected pass inspection, it is assumed that the entire lot is within the specifications and it is passed as such. If any of the inspected parts fail to pass, the entire lot is rejected. Random inspection is used by the traveling inspectors for the inspection of parts at the machines, and also as a final check upon parts which have received complete inspection for each operation. On certain classes of work, such as machine tools, where the accuracy of the product is more essential than the interchangeability of the parts, random inspection is used almost entirely as a final test on the completed machines.

52. Complete or one hundred per cent inspection requires that each part be inspected after each operation. It is used extensively on parts requiring a high degree of interchangeability not only for the operation inspection but also for the final inspection.

53. Combination inspection. — This system of inspection combines the random method with the one hundred per cent method.

When an operation is started, the first piece or pieces are inspected to test the accuracy of the set-up to insure the proper placing of tools and fixtures. The production is then started and the first few pieces watched for errors caused by the improper adjustment of tools, dull tools, or tools slipping. If errors are present, they are corrected immediately. When production is at full speed, the traveling inspectors keep a check on the quality and see that the proper standards are maintained. After a certain number of parts have had an operation completed, they are taken to the central inspection crib or room to receive one hundred per cent inspection for that operation by the bench inspectors. These parts are not

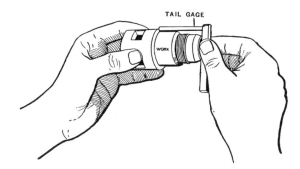

FIG. 37. — TESTING SHOULDER–DISTANCE WITH TAIL GAGE.

allowed to receive another operation until approved by the foreman inspector. When a group of parts has received all of the machining operations, a few are given complete inspection. The entire group is accepted or rejected by the result of this random inspection.

This method of inspection keeps a close check on quality of production, enables a careful watch to be maintained of each operator's work, and allows parts to be rejected at once without having further work wasted on them.

Attention. — If excessive rejections are made at the final random inspection, an immediate change to the one-hundred per cent method of inspection, is often made.

Warning. — Inspection may be carried to extremes, prove uneconomical, and be a hindrance to manufacturing. An inspection system should be so organized and conducted that it will coöperate with the management and facilitate rapid production.

54. Sets of gages for inspection. — Four sets of gages are often used for the inspection of machine parts — master, checking, working, and tolerance.

Master gages are not used for general inspection work. They are only used to set or test the checking gages.

Checking gages are the reverse of the master gages; that is, if the master gage is of the caliper type, the checking gage would be of the plug type. These gages are used to test, check, and establish the sizes of working and tolerance gages.

" **Working** " gages is the name given to all gages used by the traveling and bench inspectors and by the machine operators, with the exception of the gages used on the final inspection. The tolerances of these gages are usually made slightly smaller than those specified for the work in order to give the machine operators a small margin of safety.

Tolerance gages are those used on the final inspection by the bench inspectors. They are made with the specified tolerances.

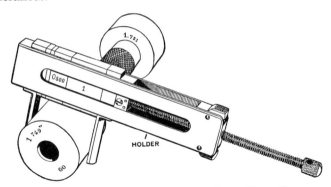

Fig. 38. — Testing "Go" End of a Limit Plug Gage.

55. Testing inspection gages. — The gages used by the bench and traveling inspectors and the machine operators, should be tested at regular intervals to see if they are still within their tolerances for wear. See Figs. 4, 17, 38.

Warning. — Snap gages or other measuring tools should never be used to measure revolving or moving work.

56. Reinspection of machine parts. — Parts which have been rejected during the operation inspection are often reinspected with tolerance or final inspection gages. The parts accepted by reinspection are allowed to go on and receive further operations; but the rejected parts are either discarded or, in many cases, salvaged by remachining.

SPECIAL FORMS OF INSPECTION

57. Functional or partial assembly inspection. — Besides inspection for accuracy and interchangeability, some of the component parts are often inspected to see if they function or operate properly. For this purpose a number of the component parts are placed in a fixture, improvised machine frame or jig, where they may be tested for operation, eccentricity, alinement and adjustment, with indicators or other instruments.

58. Profile or contour inspection. — Work of irregular outline is inspected with profile or contour gages. They are made in the form of templets. The gaging surfaces are either hardened and ground, or else hardened and ground plugs are inserted at intervals along the surface to prevent excessive wear.

If the profile is only on one side of the work, a single gage is used and its fit determined by the *light line* between gage and work. Where the profile extends nearly or completely around the work, " go " and " no-go " profile gages are used. The work should drop through one gage and not through the other.

Indicating gages are sometimes attached to profile gages to test the accuracy and location of irregular surfaces.

59. Visual inspection for finish is made on surfaces which have been cleaned, polished, ground, lapped, scraped, plated, painted, japanned, etc.

Machine parts, as well as assembled machines, are also often given a visual inspection; and if any part has large or noticeable defects, such as burrs, poorly finished surfaces, rough threads, scars from errors in tool settings, etc., it may be rejected without further handling or inspection.

Hardened, tempered, and case-hardened work is often

visually inspected by observing its color, shade or tone, in addition to standard tests. In fact, consciously or unconsciously, all work is given visual inspection in a greater or less degree.

60. Magnifying-glass inspection is often used for tools, gages, instruments, dies, punches, hardened and tempered work, and ground and lapped surfaces.

Fig. 39 shows the method of inspecting the inner surface of a drawing die to determine the quality of finish and to detect checks or cracks from hardening.

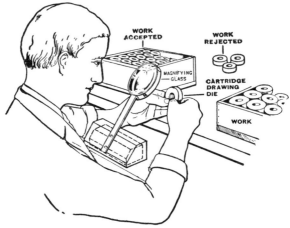

FIG. 39. — INSPECTING A CARTRIDGE DRAWING DIE, WITH MAGNIFYING GLASS.

61. Inspection for hardness. — Metals used in machine construction as well as machine parts, are inspected for hardness of material, and hardness after hardening and tempering, or heat treating. Comparative and approximate hardness may be determined by the use of a smooth file or a dead-smooth file. To measure the hardness of metal accurately, the scleroscope, No. 51, Fig. 12, or Brinell testing machine, No. 53, Fig. 12, is used. See *Principles of Machine Work*.

62. Weight inspection. — Parts of some classes of machines are inspected for weight when the weight of the part has a definite bearing upon its use or operation. Small machine parts, such as screws, bolts, nuts, and parts of small mechanisms, are often counted by weighing.

DEVICES AND MACHINES FOR INSPECTING SMALL PARTS

63. Methods of increasing the speed of hand inspection. — To permit action of both hands, avoid waste motion, conserve time, and promote general efficiency; various improvised arrangements and fixtures are often devised which increase the rapidity of hand inspection.

Both hands used to handle work for inspection. — A method of increasing the speed of hand inspection is shown in Fig. 40. The fuse bodies are inspected to insure that the

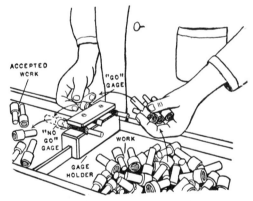

FIG. 40. — INSPECTING SMALL PARTS WITH GAGES FASTENED TO BENCH.

diameter of the threaded hole is not reduced by the threading operation. The " go " and " no-go " gages are fastened to a block or holder, leaving both hands free to handle and gage the work.

Several parts are picked up in the left hand. The right hand takes a single part and tests it on the " go " gage and then on the " no-go " gage. While the last part is being inspected, the left hand picks up another handful of parts.

Belt conveyors are sometimes used to carry parts of the work from one inspector to another during the bench inspection.

64. Hand machines for inspecting machine parts. — Some machine parts may be more rapidly inspected for a single operation by improvising power-driven devices or by adapting light power-driven machines to the work. Fig. 41 shows the final or functional inspection of a fuse for a shell by means of a light frictionally-driven tapping machine. This inspection

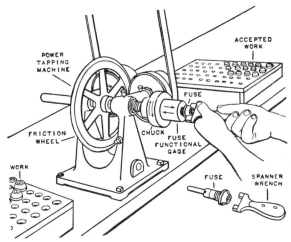

Fig. 41. — Functional Inspection with a Machine.

checks the fit of the thread, length, contour and concentricity of fuse to insure that the fuse will fit adapters made elsewhere. A single functional gage is held in a chuck. The gage consists of a threaded ring for testing the threaded portion of the fuse, and a socket which is a duplicate of the adapter. The inspector takes a fuse in the left hand, places the spanner wrench in position and inserts the fuse in the rotating gage. The fuse and spanner wrench are shown in detail, Fig. 41. The rotation carries the fuse inward. When the fuse is all the way in, or part way in and binds, the inspector pushes on the work and forces the friction wheel against the rear pulley which reverses the rotation of the gage and carries the fuse out.

Work that will screw into the gage freely is accepted; work that binds or does not enter is rejected. Some machines are arranged to reverse by using a treadle or shipper.

Information. — For "go" and "no-go" inspection, two machines may be used, one with the "go" gage, the other with the "no-go" gage.

65. Automatic machines for inspecting machine parts are used to some extent on small plain work. Balls and rollers for bearings are usually inspected and sorted, and often counted by automatic machines.

The machines for this work are so constructed that the parts move, by gravity or otherwise, past or through a series of gages which act automatically; accepted parts drop into one receptacle and rejected parts into another.

ADVANCED MACHINE WORK

SECTION 14

INSPECTION OF MACHINE TOOLS, PRECISION MEASUREMENT OF THREADS, HEAVY DUTY TURNING AND BORING

Inspection of Machine Tools: Engine Lathe, Drilling Machine, Milling Machines, Planer. Inspection of Threads. Limits Applied to Threads. Thread Lead Indicators and Testing Machines. Wire Systems for Precision Measurement of Pitch Diameters and Angles of Threads. Heavy Duty Boring and Turning. Vertical Turret Lathes.

INSPECTION OF MACHINE TOOLS

1. Introduction. — During the building and assembly of machine tools, tests are made with proof bars, surface plates, parallels, levels, straight edges, templets, and various forms of test and indicating gages to insure the alinement and accuracy of the component parts. Tests made *before and during* assembly may be repeated *after* assembly with the addition of different cutting operations to prove the accuracy and alinement of the machine, and to insure that the machine will produce accurate work. Some parts are often given their final tests and inspection during scraping and assembly, to eliminate the necessity of taking down and reassembling the machine should corrections be necessary.

Small errors are usually corrected by scraping; *excessive errors* often require remachining and scraping or the replacing of the inaccurate parts. The general procedure of inspection is practically the same for all machine tools and differs only with the type of machine and the class of work which it produces. See Scraping, *Principles of Machine Work.*

Information. — Wooden shims, or shingles, are commonly used to level up machines. One leg or corner of the machine base should remain on the floor and as few thicknesses as possible used. After the machine is set in place and leveled, additional shims, or shingles, should be driven in to fill open spaces under the legs or base to give a solid bearing.

2. The planing of beds, tables, frames, and other component parts of machines is the fundamental basis of accurate machine building. The alinement of a machine depends, to a large extent, upon the accuracy of the planer upon which the bed or frame of the machine is planed. An expert scraper can correct small errors, but depends largely upon the planer for accurate alinement since straight edges, scraping jigs, or surface plates are not always available or convenient to use on large work. Large errors cannot be corrected economically by scraping. Surfaces, especially of large work, are often scraped to fit each other. See Planers, p. **9**01, and Inspection of a Planer, p. **14**16.

3. Proof bars, except for boring tests, should be hardened and accurately ground. The tapers should fit the spindle holes properly, and the straight portions, while not necessarily of any exact dimension, should not vary over 0.0001″ in diameter along the entire lengths. The bars should be inspected periodically by testing on centers with an indicator. See p. **12**10. The proof spindles should be made with an extension on the rear end to counterbalance the weight of the test end which extends over the ways.

INSPECTION OF AN ENGINE LATHE

4. Requirements for engine lathe accuracy. — To produce accurate work a lathe should comply with the following requirements:

I. All parts of the mechanism should run smoothly and accurately.

II. Parts that travel on the bed must move parallel to the ways.

III. The spindle axis should be parallel to the ways for

turning work on centers, holding work in chuck or on face plate.

IV. A line from live center to dead center should be parallel to the ways, and be a continuation of the center line of the spindle.

V. The carriage slide and compound rest should be at right angles to the ways and axis of spindle.

VI. The adjusting screws to set over the footstock (tailstock) should be fitted free enough to be operated with the thumb and finger.

VII. The centers should be true, and the dead center hardened, tempered and ground.

5. Testing a lathe during assembly. — The bed is carefully leveled and kept so during assembly and inspection. This will insure proper alinement of the ways and prevent distortion.

To level a lathe, an accurate machinists' level is set across lathe on top of ways at each end, to test the wind of the bed. The level is then placed parallel to ways to level lengthwise. To support the level, V blocks which fit the ways of the lathe are sometimes used. If the lathe is long, tests are made at several intermediate points. Small lathes are usually bolted to the floor; large lathes may or may not be. When bolting a lathe to the floor, care should be taken not to distort or warp the bed. See Engine Lathes, p. 101.

6. To aline lathe carriage. — After scraping the loose metal from both the inner and outer ways of the bed, the carriage is scraped to a bearing with the outside ways. See Scraping, *Principles of Machine Work*. The outside ways are scraped to give accurate bearing and alinement, using the carriage as a standard or templet. The carriage V ways are then rescraped to give bearing surfaces at the ends of the V ways only, after which the carriage is tested to see if the center line of the cross slide, or compound-rest base, is at right angles to the ways.

To test carriage, set special jig *A*, Fig. 1, on the ways. Clamp dial test indicator *B* to the slide-rest base *C* with point

D of indicator touching the jig surface.   Move indicator back and forth across the carriage on the slide-rest base, and note any variation in reading of indicator.   Make corrections by scraping V ways in the carriage.

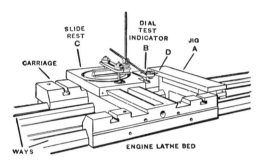

Fig. 1. — Testing Alinement of Carriage with the Ways.

It is well to make this test at two or three points along the lathe bed, and if there is any distortion in the bed, corrections may be made when alining the carriage.

To insure that the lathe will face slightly concave, the ends of the carriage V ways are scraped so that the carriage will be about 0.0005″ out of alinement toward the back of lathe.

Information. — The dovetail is scraped to a straight edge.   The slide or compound-rest base, to which the indicator is attached, is scraped to a bearing with the dovetail.

7. The cricket or footstock base is first scraped to a good bearing with the ways.   The ways are next scraped using the cricket as a standard or templet.   The cricket V ways are then scraped to give a bearing at the ends only.   The flat surfaces of cricket and footstock are scraped to a standard surface plate after which they are scraped together to insure a true bearing.

8. To aline lathe spindle with ways. — The spindle is fitted to its bearings by scraping, and the headstock V ways are scraped to a bearing with the inside ways of the bed.   See To Make Small Lathe Spindle, p. 622.

To test, place proof or dummy spindle A, Fig. 2, on head-stock bearings B, B'. The proof spindle consists of a bar a portion of which is ground to fit spindle bearings. The test part has enlarged portions C and C' at each end, 8″ to 18″ apart, which are accurately ground to the same diameter. Take readings with indicator D on top of bar, as at D and D'. A difference in the two readings not exceeding 0.00025″ is permissible.

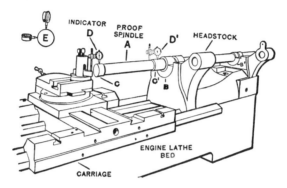

FIG. 2. — TESTING ALINEMENT OF SPINDLE WITH THE WAYS.

Correct errors by scraping the V ways in the headstock. Make additional tests at the side of bar, as shown in detail at E.

To insure that the lathe will bore slightly larger at the outer end of hole, the V ways in headstock are scraped to throw the outer end of proof spindle out of alinement 0.0005″ toward the back of lathe.

Information. — For some types of indicators, it is more convenient to take readings on the under side of spindle than on the top.

9. Inspection of lathe after assembly. — Level machine carefully. Test live and dead centers in their respective spindles by the use of artists' blue, to determine bearings. Oil and adjust all bearings; put on belts and allow lathe to run for a short time. The mechanism should operate without excessive vibration, and speed and feed changes be easily made.

10. To test accuracy of taper hole in lathe spindle. — Insert proof bar *A*, Fig. 3, in taper hole in headstock spindle; revolve, and test at spindle end with indicator *B* held in tool post, as at *C*.

An error of 0.0005″ is permissible. Move carriage along lathe bed and test proof bar at outer end, as at *C′*. The error at outer end should not exceed 0.0015″. If greater errors exist, the spindle hole may be reamed to alinement or, in extreme cases, rebored.

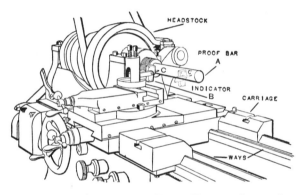

FIG. 3. — TESTING ACCURACY OF TAPER HOLE IN LATHE SPINDLE.

11. Testing alinement of lathe centers. — Duplicate plugs are inserted in the taper holes in the headstock and footstock spindles. These plugs have short, duplicate, and accurately ground gaging ends. The footstock, with the spindle run back, is moved toward headstock so that the plugs nearly touch. The distances of the plugs above the ways are then measured with an indicator. The dead center should not be over 0.0015″ higher than the live center to allow for rigid clamping and wear in the V ways. Errors may be corrected by scraping the headstock or cricket. As the headstock and complete footstock are bored and planed together in jigs, very little correction is usually necessary.

The footstock may now be set in alinement by testing the plugs at the side and making zero lines on footstock and cricket.

Attention. — Alinement of centers may also be tested by mounting a long proof bar on centers and testing with an indicator. See To Set Dead Center in Alinement to Turn Straight, p. 116.

Information. — Before the plug is removed from the footstock, the spindle is extended and the plug again measured to test the accuracy of the footstock spindle alinement.

12. To make lathe boring test. — Insert proof bar *A*, Fig. 4, with taper shank in the spindle hole. Take trial cuts on the enlarged portion at each end of the bar with the same tool

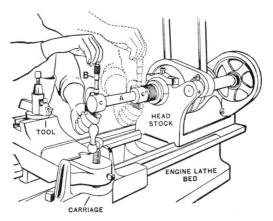

FIG. 4. — MEASURING PROOF BAR FOR BORING TEST.

setting. Measure these diameters with micrometer calipers *B*, as at *C* and *C'*. A difference in diameters of 0.00025″ in eight inches is permissible, but preferably with the outer end of bar the larger so that when the lathe is used for boring, the outer end of the hole will be slightly larger.

Information. — Another boring test may be made by holding a hollow piece in a chuck, taking several boring cuts, and measuring the diameter of the hole with inside micrometers.

13. To make lathe facing test. — Screw face plate *A*, Fig. 5, on spindle and take two facing cuts, one inward and one outward. Test face plate with straight edge *B* and three pieces of 0.001″ tissue paper, *C*, *C*, *C*. The surface should be either flat or not over 0.0005″ concave. The face should never be convex. Correct any errors by scraping the carriage V ways.

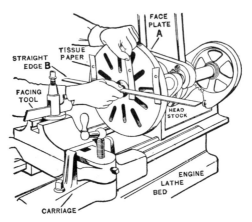

Fɪɢ. 5. — Tᴇsᴛɪɴɢ Fᴀᴄᴇ Pʟᴀᴛᴇ ғᴏʀ Fᴀᴄɪɴɢ Tᴇsᴛ.

14. Special lathe tests. — When lathes are to be used for special purposes, or when extreme accuracy is desired, the lead screws, gearing, etc., are given indicator and operation tests to determine their accuracy. See Lead Screw Testing Machine, p. **14**29.

INSPECTION OF A VERTICAL DRILLING MACHINE, DRILL PRESS OR UPRIGHT DRILL

15. Setting up drilling machine for inspection or operation. — The machine is leveled by setting the face of the column plumb. It is tested with a machinists' level held vertically against the face of column. All drilling machines should be bolted to floor to prevent vibration. See Drilling, *Principles of Machine Work.*

The principal tests are made to find out if the drilling table and base are at right angles to the spindle, and to determine the accuracy of rotation of the spindle hole.

16. To test accuracy of table of drilling machine. — Insert special taper plug A, Fig. 6, in spindle hole. The lower end is made to hold rod B which has indicator C at outer end. Place small parallel block D on table, lower spindle, so that indicator point touches block, and take reading. Test the table at several points by rotating spindle by hand and use the parallel block to span the table slots. The front of the table should be about 0.003″ high to allow for pressure of cut and sag of table. Correct alinement of table by scraping the face of column end of arm until table is in proper alinement.

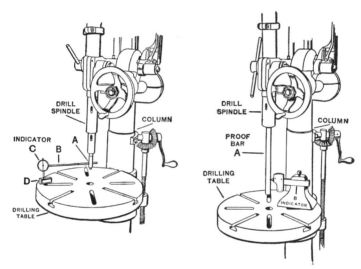

FIG. 6. — TESTING ACCURACY OF DRILLING MACHINE TABLE.

FIG. 7. — TESTING ACCURACY OF TAPER HOLE IN DRILL SPINDLE.

17. To test accuracy of drilling machine spindle hole. — Insert proof bar A, Fig. 7, in spindle; revolve, and test with indicator B. Make tests at spindle end and at lower end of bar. For ordinary work, the proof bar may run out 0.00025″ at the spindle end while the lower end may be 0.00075″ out of alinement. Make corrections by scraping the face of flange when the column is bolted to the base, and by scraping the inside surface of column socket when it is clamped in the socket.

Information. — The base may be given tests for alinement similar to the table tests.

INSPECTION OF A PLAIN OR UNIVERSAL MILLING MACHINE

18. To test a milling machine. — The spindle, column face, and knee are made to fit and aline within certain limits by careful machining and scraping; they are usually given their final inspection during scraping and assembly to eliminate the necessity of testing again when the machine is fully assembled.

19. Setting up a milling machine for inspection or operation. — To set up, an accurate machinists' level is held vertically against face of column to see if it is plumb. The column should also be tested with the level held against edge of column. Milling machines should be bolted to the floor to prevent vibration. See Milling Machines, p. **100**1.

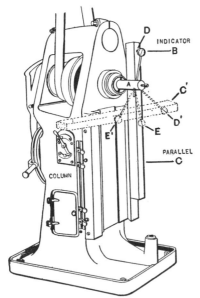

20. To test alinement of spindle axis with the face of the machine column. — Remove knee to leave face of column free for testing. Insert special plug *A*, with indicator *B* attached, Fig. 8, in spindle hole. Place parallel *C* in a vertical position against the face of column, as in figure. Rotate spindle by hand until

FIG. 8. — TESTING ALINEMENT OF SPINDLE WITH FACE OF COLUMN.

measuring point of indicator touches front surface of parallel. Take readings at *D* and *E*. Assuming the distance between

D and E to be three feet, the reading at D may be from 0.004″ less to 0.002″ more than the reading at E.

Then place parallel in a horizontal position, as shown at C' and take readings at D' and E'. The reading at D' may be 0.002″ less than E' but not greater. Correct errors in both cases by scraping the spindle bearings.

21. To test alinement of milling machine knee with the face of column. — With table and saddle removed, as in Fig. 9, place square or angle plate A on knee B. Hold indicator C, attached to stand, against face D of column so that the measuring point touches face of square. Make tests at top and bottom of square, as at E and E'. The reading at E may be 0.001″ greater than the reading at E' in a distance of 18″,

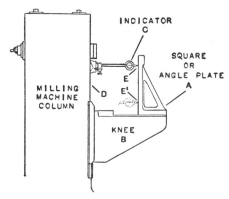

Fig. 9. — Testing Alinement of Knee with Face of Column.

to allow for pressure of cut and sag of knee. Correct errors by scraping face of knee which bears on face of column. Make similar tests between edge of column and top of knee, and between side of knee and face of column. Small errors which will offset the cutting strains of the machine are permissible.

22. To test accuracy of rotation of spindle hole. — Insert proof bar A, Fig. 10, in spindle hole. Place indicator B, attached to stand C, on top of knee D so that measuring point touches top of proof bar. Rotate proof bar and take readings at spindle end, as at E, and at outer end as at E'.

An eccentricity, not to exceed 0.0005″ at the spindle end, and 0.001″ at twelve inches from spindle end is permissible. If larger errors exist, the spindle hole may be reamed to alinement, or, in extreme cases, rebored.

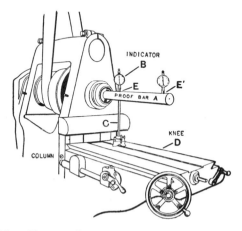

FIG. 10. — TESTING ACCURACY OF TAPER HOLE IN MILLING MACHINE SPINDLE.

23. To test alinement of milling machine table. — Place special proof bar A, Fig. 11, in spindle hole. Set indicator B, attached to stand C, on table D so that measuring point bears on top of proof bar. Run the table in and out, and take readings at E and E'. The height at E may be 0.001″ less than the height at E', but not greater. Make corrections by scraping top of saddle.

24. To test accuracy of table with knee. — Move table A, Fig. 12, inward on knee B far enough to give sufficient bearing on top of knee for indicator stand C. Set indicator D so that measuring point rests on block E which is used to span the table slots. Take a reading at F; then move indicator to position F' and take another reading. There should not be over 0.00025″ difference in the two readings. Correct errors by scraping top of saddle. If desired, the table may be moved along in the direction of travel and other readings taken.

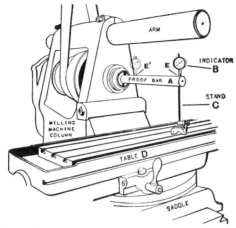

FIG. 11. — TESTING ALINEMENT OF TABLE WITH SPINDLE.

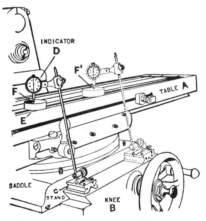

FIG. 12. — TESTING ALINEMENT OF TABLE WITH KNEE.

25. To test alinement of overhanging arm. — Move arm *A*, Fig. 13, to its working position. Raise knee *B* until special micrometer gage *C*, which rests on stand *D* on table *E*, touches lower surface of arm. Take a reading at outer end of arm, as at F. Move table back, and gage over, and take another reading, as at F'.

The reading at F should not differ more than 0.001″ from the reading at F'. Correct errors by scraping arm bearings.

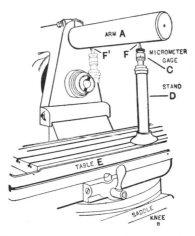

FIG. 13. — TESTING ALINEMENT OF ARM WITH TOP OF TABLE.

26. Additional inspection tests. — The dividing head, vises, and other attachments and fixtures, also the gearing, feed screws, table slots, etc., are usually given some form of inspection.

INSPECTION OF A VERTICAL MILLING MACHINE

27. The principal tests applied to a vertical milling machine are similar to those made on a plain or universal machine. Surfaces which must be parallel to the spindle axis in the plain or universal machine are perpendicular to that axis in the vertical machine, and vice versa. These machines are also usually inspected during scraping and assembly to see that the spindle knee and column fit and aline properly.

Attention. — Before inspection or assembly, the machine should be properly set up by making face of column vertical, see Vertical Milling Machine, p. 1043.

28. To test knee for alinement with spindle. — Place accurate parallel *A*, Fig. 14, on knee *B*, parallel to and near face of column.

Insert special plug *C* with radius arm and indicator *D* attached, in spindle hole. Raise knee until measuring point of indicator touches block *E* on parallel. Take a reading at *F*. Turn spindle and take another reading at other end of parallel. Move the parallel to front of knee and take similar readings.

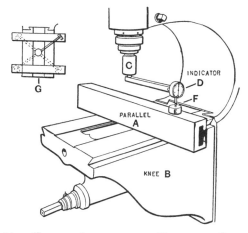

FIG. 14. — TESTING ALINEMENT OF KNEE WITH SPINDLE.

This will give four test points, as shown in detail at *G*. The front of the knee may be 0.002″ high to allow for pressure of cut and sag when saddle and table are mounted. Correct errors by scraping the face of the knee.

Information. — The spindle bearing holes are bored in a jig after planing but before scraping face of column. The holes are usually alined accurately enough for ordinary work. A few machines, however, selected at random, may be tested to check the quality of production.

29. To test alinement of table with spindle at different positions. — Mount saddle and table in position and connect the various screws, gears, and attachments. Insert special plug A, Fig. 15, with indicator B attached, in hole of spindle C. Place accurate square D, with blade vertical, on table E so that measuring point of indicator touches edge of square. Lower spindle to its extreme position and take a reading at F. Then raise spindle and take another reading at F'. Place square at 90° from first position and take similar readings. An error not to exceed $0.001''$ in $24''$ at the top in either direction is permissible. Should greater errors exist, rescrape spindle bearings or install a new spindle equipment.

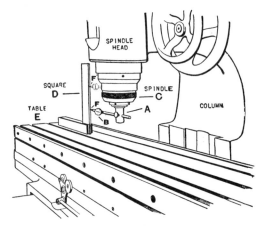

FIG. 15. — TESTING ALINEMENT OF TABLE WITH SPINDLE.

The table is also tested for alinement with the spindle by the same process used for testing the knee. See p. **14**15.

INSPECTION OF A PLANER

30. To determine the accuracy of a planer. — Since planing is the basis of the accurate alinement of many machines it is necessary that the component parts of a planer aline and operate properly. Careful tests must be made during and after assembly.

Before inspection or assembly, the bed should be carefully leveled to prevent distortion. It may or may not be bolted to the floor; but if bolted to the floor, care should be taken not to distort it by excessively tightening the floor bolts. See Planers, p. 901.

31. To level planer bed. — When the surfaces of V ways are not finished true with the face, accurate rollers or V blocks should be used to support the parallels and level.

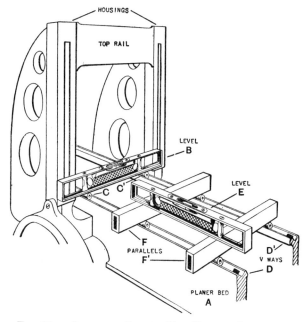

FIG. 16. — LEVELING PLANER BED BEFORE ASSEMBLY.

To level bed A, Fig. 16, set level B crosswise on rollers C, C' which rest in V ways D, D' at center of bed; then test the bed lengthwise by placing level E lengthwise on parallels F and F', as shown. Correct errors in both tests by placing shims under the legs. Repeat these operations at intervals of from 2′ to 4′ along the bed in both directions from the center, depending upon the size of the planer.

The bed may again be leveled after housings and top rail are in place, to correct any distortion of the bed due to the weight of added parts.

32. To test planer housings. — The fronts of the housings must be parallel to each other and perpendicular to the ways of the bed for the planer to produce accurate work.

The sides of housings or cross-rail guides must also be parallel to each other and perpendicular to the planer ways, so that the cross rail may be raised or lowered parallel to the table, or ways, and at the same time have no lateral motion.

To test front of housings, hold level square A, Fig. 17, against the broad face of housing B and note the position of the bubble in the level; then move level square over to housing

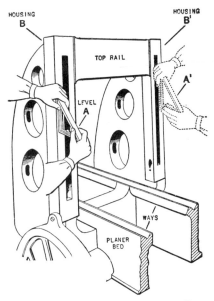

Fig. 17. — Testing Planer Housings with Level Square.

B' and hold, as at A'. Correct errors by scraping or planing the base of the housings. The level square may be placed lengthwise on the ways and the position of the bubble noted, since the location of the bubble when level square is held against housing should coincide with the position when on ways.

To test sides of housings or cross-rail guides, place parallel
A, Fig. 18, across bed on rollers *B*, *B'*. Set accurate square
C on parallel with blade near side of housing *D*. Mount indicating fixture *E* on housing *D* and guide it by edge *F*.
Adjust square so that the blade will touch measuring point of

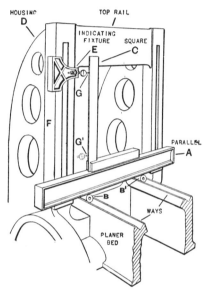

FIG. 18. — TESTING ALINEMENT OF CROSS-RAIL GUIDE TO BED.

indicator. Move the fixture up and down and take readings
at *G* and *G'*. A difference in the readings not to exceed
0.0015″ in 36″, is permissible. Repeat the operation on the
other housing. Make corrections by scraping or planing the
base of housing.

Information. — The indicating fixture is designed to bear on the
cross-rail guide. Some planers have this surface on the inside of the
housings and others on the outside.

Attention. — This method is frequently used to test the front faces
of the housings instead of the level square method.

33. Leveling planer table. — The table is first scraped to a bearing with the V ways of bed, and the V ways of bed scraped to a finish bearing using the table as a standard or templet. They should again be leveled to correct distortion due to the additional weight. The table is run backward and

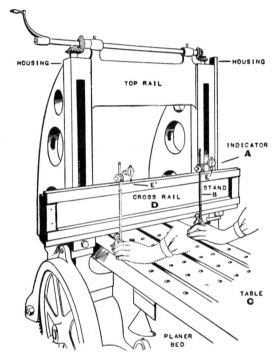

Fig. 19. — Testing Cross Rail to see if Parallel with Table.

forward as far as possible and the ways tested. As a further test, the table may be run back and a level placed across it at the front end. The table is then run forward slowly and any change in the position of the bubble noted. The operation is repeated with the level at the other end of the table. Errors are corrected by placing shims under the legs or base of bed A final leveling test should be given with the table in the middle of the bed. A parallel or straight edge is placed lengthwise on the table supported by three accurate blocks, with tissue paper between the straight edge and the blocks, to determine the bearing. Any difference in the bearing on the blocks, or difference in the number of thicknesses of the tissue paper, will show that the table is curved, due to the ways not being level. Errors are corrected by adjusting the shims under the base of bed. For long planers, this test may also be made with table alternately at both ends of bed as well as in center.

34. To test cross rail. — The cross rail after being scraped to a bearing is set in position and clamped to the housings. The elevating screws and bevel gears are then attached. The cross rail is adjusted and tested to see if it is parallel to table so that the planer will produce parallel work.

To test, set indicator A, Fig. 19, fastened to stand B, on Table C so that measuring point touches top surface of cross rail D. Make tests near each end of the cross rail, as at E and E'. The difference in the readings should not exceed $0.00025''$. Correct errors by changing the position of one bevel gear on shaft and adjusting the elevating screw. As a final test, take a light cut across the planer top after cross rail has been adjusted.

Information. — When adjusting cross rail the final movement should be upward, to remove the backlash of the side screws. See To Set Planer in Alinement, p. **9**09.

INSPECTION OF THREADS

35. Introduction. — All threads should first be given a visual inspection for smoothness.

Taps and thread gages must have smooth threads. It is also desirable that screw threads for general machine work be smooth. Screw threads that are slightly rough may be accepted for coarse and rough work.

The pitch diameter, and outside and inside diameters are the elements usually inspected with gages, but for accurate work the lead is also inspected.

36. Limit gages for screws. — External " go " and " no-go " thread gages are made circular and rectangular in form and of the same thickness as the nut which is to fit the thread. They may be solid but are usually split for adjustment and supplied with locking screws that are set and sealed with wax to detect tampering. See Adjustable Limit Snap-Thread Gages, p. 225.

37. Limit gages for nuts or threaded holes are similar to double-ended limit plug gages, but threaded. The " go " end is made with its pitch diameter to the minimum limit and the " no-go " end to the maximum limit, see No. 30, Chart, p. **13**12. Two separate single-ended gages may also be used.

Attention. — In addition to the usual limit gages for threads, plain ring and plug gages are used to measure the diameter at the tops of the threads on the screw and in the nut.

Information. — Ring and plug thread "go" and "no-go" gages not only test the pitch diameter of a thread but also check the lead.

LIMITS APPLIED TO THREADS

38. Lead and pitch diameter of threaded work. — Many times when the outside and pitch diameters are correct, a screw and nut will not fit because of the difference in the lead of the threads. For rough work this is not so important, but for work requiring accuracy, strength, long wear and inter-changeability, such as precision screws, lead screws, feed screws, jack screws, cap screws, bolts and nuts, it should be considered. The lead of the average tap is often short, owing to the use of steel which shrinks excessively when hardened. The lead of the average screw is often long, due to the in-accuracies and wear of the thread-cutting mechanism.

See Indicators and Testing Machines for the Precision Measurements of Thread Leads, p. **14**27; Thread Micrometer, p. **12**14, and the Wire Systems for Precision Measurements of Pitch Diameters and Angles of Threads, p. **14**31.

39. Evolution of a threaded fit and the effect of errors in lead. — When a nut with a short lead is turned halfway onto a screw with a long lead, as in Fig. 20, it may fit loosely for the whole length of engagement as at A, B, and C; but when the nut is turned three-quarters of the way onto the screw, as

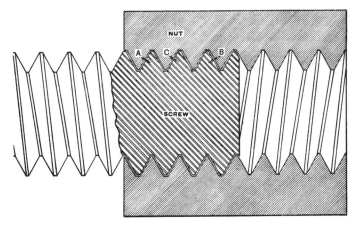

FIG. 20. — EVOLUTION OF A THREADED FIT.   HALF ENGAGEMENT. NO INTERFERENCE.

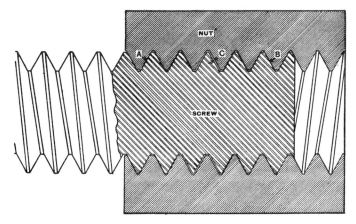

FIG. 21. — EVOLUTION OF A THREADED FIT.   THREE–QUARTER ENGAGEMENT.   INTERFERENCE.

in Fig. 21, it may bind from interference, as at A and B, and be loose, as at C, due to the differences in the lead.

40. Finger fit — Cutting the screw to fit the nut. — To allow the nut to be turned onto the screw for full engagement, the pitch diameter of the screw may be slightly reduced by taking additional cuts or by adjusting the die and running it on again; or, the hole may be made slightly larger by an adjustable tap. If the screw or nut has been thus corrected, the nut will turn all the way on, as in Fig. 22, and give complete engagement; the screw and nut will fit closely at the ends, as at *A* and *B*, but have *no contact* in the center, as at *C*.

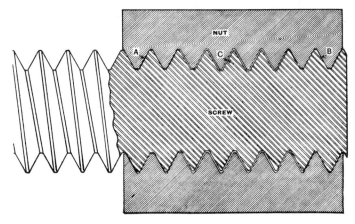

Fig. 22. — Evolution of a Threaded Fit. Full Engagement. Interference Removed.

41. Wrench fit — Forcing a threaded fit. — Nuts and screws with excessive errors in lead which can be engaged for two or three threads only with the fingers, are often forced with a wrench to a full engagement since the pressure will cause the threads to bend or flow. While such a fit is better than a loose one, if excessive force is necessary to screw the parts together the threads may be weakened and abrasion or stripping of the threads may result.

Warning. — When a nut is forced onto a screw, a lubricant should be used which will prevent abrasion, if the fit is not excessively tight.

Information. — Screws with the excessive errors of lead when used to fasten parts together in this manner, are liable to strip the threads of the nut as there are not enough threads, or surface, in contact. Screws for transmitting motion, such as lead screws, when fitted in this manner, cause excessive wear in the nut and soon become loose, due to the few threads in contact.

42. Clearance at top and bottom of threads. — To obtain a clearance at the top of threads between the nut and the screw, the outside diameter of the tap is made slightly larger than the nominal size. The amount of oversize will vary from 0.001″ to 0.020″ depending upon the diameter of the screw and the pitch of the thread. The clearance at bottom of thread is obtained by using a tap drill slightly larger than the root diameter of the tap or screw.

43. Relations between pitch diameter and lead of threads. — Errors in pitch diameter will, to a certain extent, modify errors in lead so that the greater the difference between the actual and basic pitch diameters the greater the errors in lead may be and apparently give a fair fit.

Example of a small tap. — The actual pitch diameter of a ¼″ × 20 U. S. S. tap by measurement is found to be 0.2193″. The error over basic pitch diameter (0.2175″) is 0.0018″ for which the error per inch of lead for the tapped hole is ± 0.0040″. The screw with this error of lead would be 0.0018″ under basic or have an actual pitch diameter of 0.2157″.

While the total maximum error of lead may be 0.0080″ per inch, the actual error per fit is only 0.002″ which is permissible for ordinary work.

Example of a large tap. — The actual pitch diameter of a 1″ × 8 U. S. S. tap by measurement is found to be 0.9228″. The error over basic pitch diameter (0.9188″) is 0.0040″ for which the error per inch of lead for the tapped hole is ± 0.0023″. The screw with this error of lead would be 0.004″ under basic or have an actual pitch diameter of 0.9148″. While the total maximum error of lead may be 0.0046″ per inch, the actual error per inch would be approximately 0.004″ which, though larger, would be permissible for certain classes of work.

Attention. — The actual pitch diameter of taps should be larger than the basic pitch diameter, while the actual pitch diameter of screw threads should be smaller than the basic. See pp. **12**14, **14**31.

Information. — For small screws, the error in lead per inch may be relatively large since the actual error for the length of engagement will be only a fraction of the lead per inch. For large screws, the error in lead per inch should be relatively smaller since the actual error for the length of engagement will be a multiple of the lead per inch.

44. To remedy errors in the lead of taps. — Steel which changes the least when hardened and tempered should be used in making taps. The thread should be cut in machines with precision thread-cutting mechanisms, accurately scraped ways, correct alinement, and accurate threading tools. Errors in lead due to shrinkage in hardening are sometimes compensated when cutting the thread, by using a lead screw variator.

45. To remedy errors in lead of screws they should be cut with machines having accurate screw-cutting mechanisms or with accurate dies.

46. Allowances for grinding and lapping thread gages. — To obtain the necessary accuracy and smoothness, thread gages are ground in special machines or lapped to size after hardening and tempering. Gages less than $\frac{3}{8}''$ in diameter are usually lapped as the threads are too small to grind accurately.

Gages which are only lapped are made with an allowance of from 0.0002″ to 0.001″ oversize in pitch diameter, with thread angle from 10 to 20 minutes small. An error not to exceed 0.0005″ in lead may be corrected by lapping. See p. **122**8.

Allowances for grinding are from 0.005″ to 0.010″ oversize in pitch diameter which are sufficient to correct usual errors in lead. The gages are ground to within the usual allowances for lapping.

47. Expansion and shrinkage from hardening and tempering. — Thread gages and taps may expand from 0.0005″ to 0.020″ per inch of diameter, and shrink from 0.0005″ to 0.020″ per inch of length.

INDICATORS AND TESTING MACHINES FOR THE PRE-
CISION MEASUREMENTS OF THREAD LEADS

48. Thread lead indicator A, Fig. 23, is used for the accurate measurement of threads of taps and thread gages. It
consists of fixed measuring point B, movable measuring point
B', needle scale C, and nurled nuts D and D' on barrel E.

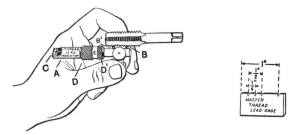

FIG. 23. — TESTING LEAD OF A TAP   FIG. 24. — MASTER THREAD LEAD
    WITH THREAD LEAD INDICATOR.   GAGE.

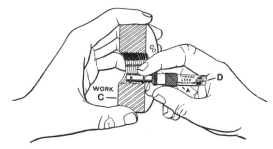

FIG. 25. — TESTING LEAD OF THREAD RING GAGE WITH THREAD
LEAD INDICATOR.

49. To test the lead of thread of a tap, Fig. 23. — Set the
points B, B' to the desired distance apart either 0.25″, 0.50″,
or 1.00″ with the master thread lead gage, Fig. 24. Adjust
needle of indicator at zero on scale C by nuts D, D'. Hold
measuring points B, B' accurately in the thread of tap and
read the variation of the actual lead on scale C in thousandths
of an inch. Half, quarter, and tenths of thousandths may be
readily estimated.

For precision work the lead of a tap should not vary over 0.0002″ to 0.0005″ per inch. Taps, however, for ordinary work often have a variation of 0.0040″.

50. To test the lead of the thread in a ring gage with lead indicator A, Fig. 25. — Hold measuring points B, B' accurately in threaded work C and take reading on scale D.

Information. — Sometimes it is more convenient to hold indicator on a stand and place work on a guide block.

51. Lead testing machine for taps and thread gages, Fig. 26.

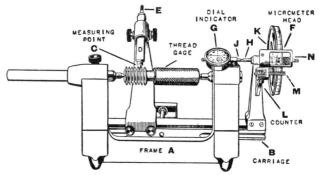

FIG. 26. — TESTING LEAD OF THREAD GAGE WITH TESTING MACHINE.

SCHEDULE OF PARTS

A — Frame with three-point bearings.	G — Dial indicator.
B — Ball-bearing carriage. Supports thread-measuring point and thread-measuring mechanism.	H — Micrometer spindle.
	J — Dial indicator spindle.
C — Thread-measuring point.	K — Graduated micrometer gear wheel reading to 0.00001″.
D — Sleeve containing spring plunger.	L — Counter for revolutions of K.
E — Plunger nurled handle.	M — Counter pinion.
F — Micrometer head.	N — Zero line.

52. To test lead of thread plug gage with lead testing machine, Fig. 26. — Mount thread gage on centers and place measuring point C in thread groove by raising and then lowering handle *E*. Rotate spindle *H* of micrometer head *F* by means of wheel *K* until it presses dial indicator spindle *J* so that dial indicator *G* registers at zero. Take reading on counter *L* and at zero line *N*. Withdraw point *C* and move it one or more threads to the right. Again adjust micrometer until dial indicator *G* registers at zero. Take reading from counter and zero line. Compute actual thread lead from two readings and compare with theoretical or calculated lead.

For fine tool work the lead of thread gages should not vary more than 0.0002" per inch.

Important. — Drunken threads may be detected by turning the gage, or tap, quarter or half-way around and testing again.

Attention. — The actual lead per inch may be measured by using standard gage blocks between point of dial indicator and micrometer spindle instead of computing from the lead per thread and the number of threads per inch.

Information. — The dial indicator is not generally used for measuring but only to obtain the same pressure on the micrometer points each time. It may, however, be used to give the exact error by moving the graduated wheel to correspond with the calculated lead, inserting the measuring point in the thread groove and then taking reading on the dial indicator. The dial indicator is graduated to read to 0.0001".

53. The lead screw testing machine, Fig. 27, is used principally to test the accuracy of Acme Standard or 29° thread lead screws.

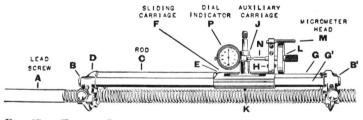

FIG. 27. — TESTING LEAD OF A LEAD SCREW WITH TESTING MACHINE.

To set, place lead screw *A* in lead testing machine and clamp securely in place in V blocks *B* and *B'*. Insert rod *C* between stop *D* on V block *B* and stop *E* on sliding carriage *F*. Move carriage F to left on carrier bars G, G' until it touches rod, and clamp with thumb nut H. Move auxiliary carriage J in sliding carriage F until plunger point K will enter a thread groove. Turn graduated wheel L of micrometer head M until micrometer spindle N touches plunger of dial indicator P and the dial reads at zero. Take reading on graduated wheel L. This wheel may be read to 0.0001″. Withdraw plunger point K and move auxiliary carriage J one or more thread grooves to right or left and set plunger point in position. Adjust micrometer again until dial indicator registers at zero and take reading on graduated wheel L. Compute the difference between readings and compare it with the theoretical or calculated lead.

If the lead per inch or per foot is desired, move carriage F to left end of machine until stops D and E touch. Clamp in position and take reading. Move carriage to right and insert standard combination gage blocks between stops D and E. Clamp in position and take second reading. These blocks may be 1″, 3″, 6″, 12″, etc., depending upon the distance desired.

For precision lead screws, the lead should not vary over 0.001″ per foot. In lead screws, for ordinary purposes such as lathe lead screws, a variation of 0.003″ to 0.005″ per foot is often allowed.

54. Optical principle for the precision measurement of threads makes use of projected and reflected rays of light in connection with special precision machines.

The enlargement of errors in the thread obtained by the multiplying effect of the instrument makes extreme accuracy possible.

The leads of threads are measured with machines which have extremely accurate micrometer screws and heads, and which use a beam of reflected light as an indicator for determining the position of the measuring point and the proper pressure on the micrometer spindle.

The forms and angles of threads are measured by the projected light principle, using a projection machine and templets or gages.

These methods while used chiefly for laboratory and research work may be used whenever extreme precision is desired.

WIRE SYSTEMS FOR PRECISION MEASUREMENT OF PITCH DIAMETERS AND ANGLES OF THREADS

55. The three-wire system is used for the accurate determination of the pitch diameter and thread angle of screw threads. It utilizes a series of groups of three wires of the same diameter, and a micrometer caliper, or similar measuring

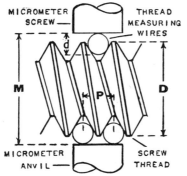

FIG. 28. — PRINCIPLE OF MEASUREMENT BY THE THREE-WIRE SYSTEM.

instrument, as shown in Fig. 28. The measurement over the wires is used in an empirical formula, together with the nominal diameter of the screw, the pitch of the thread, and the diameter of the measuring wires.

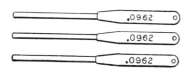

FIG. 29. — SPECIMEN WIRES.

The measuring wires, Fig. 29, are hardened, tempered, and accurately ground and lapped to size. The part of the wires finished for measuring is about 3″ to 4″ long. Wires are obtainable with the handle or grip portion made in different shapes.

To save unnecessary calculation, tables are obtainable which give the proper sizes of wires for different sizes and styles of threads. See p. 1434.

Holder for measuring threads with three wires. — To measure work conveniently, a holder or special stand, Fig. 30, may be used which consists of base A and V block B. Place thread gage C in V block B and secure by clamp screw D.

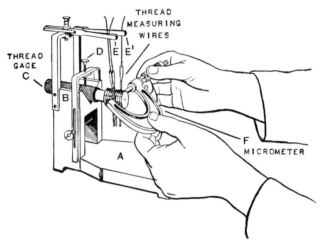

Fig. 30. — Method of Holding Work and Wires while Measuring.

Suspend measuring wires by fine wires or strings from supports E, E'. Take measurement over wires with micrometer F.

Attention. — An improvised holder can be made by placing work in a vise and supporting the measuring wires with strings and a bent wire held in the vise, or by holding the wires in position by means of rubber bands placed over their ends.

Information. — While measurements made at one point are usually sufficient, additional readings may be taken with the wires in a different set of thread grooves.

56. Principle of measuring the pitch diameter. — By trigonometry it can be proved that the measurement over the wires M, Fig. 28, equals $D - KP + 3d$ where D = nominal screw diameter, K a constant, P = pitch of thread and d = diameter of pitch wires used. Each style of thread has its own constant and wire sizes.

57. Determination of pitch diameter of a U. S. S. outside thread from measurement over wires. —

Example. — To find correct measurement over wires and pitch diameter of a $\frac{3}{4}'' \times 10$ U. S. S. thread gage.

Formula. — $M = D - KP + 3d$

where $\qquad D = 0.7500''$

$\qquad\qquad K = 1.5155$

$\qquad\qquad P = \dfrac{1}{\text{number of threads}} = \dfrac{1}{10} = 0.1000''$

$\qquad\qquad d = 0.05773.$ See Table, p. **14**34.

Solution. — $M = 0.7500 - 1.5155 \times 0.1000 + 3 \times 0.5773.$

$\qquad\qquad = 0.77164''.$

If the actual pitch diameter is correct or, in other words, equals the theoretical or basic pitch diameter, the measurement over the wire should be $0.77164''$. Any variation in the measurement over wires from the theoretical will be the same as the difference between actual and theoretical pitch diameters.

Information. — The pitch diameter is often called the "angle" or "effective" diameter.

58. Table of Seaboldt wires for measuring U. S. S. threads. — The column marked " Pitch Wire " gives sizes of wires for measuring pitch diameters, columns marked " Large Wire " and " Small Wire " give wire sizes for measuring thread angles.

Threads per Inch.	Pitch Wire.	Large Wire.	Small Wire.	Single Depth.	Pitch	Threads per Inch.
50	.01154	.01953	.01076	.01299	.02000	50
48	.01202	.02000	.01100	.01353	.02083	48
46	.01254	.02145	.01147	.01411	.02173	46
44	.01311	.02195	.01197	.01476	.02272	44
42	.01374	.02304	.01254	.01546	.02380	42
40	.01443	.02425	.01312	.01623	.02500	40
38	.01519	.02558	.01379	.01709	.02631	38
36	.01603	.02706	.01453	.01804	.02777	36
34	.01698	.02871	.01535	.01910	.02941	34
32	.01804	.03067	.01628	.02029	.03125	32
30	.01924	.03267	.01733	.02165	.03333	30
28	.02061	.03508	.01854	.02312	.03571	28
26	.02220	.03786	.01993	.02498	.03846	26
24	.02405	.04109	.02154	.02706	.04166	24
22	.02624	.04492	.02346	.02952	.04545	22
20	.02886	.04951	.02575	.03247	.05000	20
18	.03207	.05512	.02856	.03608	.05555	18
16	.03608	.06214	.03207	.04059	.06250	16
14	.04123	.07116	.03738	.04639	.07142	14
13	.04441	.07672	.03936	.04996	.07692	13
12	.04811	.08319	.04259	.05412	.08333	12
11	.05248	.09085	.04642	.05904	.09090	11
10	.05773	.10003	.05101	.06495	.10000	10
9	.06415	.11126	.05663	.07216	.11111	9
8	.07216	.12529	.06364	.08118	.12500	8
7	.08247	.14333	.07266	.09278	.14285	7
6	.09622	.16739	.08469	.10825	.16666	6
5½	.10497	.18270	.09235	.11809	.18181	5½
5	.11547	.20107	.10153	.12990	.20000	5
4½	.12830	.22352	.11276	.14433	.22222	4½
4	.14433	.25159	.12679	.16237	.25000	4
3½	.16495	.28652	.14426	.18557	.28571	3½
3	.19233	.33578	.16889	.21650	.33333	3

Attention. — Wires of any size may be used that will rest on the sides of the thread and project above the top.

59. Principle of measuring thread angle. — By trigonometry the cosecant of $\frac{1}{2}$ thread angle $= \dfrac{M - m}{W - w} - 1$, Fig. 31; M = measurement over large wires, m = measurement over small wires, W = diameter of large wires and w = diameter of small wires.

To measure, the first reading is taken over the larger wire with micrometer, as in Fig. 31. The second reading is taken over the smaller wire which is placed in the same groove as the larger ones, as shown by dotted lines, Fig. 31.

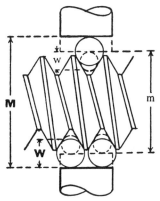

FIG. 31. — THREE-WIRE METHOD OF MEASURING THREAD ANGLE
OF U. S. S. THREAD.

60. Determination of thread angle of a U. S. S. outside thread from measurements over the wires. —

Example. — To find thread angle of $\frac{3}{4}'' \times 10$ U. S. S. thread gage.

Formula. — Cosecant of $\frac{1}{2}$ angle $= \dfrac{M - m}{W - w} - 1$. See Fig. 31.

By measurement, $M = 0.89854$; by table, $W = 0.10003$
 $m = 0.75148$; $w = 0.05101$

Difference, $M - m = \overline{0.14706}$; $W - w = \overline{0.04902}$

Solution. $-\dfrac{M - m}{W - w} - 1 = \dfrac{0.14706}{0.04902} - 1 = 3.000 - 1 = 2.0000.$

From table of trigonometrical functions the angle, the cosecant of which is 2.0000, is 30°. Therefore, the thread angle is 60°.

Information. — The thread angle in common practice, does not vary more than one-half a degree ($\frac{1}{2}$°) above or below normal. The usual thread angles and the values of $\dfrac{M - m}{W - w}$ are given in Table, p. **14**36

61. Table of common variations of thread angle of a U. S. S. thread, with corresponding quotients. —

Degrees.	Minutes.	Quotients.
59	30	3.0152
59	40	3.0101
59	50	3.0050
60		3.0000
60	10	2.9950
60	20	2.9900
60	30	2.9850

62. Three-wire system applied to other threads. — The method of measuring other types of threads is the same as for U. S. S. threads. The formulas and their values, however, are changed for the difference in the shapes and angles of the threads.

63. Determination of the pitch diameter of a Whitworth outside thread from measurement over wires. —

Example. — To find correct measurement over wires and pitch diameter of a $1\frac{1}{4}'' \times 5$ Whitworth tap.

Formula. — $M = D - KP + 3.1657\,d$
where
$D = 1.2500''$
$K = 1.6008$
$P = \frac{1}{5} = 0.2000''$
$d = 0.11273.$ See Table, p. **14**37.
Solution. — $M = 1.2500 - 1.6008 \times 0.2000 + 3.1657 \times .11273$
$= 1.28668''.$

If the pitch diameter is correct the measurement over wires should be 1.28668''. Any variation in this measurement will be the same as the variation in the pitch diameter.

64. Table of Seaboldt wires for measuring Whitworth threads. —

THREADS PER INCH.	PITCH WIRE.	LARGE WIRE.	SMALL WIRE.	SINGLE DEPTH.	PITCH.	THREADS PER INCH.
40	.01409	.02031	.01314	.01600	.02500	40
38	.01483	.02144	.01380	.01685	.02631	38
36	.01565	.02268	.01454	.01777	.02777	36
34	.01657	.02408	.01537	.01883	.02941	34
32	.01761	.02564	.01630	.02001	.03125	32
30	.01878	.02742	.01735	.02134	.03333	30
28	.02013	.02943	.01855	.02286	.03571	28
26	.02168	.03179	.01994	.02462	.03846	26
24	.02348	.03453	.02156	.02668	.04166	24
22	.02562	.03756	.02348	.02900	.04545	22
20	.02818	.04163	.02578	.03201	.05000	20
18	.03131	.04637	.02859	.03557	.05555	18
16	.03523	.05229	.03210	.04002	.06250	16
14	.04026	.05991	.03661	.04573	.07142	14
12	.04697	.07006	.04263	.05336	.08333	12
11	.05124	.07652	.04647	.05821	.09090	11
10	.05636	.08426	.05106	.06403	.10000	10
9	.06263	.09375	.05668	.07114	.11111	9
8	.07046	.10559	.06370	.08004	.12500	8
7	.08052	.12082	.07273	.09147	.14285	7
6	.09394	.14113	.08477	.10671	.16666	6
$5\frac{1}{2}$	.10248	.15405	.09244	.11642	.18181	$5\frac{1}{2}$
5	.11273	.16955	.10163	.12806	.20000	5
$4\frac{1}{2}$	.12526	.18850	.11287	.14229	.22222	$4\frac{1}{2}$
4	.14092	.21219	.12691	.16008	.25000	4
$3\frac{1}{2}$	.16105	.24265	.14497	.18295	.28571	$3\frac{1}{2}$
$3\frac{1}{4}$	.17344	.26139	.15609	.19702	.30769	$3\frac{1}{4}$
3	.18789	.28326	.16905	.21344	.33333	3
$2\frac{7}{8}$	.19606	.29562	.17638	.22272	.34782	$2\frac{7}{8}$
$2\frac{3}{4}$	.20497	.30910	.18438	.23284	.36363	$2\frac{3}{4}$
$2\frac{5}{8}$	.21473	.32387	.19313	.24393	.38095	$2\frac{5}{8}$

65. Determination of thread angle of a Whitworth outside thread from measurements over the wires. —

Example. — To find thread angle of $1\frac{1}{4}'' \times 5$ Whitworth tap.

Formula. — Cosecant of $\frac{1}{2}$ angle $= \dfrac{M - m}{W - w} - 1$. See Fig. 32.

By measurement, $M = 1.46658$;   by table $W = 0.16955$

$\qquad\qquad\qquad\quad m = 1.25157$; $\qquad\qquad w = 0.10163$

Difference, $M - m = 0.21501$; $W - w = 0.06792.$

Solution. — $\dfrac{M - m}{W - w} - 1 = \dfrac{0.21501}{0.06792} - 1 = 3.1656 - 1 = 2.1656.$

From table of trigonometrical functions the angle, the cosecant of which is 2.1656, is 27° 30'. Therefore the thread angle is 55°.

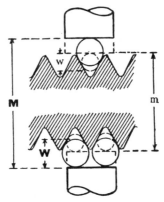

FIG. 32. — THREE-WIRE METHOD OF MEASURING THREAD ANGLE OF WHITWORTH THREAD.

Information. — Common variations of thread angle and corresponding quotients are shown in Table, Art. 66.

Attention. — Both measurements should be taken with the wires placed in the same thread grooves, as shown in Fig. 32.

66. Table of common variations of thread angle of a Whitworth thread, with corresponding quotients. —

DEGREES.	MINUTES.	QUOTIENTS.
54	30	3.1840
54	40	3.1778
54	50	3.1717
55		3.1657
55	10	3.1596
55	20	3.1536
55	30	3.1477

67. One- and two-wire systems are sometimes used to measure the pitch diameter of threads, but not so commonly used as the three-wire system.

68. Drunken threads are not detected by any of these wire systems. They may be found by means of a thread micrometer, p. **12**14, or by a lead testing machine, p. **14**28.

HEAVY DUTY BORING AND TURNING

69. Heavy duty boring, facing and turning. — Heavy castings and forgings may be bored, faced, and turned in engine lathes.   See Boring, Boring Bars, and Boring Machines, p. **6**10.

Boring and facing when no turning is required, may be more rapidly performed in a horizontal boring machine.   When the length of the work is small as compared with the diameter, and more especially if the work is unusually heavy, it may be machined in a vertical boring and turning mill.

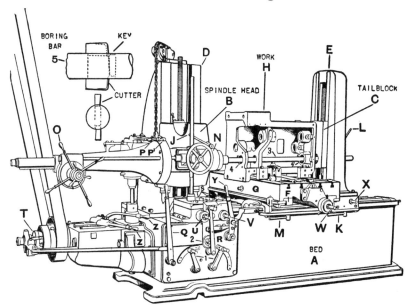

Fig. 33. — Boring a Gear Box with Horizontal Boring Machine.

70. Horizontal boring machines are arranged to bore, ream, counterbore, and mill any work that can be held conveniently on a horizontal table.   They are made in two types, — one with a sliding head and tailblock, as in Fig. 33, the other, with a fixed spindle head and tailblock, but having a table that may be raised and lowered. See Horizontal Boring Machines, p. 616.

71. Horizontal boring machine with sliding spindle head, described. Fig. 33.

SCHEDULE OF PARTS

A — Bed.	S — Interlocking feed selecting levers.
B — Spindle head.	
C — Tailblock.	T — Friction clutch.
D — Column.	U — Spindle hand **adjustment,** with micrometer dial.
E — Back rest.	
F — Saddle.	V — Saddle hand adjustment, with micrometer dial.
G — Table or platen. Circular table may be added.	W — Table hand adjustment, with micrometer dial.
H — Work.	
J — Spindle-head clamp.	X — Back rest longitudinal adjustment.
K — Back-rest clamp.	
L — Tailblock clamp.	Y — Feed and quick return lever.
M — Saddle clamp.	
N — Slow spindle movement hand wheel.	Z — Speed change levers.
	1 — Feed change levers.
O — Quick spindle movement hand wheel.	2 — Adjustable feed safety friction.
P, P' — Interlocking back gear levers.	3 — Boring bar.
	4, 4' — Boring bar cutters.
Q — Starting and stopping lever.	5 — Boring bar cutter and key.
R — Reversing lever for feed and quick return.	

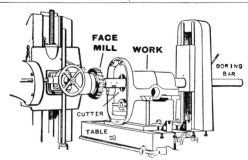

FIG. 34. — BORING AND MILLING IN HORIZONTAL BORING MACHINE.

72. Boring and face milling at one setting of work may be performed in a horizontal boring machine, as in Fig. 34.   The construction of the spindle permits the use of a milling cutter with the boring bar.

73. Heavy duty vertical boring and turning mills. — These machines are especially adapted for heavy work such as large fly wheels, pulleys, engine cylinders, etc. The work may be held on the table by chuck jaws, or by bolts, clamps, and blocking. The weight of the work assists in holding it in place.

The two tool posts permit two cutting operations at once thus effecting a saving of time. These machines are obtain-

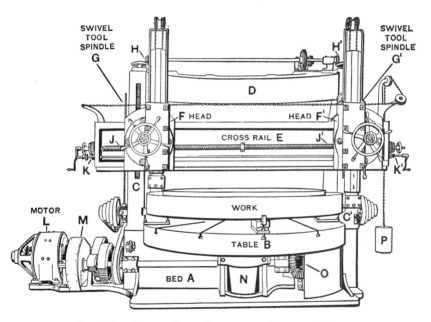

Fig. 35. — Heavy Duty Boring and Turning Mill.

able in sizes of from 4 feet to 8 feet swing, and from 3 feet to 4 feet under the rail. On the heavier sizes the tool spindles and cross rail are usually moved by individual motors to assist in the setting of the tools. The motor-driven mill, Fig. 35, is turning and facing a large casting.

74. Heavy duty vertical boring and turning mill described, Fig. 35.

SCHEDULE OF PARTS

A — Bed.	J, J' — Head feed screws, —
B — Revolving table.	one threaded right-handed and
C, C' — Uprights or housings.	the other threaded left-handed for
D — Top rail.	moving heads independently in
E — Cross rail. May be raised	the same or opposite directions.
or lowered and clamped in position.	K, K' — Tool-post feed shafts.
F, F' — Heads with hand and	L — Motor. When belt-driven,
power feed.	a six-step cone pulley.
G, G' — Swivel tool spindles	M — Driving gears.
with hand and power feed,	N — Table bearing.
counterbalanced.	O — Table-drive gears.
H, H' — Side-screw bevel gear	P — Tool spindle balance
for raising and lowering cross rail.	weight.

VERTICAL TURRET LATHES

75. Vertical turret lathes are often used for boring, reaming, turning, facing, and also for tapping and chasing threads on such work as cone pulleys, gear blanks, car and truck wheels and similar face-plate work. The turret heads carry a large number of tools which permit several operations at one setting of the work.

76. Operation of a vertical turret lathe. — Fig. 36 shows a vertical turret lathe turning and facing gear blank A. Work is held by chuck jaws on table B which revolves on bed C. Main turret D has five tool holders, E, F, G, H, J, and is revolved by handle K. An indexing pin which is operated by handle L brings the tools into accurate alinement. Auxiliary turret M has four tool positions and is used for facing and turning. Both turrets can be adjusted vertically and horizontally, and have hand and power feeds. Two cutting tools can be arranged in a single tool holder in both turrets so that four cutting operations may be performed at once. Fig. 36 shows three tools cutting at once, — two by the main turret cutting tools and one by an auxiliary head. Another feature is that during a long boring or turning operation, two or three other shorter operations may be performed by the other turret. A single tool may also be used for several operations.

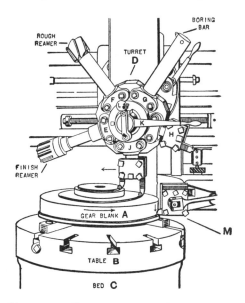

FIG. 36. — MACHINING WORK HELD IN A VERTICAL TURRET LATHE.

77. To machine a steel motor-truck wheel. — To facilitate the setting of the machine and operations, a graphic schedule, as shown in Figs. 37, 38, and described in Schedule of Operations, p. **14**46, may be used.

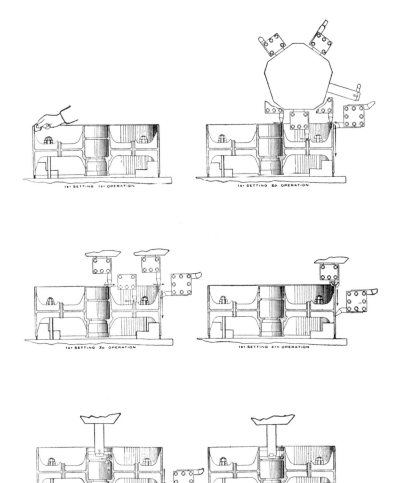

FIG. 37. — GRAPHIC SCHEDULE OF OPERATIONS FOR FIRST SETTING
OF A STEEL TRUCK WHEEL.

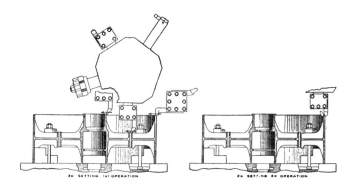

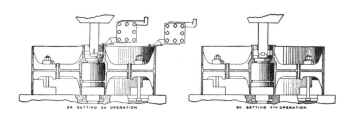

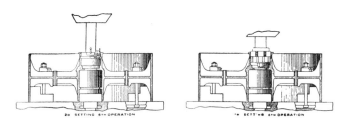

FIG. 38. — GRAPHIC SCHEDULE OF OPERATIONS FOR SECOND SETTING
OF A STEEL TRUCK WHEEL.

SCHEDULE OF OPERATIONS

OPERA-TION.	FIRST SETTING. FIG. 37.	SECOND SETTING. FIG. 38.
1st	Chuck and true up casting.	Rough face rim and hub and with one cut, finish face bolt bosses.
2d	Rough tu.n rim, and rough face bolt, bosse , and rim.	Chamfer rim.
3d	Finish face hub, rim, and bolt bosses. Start rim finish-turning cut.	Finish face rim and hub, and start boring hub.
4th	Chamfer rim while continuing the rim finish-turning cut.	Finish bore hub.
5th	Rough bore and square inside of hub bearing case while terminating finishing cut on rim.	Counterbore and recess for thread.
6th	Finish bore hub bearing case.	Tap.

Information. — By making a schedule of operations and settings, idle tools are eliminated and operations will not overlap.

Attention. — Attachments are obtainable for use of cutting and cooling compounds. The tank, guards, pumps and piping are independent of the machine and can be detached at will.

ADVANCED MACHINE WORK

SECTION 15

MECHANICAL MEASUREMENTS WITH LIGHT WAVES

Introduction. Light Waves and the Standard Inch. Equipment for Measuring with Light Waves. Measuring Flat Surfaces. Measuring Lengths. Measuring Plug Gage. Problems in Light Wave Measurements. Theory of Light Waves. Measuring the Length of Light Waves.

INTRODUCTION

1. Evolution of mechanical measurements. — In the construction of the earlier machines and machine parts measurements were made with a *wooden* rule and calipers. The next forward step was the introduction of the *steel* rule. Machine parts made by these measurements were not interchangeable.

This was followed by the vernier and micrometer which measured to a thousandth of an inch and permitted such dimensions to be specified on drawings. The use of these instruments eliminated the existing guess work and to a large extent made possible the manufacture of interchangeable machine parts. Great good has resulted from the use of micrometers and verniers which at first were considered unnecessary refinements.

2. The development of the Limit System of manufacturing, and of the interchangeability of machine parts, showed the necessity for more precise and reliable standards with which to promote quantity production. These standards are supplied in large degree by limit gages and also by combination gage blocks which furnish a degree of mechanical accuracy that exceeds the limit of mechanical measuring machines.

To measure new gages and instruments and to determine the wear on old gages, a still finer unit was necessary. This is now supplied by light waves.

Light waves are also used in industrial work for measuring gage blocks, plug gages, thread gages, precision flat surfaces, measuring wires, steel balls, hair springs, contraction of dental amalgams, coefficient of expansion of metals, etc.

Wooden rule to light waves. — The wonderful progress in the development of machinery from the age of the wooden rule to light waves, proves that the accuracy of man's work is limited only by the measuring instruments available.

3. The micrometer unit is .001″ (one thousandth of an inch) but .0001″ (one ten-thousandth of an inch) may be estimated.

4. The light-wave unit (one dark band or one-half wavelength) obtained with red monochromatic light, and generally used for mechanical measurements, is .0000125″ (twelve and one-half millionths of an inch) but .000001″ (one millionth of an inch) may be estimated. See Fig. 1 and Arts. 11, 12 and 14.

LIGHT WAVES AND THE STANDARD INCH

5. Light waves give a reliable and permanent unit of measurement. If all the standard inches, yards and meter bars (39.37 inches) distributed throughout the world, from which the inch has been derived, were destroyed, the standard could be duplicated by means of light waves.

For the different radiations of cadmium light, the International Meter contains 1,553,163.5 red waves; 1,966,249.7 green waves; and 2,083,372.1 blue waves. While the standard inch, yard, or meter bar is subject to minute variations, light waves furnish an unchanging standard of measurement.

The United States Bureau of Standards measures its primary standard inches with light waves.

EQUIPMENT FOR MEASURING WITH LIGHT WAVES

6. Set-up for measuring with light waves, Fig. 1.

 I. For measuring flat surfaces, set-up consists of an optical glass flat A and monochromatic light E.

II. For length measurements, set-up consists of an optical glass flat A, metal flat B, standard gage block CC', or a number of combination gage blocks, and monochromatic light E.

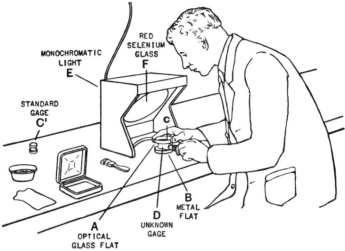

RED SELENIUM GLASS **F**

MONOCHROMATIC LIGHT **E**

STANDARD GAGE **C'**

C

B METAL FLAT

D UNKNOWN GAGE

A OPTICAL GLASS FLAT

Fig. 1. — Equipment for Measuring with Light Waves.

7. The optical glass flat A, Fig. 1, is a flat piece of glass with both sides polished so as to be transparent. One surface only need be lapped optically flat and should be accurate within three millionths of an inch.

8. The metal flat B, Fig. 1, must also have one surface accurate within three millionths of an inch. Glass is sometimes used for the lower flat in measuring; but tool steel, hardened, ground and lapped, is generally used as it has better wearing qualities.

9. Optical and metal flats are originated by making three of a kind and correcting by the process of grinding and lapping and testing with each other, in binary combinations, the same as surface plates and straight edges are originated.

Information. — Glass parallels are obtainable.

10. The standard gage block, as shown at C, C', Fig. 1, must be of a known length, and have the end surfaces parallel and flat. The blocks may be round, square, or rectangular in cross-section. A single block or a combination of blocks may be used, as required.

Small gage blocks are made within .000005″ (five millionths of an inch), and the large sizes within .000010″ (ten millionths of an inch). See pp. 1322–1329. The unknown plug gage *D*, Fig. 1, is to be measured by light waves.

11. Monochromatic light *E*, Fig. 1, or light of one color, is composed of waves of one length only, or with such waves predominating. For mechanical measurements daylight or electric light, is transmitted through red selenium glass *F*.

The light box is made of wood in which is located an electric-light bulb with an opening for the red selenium glass *F*.

Information. — Good colored bands may also be obtained by transmitting daylight or electric light through a sheet of tissue paper or ground glass.

12. Interference bands — half-wave lengths —, appear on a polished metal surface when an optical glass flat is placed upon it. The bands are produced by the light reflected from one surface interfering with the light reflected from the other.

If daylight is used, a series of colored interference bands or fringes is seen; the value of each purple band (one-half wave length) being approximately .000010″ (ten millionths of an inch).

If red monochromatic light is used, a series of dark bands and light spaces is seen; the value of each dark band (one-half wave length) being .0000125″ (twelve and one-half millionths of an inch). See Art. 11.

By estimating the straightness of the bands, errors in flat surfaces may be determined; and by counting the number of bands and fractions of bands, lengths and diameters can be obtained. See Theory of Light Waves, Art. 37, and Measuring Length of Light Waves, Art. 46.

Note. — The bands are not light waves but simply show where interference is produced by reflections from two surfaces.

MEASURING OR TESTING FLAT SURFACES

13. Reading hundred-thousandths and millionths. — After cleaning with cloth and camel's-hair brush, place optical flat *A*, Fig. 2, on gage block *B* and note the appearance of interference (dark) bands *C* under red monochromatic light. Press downward slightly at line *D* of flat to give contact along edge *EF*. The film of air *G* becomes wedge-shaped and interference bands will appear at *HJ*, *KL*, *MN*, etc. Band *HJ*, according to the theory of light waves shows that the vertical distance *HH'* and *JJ'* is one band or unit (one-half wave length) or .000012″ (twelve millionths of an

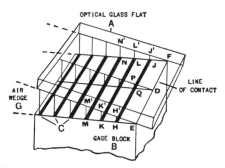

inch). Similarly, *KK'* and *LL'* are two bands or .000025″ (twenty-five millionths of an inch) and *MM'* and *NN'* are three bands or .000038″ (thirty-eight millionths of an inch). The number of bands seen on the gage indicates the amount of slope of optical flat and thus the distance between back edge of

Fig. 2. — Measuring Flat Surface with Light Waves. Optical Flat Making Angular Contact with Gage Showing Air–Wedge.

gage and glass flat is six bands or .000075″ (seventy-five millionths of an inch). See the following table also Arts. 11 and 12.

14. Table of light-wave equivalents when red monochromatic light is used.

No. of Bands	Value in Millionths of an Inch	No. of Bands	Value in Millionths of an Inch
$\frac{1}{10}$	.000001	6	.000075
$\frac{1}{4}$	.000003	7	.000088
$\frac{1}{2}$	.000006	8	.000100 or one ten-thousandth
$\frac{3}{4}$	.000009	10	.000125
1	.000012	12	.000150
2	.000025	16	.000200 or two ten-thousandths
3	.000038	24	.000300 or three ten-thousandths
4	.000050 or five hundred-thousandths	32	.000400 or four ten-thousandths
5	.000062	40	.000500 or five ten-thousandths

Attention. — In the above table equivalents of fractions, or multiples of bands, are given to the nearest even millionth.

The value of each dark band is .0000125″ (twelve and one-half millionths). Thus there are eight bands for every .0001″ (one ten-thousandth of an inch).

15. Straight bands indicate a flat surface since by geometry, triangles $H'HE$, $J'JF$, PQD, Fig. 2, are similar.

16. Line of contact. — Using daylight the line of contact is located by the absence of color; the grayish surface of the steel gage is seen. With red monochromatic light the bands are more distinct near the line of contact.

Fig. 2 shows that line of contact between the optical flat and the surface being tested, is at EF. This must be located before surface readings are taken.

17. The air-wedge G, Fig. 2, (enormously exaggerated to illustrate it), which is formed by placing the optical glass flat A upon gage block B, may be called the micrometer of this system.

18. The number of bands depends upon the angle of the air-wedge. If the angle of the air-wedge is increased, the number of bands will increase and they will be closer together; and, inversely, if the angle is decreased, there will be fewer bands but farther apart.

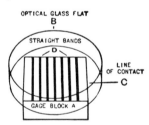

FIG. 3. — TESTING SURFACE OF GAGE BLOCK. STRAIGHT BANDS SHOW THAT SURFACE IS FLAT WITHIN ONE MILLIONTH OF AN INCH.

Fig. 3 shows the surface of a square gage block A being tested with an optical glass flat B. The line of contact is at C. The bands D on this surface are straight, parallel, and equidistant showing that the surface is perfectly flat.

Attention. — It is best to test flat surfaces twice making the second test with the bands at right angles to their first position.

19. Curved bands indicate curved surfaces because triangles ABC, DEF, GHJ, Fig. 4, are not equal. The amount of deviation in a flat surface is indicated by the amount the bands curve.

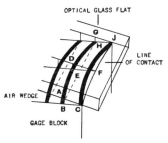

FIG. 4. — CURVED BANDS INDICATE CONVEX OR CONCAVE SURFACES.

20. Convex surface. — In Fig. 5 the optical glass flat A rests on gage B and the line of contact is at C. The curved bands D show that the surface of the gage block B is not flat but convex. When the bands curve around the line of contact, the surface tested is convex, as in Fig. 5. The dotted lines show that the bands curve an amount equal to the distance from one band to the next, indicating that the surface is convex and the side edges are low one band or .000012″ (twelve millionths of an inch).

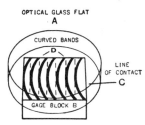

FIG. 5. — TESTING SURFACE OF GAGE BLOCK. SURFACE CONVEX ONE BAND OR TWELVE MILLIONTHS OF AN INCH.

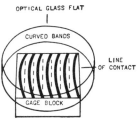

FIG. 6. — TESTING SURFACE OF GAGE BLOCK. SURFACE CONVEX ONE – HALF BAND OR SIX MILLIONTHS OF AN INCH.

In Fig. 6 the bands curve around the line of contact an amount equal to one-half the distance from one band to the next (see dotted lines), and show that the surface of the gage block is convex one-half a band or .000006" (six millionths of an inch).

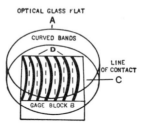

FIG. 7. — TESTING SURFACE OF GAGE BLOCK. SURFACE CONCAVE ONE-HALF BAND OR SIX MILLIONTHS OF AN INCH.

21. Concave surface. — Fig. 7 shows an optical glass flat A resting on the square gage block B with line of contact at C. The bands curve away from the line of contact C an amount equal to one-half the distance from one band to the next (see dotted lines), and show that the surface of the gage block is concave one-half a band or .000006" (six millionths of an inch).

Information. — Most surfaces that have to be measured and corrected are found to be convex rather than concave.

22. High spots are indicated by circular points of contact with concentric bands around them.

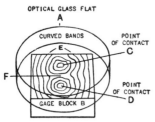

FIG. 8. — TESTING SURFACE OF GAGE BLOCK. SURFACE IRREGULAR.

In Fig. 8 the optical glass flat A rests on gage block B and makes contact at the high spots C and D. The bands curve around these high spots C and D showing an irregular surface.

The surface slopes downward from the high spots one unit for each band. There are three and one-half bands indicating that the depression between the high spots is three and one-half bands or .000044" (forty-four millionths of an inch) deep.

MEASURING LENGTHS

23. Measuring lengths with light waves is the determining of the unknown from the known. It requires an optical glass flat A, Fig. 9, a master metal flat B, a gage of known standard length C, and red monochromatic light. See Fig. 1 and Arts. 11, 12 and 14.

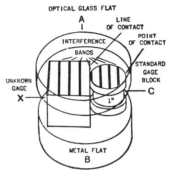

24. To wring the gages to the metal flat. — With a clean soft cloth or tissue paper, wipe the surfaces of the standard gage C, Fig. 9, and the unknown gage block X which is to be measured, and the metal flat B.

To supply the film for the surfaces, the moisture of the breath may be used, or any liquid such as alcohol or benzine.

Fig. 9. — Showing Arrangement of Metal Flat, Standard and Unknown Gages and Optical Flat for Measuring Lengths.

Standard gage C is started at edge of metal flat B and brought to its position with a sliding movement which distributes the film and excludes dust particles. It is located at the right of metal flat. The unknown gage block X, Fig. 9, is wrung on to the metal flat the same way and located at the left. The optical glass flat A and the upper surfaces of the gages are wiped with a clean cloth and then dusted with a camel's-hair brush. The optical flat is then placed on top of the gages for measuring and with a little pressure on the optical flat the measuring bands will appear.

Warning. — The gage blocks and flats should not be left wrung together after using but should be separated and each surface wiped with a clean soft cloth and then oiled.

Information. — If gage blocks and metal flat are cleaned with soap and water and thoroughly dried, no film will remain and they cannot be wrung together.

Important. — To measure accurately with light waves, the equipment must be clean, carefully assembled and the temperatures uniform.

25. The thickness of the film which holds a gage to the flat is from .000001″ (one millionth of an inch) to .000003″ (three millionths of an inch) depending on the nature of the film and the accuracy with which the gage is wrung to the metal flat. It is negligible when measuring ordinary gages, but must be considered when making measurements of great precision and particularly when using a combination gage consisting of several blocks wrung together.

26. Temperature of gages. — The gages are standard at 68° F. It is essential that work to be measured be the same temperature as the standard gage when measured. In summer, at 80°, the work will expand as much as the standard gage. Upon cooling to 68° the standard temperature, the work will contract.

Expansion of steel. — A 1″ gage expands for a rise of temperature of 1° F., about .000007″; a 4″ gage expands for a rise of 10° F., more than one-quarter thousandth of an inch.

Information. — In all cases of measurement, the instruments should rest untouched for a few minutes after adjustment before taking final measurements, so that the parts may all assume the same temperature.

27. Set-up for measuring length. — The standard gage

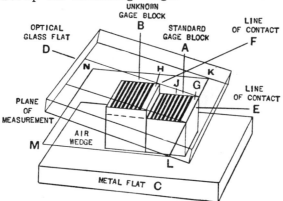

Fig. 10. — Principle of Measuring Length of Gage Block with Light Waves.

block *A*, Fig. 10, and the unknown gage *B* to be measured, are wrung on to the true side of metal flat *C* and placed side by side for precision measurements, with the standard gage at the right. The true side of the optical glass flat D is then placed on the top of the two gages and pressed down upon both gages.

28. Two lines of contact. — There are two lines of contact, E on the standard gage and F on the unknown gage, which are indicated by the light spaces G and H.

The optical glass flat D slopes upward from the line of contact E on the standard gage A, eight (8) bands (units), to the line of contact F on the unknown gage B. If the same number of bands appears on the unknown gage B as on the standard gage A, the surfaces of the two gages are parallel, and the measurement may be considered complete.

29. Number of bands on the standard gage show height of unknown gage. — The unknown gage is eight bands or .000100″ (one hundred millionths of an inch) large. The unknown gage block is then lapped and measured until it is within three or four bands of standard. As the unknown gage comes nearer standard the number of bands on both gages becomes less.

30. To take final measurement, arrange the gages side by side, as in Fig. 11, then place the optical glass flat on the gages with points of contact as at A and B.

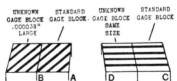

Fig. 11. — Taking Final Measurement of an Unknown Gage Block.

Begin at the lowest point of contact A of standard gage; count the bands to the point of contact B of the unknown gage which are three, indicating that the unknown gage is three bands or .000038″ (thirty-eight millionths of an inch) large.

Continue to lap the unknown gage until it is the desired size. It is good practice to leave the unknown gage one band large for wear allowance. If the bands on the unknown gage exactly match those on the standard gage, as shown at C and D, Fig. 11, the two are the same size.

MEASURING PLUG GAGE

31. To measure diameter of 1″ plug gage with light waves, using red monochromatic light. See Figs. 1, 12–15, and Arts. 11, 12 and 14.

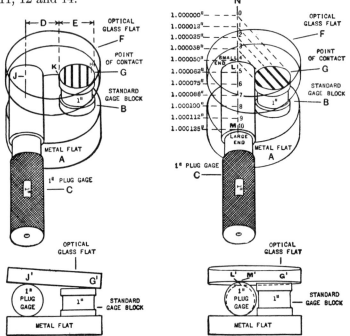

Fig. 12. — Measuring 1″ Plug Gage. Reading Ten Bands Large, Or 1.000125″.

Fig. 13. — Measuring 1″ Plug Gage. Reading Small End, 1.000050″; Large End, 1.000125″; Taper, .000075″.

Specifications: Hardened-steel plug gage ground and lapped to limit $^{1.000200″}_{1.000100″}$; measurement to be obtained with measuring machine or ten-thousandth micrometer. See pp. 1236 and 213.

Tools: One 2″ optical glass flat, one 2″ metal flat, one 1″ standard gage block, light box, camel's-hair brush, clean cloth, and paper.

SCHEDULE OF OPERATIONS

I. Set up apparatus. Wipe true surface of metal flat A, Fig. 12, and both surfaces of 1″ standard gage B with clean cloth and paper.

Wring 1″ standard gage block B on metal flat and locate it at right edge.

Dip 1″ plug gage C in soap and water or gasolene to clean, and wipe dry with cloth and paper.

Place plug gage on metal flat A so that distance D from center line J of plug gage equals diameter E of standard gage.

Clean optical flat F with camel's-hair brush and place it upon gages and press down to secure contact on both the standard gage block and the plug gage.

II. Reading for oversize gage. Look for point of contact. If point of contact is at G, as indicated by light space H, plug gage is large and the optical glass flat slopes from standard gage at point G to top of plug gage at J, as in detail at G′,J′. Count number of bands appearing on top of standard gage, as 1,2,3,4,5. As center line of plug gage is two diameters of standard gage away from contact point G, the top of plug gage is twice as far above point G as point K. Since there are five bands from G to K, there would be 10 from G to J, therefore the plug gage is ten bands (units) or .000125″ large. See To Lap Standard Plug Gage, p. 1229.

If bands are parallel to center line, gage is straight; if at angle, gage is taper. See Fig. 13.

III. Reading when gage is taper. If plug gage C, Fig. 13, is taper, the bands if prolonged will intersect the vertical plane of the axis of the plug gage. Where the first band prolonged intersects, it is one unit (band) large. In a similar manner where the second band intersects, it is two units (bands) large, etc. Thus end L of gage, where the fourth band prolonged intersects M N, is four units (bands) large or 1.000050″. By estimation it can be easily determined that a tenth band would intersect at M.

Thus end M of gage is ten bands (units) large or 1.000125″ or .000075″ larger than end L.

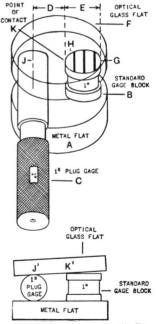

Fig. 14. — Measuring 1″ Plug Gage. Four Bands Small.

IV. Reading when gage is small.
If contact is at point K, Fig. 14, as indicated by light space H, plug gage is small and the optical flat F slopes downward from top of standard gage B at K through point J, as in detail at K',J'.

Since there are four bands from K to G, there would be four from K to J, therefore the plug gage is four bands (units) or .000050″ small. If bands are parallel to center line of gage, the gage is straight.

V. To take final precision measurement of plug gage. Place plug gage C, Fig. 15, on metal flat A with end near standard gage B and back of front edge of gage G.

Place optical flat F on gage and

press down to exclude air and show point of contact G.

Bands will appear as at 1 and 2. Roll plug gage forward in direction of arrow P until it touches the optical flat, as at Q.

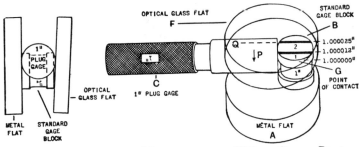

FIG. 15. — TAKING FINAL MEASUREMENT OF 1″ PLUG GAGE. READING TWO BANDS LARGE, OR 1.000025″.

After plug gage C touches optical flat F, further movement will disturb position of bands 1 and 2.

Count number of bands from point of contact G to where plug gage touches optical flat. Thus in Fig. 15, gage has stopped at 2 and therefore gage is two bands

(units) or .000025″ (twenty-five millionths) large. Every point on the flat F along band 1 (one) is one unit over 1″ and similarly with band 2, etc.

Important. — This is using the two flats as a taper gage with the length bands as graduations.

Attention. — In the final measurements, it is best to take several readings.

" Distance " or " close " set-up. — If the unknown gage which is to be measured is from .0002″ to .0005″ large, it is best to use " distance " set-up and process with unknown gage the width of the standard gage away, as in Fig. 12, to avoid a large number of bands appearing on the surface of the standard gage block which would be difficult to count; but if the unknown gage is about .0001″ large, use " close " set-up and process, as in Fig. 15.

PROBLEMS IN LIGHT WAVE MEASUREMENTS

32. To measure diameter of steel ball with light waves. — Select standard gage A, Fig. 16, (or several combination gage blocks wrung together as a unit) equal to the desired diameter of steel ball B. Wring standard gage A to metal flat C.

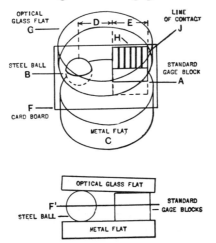

FIG. 16. — MEASURING DIAMETER OF STEEL BALL.

Place ball B on metal flat C so that distance D of center of ball B from standard gage A equals width E of standard gage A. To hold ball in proper position, make cardboard retainer, FF', to fit both ball and gage tightly. Place optical flat G on standard gage block. Press down to exclude air and to produce interference bands on surface H. Note line of contact J. Determine error in ball by method given in Art. 31,

33. To measure outside diameter of ball bearing with light waves. — Select combination of gage blocks A, Fig. 17, 1,2,3,4,5, to equal the given diameter of ball bearing B.   Clean surfaces of

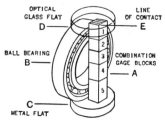

FIG. 17. — MEASURING OUTSIDE DIAMETER OF BALL BEARING.

gage blocks, metal flat C and optical flat D.   Wring gage blocks together and then wring unit on to metal flat.   Hold ball bearing in position and place optical flat on top of gage blocks and bearing, and locate line of contact at E.   Determine error by method given in Art. 31.

34. To measure thread wires, hair springs, and other wires of precision. — Select a combination of gages, as at A, Fig. 18, and another combination of gages B, which when combined with the wire C will equal height of gages A.

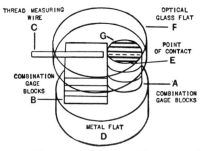

FIG. 18. — MEASURING THREAD–MEASURING WIRE.

Assemble gages and measuring wire C on metal flat D.   Point of contact is at E.   See Three-Wire System, pp. 1431–1438.

Place optical flat F in position and take reading from bands appearing on top of gage G.

Exception. — If height of gage blocks *B* plus diameter of measuring wire *C* is nearly the same height as combination gages *A*, use " close " set-up and process, as in Fig. 18; but if there is considerable difference in these heights, use " distance " set-up and process shown in Fig. 19, making distance *D* the same as distance *E*. See Method of Measuring Gages, Art. 31.

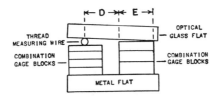

FIG. 19. — MEASURING THREAD–MEASURING WIRE.

35. To measure . pitch diameter of thread gage with light waves. — Select three thread measuring wires *A,A,A*, Fig. 20, of proper size to fit thread of gage *B*, see p. 1431. Select combination of gage blocks *C*, 1,2,3, to give desired measurement over thread-measuring wires.

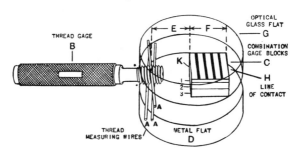

FIG. 20. — MEASURING PITCH DIAMETER OF THREAD GAGE.

Wring gage blocks *C*, 1,2,3, together and then wring them on to metal flat *D*.

Place two thread-measuring wires upon the metal flat *D*; place thread gage *B* upon them, then place one wire in thread on top. Place optical flat *G* upon gages *C* so that line of contact will be at *H*. Determine error by method given in Art. **31**.

36. To determine flatness of face of micrometer anvil by light waves. — Place optical flat A, Fig. 21, between anvil B and screw C of micrometer D. Rotate screw C with thimble E to press optical flat against anvil B.

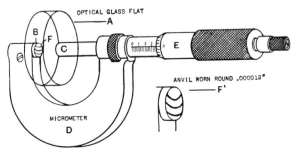

Fig. 21.—Measuring Flatness of Anvil Face of Micrometer Caliper.

The rounded condition of surface F, F', on anvil B, is shown by the curved bands enlarged at F'.

To determine the flatness of the micrometer screw, place the flat face of the optical glass flat against face C.

Information. — If an optical glass parallel is available, the parallelism of the measuring surfaces of anvil B and screw C may also be determined.

THEORY OF LIGHT WAVES
UTILIZATION IN MINUTE MEASUREMENTS OF LIGHT
INTERFERENCE PHENOMENA

37. Light is transmitted through space by means of a wave motion. It is convenient to consider the waves as related to a single plane, and that light energy is transmitted just as mechanical energy is sent forward across a water surface by a wave motion. The velocity of light waves is 180,000 miles per second.

Light can be manipulated to produce reflection, refraction, interference, etc. Minute measurements deal only with interference.

38. A wave train consists of a series of waves. The length of the waves — from crest to crest — is related to the color of the light.

39. Spectrum. — Every color in the spectrum corresponds to a different wave length, from the extreme red which is about .000030″ (thirty millionths of an inch) to the extreme violet, about .000015″ (fifteen millionths of an inch). Different shades of color have different wave lengths.

40. Monochromatic light has a single wave length. It may be of any color. The red light which passes through selenium glass, for example, has a wave length of .000025″ (twenty-five millionths of an inch).

Daylight is composed of a mixture of all the different colors in the spectrum and if it is allowed to fall on selenium glass, all the colors are absorbed except the red.

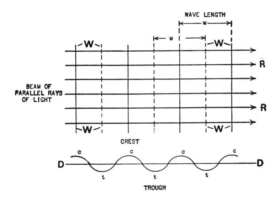

Fig. 22. — A Train of Light Waves.

41. To understand the interference of light a beam of parallel rays *RR*, Fig. 22, must be thought of as a series or *train* of waves as represented by the full and dotted cross-lines *WW*. The full cross-lines correspond to the crests, and the dotted cross-lines to the troughs, as represented by the wave diagram *DD*. The wave length is the distance *w,w* from crest to crest or from trough to trough.

Different positions of reflecting surfaces affect interference of light which is the phenomenon used in minute measurements.

42. When light waves are reflected from the front face of glass in air, as shown in Fig. 23, crests are reflected as crests

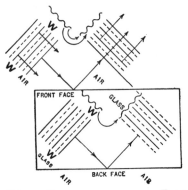

and troughs are reflected as troughs; and

43. When light waves are reflected from the back face of a glass in the glass itself, as shown in Fig. 23, crests are reflected as troughs and troughs are reflected as crests. This effect is called *phase reversal.*

The parallel equidistant full lines WW in Fig. 24 are the crests in a train of light waves falling normally on glass flat AB and steel plate CD.

Part of the wave train WW is reflected or turned back by the front face FF of the glass. This reflection produces general illumination and does not concern the phenomena of light interference.

Part of the wave train WW is reflected or turned back by the back face AB of the glass and this reflection converts *every crest* into a *trough* and *every trough* into a *crest* as stated in connection with Fig. 23. Inasmuch as we wish to follow the *identical waves* which are represented by the full lines WW in Fig. 24, we must consider where these identical lines are at a later instant, and the dotted lines P in Fig. 25 show the positions of these identical lines as troughs.

Part of the wave train WW, Fig. 24, is reflected or turned back by the steel surface CD in Fig. 24, and the full lines Q in Fig. 25 show the positions of the identical lines WW as crests at a later instant. The wave crests Q have had to travel twice across the air space d and therefore these wave crests are a distance $2d$ behind the wave troughs P. Now the wave system including trains P and Q, fills the whole region above the glass, and of course each wave train consists of crests and troughs, and two particular positions are of importance; namely,

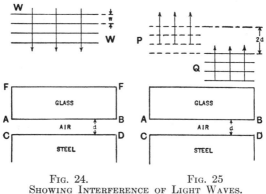

FIG. 24.   FIG. 25
SHOWING INTERFERENCE OF LIGHT WAVES.

I. When the crests of the wave system Q coincide with the troughs of the wave system P, and

II. When the crests of the wave system Q coincide with the crests of the wave system P.

The first condition occurs when the distance $2d$ is a multiple of w, that is, when $2d$ divided by w is an integer; and in this case the wave systems P and Q destroy each other; or in other words, no light of the particular wave length is reflected from AB and CD together.

The second position occurs when $2d$ divided by w is an odd number of halves, $\frac{1}{2}$, or $\frac{3}{2}$, or $\frac{5}{2}$, or $\frac{7}{2}$, etc.; and in this case a great deal of light is reflected from AB and CD together.

44. Apparatus for measuring air-wedge. — The simplest arrangement theoretically for measuring the thickness of a very thin layer of air a between glass and steel, is shown in Fig. 26.

Plate of selenium glass SS, is used for obtaining red monochro-matic light; and plate of un-silvered glass GG, to reflect light from the lamp downward to the glass and steel flat. This allows the light reflected by AB and CD to pass through to the observer's eye.

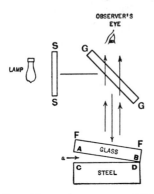

FIG. 26. — APPARATUS FOR MEASURING AIR - WEDGE WHEN LIGHT IS INCIDENT PERPENDICULAR.

When the thickness of the air layer d, Fig. 25, is such that $2d$ is a multiple of the wave length w of the light used, the observer sees no light upon the surface and a black band appears, as in Fig. 27, and when the thickness d is such that $2d$ divided by w is an odd number of halves, the observer sees bright light upon the surface, and light bands appear.

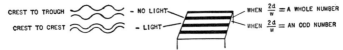

FIG. 27. — CONDITIONS NECESSARY FOR DARK, AND LIGHT BANDS.

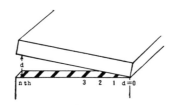

2d = NUMBER TIMES WAVE LENGTH

FIG. 28. — SHOWING METHOD OF DETERMINING THICKNESS OF AIR-WEDGE.

Therefore in case of a wedge-shaped air layer, as shown in Figs. 26 and 28, the observer sees a series of parallel bright and dark bands. Counting the first dark band from the line of contact, see Fig. 28, as No. 1, the next as No. 2, the next as

No. 3, etc., the thickness d at the nth dark band is such that $2d = nw$, so that if w is known and if n is counted, the value of d becomes known.

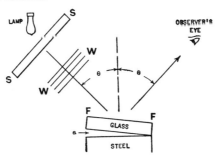

FIG. 29. — APPARATUS FOR MEASURING AIR–WEDGE WITH LIGHT IN-
CIDENT AT AN ANGLE.

The apparatus that is generally used in the measurement of the thickness of a thin layer of air by interference is shown in Fig. 29. The light through selenium glass falls on the measuring device of glass and steel flats and is seen by the observer's eye. A detail of the path of the light through the glass, Fig. 30, which shows that $2d = nw$ is slightly in error, as $2l - e = nw$. If the angle θ is small, however, the error is negligible.

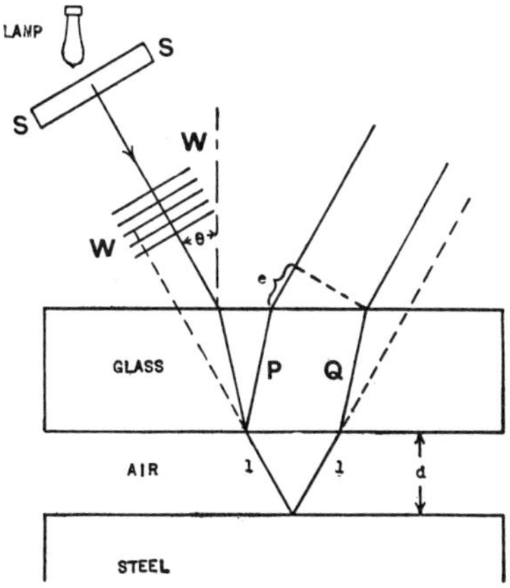

In making actual measurements of the thickness

FIG. 30. — DETAIL WITH LIGHT INCIDENT
AT AN ANGLE.

of the thin air layer a, Fig. 26, or in Fig. 29, the light from a convenient source is transmitted through selenium glass and a red monochromatic light is obtained whose wave length may be taken as .000025″ (twenty-five millionths of an inch).

45. When white light is used the observer sees a succession of colored bands instead of light spaces and dark bands. These bands repeat themselves and if one wishes to make a measurement of the thickness of the air layer, as in Figs. 26 and 29, using white light, one should start from the line of contact of glass and steel and count the number of bands of purple color (the " sensitive purple," as it is called) up to the chosen spot. Then the thickness d of the air layer at this spot is such that $2d = nw$ where w the value of the wave length to be used is .000020''. It is better to use selenium glass.

There is, however, one important use of daylight or white light. The line of contact of glass and steel is a line of no color, and is very easily distinguished from the colored bands.

MEASURING THE LENGTH OF LIGHT WAVES

46. Interferometer method of measuring the length of light waves, Fig. 31. — The wave length of light waves may be measured by means of the Michelson interferometer as shown

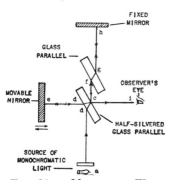

FIG. 31. — MEASURING WAVE LENGTHS WITH THE INTERFEROMETER.

by the diagram, Fig. 31. Light from source a is divided into two beams at the surface of the half-silvered glass parallel. One part travels the path $cdedci$ and the other part travels the path $cfghgfci$. These two beams interfere along the path ci and may meet crest to crest (light), or crest to trough (no light). If the two cross-section mirrors are not exactly at right angles the eye will see a series of light spaces and dark bands. If the mirror is moved along the line ed, the bands will move across the field of view and if the number of bands which pass a given view point is counted and the distance the mirror travels is measured by a micrometer screw, the wave length of the particular monochromatic light used can be calculated.

THIS BOOK CONTAINS
ONE HUNDRED SCHEDULES
OF
PROBLEMS AND PROCESSES IN MACHINE WORK

SPECIAL INDEX
ARRANGED TO ENABLE STUDENTS TO FIND THE PROBLEMS AND PROCESSES QUICKLY
Problems in Black-face Type — Processes in Light-face Type

1

GENERAL INDEX

HOW TO FIND A SUBJECT

Each Section is a unit — a book in itself. In the General Index, Section numbers are joined with Page numbers, thus: 2₂₄. To find a subject, for example, "Limit gages": turn to Index and find "Limit gages . . . . 2₂₄" which means that limit gages are in Section 2, page 24. Then find 2₂₄ (read **two twenty-four**) at top of page.

1

SUPPLEMENT

TO

GENERAL INDEX

Sections 13 and 14

14

Section 15

LIGHT-WAVE MEASUREMENTS